Fundamentals of Mathematics

Fundamentals of Mathematics
An Introduction to Proofs, Logic, Sets, and Numbers

Bernd S. W. Schröder

Louisiana Tech University
Program of Mathematics and Statistics
Ruston, LA

A JOHN WILEY & SONS, INC., PUBLICATION

For general information on our other products and services or for technical support, please contact our Customer Care Department within the United States at (800) 762-2974, outside the United States at (317) 572-3993 or fax (317) 572-4002.

Wiley also publishes its books in a variety of electronic formats. Some content that appears in print may not be available in electronic formats. For more information about Wiley products, visit our web site at www.wiley.com.

Library of Congress Cataloging-in-Publication Data:

Schröder, Bernd S. W. (Bernd Siegfried Walter), 1966–
 Fundamentals of mathematics : an introduction to proofs, logic, sets, and numbers / Bernd S. W. Schröder.
   p. cm.
 Includes bibliographical references and index.
 ISBN 978-0-470-55138-7 (hardback)
 1. Set theory. 2. Logic, Symbolic and mathematical. I. Title.
 QA248.S358 2010
 510—dc22   2010010589

10 9 8 7 6 5 4 3 2 1

Contents

Preface

Mathematics is a general way of thinking in a very rigorous logical fashion. This text is written especially for students who are about to make first contact with this way of thinking.[1] No matter what the first "proof class" in your curriculum is, you may have heard stories from upperclassmen about how hard it is. The bad news: They are right. Proofs are hard. The good news: You can adjust to this new way of thinking, and it will make you more capable in general.

The biggest challenge in a first proof class is that you need the mental discipline to not use mathematics that you already know (all that calculus you took) to prove some rather simple-looking results. In fact, for some early theorems, it is very common to ask yourself "Why do we even need to prove this?" or to realize "I already know this, but why is it true?" Check out page x for some simple-looking questions (and some hard ones) that we will consider in this text.

To keep the temptation to use "known stuff" to a minimum, it is appropriate to start with nothing but formal logic and the axioms of set theory. But we must assure that we do not get too tangled in the technical details of these subjects. So the trick to this introduction (hopefully) is to give you enough of everything, but not too much of anything: Because mathematics is pure logical reasoning, the text is about logic. But formal logic is also a branch of mathematics, and this text is not a logic text. Because everything in mathematics is constructed from sets, the text is about set theory. But formal set theory is also a branch of mathematics, and this text is not a set theory text. Because mathematics deals with numbers, the text is about numbers. But formal number theory is also a branch of mathematics, and this text is not a number theory text. Because a lot of mathematics is algebraic, the text is about algebra. But formal algebra is also a branch of mathematics, and this text is not an algebra text.

So what is this text? Chapters 1-5, which I recommend for an introductory course, provide a rigorous, self-contained construction of the familiar number systems (natural numbers, integers, rational, real, and complex numbers) from the axioms of set theory. The construction in itself is quite beautiful, and it will train you in many of the proof techniques that mathematicians use almost subconsciously. But learning anything just for the sake of doing it is typically less efficient than learning something for a known purpose or within a known context (see [4]). Whenever possible, to increase motivation, I have included important applications of the fundamentals: Among other things, we will discuss the scientific method in general (which is the reason why civilization

---

[1] But nonetheless, I hope that advanced students, and even experts, will enjoy it, too.

has advanced to today's highly technological state), the fundamental building blocks of digital processors (which make computers work), and public key encryption (which makes internet commerce secure). To attach more familiar meanings to the entities with which we work in this text, and to specifically serve education majors, I have also included examples and exercises on a lot of the neat mathematics that we have learned in elementary and high school.

Historically, the various number systems were developed to better solve certain problems, most notably equations. So it is natural to conclude the construction of the number systems with the solution of cubic and quartic equations in Section 5.8 and with the unsolvability of the quintic by radicals in Chapter 6. I don't expect Chapter 6 to fit into a standard one-term introductory course. But it makes for fascinating, if challenging, reading, and it can be used in a follow-up course or seminar.

Finally, Chapter 7 puts the finishing touches on our excursion into set theory. The axioms presented there do not directly impact our elementary construction of the number systems. But once they are needed in an advanced class, you will appreciate them.

Once you have gone through the majority of this text, I am confident that you will be prepared for your first "targeted" proof class, no matter if it is analysis, algebra, linear algebra (these are "the usual suspects"), or another subject altogether.

That leaves us with the subject of proofs themselves. I have tried to present standard proof methods and to make the transition into proofs and abstract reasoning as smooth as possible. But that's the same as saying that a successful cross country running coach made the start of spring training as smooth as possible: The overall effort to go from untrained to competition level remains considerable. All we can do is make the preparation more effective. Moreover, unlike in, say, calculus, we will not be able to mimic examples. The only good example for a proof pretty much amounts to the proof itself. This is a radical shift for many of us, whether we are consciously aware of it or not. I had a hard time with it myself, because I did not even realize that the comparatively small number of techniques (examples if you will) in calculus had become second nature. They sustained me through a lot of physics, but proofs were an entirely new game.

To help us understand mathematical arguments, in the early chapters many additional details and remarks are provided in **handwriting** and the reasons why proofs are set up in a certain way are discussed frequently. You should fill in any necessary details and think about why a proof ran the way it did for *every* proof you read. The effort spent thinking about these ideas will help settle the necessary structures in your mind. (Consider [3] for the value of training that "feels hard.") In later chapters, proofs are presented as you would see them in an advanced text. So as we progress, you will need to fill in more and more logical steps. This is quite customary in mathematics. If we were to fill in every detail of every proof, communication would be hard, because a lot of unnecessary detail would be included every time. In the video presentations (see [32]) many details are filled in verbally.

It is virtually certain that at some point you will wonder how your teacher or your classmates can produce proofs with seeming ease, while you cannot. This can be very frustrating, but just about everyone I know has gone through this struggle.[2] Think of it

---

[2]The only possible exception is an individual who started taking college classes at 12, graduated with a

this way: Do we really know what happens when we walk or pursue any daily activity that we may take for granted? Naturally, the task of walking is simple for a healthy, fully developed, non-inebriated adult.[3] But this ability must be acquired, too. Children spend hours, days, and weeks learning to stand up, stand without support, take one step, two steps, and so on. Just because we can do it now does not mean it was always easy. So what exactly happens when we walk? Few people, and I am not one of them, even know all the muscle groups and neural connections that have to work in a very coordinated fashion as people walk. Moreover, an intellectual understanding of all the underlying processes will not make anyone, say, a better soccer player. On the other hand, a child who grew up playing the game can be a very good soccer player, even with no knowledge of physiology. Similarly, I do not think anyone really knows what happens when a mathematician conceives a proof. Attempting to do proofs ("play the game") as much as we can is our best chance to make them second nature, just like walking.

Acknowledgments. Constructing the real numbers "from nearly nothing" is a natural start. My first analysis teacher, Professor Wegener, constructed $\mathbb{R}$ from $\mathbb{N}$ at the start of our class. At Tech, we used Clayton Dodge's out of print book [7], which constructs $\mathbb{R}$ in set theory, in the class for which I developed this text. Halmos' classic text [12] helped me with set theory, the Encyclopaedia of Mathematics [13] was an invaluable resource for quick references as well as some historical notes, my background in logic is from Hurd and Loeb [15], the idea to include a proof of the unsolvability of the quintic, and the proof itself, stem mostly from Maxfield and Maxfield [19], with Meyberg's texts [20] and [21] further helping with the presentation of Galois Theory, and [2] helping with Descartes' Rule of Signs. Finally, I used [13] and Wikipedia to refresh my memory on the few Latin words I know, on RSA encryption and as a general resource. The students in my Spring and Summer 2009 classes provided good feedback on the text and they were patient with typos. Special thanks go to James Sims, who inspired Exercise 3-82, Wei Wang, whose ideas greatly improved Exercise 6-8 on Descartes' Rule of Signs, and Nichamon Naksinehaboon and Narate Taerat who corrected Exercise 6-43 on commuting cycles. At Wiley, Susanne Steitz-Filler, Jacqueline Palmieri, Melissa Yanuzzi, Lisa Vanhorn, and Kristen McGowan continue to be wonderful to work with.

But most importantly, this work would not have been possible without the love and understanding of my family. They continue to live with me and my obsession for mathematics, and I am eternally grateful for that.

BERND S. W. SCHRÖDER

Ruston, Louisiana
June 3, 2010

BS in mathematics at 18 and had a Ph.D. in mathematics from a world-leading institution by age 23. He is an exceptional talent. Do not be disappointed if your progress is slower. Do not get discouraged when you get bogged down. That is simply normal.

[3] In case of inebriation, walking may be difficult, but it must be preferred over driving.

Questions

Human inquisitiveness is driven by questions. In high school, Einstein once asked himself a very simple question: If he had the appropriate means of transportation, could he catch up with the light emitted from a match he just lit? The answer he found about a decade later, relativity theory, has revolutionized our understanding of the world. So it is good to ask questions, even questions that sound simple or strange. Of course, not all of us can ask and answer questions of Einstein's caliber. But answers to our questions can change *our* understanding of any subject, including mathematics. Think about the questions below, then start reading. References to the answers, if they exist as this text is published, are provided in parentheses.

1. Why is $1 + 1 = 2$? (Definition 3.7.)

2. Why is $3 < 9$? (Comment after Definition 3.20.)

3. Why can we not divide by zero? (Comments after Proposition 4.17 and after Proposition 5.1, as well as Proposition 5.11 itself.)

4. Is 3 a solution of $x^3 - 5x^2 + 3x + 9 = 0$? Why? Are there more solutions? If so, what are they? (Definition 4.55, Theorems 5.65, 5.66 and Exercise 5-66d.)

5. What is a set? (Sections 1.8 and 2.1.)

6. What is a digital computer made of? (Example 1.22 and Exercise 1-10.)

7. What is the scientific method? (Example 1.58.)

8. Why are publicly encrypted internet transactions safe? (Section 3.9.)

9. Why is there no quintic formula? (Chapter 6 – all of it!)

10. Why can angles not be trisected with straightedge and compass? (Exercises 6-5, 6-16, 6-31, 6-32 and 6-33.)

11. Can we count past infinity? (Section 7.2.)

12. Are there infinitely many twin prime numbers? (Unsolved, see Example 1.5 for the statement.)

13. Is there an efficient factorization algorithm for integers? (Unsolved, see the discussion after Theorem 3.109 for the relevance of this question.)

Chapter 1

Logic

Mathematics is built upon two fundamental disciplines: Logic and set theory. Logic provides the language and rules that are needed to discuss and explore mathematical objects, and set theory provides the objects. When introducing these two areas, we are faced with a "chicken and the egg" problem. If we introduce logic first, then we have no formally defined objects to talk about in the introduction. If we introduce set theory first, then the language we use to describe sets is not yet formally defined. We break this gridlock by introducing logic first. To have something to talk about, in this first chapter we will refer to some standard mathematical concepts without formally defining them. Specifically, we assume the reader is familiar with fundamental mathematical concepts such as natural numbers $\mathbb{N} = \{1, 2, 3, \ldots\}$, integers $\mathbb{Z} = \{\ldots, -2, -1, 0, 1, 2, \ldots\}$, divisibility, prime numbers, rational numbers $\mathbb{Q}$ (fractions with integer numerator and denominator), real numbers $\mathbb{R}$ (rational numbers, roots and further numbers that require a technical explanation) and complex numbers $\mathbb{C}$ (real numbers with an additional number i, which satisfies $i^2 = -1$, adjoined), and we also assume that the reader has an intuitive idea that a set is a collection of objects. Except for sets themselves, we will give formal definitions for all these (ultimately set theoretical) notions later in the text.

1.1 Statements

Mathematics differs from other scientific disciplines in that it has a very rigid notion of truth. Once a result has been derived from a set of ground rules (axioms) that are accepted as true, the result itself is considered to be true. Within the framework given by the axioms, no further investigation, no change of scales, no refinement of knowledge can change the result from true to false.

This situation is fundamentally different from the sciences, because in the sciences results are tested by whether they are consistent with experimental observations. In-

Fundamentals of Mathematics: An Introduction to Proofs, Logic, Sets, and Numbers.
By Bernd S. W. Schröder.
Copyright © 2010 John Wiley & Sons, Inc.

ternal logical consistency of theories is important, and it has allowed science to make great strides. But the experiment is the ultimate arbitrator, and scientific results can be true in one context and invalid in another.

Consider, for example, the assumption that matter is made up of continuous substances. This assumption is "true" in classical mechanics in the sense that the assumption that fluids, solids, etc. are made up of continuous matter (instead of individual particles) leads to verifiable results. In fact, this model of matter as a continuum is so successful that computer programs can be used to design functional airplanes, which can be put into service without first building and testing prototypes. Now consider that the behavior of matter is different under high magnification. Observations that an electron beam can penetrate a thin, but solid, metal foil without damaging the foil, show that there must be emptiness inside matter. Such observations ultimately led to the realization that matter is made up of individual atoms, which each contain a lot of empty space.

This means that in science a result can be "true" (meaning it leads to verifiable results) in one context, such as the macroscopic modeling of matter, and at the same time it can be problematic (meaning it leads to predictions that contradict experimental results) in another context, such as phenomena that occur on an atomic scale. Consequently, in science we don't just need internal consistency, but we also need experimental validation that our underlying assumptions can be trusted.

Given that all science ultimately exists in the same context (our universe), the preceding paragraph may be a little disconcerting. But rather than worrying about whether it is or is not "true" that matter is continuous, science focuses on whether the model provides verifiable results. If it does, the model will be used, otherwise the model must be refined or discarded. These considerations lead to extremely interesting work in all areas of science, and intrepid souls can also choose to work in areas where behavior is best modeled by a mixture of traditional models. For example, the description of the behavior of matter on the nanoscale can require a mixed model that uses the fact that matter consists of individual particles as well as the fact that matter can behave like a continuum.[1]

The role of mathematics in science is that it provides the language and the rules of inference that govern theoretical science. Although experimental verification is the ultimate scientific validation, theoretical derivations have driven science and engineering forward as well. The above mentioned design of functional airplanes without needing to build and test prototypes is an astounding achievement that requires sound mathematics, sound science and sound engineering. Similarly, there are numerous phenomena that were predicted by theory, that is to say, mathematics, before they could be experimentally verified. For example, James Clerk Maxwell predicted the existence of electromagnetic waves. These electromagnetic waves are what makes radios, cell phones, bluetooth devices, and so on, function, and they were first experimentally observed after Maxwell had already died. So, even though mathematical truth must be interpreted very carefully when applied in science, it has great predictive power. This predictive power led Galileo Galilei to his famous statement that "Nature's great book is written in mathematical symbols."

---

[1] Some models must even take quantum effects into account.

By studying mathematics, the reader will learn to reason in logically correct fashion and to prove results from earlier results or axioms. This very structured reasoning, often combined with computation, which is nothing but another form of reasoning, allows the prediction of the behavior of designed and natural systems. The examples above have already shown that these predictions can be correct independent of whether they are verifiable at the time (the designed airplanes can be built and tested) or not (Maxwell was unable to test his prediction of electromagnetic waves). The connection between mathematics and science, if built from the foundations of mathematics without gaps, takes a while to develop. The author has attempted to do so in this text, as well as in [30] and [31]. This text provides the introduction to mathematical communication and to the fundamental building blocks of mathematics.

> The reader may notice that the narrative in the introduction is a bit technical. Many sentences are lengthened by added comments. These added comments make the statement in question more precise, but they also make it a little harder to read. One skill that is acquired in doing mathematics at this level is the ability to take a large amount of information and make sense of it. Hence the attempt at high precision in the introduction is the first step towards learning how to read mathematics. If there are parts that sounded confusing, read them again and think about them until the fog lifts.

Although mathematics allows a very rigid and reliable sense of truth, it will not provide absolute truth. Truth of a result will always depend on the axioms we are willing to accept. Therefore, we cannot define what truth is. Instead, we must limit our language to only allow phrases that are (at least in theory) verifiable. Such phrases are called statements.

Definition 1.1 *A* **statement** *or* **proposition** *is a sentence that is either true or false.*

Ideally, everything communicated in formal mathematics is a statement. Of course not every sentence in a mathematics text is a statement. For example, the introduction above contains several sentences for which it does not make sense to ask if they are true or false. Theorems and their proofs, however, should consist of statements only, with maybe some explanatory prose thrown in to make them easier to read.

Example 1.2 The sentence *"Einstein once said 'Do not worry about your difficulties in mathematics. Mine are greater yet.'"* is a statement. It is a statement, because we can all agree that it must be either true or false: Either Einstein once said it or he did not. From what the author has heard, the above is an actual quote, so the sentence would be a true statement. □

Example 1.3 The sentence *"Einstein was awarded the 1921 Nobel prize in Physics for his pioneering work on the theory of relativity."* is a statement, too. After all, we can check (see, for example, [24]) to determine if the statement is true. Surprisingly enough, the statement is false. Einstein was awarded the Nobel prize that year, but it was "for his services to Theoretical Physics, and especially for his discovery of the law of the photoelectric effect," which is a quantum mechanical effect, not a relativistic one.

This example shows another feature of mathematical statements: There are no half-truths. A single inconsistency invalidates a whole statement. Similarly, even lengthy arguments can be invalidated by a single mistake. This merciless insistence on validity in every detail is what makes mathematics so highly useful, but it is also what makes it so challenging. There is no way to cut corners, there is no partial credit. Either it's right or it's not. □

Example 1.4 The sentence *"Consider, for example, the assumption that matter is made up of continuous substances."* from the introduction (see page 2) is not a statement. How could we assign a truth value to a recommendation? If the reader did consider the assumption, well, that is what the author wanted. But this reaction does not assign truth to the sentence *"Consider, for example, the assumption that matter is made up of continuous substances."* □

Example 1.5 A **prime number** is a natural number that is only divisible by 1 and by itself. Examples are $2, 3, 5, 7, 11, 13, 17, 19, 23, 29, 31$, etc. A **twin prime** number is a prime number that differs by 2 from another prime number. Examples are $3, 5, 7, 11, 13, 17, 19, 29, 31$, etc., but not 23. Now consider the sentence *"There are infinitely many twin prime numbers."* This sentence is a statement, because it is either true or false. Either the sequence of twin prime numbers always has a next element, or it stops with a final, largest twin prime number. The interesting thing here is that it is unknown whether the statement is true or false. To date, the statement has neither been proved nor disproved. (Therefore, as this is written, question 12 on page x cannot be answered.)

This example shows that the idea of a statement is independent of whether we have a mechanism to establish the truth of the statement. □

Example 1.6 The sequence of symbols "$a^2 + b^2 = c^2$" is *not* a statement and it is *not* **Pythagoras' Theorem**. We do not know what a, b and c are, and depending on the values of a, b and c, the equality may or may not hold. We will state Pythagoras' Theorem in Example 1.15 and we will introduce ways to handle variables in Sections 1.2 and 1.5. □

We should mention briefly that it has been proved by Gödel that every sufficiently rich axiomatic system allows the formulation of statements that cannot be verified with the axioms at hand. More formally, Gödel's first incompleteness theorem says that a formal system that supports arithmetic of natural numbers and contains the universal and existential quantifiers (see Section 1.5) contains a statement so that neither it, nor its negation can be deduced within the system. We will not run into any problems with undecidable statements in this text, but it is only prudent to warn the reader that even in mathematics not everything is simply true or false. Interested readers are advised to consider texts that present logic more formally, such as, for example, [1] or [6], which contain proofs of Gödel's theorems.

Exercises

1-1. For each of the following sentences, determine if it is a statement. For those that are statements determine if they are true or false.

    (a) In the year 2008, the US housing market was in a serious crisis.

    (b) The number 91 is a prime number.

    (c) Do your homework.

    (d) *Archon* is my favorite video game.

    (e) The number 3 is the only prime number p so that p, $p + 2$ and $p + 4$ are prime numbers.

    (f) Is there still time left to change my bid on eBay?

    (g) Say Boudreaux here come.

    (h) The mathematician Kronecker once said "The natural numbers were made by the good Lord. The rest is man's invention."

    (i) This sentence is a false statement.

    (j) I like my new mp3 player.

    (k) Chagrinned, Dr. Schröder said "I don't like losing my wizard to a basilisk."

    (l) OMG

1-2. For each of the following statements, determine if the statement is true or false and prove it if possible. For some of the true statements, a proof is currently out of reach, but think about how the statement could be justified or shown to be incorrect.

    (a) $5 < 9$

    (b) The square of every complex number is positive.

    (c) The square of every real number is positive.

    (d) Given natural numbers k, $k + 2$ and $k + 4$, one of them is divisible by 3.

1-3. For each of the following statements, determine if the statement is true or false.

    (a) Every differentiable function is continuous.

    (b) Every continuous function is differentiable.

    (c) The harmonic series $\sum_{n=1}^{\infty} \frac{1}{n}$ diverges.

    (d) The alternating harmonic series $\sum_{n=1}^{\infty} \frac{(-1)^{n+1}}{n}$ diverges.

    (e) Every integrable function has an antiderivative.

1.2 Implications

Mathematical reasoning, indeed all sensible reasoning, is about deriving new true results from established true results or from axioms. Therefore, we must carefully analyze the nature of implications or "if-then" statements. In an implication, we connect two statements to each other. Because we will frequently connect two statements to each other, using a variety of so-called connectives, we will first introduce the general language and then continue with implications.

p	q	$p \Rightarrow q$
FALSE	FALSE	TRUE
FALSE	TRUE	TRUE
TRUE	FALSE	FALSE
TRUE	TRUE	TRUE

p	q	$p \Leftrightarrow q$
FALSE	FALSE	TRUE
FALSE	TRUE	FALSE
TRUE	FALSE	FALSE
TRUE	TRUE	TRUE

Figure 1.1 Truth tables for implication and biconditional.

Definition 1.7 *Statements symbolized by a single letter are called* **primitive proposi-tions***. A symbol* ∘ *that can be put between two primitive propositions p and q so that* *p* ∘ *q is a statement is called a* **connective** *(for examples, see Definitions 1.8, 1.10 and 1.16). A statement that is constructed from other statements using connectives is called a* **compound statement***. Because compound statements are very similar to functions of several variables, we will also sometimes call them* **logical functions***.*

Our first example for connectives, as discussed in Definition 1.7, is the implication. Of course, the implication is much more than just an example of what can be done in a logical framework, which is why we investigate its properties in detail.

Definition 1.8 *Let p and q be primitive propositions. The compound statement p* $\Rightarrow$ *q* *(*"*p* **implies** *q*" *or* "**if** *p,* **then** *q*"*) is false if and only if p is true and q is false. Such a statement is called an* **implication** *or a* **conditional statement***. The statement p is also called the* **hypothesis** *and the statement q is called the* **conclusion***. Moreover, when p* $\Rightarrow$ *q is a true statement, then p is called a* **sufficient condition**[2] *for q and q is called a* **necessary condition**[3] *for p.*

The definition of an implication does not state when an implication is true, but instead it focuses on when an implication is false. This is because in logical reasoning, true statements will always imply true statements, true statements *cannot* imply false statements and false statements can imply anything, be it true or false.

A good way to remember how logical connectives, such as ⇒, work is to remember their **truth table**. A truth table of a compound statement provides the truth value of the compound statement for each possible combination of truth values of the primitive propositions. The representation of logical connectives via truth tables also makes it clear why logical connectives are so similar to functions of several variables. The truth table is nothing but a chart that assigns output values for the various inputs of the function. Figure 1.1 gives the truth tables for implication and biconditional. (The biconditional will be tackled in Definition 1.10.) The truth table for the implication reflects the definition as well as the earlier discussion. Truth implying falsehood is excluded and everything else is possible.

Aside from just stating that Definition 1.8 and the truth table in Figure 1.1 are "how implications work," we should make as much sense of the definition as possible. When we think of an implication, we typically think of an implication with a true

---

[2]Because knowing that p is true is *sufficient* for knowing that q is true.

[3]Because whenever p is true, it is *necessary* that q is true, too.

hypothesis. This is because, in everyday life, people do not deliberately start a logical argument with a false hypothesis. Something does not feel right about that. So, if we first consider implications with a true hypothesis, we see that the truth table makes sense. True statements should imply true statements, so "$TRUE \Rightarrow TRUE$" is true, and true statements should never imply false statements, so "$TRUE \Rightarrow FALSE$" is false. Any alteration to this part of the truth table leads to nonsense.

That leaves us to make sense of the demand that *both* "$FALSE \Rightarrow TRUE$" and "$FALSE \Rightarrow FALSE$" are true. The key to understanding these assignments is to realize that we are concerned with the truth value of the implication itself, not with the truth values of hypothesis and conclusion. In everyday usage, we often doubt the validity of an argument if the conclusion is incorrect. This is indeed good practice. For example, if an integral was computed incorrectly in calculus, then the problem is typically with the computation (which is a derivation or a chain of implications) that led to the result. When we work with logic, we will typically assume that all our implications are correct. That is, we assume that somehow we have made sure that there are no mistakes in going from the hypothesis to the conclusion. Indeed, the main task when doing proofs can be summarized in three words: "Make no mistakes." (Recall Example 1.3 for how subtle mistakes can be.)

But the next example shows that, when no mistakes are made, a false hypothesis can lead to true conclusions as well as to false conclusions, each time using a valid set of implications.

Example 1.9 Consider the statements $p := $ "$0 = 1$", $q := $ "$0 = 2$" and $r := $ "$1 = 1$". By definition, both implications $p \Rightarrow q$ and $p \Rightarrow r$ are true. That means, we should be able to prove both of them.

Proof that $p \Rightarrow q$ is true. Our hypothesis is that $0 = 1$. Adding the equation to itself (which is allowed) we obtain $0 = 0+0 = 1+1 = 2$. Thus the implication $p \Rightarrow q$ has been proved. The *implication* must be true. But note that we have only shown that our *argument* is valid. We have *not* validated the hypothesis or the conclusion.

Proof that $p \Rightarrow r$ is true. Our hypothesis is that $0 = 1$. Consequently, it must be true that $1 = 0$ (we can switch sides in an equation). Adding the two equations (which is allowed) we obtain $1 = 1 + 0 = 0 + 1 = 1$. Thus the implication $p \Rightarrow r$ has been proved. The *implication* must be true. But, once more, we have only shown that our *argument* is valid. We have *not* validated the hypothesis or the conclusion.

Because there are valid logical arguments that lead from $FALSE$ to $FALSE$ and from $FALSE$ to $TRUE$, we are right to define both "$FALSE \Rightarrow TRUE$" and "$FALSE \Rightarrow FALSE$" as valid implications. $\qquad\square$

The logical rules that implications with false hypotheses are always true can be seen as a precaution against logical inconsistencies. False hypotheses are also often dismissed by saying that "Anything can be derived from a false hypothesis." Nonetheless, although mathematics is concerned with deriving true conclusions from true hypotheses, implications that start with a false hypothesis have an important place in mathematics. They are the foundation for the very important proof method of proving a result by contradiction (see Standard Proof Technique 1.56). So we will frequently encounter situations in which we start a correct chain of implications from a hypothesis that ultimately turns out to be false. In such situations, it is important to focus on the *internal validity* of the argument. As long as that is given, a proof by contradiction works very well.

Next, we introduce the logical connective that takes the place of equality in logic. Two statements should be "equal" ("equivalent" in logic) when they have the same truth values. That is, an equivalence should be true when both statements involved are true or when both statements involved are false and that's it.

Definition 1.10 *Let p and q be primitive propositions. The compound statement $p \Leftrightarrow q$ ("p **if and only if** q" or "p **iff** q") is true if and only if p and q are both true or both false. Such a statement is also called a* **biconditional** *or an* **equivalence**.

The truth table for equivalence is given in Figure 1.1. The symbol "$\Leftrightarrow$" suggests that the equivalence $p \Leftrightarrow q$ is true if and only if $p \Rightarrow q$ is true and $q \Rightarrow p$ is true. This is indeed the case.

Proposition 1.11 *Let p and q be primitive statements. The equivalence $p \Leftrightarrow q$ is true if and only if the implication $p \Rightarrow q$ is true and the implication $q \Rightarrow p$ is true.*

Proof. The statement $p \Leftrightarrow q$ is true if and only if p and q have the same truth values. We claim that this is the case if and only if $p \Rightarrow q$ is true and $q \Rightarrow p$ is true. An explanation of the claim follows below.

If p and q both have the same truth value, then $p \Rightarrow q$ is true and $q \Rightarrow p$ is true. Conversely, let $p \Rightarrow q$ and $q \Rightarrow p$ both be true and consider p. The primitive proposition p is either true or false. If p is true, then truth of $p \Rightarrow q$ forces q to be true. On the other hand, if p is false, then truth of $q \Rightarrow p$ forces q to be false. ∎

Note also that we use the phrase "if and only if" in definitions of terms. This is because a definition is a biconditional. If a term applies, then whatever the definition says must be true. Conversely, if what the definition says is true, then the term from the definition can (and should) be used.

For a given implication $p \Rightarrow q$, the implication $q \Rightarrow p$ has a special name.

Definition 1.12 *Let p and q be primitive propositions. Then the implication $q \Rightarrow p$ is called the* **converse** *of the implication $p \Rightarrow q$.*

In this language, Proposition 1.11 simply says that a biconditional statement $p \Leftrightarrow q$ is true if and only if the conditional statement $p \Rightarrow q$ and its converse are both true. However, an implication $p \Rightarrow q$ should never be confused with its converse $q \Rightarrow p$, because the two need not have the same truth value and they need not say the same thing.

Example 1.13 Let n be a natural number greater than 3. The implication "If n is divisible by 3, then n is not prime." has as its hypothesis $p =$"n is divisible by 3," and as its conclusion $q =$"n is not prime," and it is a true statement. But the converse $q \Rightarrow p$, reads "If n is not prime, then n is divisible by 3," which is certainly not true. So the implication and its converse cannot say the same thing, because they need not even have the same truth values. □

It is important to keep in mind that implications are "one-way streets" and that validity of the converse does not say anything about the validity of the original implication. When writing a proof of an implication, it is easy to get lost in the details and to end up proving the converse. Once we have accepted that the two are not the same, we simply must safeguard against accidentally reversing the direction of our reasoning.

Remark 1.14 Note that in Example 1.13 we have also used the common practice of introducing variables to refer to certain objects. In this fashion, we avoid statements such as "If a natural number n that is greater than 3 is divisible by 3, then it is not a prime number." Although this is a perfectly fine mathematical statement, it is a little harder to see that, because the description of the number must stay with the number at the start of the sentence, the converse is "If a natural number n that is greater than 3 is not a prime number, then it is divisible by 3." By deliberately making language a bit more schematic, we achieve greater clarity at the expense of prose becoming a little less rich. As clarity and exactness are our main goals, this is a small price to pay. Once we are sufficiently familiar with the language of mathematics, our prose will automatically get richer again. But, for starters, it's better to be safe than sorry.   □

Further remarks about language are in order. Proposition 1.11 shows that calling the connective $\Leftrightarrow$ "if and only if" comes from the fact that $p \Leftrightarrow q$ says that $p \Rightarrow q$, which can be read "p **only if** q," and $p \Leftarrow q$, which can be read "p **if** q."[4] Combining the two gives "if and only if." Instead of writing "if and only if," in mathematics this phrase is abbreviated with the artificial word "**iff**." Note that we have used the phrase "if and only if" before "$\Leftrightarrow$" was defined. In fact, it was even used in the definition of "$\Leftrightarrow$" itself. This means our use of language is a bit circular. For a more formal approach to building languages in logic, the reader should consider texts that cover formal logic, such as [1] or [6]. For the purposes of this exposition, let us accept that we have no choice but to use words to define words. Hence, although formally a lot can depend on "what your definition of the word 'is' is," we will assume that the commonly accepted rules of the English language are understood and we will use them without any qualms, formal or otherwise.

The final example in this section shows another typical use of implications in mathematical language. It also re-emphsizes that we do not want to become too mechanical (or pedantic) in our use of language.

Example 1.15 The sentence *"If c is the length of the hypothenuse of a right triangle and a and b are the lengths of the other two sides, then $a^2 + b^2 = c^2$."* is a statement. It is a true statement and it is called **Pythagoras' Theorem**. However, we must be careful. We might say that the statement formally is not a compound statement, because "$a^2 + b^2 = c^2$." is not a statement. Then again, once a, b and c are determined, "$a^2 + b^2 = c^2$." is a statement, and the hypothesis determines a, b and c. So, we should consider the sentence a compound statement.

---

[4]Although grammatically the uses of "if" and "only if" are correct, the author recommends staying with straight "if-then" statements. There is too much possible confusion with "p if q" being the converse of "if p then q," which is equivalent to "p only if q." Compare with Remark 1.14: It's better to be safe than sorry.

The precise definition of language is best left to texts that specialize in formal logic. Instead of worrying about whether a sentence fits a certain mold, we should here simply use our command of the English language to understand it. □

Exercises

1-4. For each statement below, determine if it is true or false.

    (a) If you are reading these words, then the book must be opened to page 10.

    (b) If $5 + 3 = 8$, then $2 + 4 = 6$.

    (c) If $5 + 3 = 9$, then $2 + 4 = 6$.

    (d) If $5 + 3 = 8$, then $2 + 4 = 7$.

    (e) If $5 + 3 = 9$, then $2 + 4 = 7$.

    (f) $5 + 3 = 8$ if and only if $2 + 4 = 7$.

    (g) $5 + 3 = 8$ if and only if $2 + 4 = 6$.

    (h) $5 + 3 = 9$ if and only if $2 + 4 = 7$.

    (i) $5 + 3 = 9$ if and only if $2 + 4 = 6$.

    (j) If the Sun revolves around the Earth, then there is life on Io.

1-5. For each of the following implications, state the converse. Then determine if the implication is true and if the converse is true.

    (a) Let n be an integer. If $n > 0$, then n is a natural number.

    (b) Let $n \neq 3$ be a natural number. If n is prime, then n is not divisible by 3.

    (c) Let n be a natural number. If n is prime, then n is not divisible by 3.

    (d) If a rational number cannot be written as a fraction with denominator 1, then the number is not an integer.

    (e) If there is lightning in the area, then the football game will be stopped.

    (f) If a real number can be represented as an infinite nonrepeating decimal, then the number is not rational.

    (g) Let f be a function. If f is differentiable, then f is continuous.

    (h) Let n be a natural number. If n is prime and $n + 2$ is prime, then n is a twin prime number.

    (i) Let n be a natural number. If $n = pq$, where $p \neq 1$ and $q \neq 1$, then n is not a prime number.

    (j) Let C be a geometric figure. If C is a circle, then every tangent of C touches C in exactly one point.

1-6. Rewrite each of the following statements as an implication in the form "if p then q."

    (a) The sum of the angles in a triangle is $180°$.

    (b) Every horizontal line has slope 0.

    (c) The graph of a differentiable function has no corners.

    (d) Birds have feet.

p	q	$p \wedge q$
FALSE	FALSE	FALSE
FALSE	TRUE	FALSE
TRUE	FALSE	FALSE
TRUE	TRUE	TRUE

p	q	$p \vee q$
FALSE	FALSE	FALSE
FALSE	TRUE	TRUE
TRUE	FALSE	TRUE
TRUE	TRUE	TRUE

p	$\neg p$
FALSE	TRUE
TRUE	FALSE

Figure 1.2 Truth tables for conjunction (AND), disjunction (OR) and negation (NOT).

1.3 Conjunction, Disjunction, and Negation

Statements can be combined in other ways than just through implications and biconditionals. For example, to remain eligible for certain types of financial aid, a student must be enrolled at a university *and* have a GPA greater than a certain cutoff *and* may need to satisfy certain other requirements. That is, the following statements must all be true.

- $p =$"The student is enrolled at a university."

- $q =$"The student's GPA is greater than ..."

- $r =$"⟨Statements encoding other conditions that need to be satisfied.⟩"

Note that we already used an "and" in Proposition 1.11. So once more, we rely on our familiarity with language before we introduce formal definitions. Similarly, in some of the exercises we have used the word "not," which will also be formalized below. The following three connectives give precise meaning to words that are commonly used in mathematical reasoning as well as in everyday language.

Definition 1.16 *Let p and q be statements.*

1. *The statement $p \wedge q$ (read "p **and** q"), also called the* **conjunction** *of p and q, is true iff p is true and q is true.*

2. *The statement $p \vee q$ (read "p **or** q"), also called the* **disjunction** *of p and q, is true iff p is true or q is true, where the "or" also allows for both statements to be true.*

3. *The statement $\neg p$ ("**not** p"), also called the* **negation** *of p, is true iff p is false.*

Figure 1.2 gives the truth tables for conjunction, disjunction and negation. The following examples provide some settings in which these connectives can be encountered.

Example 1.17 Many internet search engines allow the user to include AND, OR and NOT operators when searching for content. This feature is important when a user is looking for specific content. For example, a search for "Schroeder" yields long lists of links to events involving people named "Schroeder." A search for "Schroeder AND analysis" yields long lists of links to analyses of all kinds that were done by people named "Schroeder." A search for "Schroeder AND mathematical analysis" still yields a surprising amount of content, but the first few links contain a link to [30].

This means that, when looking for specific content on the web (such as [30], which would really please the publisher), logical connectives can simplify the search.   □

Example 1.18 The operators AND, OR and NOT are also available in spreadsheets. They can be used to filter data.

Professional societies exist for just about all professions. One such society for mathematics is the American Mathematical Society. A professional society may be interested in the demographics of the next cohort of professionals who enter the profession. In mathematics, every year the American Mathematical Society surveys the new mathematics Ph.D.s in the United States and the results are published (see, for example, [26]). When this data comes in, the demographic data for each respondent can be recorded in separate columns of a spreadsheet. For example, in a "gender column" it can be recorded if the respondent is male (M) or female (F), and in a "citizenship column" it can be recorded if the respondent is a US citizen (C), a non-US-citizen on a permanent visa (P), a non-US-citizen on a temporary visa (T), or a non-US-citizen of unknown visa status (U).

Once such data is recorded, it can be determined, for example, how many new Ph.D.s were female US citizens. One way to do this would be to open a new column in the spreadsheet that records a 1 iff there is an "F" in the gender column AND a "C" in the citizenship column. (The specific encoding depends on the type of spreadsheet used.) Add the values in this column to obtain the number of new mathematics Ph.D.s who are female US citizens.

The author should say here that he does not know how the American Mathematical Society actually aggregates their data. Most likely they use more sophisticated database software. But the above describes how it can be done (and sometimes is done) using the elementary functions of a spreadsheet.   □

The words "and," "or" and "not" are also used extensively in legal documents, where they are used in exactly the same way as in mathematics. For an example from tax instructions, consider Exercise 1-7, which shows that, although taxes may be a hot button issue for some individuals, the instructions can be analyzed and understood.

It usually takes a little while to really get used to the mathematical definition of "or." This is because, in everyday language, the word "or" is often used in an exclusive fashion, as in "I can take the MWF 11:00 class on differential equations or I can take the MWF 11:00 class on set theory and logic." The mathematical use of "or" is that "p or q" is also true when *both* p and q are true. So when your plan for this afternoon is "I will read the book or do homework." then the statement will also be true when you do both. (**And you should.**) The exclusive or operator will be constructed in Example 1.21.

Example 1.19 Logical connectives can also be visualized with **switch diagrams**. Figure 1.3 shows the switch diagrams for conjunction and disjunction. Every switch in a switch diagram represents a primitive proposition, and the whole circuit represents a compound statement. A primitive proposition is true if its switch is closed, and it is false if its switch is open. The compound statement is true iff there is a conducting connection from the starting point A to the finishing point B.

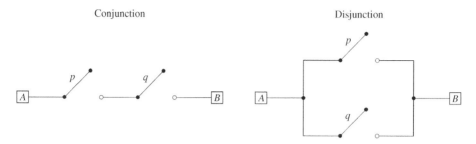

Figure 1.3 Visualization of conjunction and disjunction with simple electric circuits ("switch diagrams"). On the left, current will flow from point *A* to point *B* iff both switch *p* *and* switch *q* are closed. On the right, current will flow from point *A* to point *B* iff switch *p* *or* switch *q* (or both) are closed.

It should be emphasized, however, that switch diagrams are visualizations only. Yes, we could build circuits to the specifications of the switch diagrams, but they would serve no other purpose than to model a specific logical statement. Logical circuits that are used in building computers are discussed in Example 1.22. □

Before we can talk about building actual circuits, we need to prove that conjunction, disjunction and negation are "enough" to describe arbitrary compound statements.

Theorem 1.20 *Let ○ be a connective and let p, q be primitive propositions. Then p ○ q can be represented as a disjunction of conjunctions of statements p, ¬p, q, or ¬q.*

Proof. For any primitive propositions u and v, the conjunction $u \wedge v$ is true iff u is true and v is true. Thus any logical function of two primitive propositions u and v which is true for exactly one combination of truth values of u and v can be represented using a conjunction of u or $\neg u$ with v or $\neg v$: Indeed,

- $\neg u \wedge \neg v$ is true iff u is false and v is false.

- $\neg u \wedge v$ is true iff u is false and v is true.

- $u \wedge \neg v$ is true iff u is true and v is false.

- $u \wedge v$ is true iff u is true and v is true.

Now consider an arbitrary logical function ○ of two primitive propositions p and q. Then the column for $p \circ q$ in the truth table of $p \circ q$ has between 0 and 4 entries that read "TRUE." If $p \circ q$ is always false, then $p \circ q$ is $p \wedge \neg p$. Otherwise, for each "TRUE" entry in the column for $p \circ q$, choose the compound statement from the list above which is true for exactly the input values in the entry's row. The compound statement $p \circ q$ is equal to the disjunction of the thus collected statements. ■

Regarding the *exact* statement of Theorem 1.20, we note that, formally, a statement such as $p \wedge q$ is not a disjunction of conjunctions. But because we can turn it into the rather trivial disjunction $(p \wedge q) \vee (p \wedge q)$, we can safely consider it a disjunction of conjunctions, too.

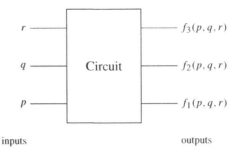

Figure 1.4 Visualization of a simple component of a digital computer. Every output f_1, f_2 and f_3 on the right side depends on the inputs p, q and r. That is, every individual output is a logical function of the inputs.

Example 1.21 The **exclusive or** of statements p and q, denoted $p \oplus q$ is true iff *exactly one* of p and q is true. By the above we note that $p \oplus q = (p \wedge \neg q) \vee (\neg p \wedge q)$. Double checking with a truth table confirms the claimed identity (Exercise 1-8). □

Example 1.22 Theorem 1.20 has much further reaching consequences than the simple ability to represent the "exclusive or" connective in Example 1.21. By Exercise 1-9b, any compound statement can be represented using AND, OR and NOT.

Now think of a compound statement as a device that turns inputs that are combinations of $TRUE$ and $FALSE$ into an output that is either $TRUE$ or $FALSE$. In a digital computer, a voltage near zero volts is associated with $FALSE$ (also denoted 0 in digital circuits) and a voltage near five volts is associated with $TRUE$ (also denoted 1 in digital circuits). Every component of a digital computer is a hardware implementation of a logical function. That is, every component takes a set of voltages that are either near zero volts or near 5 volts and turns these inputs into output voltages that are either near zero volts or near five volts. See Figure 1.4 for a simple visualization of a component with three inputs and three outputs. Theorem 1.20 and its extension through Exercise 1-9b now guarantee that every component of a digital computer can be built from components that model the logical operations of conjunction, disjunction and negation. Such components actually exist, and they are called AND-gate, OR-gate and NOT-gate (or inverter), respectively.

Figure 1.5 shows the symbols used in digital circuits for the respective gates as well as a circuit that can be used to encode the action of the "exclusive or" connective. Example 1.26 will show how the circuit was conceived. □

The above is a first, very impressive, example of the effectiveness of mathematics. Our discussion started with the goal of formalizing language so that we can communicate unambiguously. It led us to the answer to question 6 on page x: All digital devices are, in theory, made up of three simple components: AND-gates, OR-gates and NOT-gates. See Figure 1.5 for their symbols and a simple digital circuit. For more on the gates that are actually used most often on chips, consider Exercise 1-10.

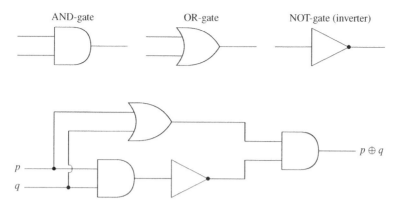

Figure 1.5 AND-, OR- and NOT-gates in digital circuits and an implementation of the exclusive OR function as a digital circuit.

When designing digital devices, it is important to come up with the most efficient design. The more efficient the design, the smaller/less expensive/more energy efficient/etc. the device will be. Although chip design is a highly sophisticated engineering process, we can describe a naive approach to designing a component. Once the desired truth table has been written down, use Theorem 1.20 and its extension through Exercise 1-9b to obtain an expression for the compound statement that is to be encoded. Then simplify the statement. As simplification reduces the number of connectives $\wedge$, $\vee$ and $\neg$ in the statement, the number of gates needed to build the device decreases.

Simplification of expressions involving the connectives $\wedge$, $\vee$ and $\neg$ is governed by laws that are similar to the familiar laws of algebra. Theorem 1.24 and Theorem 1.25 below provide these laws. The algebraic process described above will be used in the algebraic proof of Theorem 1.25 on page 20. Because of the algebraic flavor of what follows, we introduce the equal sign.

Definition 1.23 *Two compound statements are* **equal**, *denoted with "=," iff for all assignments of truth values to the primitive propositions, both statements give the same value.*

Definition 1.23 simply gives another name for the situation in which two compound statements are equivalent. The use of "=" is simply more common when something is of an algebraic nature than the use of "$\Leftrightarrow$." Moreover, because the compound statements may involve the symbol $\Leftrightarrow$, it is sensible to use a symbol, such as =, that does not occur in a compound logical statement.

Next, we can introduce some properties of the logical operations. To not have too many separate definitions, we introduce the vocabulary at the same time as the property (which is still to be proved). The introduction of terminology "as we go" is quite common in mathematics.

Theorem 1.24 Boolean algebra, *part I.*

1. *The conjunction/AND operation is* **commutative**, *that is, for all primitive propositions p, q we have that $p \wedge q = q \wedge p$.*

2. *The disjunction/OR operation is* **commutative**, *that is, for all primitive propositions p, q we have that $p \vee q = q \vee p$.*

3. *The conjunction/AND operation is* **associative**, *that is, for all primitive propositions p, q, r we have that $(p \wedge q) \wedge r = p \wedge (q \wedge r)$.*

4. *The disjunction/OR operation is* **associative**, *that is, for all primitive propositions p, q, r we have that $(p \vee q) \vee r = p \vee (q \vee r)$.*

5. *The conjunction/AND operation is* **distributive** *over the disjunction/OR operation, that is, for all primitive propositions p, q, r we have the equality $p \wedge (q \vee r) = (p \wedge q) \vee (p \wedge r)$.*

6. *The disjunction/OR operation is* **distributive** *over the conjunction/AND operation, that is, for all primitive propositions p, q, r we have the equality $p \vee (q \wedge r) = (p \vee q) \wedge (p \vee r)$.*

Proof. Any claim of equality of two compound statements can be proved by showing that the compound statements have the same truth table. For example, the following truth table proves part 5.

p	q	r	$q \vee r$	$p \wedge (q \vee r)$	$p \wedge q$	$p \wedge r$	$(p \wedge q) \vee (p \wedge r)$
FALSE	FALSE	FALSE	FALSE	FALSE	FALSE	FALSE	FALSE
FALSE	FALSE	TRUE	TRUE	FALSE	FALSE	FALSE	FALSE
FALSE	TRUE	FALSE	TRUE	FALSE	FALSE	FALSE	FALSE
FALSE	TRUE	TRUE	TRUE	FALSE	FALSE	FALSE	FALSE
TRUE	FALSE	FALSE	FALSE	FALSE	FALSE	FALSE	FALSE
TRUE	FALSE	TRUE	TRUE	TRUE	FALSE	TRUE	TRUE
TRUE	TRUE	FALSE	TRUE	TRUE	TRUE	FALSE	TRUE
TRUE	TRUE	TRUE	TRUE	TRUE	TRUE	TRUE	TRUE

The proof of part 6 is left as Exercise 1-11, and the proofs of the other parts are so simple that the reader should verify them "by inspection." ∎

Regarding distributivity, in verbal statements, parentheses are not available to clarify which of "and" and "or" takes precedence. Therefore, the following convention has been adopted: In verbal statements, **"and" takes precedence over "or."** That is, a verbal statement "*A* or *B* and *C*" means "$A \vee (B \wedge C)$."

Theorem 1.25 will provide the remaining laws of Boolean algebra, which involve negation. Because we need to be very familiar with negation as we embark into more formal mathematics, we conclude this section here, and we devote the entire next section to negation.

Exercises

1-7. (Adapted from the instructions for the 2007 US income tax form 1040.) Under the US tax code, a qualifying child may entitle a taxpayer to certain deductions or even tax credits. Therefore, any taxpayer and any child of a taxpayer has an interest in determining if they satisfy this status.

For the tax year 2007, a qualifying child is a child who

(I) Is the taxpayer's son, daughter, stepchild, foster child, brother, sister, stepbrother, stepsister, or a descendant of any of them (for example, a grandchild, niece, or nephew), and

(II) Is under age 19 at the end of 2007, or, under age 24 at the end of 2007 and a student, or, any age and permanently and totally disabled, and

(III) Did not provide over half of his or her own support for 2007, and

(IV) Lived with the taxpayer for more than half of 2007. (Time in which the child attended school is counted as time lived with the taxpayer.)

Use the above rules to determine if the person is a qualifying child. Justify your answer.

(a) A 20 year old son who goes to college, has no income and lives on campus year round.

(b) A 23 year old daughter who is attending graduate school, lives on campus, who has a graduate assistantship that pays $18,000 per year and whose living expenses are $20,000 per year.

(c) A 3 year old grandchild living with the grandparents since April 2007.

(d) A 5 year old child living with its father from July 10th, 2007 to the end of the year.

(e) A son who was born on 12-30-1988, lives at home, but who works and who pays for all his living expenses from his salary.

1-8. Double check the claimed identity in Example 1.21 with a truth table.

1-9. Extending Theorem 1.20.

(a) Let $o(p, q, r)$ be a compound statement that depends on the primitive propositions p, q, and r. Prove that there is a formula for o which consists solely of $\land, \lor, \neg$ and p, q, r.

(b) Can you extend part 1-9a to compound statements with more than three entries?

Note. Presently, a description of how the formula would be constructed will do. A formal proof will be given in Exercise 3-40.

1-10. It can be proved (see Exercises 1-9b and 3-40) that every logical function can be represented with the operations $\land, \lor, \neg$. Therefore, as long as enough AND-gates, OR-gates and inverters are available, any logical function can be encoded on a chip. However, because of the way transistors[5] behave, the easiest logical gate to put on a chip is a "NAND"="NOT-AND" gate. A NAND gate returns a value of 0 ($FALSE$) iff both entries are set to 1 ($TRUE$), hence the name. The corresponding operation in logic is $p|q$, where the connective "|" is also called the **Sheffer stroke**.

(a) Prove that $\neg p$ can be represented by using only the Sheffer stroke.

(b) Prove that $p \land q$ can be represented by using only the Sheffer stroke.

(c) Prove that $p \lor q$ can be represented by using only the Sheffer stroke.

(d) Conclude that any logical function can be represented by using only the Sheffer stroke.

Hint. You may assume that the answer in Exercise 1-9b is affirmative.

---

[5] A transistor is the fundamental electrical device that makes digital circuits possible.

Note. This is our final answer to question 6 on page x. In principle, computers can be made of NAND-gates, which function like the Sheffer stroke.

1-11. Prove part 6 of Theorem 1.24.

1-12. Prove that the logic circuit in Figure 1.5 encodes the "exclusive or" operation.

Hint. Use the figure to write the corresponding statement involving $p, q, \wedge, \vee$ and $\neg$.

1-13. **Binary arithmetic.** The only arithmetic that a digital computer can perform is binary arithmetic, that is, arithmetic with zeroes and ones. In this arithmetic, $0 + 0 = 0$, $1 + 0 = 0 + 1 = 1$, and $1 + 1 = 0$, with a carryover of 1 to the next higher digit. So to add two binary numbers, we need to add corresponding digits and a potential carryover from the previous digit. The place value in binary arithmetic is that the rightmost digit has a value of $2^0 = 1$ and the place value increases as powers of 2. That is, from right to left, the values of the digits in a binary number are $2^0 = 1$, $2^1 = 2$, $2^2 = 4$, $2^3 = 8$, and so on. (This is just like regular decimals, for which the place values are, from right to left, $10^0 = 1$, $10^1 = 10$, $10^2 = 100$, and so on.) For the following, use $0 = FALSE$ and $1 = TRUE$.

   (a) Warm-up. Add the two binary numbers 0110 1011 and 1000 1111 using binary arithmetic. (It is customary to group binary digits in groups of four, similar to how decimal digits are grouped in groups of three.) Then convert the numbers into regular base 10 numbers and verify your computation.

   (b) Find a compound statement $a(p, q, c)$ that encodes the sum of the three binary digits p, q and c, but not the carry over to the next digit. (That is, for example, the sum of two ones and a zero will be zero.)

   (c) Find a compound statement $n(p, q, c)$ that encodes the carry over to the next digit of the sum of the three binary digits p, q and c. (That is, for example, the carry over of the sum of two ones and a zero will be one.)

   Note. The digital circuit that manifests the two compound statements $a(p, q, c)$ and $n(p, q, c)$ in a computer is called a **half adder**.

   Note. To add binary numbers in a computer, several half adders are connected to each other in such a way that the ps and qs are the digits of the numbers to be added and the c is the carry over n from the digit immediately to the right. The digits of the result are the outputs of functions a as above. To add two 64-bit numbers, 64 linked half adders are needed and the carry over from the last one is also called the "**overflow flag.**"

1-14. For each switch diagram in Figure 1.6, state a logical formula that expresses when current can flow from point A to point B.

1-15. Let p, q, r be primitive propositions. Construct the truth table for each of the following compound statements.

   (a) $(p \vee q) \Rightarrow r$

   (b) $(\neg r) \Rightarrow (p \wedge q)$

   (c) $\neg \left(r \Rightarrow (p \wedge q) \right)$

   (d) $(p \vee q) \Rightarrow (p \wedge q)$

1-16. **Representation of implications with or and not** Let p, q be primitive propositions. Prove that $[p \Rightarrow q] = \left[(\neg p) \vee q \right]$.

1-17. Let q be a primitive proposition. Prove that $q \vee (\neg q)$ is always true.

1-18. Let p be a primitive proposition.

   (a) Prove that $p \wedge TRUE = p$.

   (b) Prove that $p \vee FALSE = p$.

1-19. Let p be a primitive proposition.

   (a) Prove that $p \wedge FALSE = FALSE$.

   (b) Prove that $p \vee TRUE = TRUE$.

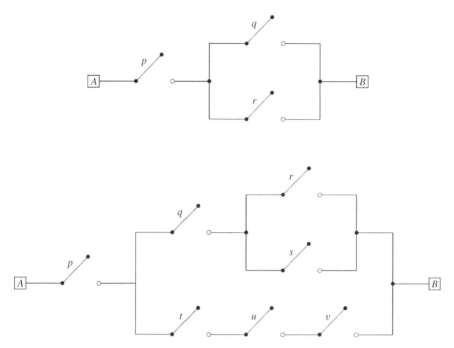

Figure 1.6 The switch diagrams for Exercise 1-14.

1.4 Special Focus on Negation

The negation of a statement is a very important operation in mathematics. Proofs by contradiction (see Standard Proof Technique 1.56) and proofs by contraposition (see Standard Proof Technique 1.66) both require the correct negation of a statement. Hence it is only appropriate that we devote special attention to negation. In this section, we show how to correctly negate the connectives $\wedge$, $\vee$, $\neg$, $\Rightarrow$ and $\Leftrightarrow$.

It was mentioned in the proof of Theorem 1.24 that any equality of compound statements can be formally verified using truth tables. Although such a verification shows beyond a doubt *that* the equality is true, it may not really convey *what the equality means*. But to correctly use negations in proofs, we must be able to assign meaning to the equalities in this section, to the point where we can use them instinctively. **Think of what happens in your mind when you read the word "horse:" You picture a large animal that looks like a horse, and you do not try to come up with any other verbal or conscious description. This is the level of instinctive familiarity that we are aiming for.**

Of course, this familiarity will take time and practice to develop. But an initial verbal explanation would contribute more to the reader's development than a mechanical verification via truth tables.[6] Moreover, in some ways, proofs are where mathematics merges with essay writing, and we might as well get started with that. Therefore, we first present a verbal proof for the next theorem. Subsequently, we present an algebraic

---

[6]Truth tables are a wonderful tool, but they must not turn into a crutch.

proof, which shows why the word "algebra" occurs in "Boolean algebra." It is quite instructive to work with rules of algebra that are slightly different from the familiar rules for numbers. In terms of developing mathematical ability, working with different rules of algebra accustoms us to the idea of following rules *as they are*, not hoping for rules to be as we *want* them to be. This is important, because, although some results in mathematics are quite surprising, we must embrace any result that can be proved, no matter how surprising it is. Moreover, the algebraic proof shows that some proofs can be "simply computations." In fact, a lot of good mathematics has been done with such computations. Yet we must be warned again: Not all proofs are computations. Although the algebraic proof develops important computational facility, it has only limited potential to explain what the equality means. Finally, for completeness, we include a proof via truth tables. Although work with truth tables has its place and should be practiced, in the context of proof writing, we want to use it only as a last resort.

Another valuable lesson here is that a statement can be proved in different ways. Throughout this text, the reader's first priority should be to produce *correct* proofs. Once a correct proof is found, the reader should analyze the proof and determine if there is a better way, say, a shorter or simpler argument. Sometimes a different proof produces different insights, too.

Theorem 1.25 Boolean algebra, *part II.* **DeMorgan's Laws.** *Let p, q be primitive propositions.*

1. *The negation of the statement $p \wedge q$ is $\neg(p \wedge q) = \neg p \vee \neg q$.*

2. *The negation of the statement $p \vee q$ is $\neg(p \vee q) = \neg p \wedge \neg q$.*

Verbal proof of Theorem 1.25. To prove part 1, note that the statement $\neg(p \wedge q)$ is false iff both p and q are true. Similarly, the statement $\neg p \vee \neg q$ is false iff both p and q are true. Hence $\neg(p \wedge q)$ is false iff $\neg p \vee \neg q$ is false, and, consequently, $\neg(p \wedge q)$ is true iff $\neg p \vee \neg q$ is true, which means that the two statements are equal.

The corresponding proof of part 2 is left to Exercise 1-20a. ∎

Algebraic proof of Theorem 1.25. The statement $\neg(p \wedge q)$ is true in three cases: when p and q are both false, when p is false and q is true, or, when p is true and q is false. Therefore, $\neg(p \wedge q) = (\neg p \wedge \neg q) \vee (\neg p \wedge q) \vee (p \wedge \neg q)$ (see the proof of Theorem 1.20). Simplification of this expression, as shown below, proves part 1.

$$\neg(p \wedge q) \quad = \quad (\neg p \wedge \neg q) \vee (\neg p \wedge q) \vee (p \wedge \neg q)$$

> By the associative law we can introduce parentheses in any way we may want to.

$$= \quad \left[(\neg p \wedge \neg q) \vee (\neg p \wedge q)\right] \vee (p \wedge \neg q)$$

> The distributive law allows us to "factor out" a $\neg p$.

$$= \quad \left[\neg p \wedge \underbrace{(\neg q \vee q)}_{=TRUE}\right] \vee (p \wedge \neg q)$$

> In a conjunction, a true input can be dropped. (See Exercise 1-18a.)

$$= \quad \neg p \vee (p \wedge \neg q)$$

> Now we use the distributive law to " multiply out" the parenthesis.

$$= \quad \underbrace{(\neg p \vee p)}_{=TRUE} \wedge (\neg p \vee \neg q)$$

> Again we can drop the true input.

$$= \quad \neg p \vee \neg q.$$

The corresponding proof of part 2 is left to Exercise 1-20b. ∎

Proof of Theorem 1.25 via truth tables.

p	q	$p \wedge q$	$\neg(p \wedge q)$	$\neg p$	$\neg q$	$\neg p \vee \neg q$
FALSE	FALSE	FALSE	TRUE	TRUE	TRUE	TRUE
FALSE	TRUE	FALSE	TRUE	TRUE	FALSE	TRUE
TRUE	FALSE	FALSE	TRUE	FALSE	TRUE	TRUE
TRUE	TRUE	TRUE	FALSE	FALSE	FALSE	FALSE

The corresponding proof of part 2 is left to Exercise 1-20c. ∎

Although chip design is more complicated than can reasonably be explained here, we can at least show how the laws of Boolean algebra are useful in industry.

Example 1.26 To build a digital circuit that encodes the "exclusive or" connective $\oplus$, we first note that $p \oplus q$ is true iff p is true and q is false or p is false and q is true. That is, $p \oplus q = (p \wedge \neg q) \vee (\neg p \wedge q)$. The right side would require five gates to be built from AND-, OR- and NOT-gates: Two AND-gates, two NOT-gates and one OR-gate. To decrease the number of gates, we use Boolean algebra to simplify the expression. Similar to the boxes in the algebraic proof of Theorem 1.25, the reader should annotate each line with the laws that were used to get from one line to the next.[7]

$$
\begin{aligned}
p \oplus q \quad &= \quad (p \wedge \neg q) \vee (\neg p \wedge q) \\
&= \quad \big[p \vee (\neg p \wedge q)\big] \wedge \big[\neg q \vee (\neg p \wedge q)\big] \\
&= \quad \big[\underbrace{(p \vee \neg p)}_{=TRUE} \wedge (p \vee q)\big] \wedge \big[(\neg q \vee \neg p) \wedge \underbrace{(\neg q \vee q)}_{=TRUE}\big] \\
&= \quad (p \vee q) \wedge (\neg q \vee \neg p) \\
&= \quad (p \vee q) \wedge \neg(q \wedge p)
\end{aligned}
$$

Upon completing the simplification, we note that the new formula would require four gates to be built from AND-, OR- and NOT-gates: Two AND-gates, one NOT-gate

---

[7] The software tool "Lurch" [5] allows the user to produce such arguments including the reasons for every step, and it highlights correct reasons in green and incorrect reasons in red.

and one OR-gate. This may not look impressive, but a 20% reduction in the number of needed gates in a circuit is significant. If such savings could be achieved with all components of a processor, then up to 20% of the original size would become "spare room" for added functionality without the need for any engineering and scientific advances to make it possible. Thus, clearly, it is important in chip design to come up with the most efficient design, and logic can help minimize the number of circuits. Again, the reader should recall that the devices most easily built on a chip model the Sheffer stroke of Exercise 1-10. But if it is possible to use algebraic simplifications to save AND-, OR- and NOT-gates, then it should also be possible to use algebra to simplify expressions involving the Sheffer stroke. Some algebraic rules for the Sheffer stroke are given in Exercise 1-30.   □

Continuing with our analysis of negations, the negation of the statement $\neg p$ is so easy to verify with a truth table that we omit the proof. The discussion below is more important than the rather trivial proof.

Proposition 1.27 Law of double negation. *Let p be a primitive proposition. The negation of the statement $\neg p$ is $\neg(\neg p) = p$.* ∎

The law of double negation says that if a statement is not not true, then it must be true. There is no third option. The statement can only be true or false. So, if it is not not true, then it is not false, so it must be true.

One of the reasons we focus on negation is because it helps us reformulate implications. A common way to reformulate $p \Rightarrow q$ is its contrapositive.

Definition 1.28 *Let p and q be primitive propositions. Then the* **contrapositive** *of $p \Rightarrow q$ is $\neg q \Rightarrow \neg p$.*

The contrapositive is a reformulation of the implication because of the following theorem.

Theorem 1.29 *Let p and q be primitive propositions. Then the statement $p \Rightarrow q$ is equivalent to the statement $\neg q \Rightarrow \neg p$.*

Verbal Proof. The statement $p \Rightarrow q$ is false iff p is true and q is false. Similarly, the statement $\neg q \Rightarrow \neg p$ is false iff $\neg q$ is false and $\neg p$ is true, that is, when p is true and q is false. Hence the two are equivalent. ∎

Because the contrapositive is equivalent to the original implication, an implication is proved if we can prove its contrapositive. Sometimes the conclusion of an implication looks easier to work with than its hypothesis. So, although formally we cannot work backwards from the conclusion, we can try to start with the negation of the conclusion and work towards the negation of the hypothesis.

Example 1.30 The contrapositive of "If a is even, then ab is even" is "If ab is not even, then a is not even." This statement should be simplified to "If ab is odd, then a is odd."   □

Sometimes we will be interested in proving that a certain implication is *not* true. The ability to prove that an implication is not true hinges upon understanding when the negation of an implication is true.

Theorem 1.31 *Let p and q be primitive propositions. The negation of the statement $p \Rightarrow q$ is $\big(\neg(p \Rightarrow q)\big) = (p \wedge \neg q)$.*

Proof. (Algebraic proof using Theorem 1.20.) The statement $\neg(p \Rightarrow q)$ is true iff p is true and q is false. Hence $\big(\neg(p \Rightarrow q)\big) = (p \wedge \neg q)$. ∎

So Theorem 1.31 says that if we want to show that an implication $p \Rightarrow q$ does not always hold, we must find one object for which p is true and q is not. Such an object is called a **counterexample**.

Example 1.32 The implication "If p is a prime number, then p is odd." is not true, because $p = 2$ is a counterexample.

Note that it does not matter that $p = 2$ is the only counterexample. One counterexample is enough. □

Finally, equivalences rarely need to be negated (see Standard Proof Technique 1.63 on how equivalences usually are treated). Hence we leave the negation of $\Leftrightarrow$ to Exercise 1-29.

Exercises

1-20. Proving part 2 of Theorem 1.25.

    (a) Give a verbal proof of part 2 of Theorem 1.25.

    (b) Give an algebraic proof of part 2 of Theorem 1.25.
        Hint. Try using [5] if you have doubts in your steps.

    (c) Prove part 2 of Theorem 1.25 by giving a truth table.

1-21. Prove Theorem 1.29 using a truth table.

1-22. Other proofs for Theorem 1.31.

    (a) Give a verbal proof for Theorem 1.31.

    (b) Prove Theorem 1.31 by giving a truth table.

1-23. Find the contrapositive of each of the following implications.

    (a) Let f be a function. If f is differentiable, then f is continuous.

    (b) Let n be a natural number. If n is divisible by 3 and $n > 3$, then n is not prime.

    (c) If our team wins the next game, then the team is eligible for postseason play.

    (d) If $f(x) = x^2$, then $f'(x) = 2x$.

    (e) If p is a twin prime number, then $p > 2$.

1-24. Answer each of the following questions and justify your answer.

    (a) Can an implication and its contrapositive both be true?

    (b) Can an implication and its contrapositive both be false?

    (c) Can an implication be true while its contrapositive is false?

(d) Can an implication be false while its contrapositive is true?

1-25. For each of the following *false* implications, give a counterexample.

(a) If n is an even number, then n is divisible by 4.

(b) If f is a continuous function, then f is differentiable.

(c) If $\circ$ is a logical function of two variables, then $\circ$ is one of $\Rightarrow, \Leftrightarrow, \wedge, \vee, \oplus, \Leftarrow$.

1-26. For each implication below, determine if it is true or false. For the false statements, provide a counterexample which shows that the statement is not true.

(a) If f is a function with $f(a) < 0$ and $f(b) > 0$, then there is a c between a and b so that $f(c) = 0$.

(b) If f is a continuous function with $f(a) < 0$ and $f(b) > 0$, then there is a c between a and b so that $f(c) = 0$.

(c) If f is a function with $f(a) = f(b)$, then there is a c between a and b so that $f'(c) = 0$.

(d) If f is a differentiable function with $f(a) = f(b)$, then there is a c between a and b so that $f'(c) = 0$.

(e) If p is a polynomial, then $\lim_{x \to \infty} p(x) = \infty$.

(f) If p is a polynomial, then $\lim_{x \to \infty} \left| p(x) \right| = \infty$.

1-27. For each of the following, give a verbal proof.

(a) If p, q, r are primitive propositions, then $(p \wedge q) \wedge (p \wedge r) = (p \wedge q) \wedge r$.

(b) If p, q, r are primitive propositions, then $(p \vee q) \vee (p \vee r) = (p \vee q) \vee r$.

1-28. For each of the following, give an algebraic proof. Justify each step with a part of Theorem 1.24 or 1.25.

(a) If p, q, r are primitive propositions, then $(p \wedge q) \wedge (p \wedge r) = (p \wedge q) \wedge r$.

(b) If p, q, r are primitive propositions, then $(p \vee q) \vee (p \vee r) = (p \vee q) \vee r$.

1-29. Prove that the negation of $p \Leftrightarrow q$ is $(\neg p \wedge q) \vee (p \wedge \neg q)$.

1-30. Algebra with the **Sheffer stroke** and negation.

(a) Prove that $(p|q)|(p|\neg q) = p$.

(b) Prove that $p|(q|r) = \neg \left[(\neg r|p)|(\neg q|p) \right]$.

1-31. Let p and q be primitive propositions. The **inverse** of the implication $p \Rightarrow q$ is $\neg p \Rightarrow \neg q$.

(a) Prove that the inverse is not equivalent to the original implication.

(b) Prove that the inverse is equivalent to the converse of the original implication.

1-32. The statement "If you don't try, then you won't win." is undoubtedly true. Does that mean that if you do try, then you will win? Explain your answer.

1.5 Variables and Quantifiers

Many mathematical theorems say something about all elements in a set or they say something about the existence of an object with certain properties. In either case, the object is undetermined and it needs a name. Mathematical entities are typically named with letters. Example 1.15 already showed some of the potential pitfalls when using symbols: As soon as symbols are involved, the truth value of a sentence depends on what is substituted for the symbols. Nonetheless, sentences that involve symbols will be very important for us.

Definition 1.33 *A sentence that includes symbols (called* **variables***) like* x, y, *etc. and which becomes a statement when all variables are replaced with objects taken from a given set*[8] *is called an* **open sentence***.*

Open sentences can be turned into statements by specifying where the objects that can replace the variables come from and by specifying in what way the open sentence is supposed to be true. There are two ways in which variables in open sentences typically are used, or, "quantified:" The open sentence should either be true when x is replaced with *any* object in a set, or, it should be true when x is replaced with *some* object in a set.

Definition 1.34 *Let* $p(x)$ *be an open sentence that depends on the variable* x, *let* S *be a set and let "*$x \in S$*" denote the fact that* x *is an element of* S.

1. *The statement* $\forall x \in S : p(x)$ *(read "for all* x *in* S *we have* $p(x)$*") is true iff* $p(x)$ *holds for all elements* x *in the set* S.

2. *The statement* $\exists x \in S : p(x)$ *(read "there is an* x *in* S *so that* $p(x)$*") is true iff* $p(x)$ *is true for at least one element* x *in the set* S.

The symbols $\forall$ *and* $\exists$ *are called* **quantifiers***.* $\forall$ *is the* **universal quantifier** *and* $\exists$ *is the* **existential quantifier***. The statement of what set* x *belongs to may also be encoded by an equation or an inequality that defines the set. So, for example, instead of writing* $\forall x \in \{y \in \mathbb{R} : y > 0\} : p(x)$ *we may write* $\forall x > 0 : p(x)$, *if it is understood that* x *must be a real number.*

Quantifiers help formulate complicated statements in "mathematical shorthand." They are especially useful when finding negations of complicated statements (see Theorem 1.44 below). We start with some reasonably straightforward statements in Examples 1.35-1.38 and one peculiar one in Example 1.39. Then we investigate nested quantifications.

Example 1.35 *The sentence "*$\forall x \in \mathbb{R} : x^2 \geq 0$*" is a true universally quantified statement.*
Because "$x^2 \geq 0$" becomes a statement when x is replaced with a real number, the open sentence is $p(x) = $ "$x^2 \geq 0$." Overall, the sentence is a universally quantified

---

[8]Here we rely on the reader's intuition of what a set might be. In this section, we will use sets without qualms.

statement because it is of the form "$\forall x \in \mathbb{R} : p(x)$." At this stage of our development of mathematics we can *say* the statement is true, because we *know* that squares of real numbers are nonnegative. To *prove* that the statement is true we need to know more about the real numbers. The statement will be proved in Proposition 5.23. □

Example 1.36 *The sentence "$\forall x \in \mathbb{C} : x^2 \in [0, \infty)$" is a false universally quantified statement.*

The argument that the sentence is a universally quantified statement is similar to the argument in Example 1.35. To show that the statement is false, note that $i^2 = -1$, which is not in $[0, \infty)$. Of course, we use some knowledge about the real and complex numbers here, but the facts that $i^2 = -1$ and that -1 is not in $[0, \infty)$ will be confirmed by our construction of the real and complex numbers in Sections 5.5 and 5.7.

Note that because the complex numbers are not ordered (see Exercise 5-52 later in the text), it is safer to talk about numbers being in the interval $[0, \infty)$ than about numbers being greater than or equal to zero. □

The similarity between universally quantified statements and implications is not an accident. The statement "$\forall x \in S : p(x)$" is true iff the statement "Let x be an object. If $x \in S$, then $p(x)$." is true. So universal quantification truly is shorthand for a way to formulate mathematics that we have already seen, say, in Example 1.13.

On the other hand, existentially quantified statements typically are not reformulated using other connectives. (But consider Exercise 1-33a for a brute force way to avoid existential quantifiers.)

From now on, we will let the reader double check that sentences with quantifiers really are quantified statements. To perform the check, simply confirm that what follows the quantifier is an open sentence in the quantified variable.

Example 1.37 *The sentence "$\exists x \in \mathbb{C} : x^2 = -1$" is a true existentially quantified statement.*

Our construction of the complex numbers will show that $i^2 = -1$, so there is $x = i \in \mathbb{C}$ so that $x^2 = i^2 = -1$. □

Example 1.38 *The sentence "$\exists x \in \mathbb{R} : x^2 = -1$" is a false existentially quantified statement.*

We *know* that squares of real numbers are nonnegative, so the statement must be false. The *proof* that squares of real numbers are nonnegative will be given in Proposition 5.23. □

Some true implications can look strange because a false hypothesis can imply anything. Similarly, universally quantified statements can look strange, even though they are formally true.

Example 1.39 *The statement "Every man born in Antarctica became president of Antarctica." is a true universally quantified statement.*

The universal quantification can be recognized upon formalizing the sentence a bit. If M is the set of all men born in Antarctica, and $p(x)$ is the open sentence "x became president of Antarctica," then the statement above can be encoded as "$\forall x \in M : p(x)$."

The claim that the statement is true is a bit harder to accept. How can every man become president? Note, however, that in recorded human history, no man was born in Antarctica. Therefore, the set M is empty (that is, it has no elements). Paradoxical as it may sound, every element of the empty set has become president of Antarctica, because there is no element that could serve as a counterexample.

Let us reformulate the statement as the implication "If $m \in M$, then m became president of Antarctica." In this formulation, we see that the statement is true, because any implication with a false hypothesis (m cannot be in M because M is empty) is true. So, even though the language has changed a little, we are not running into anything new here. $\qquad\square$

Because the whole argument in Example 1.39 happens in the "vacuum" that is the empty set, statements as in Example 1.39 are called "vacuously true."

Definition 1.40 *Universally quantified statements that quantify over a set that turns out to be empty are also called* **vacuously true**.

The true power of quantifications reveals itself when we start nesting quantifiers. Many important definitions and theorems are stated as nested quantifications. For example, the all-important definition of the limit is a triply nested quantification (see Example 1.42). But before we reach this level of complexity, let us first consider a doubly nested quantification.

Example 1.41 *The sentence* $\forall x \in [0, \infty) : \exists r \in [0, \infty) : r^2 = x$ *is a true doubly quantified statement.*

Note that $p(x) := \text{"}\exists r \in [0, \infty) : r^2 = x\text{"}$ is an open sentence with one variable x. The open sentence contains another open sentence $q(x, r)$ with two variables, x and r. In the double quantification, it should be understood that the outer quantification fixes the variable x. Hence we could say that the inner quantification "$\exists r \in [0, \infty) : r^2 = x$" can be interpreted as an open sentence with one variable, because x has already been fixed. In this fashion, the structure of the inner quantification still fits Definition 1.34. We therefore avoid the need to define open sentences with multiple variables, which would be required for a formal definition of nested quantifications.[9]

The statement is true, because it says that every nonnegative real number has a square root, which we *know* to be true. We will *prove* this fact in Theorem 5.44. $\qquad\square$

For our final example on nested quantifications, we borrow sequence notation from calculus.

Example 1.42 *The statement "The real number L is the limit of the sequence of real numbers $\{a_n\}_{n=1}^{\infty}$." is formally encoded as the triply nested quantified statement "$\forall \varepsilon > 0 : \exists N \in \mathbb{N} : \forall n \geq N : |a_n - L| < \varepsilon$."*

We cannot discuss the truth value here, because we do not know what L and $\{a_n\}_{n=1}^{\infty}$ are. But we can interpret what is said above. To make the argument easier to read, let

---

[9]Again, we will not go deeper into these finer points. We are interested in the fundamentals of mathematical communication, leaving the subtleties to specialists in logic.

us assume that L and $\{a_n\}_{n=1}^{\infty}$ have been given to us so that we can talk about the statement being true or false.

The idea of the limit is that, for all sufficiently large indices n, the values a_n of the sequence must be close to the limit L. The quantified statement says that this is the case iff for all $\varepsilon > 0$ there is an $N \in \mathbb{N}$ so that for all $n \geq N$ we have $|a_n - L| < \varepsilon$. That is, no matter how small, or, "close," we choose a tolerance ε, there will always be a threshold N so that for all $n \geq N$, that is, for all "sufficiently large n," we have $|a_n - L| < \varepsilon$, that is, a_n is "close to" L for these n. Hence, the quantified statement above gives precise mathematical meaning to what otherwise involves the undefined terms "close to" and "sufficiently large." $\qquad\square$

The definition of the limit of a sequence is explored in depth in any introductory text on analysis.[10]

Notation 1.43 *For consecutive existential or universal quantifications over the same set, the quantifications are often combined, and we will do so in this text. So, instead of "$\exists p \in \mathbb{N} \setminus \{1\} : \exists q \in \mathbb{N} \setminus \{1\} : pq = 12$" we say "$\exists p, q \in \mathbb{N} \setminus \{1\} : pq = 12$," and instead of "$\forall x > 0 : \forall y > 0 : xy > 0$" we say "$\forall x > 0, y > 0 : xy > 0$," or, "$\forall x, y > 0 : xy > 0$."*

Theorem 1.44 *Let $p(x)$ be an open sentence that depends on the variable x and let S be a set.*

1. *The negation of $\forall x \in S : p(x)$ is $\neg\big(\forall x \in S : p(x)\big) = \exists x \in S : \big(\neg p(x)\big)$.*

2. *The negation of $\exists x \in S : p(x)$ is $\neg\big(\exists x \in S : p(x)\big) = \forall x \in S : \big(\neg p(x)\big)$.*

Proof. For part 1, note that $\neg\big(\forall x \in S : p(x)\big)$ is true iff $\forall x \in S : p(x)$ is false, which is the case iff there is an $x \in S$ so that $p(x)$ is false, which is the case iff there is an $x \in S$ so that $\neg p(x)$ is true, which is the case iff $\exists x \in S : \big(\neg p(x)\big)$ is true.

The proof of part 2 is left to the reader as Exercise 1-34. $\qquad\blacksquare$

So, negation of a quantified statement is easy: Simply interchange the quantifier for the respective other one and negate the open sentence inside the quantification. Example 1.45 shows that this is easy for single quantifications and Example 1.46 shows that quantifiers can help clear things up for more complicated statements.

Example 1.45 For the universally quantified statement "$\forall x \in \mathbb{C} : x^2 \in [0, \infty)$" from Example 1.36, the negation is "$\exists x \in \mathbb{C} : x^2 \notin [0, \infty)$." $\qquad\square$

Example 1.46 *The negation of "$\forall \varepsilon > 0 : \exists N \in \mathbb{N} : \forall n \geq N : |a_n - L| < \varepsilon$" from Example 1.42 is "$\exists \varepsilon > 0 : \forall N \in \mathbb{N} : \exists n \geq N : |a_n - L| \geq \varepsilon$."*

The negation is not as easy to see as in Example 1.45. For quantifications such as the one in Example 1.42, it can be helpful to work the quantifiers one at a time.

"$\neg\big(\forall \varepsilon > 0 : \exists N \in \mathbb{N} : \forall n \geq N : |a_n - L| < \varepsilon\big)$" is equivalent to

"$\exists \varepsilon > 0 : \neg\big(\exists N \in \mathbb{N} : \forall n \geq N : |a_n - L| < \varepsilon\big)$," which is equivalent to

---

[10] And the author is pretty sure that it would please the publisher if the reader would consider [30].

"$\exists \varepsilon > 0 : \forall N \in \mathbb{N} : \neg \big(\forall n \geq N : |a_n - L| < \varepsilon \big),$" which is equivalent to

"$\exists \varepsilon > 0 : \forall N \in \mathbb{N} : \exists n \geq N : \neg \big(|a_n - L| < \varepsilon \big),$" which is equivalent to

"$\exists \varepsilon > 0 : \forall N \in \mathbb{N} : \exists n \geq N : |a_n - L| \geq \varepsilon.$"

In the last step, we were able to use that in the context of real numbers the negation of $|a_n - L| < \varepsilon$ is $|a_n - L| \geq \varepsilon$. □

Notation 1.47 In some texts, the universal quantifier $\forall$ is denoted $\bigwedge$ and the existential quantifier $\exists$ is denoted $\bigvee$. This is just a different way of working mnemonics. The $\forall$ is the "A" in "For **All**" turned upside down and the $\exists$ is the "E" in "There **Exists**" turned upside down. Similarly, note that, for a finite set, a universal quantification is true iff the open sentence is true for the first element *and* for the second element *and* for the third element, etc. So the quantifier $\bigwedge$ is just a large AND symbol $\wedge$. Moreover, for a finite set, an existential quantification is true iff the open sentence is true for the first element *or* for the second element *or* for the third element, etc. So the quantifier $\bigvee$ is just a large OR symbol $\vee$.

With the quantifiers $\bigwedge$ and $\bigvee$, Theorem 1.44 looks very similar to DeMorgan's Laws (see Theorem 1.25). It is unfortunate that the universal quantifier is denoted by $\forall$ in one notation and the existential quantifier is denoted by $\bigvee$ in the other. Because the symbols are so similar, translation between the two conventions is more tedious than we want it to be.

In this text, we will use $\forall$ and $\exists$ throughout.

Exercises

1-33. Reformulating a quantifier in terms of the other quantifier.

    (a) Prove that the statement "$\exists x \in S : p(x)$" is true iff the statement "$\neg \big[\forall x \in S : \neg p(x) \big]$" is true.

    (b) Prove that the statement "$\forall x \in S : p(x)$" is true iff the statement "$\neg \big[\exists x \in S : \neg p(x) \big]$" is true.

1-34. Prove part 2 of Theorem 1.44.

1-35. State the negation of each of the following statements. Also state if the statement or its negation is true.

    (a) $\exists p \in \mathbb{N} : 2p = 6$

    (b) $\forall x \in \mathbb{R} : x^2 \geq 1$

    (c) $\exists p, q \in \mathbb{N} : pq = 12$

    (d) $\forall n \in \mathbb{N} : \exists p, q \in \{2, \ldots, n-1\} : pq = n.$

    (e) $\forall x > 0 : \exists n \in \mathbb{N} : n > x$

    (f) $\forall \varepsilon > 0 : \exists \delta > 0 : \forall x \in (-\delta, \delta) : x^2 < \varepsilon$

    (g) $\exists a \in \mathbb{R} : \forall \varepsilon > 0 : \exists \delta > 0 : \forall x \in (-\delta, \delta) : \big| x^2 - a^2 \big| < \varepsilon$

    Note. This one is challenging.

1-36. State the negation of each of the following statements.

    (a) $\forall x \in \mathbb{N} : x^2 \geq 1$

    (b) $\forall \varepsilon > 0 : \exists \delta > 0 : \forall x \in (x_0 - \delta, x_0 + \delta) : \left| f(x) - f(x_0) \right| < \varepsilon$, where f denotes one fixed function.

    (c) $\forall \varepsilon > 0 : \exists \delta > 0 : \forall x \in (-\delta, \delta) : \sqrt{x} < \varepsilon$

    (d) $\exists f \in \mathcal{F} : \forall \varepsilon > 0 : \exists \delta > 0 : \forall |x - a| < \delta : \left| f(x) - f(a) \right| < \varepsilon$, where $\mathcal{F}$ denotes a fixed family of functions.

1-37. For each of the following *false* universally quantified statements, provide a counterexample.

    (a) Every prime number is odd.

    (b) $\forall n \in \mathbb{Z} : \exists k \in \mathbb{Z} : nk = 1$

    (c) $\forall n \in \mathbb{Q} : \exists k \in \mathbb{Q} : nk = 1$

1-38. Although the results can look strange and natural language is ultimately preferable, it is a good exercise to overlay the quantifier formalism over statements in natural language. Formulate each of the following statements with quantifiers and decide if it is true or false.

    (a) Stars other than the Sun have planets.

    (b) All planets have moons.

    (c) Every member of an on-line social network owns a computer.

    (d) Every real number has a positive square root.

1.6   Proofs

Now that the formal language is in place, we can analyze the structure of mathematical proofs. In the proofs we have seen so far, there were three main approaches. With a truth table, we simply computed truth values to prove that two statements were equal. In an algebraic proof, we found the right computation to show that two statements were equal. Finally in a verbal proof, we had to put up a "convincing argument" to show that whatever we wanted to prove was true. Ultimately, we will need to write proofs in any branch of mathematics. So the utility of truth tables is limited to (elementary) formal logic. They can only be used to prove the equality of compound statements. Algebraic proofs are a bit more widely applicable, because many proofs in mathematics have a computational component. That is, there will be situations in which a computation that follows certain rules will prove an identity or an inequality.[11] Most of the time, however, a proof is a verbal argument that shows that a certain hypothesis implies a certain conclusion.

    The formalisms we have introduced so far can help us write and analyze proofs, because they provide the requisite structure for proofs. Even though it seems as if natural language allows us to do just about anything in a verbal proof, logic limits us to what is sensible: A proof typically starts with a hypothesis and, through some valid argument, the proof derives the conclusion. If we replace the hypothesis, the conclusion, as well as all intermediate statements with primitive propositions, then we

---

[11] At this stage we will not get into philosophical discussions where computer generated proofs and calculations fit in. It should be said, though, that such arguments are frequently used when a computation can be carried out "in principle" by hand, but would take an unacceptable amount of time to complete.

obtain the logical structure of the proof, which leads us from the hypothesis to the conclusion. Interestingly enough, this logical structure is independent of the context of a particular proof. That is, the structure of the proof must be so that truth of the hypothesis plus truth of any other results that are used along the way leads to truth of the conclusion. In logic notation, the structure of a proof of $p \Rightarrow q$ is always $\left[p \wedge \langle \text{argument} \rangle \right] \Rightarrow q$. Because the logical structure of a proof is context independent, the (possibly very long and complicated) statement $\left[p \wedge \langle \text{argument} \rangle \right] \Rightarrow q$ must be true no matter what the truth values of p and q are, as long as all outside results used in the argument are true. This is the idea of a tautology.

Definition 1.48 *A compound statement is called a* **tautology** *iff for all assignments of truth values to the primitive propositions, the compound statement returns the value $TRUE$. A compound statement is called an* **absurdity** *iff for all assignments of truth values to the primitive propositions, the compound statement returns the value $FALSE$. If the compound statement can return both $TRUE$ and $FALSE$ depending on the values of the primitive propositions, then the statement is called a* **contingency**.

Tautologies are the basic building blocks for proofs. No matter what the context is, an inference that is based on a tautology will always work. In the following, we will provide a set of tautologies, which we will also call **rules of inference**, that are frequently encountered in proofs. We will also analyze some sample arguments in light of these rules of inference. Throughout, the reader should keep in mind that this mechanical analysis mainly serves to illustrate the points made. After enough practice, a working mathematician does not consciously recall every rule of inference that is applied correctly in a proof, and the names of the rules also recede from memory. This level of familiarity is similar to what happens when in a computation we apply the commutative law or factor out a common term. We may not even consciously recall that we apply a law when we say $3 \cdot 5 = 5 \cdot 3$, and we may not remember that, formally, when we "factor out" a term that the law we apply is called the distributive law. Once this level of familiarity is reached, there is nothing wrong with correctly using the rules of inference in an essentially subconscious fashion. But we should remember that the rules of algebra we now use without much worry were hard won when we were younger. Rules like $3 \cdot 5 = 5 \cdot 3$ were acquired when learning multiplication tables, and the distributive law was acquired through multiplying out parentheses and very carefully (consciously) taking care of all summands in the process. Early on in our education, many algebraic rules required a conscious effort, just like our analysis of proofs will be a conscious effort. With experience, acquired through this text and others, the conscious effort will become second nature, just like elementary algebra.

Notation 1.49 *Throughout this section, p, q, r, etc. denote primitive propositions, so we will not re-state this assumption in every theorem.*

Theorems 1.50, 1.55 and 1.59 provide the tautologies that underly three fundamental techniques used to prove that a proposition p implies a proposition q: direct proof, proof by contradiction and proof by case analysis, respectively.

Theorem 1.50 *The statement $\left[p \wedge (p \Rightarrow q)\right] \Rightarrow q$ is a tautology.*

Proof. The result can be proved verbally as follows. The statement is true when q is true, because any implication with a true conclusion is true. Similarly, the statement is true when p is false, because then $\left[p \wedge (p \Rightarrow q)\right]$ is false and any implication with a false hypothesis is true. Finally, when p is true and q is false, note that $p \Rightarrow q$ is false, which means the hypothesis $\left[p \wedge (p \Rightarrow q)\right]$ is false and hence the implication $\left[p \wedge (p \Rightarrow q)\right] \Rightarrow q$ is true.   ∎

The last part of the proof of Theorem 1.50 can be interpreted as saying that when p is true and q is false, the actual proof, which is to establish that $p \Rightarrow q$ is true, cannot work.

Standard Proof Technique 1.51 *Proofs built on the rule of inference in Theorem 1.50 are called* **direct proofs** *and the inference is also called the* **law of detachment,** *or* **modus ponens,** *or, more fully, modus ponendo ponens, which is Latin for "mode that affirms by affirming."*

Modus ponens proofs are easy to explain. If we know that a certain hypothesis implies a certain conclusion and if we also know that the hypothesis is true, then the conclusion must be true.   □

In this text, we will typically stick to the English name of a tautology. To assure that the reader will not be caught by surprise if slightly different terminology is used elsewhere, for many tautologies we provide (a multitude of) other names, too.

A direct proof is nothing but the application of a known implication.

Example 1.52 *The sum of the three interior angles in a plane triangle is* $180°$.
   Proof. We will prove this statement using the following facts.

1. If L is a line and P is a point not on the line L, then there is exactly one line parallel to L that goes through P.

2. If two angles are opposite on parallel lines, then the two angles are equal.

3. If σ is a straight angle, then σ measures $180°$.

Consider an arbitrary triangle (see Figure 1.7). Label the vertices A, B and C and let α, β and γ be the angles at A, B and C, respectively. We can use modus ponens with fact 1, the line through A and B and the point $P = C$ to conclude that there is a unique line through C that is parallel to the line through A and B. (So modus ponens is quite literally the application of an implication that is known to be true.) At C, this line gives the initial and terminal sides of a straight angle. This straight angle is the sum of three angles: the angle γ plus the angles made by the sides of the triangle and the line through C that are to the left and to the right of γ. Opposite angles on parallel lines are equal. So modus ponens with fact 2, using the line through A and C shows that the angle to the left of γ is α. Similarly, modus ponens with fact 2, using the line through B and C shows that the angle to the right of γ is β. Hence $\alpha + \beta + \gamma$ is the measure of a straight angle, and by modus ponens with fact 3, we conclude that the sum is $180°$.   ∎

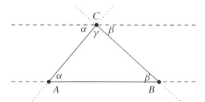

Figure 1.7 Illustration of the proof in Example 1.52.

Note that because we did not formally state our axioms for geometry, we had to state *from what* the statement in Example 1.52 would be derived. We will only face this challenge in the first chapter. In the rest of the text, we will, at least in principle, be able to reduce results to the axioms of set theory, which will be given in Chapter 2. Moreover, note that the references to modus ponens in Example 1.52 were merely included to illustrate the use of the proof technique. Outside of introductions to proof techniques (such as in this section), it is not necessary to say that modus ponens was used to get from one statement to another.

For our next rule of inference, we first have to lay some groundwork.

Proposition 1.53 *The statement $p \vee \neg p$ is a tautology and $p \wedge \neg p$ is an absurdity.*

Proof. For $p \vee \neg p$, note that the statement clearly is true if p is true. On the other hand, if p is false, then $\neg p$ is true and hence $p \vee \neg p$ is true. Thus $p \vee \neg p$ is always true.

The argument that $p \wedge \neg p$ is an absurdity is similar. (Exercise 1-40.) ∎

Remark 1.54 *The tautology $p \vee \neg p$ is also called the* **law of the excluded middle**.

This name is appropriate, because $p \vee \neg p$ being always true means that whatever happens, one of p or $\neg p$ is always true. There is no need for a third option in case both fail, because they cannot fail at the same time. Also recall that we have already used the law of the excluded middle several times in Section 1.4. This was formally o.k., because we proved it in Exercise 1-17.

Along these lines, $p \wedge \neg p$ being an absurdity shows that there is no overlap between the two options of p being true or $\neg p$ being true. □

Theorem 1.55 *The statement $\left[p \wedge \big((p \wedge \neg q) \Rightarrow FALSE \big) \right] \Rightarrow q$ is a tautology.*

Proof. The result can be proved algebraically as follows.

$$\left[p \wedge \big((p \wedge \neg q) \Rightarrow FALSE \big) \right] \Rightarrow q$$

> By Theorem 1.31 and DeMorgan's Laws, we know that the statement "$x \Rightarrow y$" is the same as "$\neg(x \wedge \neg y)$"="$\neg x \vee y$." So we can eliminate the implications in the statement. We first eliminate the implication at the end.

$$= \quad \neg \left[p \wedge \big((p \wedge \neg q) \Rightarrow FALSE \big) \right] \vee q$$

| Now we eliminate the implication in the square brackets. |

$$= \quad \neg\Big[p \wedge \big(\neg(p \wedge \neg q) \vee FALSE\big)\Big] \vee q$$

| Now we apply DeMorgan's Laws to $\neg(p \wedge \neg q)$. |

$$= \quad \neg\Big[p \wedge \big((\neg p \vee q) \vee FALSE\big)\Big] \vee q$$

| If one of the two primitive propositions in an "or" is false, then the truth value is determined solely by the other primitive proposition (see Exercise 1-18b). |

$$= \quad \neg\Big[p \wedge (\neg p \vee q)\Big] \vee q$$

| Use the distributive law. |

$$= \quad \neg\Big[\underbrace{(p \wedge \neg p)}_{=FALSE} \vee (p \wedge q)\Big] \vee q$$

| Once more, an entry of $FALSE$ in an "or" can be ignored. |

$$= \quad \neg(p \wedge q) \vee q$$

| Use DeMorgan's Laws. |

$$= \quad (\neg p \vee \neg q) \vee q$$

| Use the associative law for "or." |

$$= \quad \neg p \vee \underbrace{(\neg q \vee q)}_{=TRUE}$$

| Once one entry of an "or" is true, the whole or-statement will be true (see Exercise 1-19b). |

$$= \quad TRUE$$

Because the expression simplifies to an expression that is always true, the expression is a tautology. ∎

Standard Proof Technique 1.56 *Proofs built on the rule of inference in Theorem 1.55 are called* **proofs by contradiction** *or* **reductio ad absurdum**.

The idea for proofs by contradiction stems from the fact that a true statement cannot imply a false statement. Consider a true statement p and suppose that we can prove that $p \wedge \neg q$ leads to a contradiction; that is, suppose that we can prove that $p \wedge \neg q$ implies an absurdity. Then $p \wedge \neg q$ must be false, because a true statement cannot imply a false statement. But if $p \wedge \neg q$ is false and p is true, then $\neg q$ must be false. Therefore, by the law of double negation, q must be true.

Another way to set up proofs by contradiction is given in Exercise 1-42.

To clearly indicate that we are doing a proof by contradiction, in this text proofs by contradiction typically start with "Suppose for a contradiction ...". This clear indication is not standardized. In the literature, proofs by contradiction also start with "Assume for a contradiction ..." or, without reference to contradictions with "Suppose ..." or

"Assume ...". At the conclusion of a proof by contradiction, it is customary to indicate that a contradiction has been reached. □

A standard example of a proof by contradiction is the proof that the square root of 2 is not a rational number.

Example 1.57 $\sqrt{2}$ *is not rational.*

Proof. *Suppose for a contradiction* that $\sqrt{2}$ is rational. Then, because rational numbers are fractions that can be written in lowest terms, there are $n, d \in \mathbb{N}$ so that $\sqrt{2} = \frac{n}{d}$ and so that n and d have no common prime factors. Then $\frac{n^2}{d^2} = \left(\frac{n}{d}\right)^2 = 2$, which means that $n^2 = 2d^2$. But then 2 is a prime factor of n^2 and hence 2 is a prime factor of n. Therefore, $n = 2m$ and $2d^2 = n^2 = 4m^2$, that is, $d^2 = 2m^2$. But then 2 is a prime factor of d^2 and hence 2 is a prime factor of d, contradicting the assumption that n and d have no common prime factors. *Because we have reached a contradiction,* $\sqrt{2}$ *cannot be rational.* ∎

Note that in the proof in Example 1.57 we used that if 2 is a prime factor of a square k^2, then 2 must be a prime factor of k. Formally, we would need to prove this fact. But to prove it, we may need to establish other things. Once again we don't know precisely where to stop proving results, because we do not yet have a definite starting point in our construction of mathematics. Starting with Chapter 2, this will not be an issue any more, because we will build mathematics "from the ground up." Eventually, Exercise 3-71 will show that if a number is a prime factor of a square k^2, then it must also be a prime factor of k.

Example 1.58 The idea of a proof by contradiction is the cornerstone of the **scientific method**. Basically, the answer to question 7 on page x is that *a scientific*[12] *statement that contradicts experimental observations must be false.*

For example, consider the model of matter as a continuous substance from the introduction to Section 1.1. Based on this model, a solid, thin metal foil should completely block an electron beam from passing through. But experiments show that electron beams go through very thin metal foils without damaging them. Consequently the assumption that matter is a continuum leads to a conclusion that was experimentally shown to be false. Because the implication was valid, the hypothesis must be false.

In the notation of Theorem 1.55, we could say that $p =$"barriers in which there is no empty space block electron beams" and $q =$"matter contains empty space." □

Theorem 1.59 *The statement* $\left[p \wedge (A \vee B) \wedge \left((p \wedge A) \Rightarrow q\right) \wedge \left((p \wedge B) \Rightarrow q\right)\right] \Rightarrow q$ *is a tautology.*

Proof. The proof is similar to the proof of Theorems 1.50 and 1.55 (see Exercise 1-43). ∎

Standard Proof Technique 1.60 *Proofs built on the rule of inference in Theorem 1.59 are called* **proofs by case distinction** *or* **proofs by exhaustion***.*

---

[12] Actually, this is true for any statement, but the subject for this example is the scientific method.

Proofs by case distinction are used when it is realized that a conclusion could be proved if only we had an additional hypothesis A. If we know p is true and that $p \wedge A$ implies q, then to fully prove q, we must somehow prove q when A is not guaranteed to be true. So, if there are two possible additional hypotheses, A and B, so that one of A and B is always true (this is what the "additional hypothesis" $A \vee B$ in Theorem 1.59 guarantees: formally the theorem establishes that $p \wedge (A \vee B)$ implies q), then we could split the proof that p implies q into two parts. In one part, we prove that $p \wedge A$ implies q. In the other part, we prove that $p \wedge B$ implies q. Both parts are presumably easier to prove, because we have an additional hypothesis. Once both parts are proved, we can conclude that p must imply q, because, as long as A or B is true, adding the one that is true to the hypothesis will establish truth of q. To make sure p always implies q, case distinctions typically use additional hypotheses on A and B so that $A \vee B$ is always true. Note, however, that A and B are allowed to overlap.

Proofs by case distinction are not limited to splitting the argument into two parts (see Exercise 1-44).

Proofs by case distinction are often presented by explicitly considering the cases in separate paragraphs. But short case distinctions can also be handled in short sentences, as in "If A, then ..., and if B, then ..., so ... is true." □

Example 1.61 *Every natural number that is not a power of 2 has an odd prime factor.*

If n is a natural number that is not of the form 2^k, then *we distinguish two cases*: Either n is odd or n is even.

Case 1: n is odd. If n is odd, then n is not divisible by 2 and no factor of n is divisible by 2. Therefore all prime factors of n (if there are any) are odd. Moreover, because $n \neq 2^0 = 1$, n is at least greater than or equal to 2 and thus it must have prime factors. This proves the result in case n is odd.

Case 2: n is even. If n is even, then n is divisible by 2. But because n is not a power of 2, not all prime factors of n are 2. Because 2 is the only even prime number, n must have an odd prime factor. This proves the result in case n is even. ∎

The next two proof techniques are different from the first three in Theorems 1.50, 1.55 and 1.59 in that they are not directly starting with a hypothesis p and they do not use an argument to derive q. Instead, these proof techniques take a statement (which we may be required to prove) and translate it into an equivalent statement. This is another common approach to proving results: Translate the desired result into another version which may be easier to prove.

Theorem 1.62 *The statement* $(p \Leftrightarrow q) \Leftrightarrow \big((p \Rightarrow q) \wedge (q \Rightarrow p)\big)$ *is a tautology, called the* **biconditional law**.

Proof. Exercise 1-45. ∎

Standard Proof Technique 1.63 Proofs that use the biconditional law simply split up the work to prove a biconditional into the often simpler task of proving two implications, left to right "$\Rightarrow$" and right to left "$\Leftarrow$". *Throughout the text, the two halves of a biconditional "A iff B" will be referred to as "$\Rightarrow$," denoting "if A, then B" and "$\Leftarrow$," denoting "if B, then A."* □

The next example shows that, even for very simple biconditionals, it is typically easier to split up the proof into two implications than to try to preserve biconditionals throughout.

Example 1.64 *A natural number x is even and prime iff x = 2.*

Proof. For "$\Leftarrow$" note that 2 is even and prime.

For the converse direction "$\Rightarrow$," let x be even and prime. Because x is even, there is a natural number k so that $x = 2k$. Because x is prime, we must have that $k = 1$. Hence $x = 2$. ∎

Theorem 1.65 *The statement $(p \Rightarrow q) \Leftrightarrow (\neg q \Rightarrow \neg p)$ is a tautology, called the* **law of contraposition**.

Proof. Left as Exercise 1-46. ∎

Standard Proof Technique 1.66 *Proofs built on the law of contraposition are also called* **proof by contraposition** *or* **modus tollens** *(more fully, modus tollendo tollens, "the way that denies by denying").*

There are situations in which the negation of the conclusion makes for a better hypothesis than the hypothesis itself. When that is the case, the contrapositive can be easier to prove than the original statement. Overall, proving the contrapositive is merely shifting the focus of the argument to a translation of the original implication.

It should be noted that proofs by contraposition are sometimes disguised as proofs by contradiction. After all, if we know p and we can prove that $\neg q$ implies $\neg p$, then assuming $\neg q$ leads to the contradiction $p \wedge \neg p$. (Also see Exercise 1-47.) □

The following example may be simple, but it shows how the contrapositive can simplify the task of proving a statement, for example, by removing negations.

Example 1.67 *Let x be a natural number. If 2 is not a factor of x, then 4 is also not a factor of x.*

Proof. The contrapositive of the above statement is "If 4 is a factor of x, then 2 is a factor of x." To prove the contrapositive, let x be a natural number so that 4 is a factor of x. This means that there is a natural number k so that $n = 4 \cdot k = 2 \cdot 2 \cdot k$, which means that 2 is a factor of x. Because we have proved the contrapositive, the result must be true. ∎

Sometimes an implication actually is *not* true and needs to be *dis*proved.

Theorem 1.68 *The statement $\left[\neg (p \Rightarrow q) \right] \Leftrightarrow (p \wedge \neg q)$ is a tautology.*

Proof. Left as Exercise 1-48. ∎

Standard Proof Technique 1.69 *The statement in Theorem 1.68 does not seem to have a separate name. We'll call it* **disproof by counterexample**.

For a disproof by counterexample, we must somehow establish that p and $\neg q$ can be true at the same time. This is typically done by constructing an example that has these properties (hence the name). □

Example 1.70 *Disprove the implication "If p is a prime number, then p is odd."*
The number 2 is prime and even, which is a counterexample for this statement. □

Note that the three tautologies in Theorems 1.62, 1.65 and 1.68 could also have been stated as the three Boolean identities $(p \Leftrightarrow q) = ((p \Rightarrow q) \wedge (q \Rightarrow p))$, $(p \Rightarrow q) = (\neg q \Rightarrow \neg p)$ and $[\neg(p \Rightarrow q)] = (p \wedge \neg q)$. To retain the uniform pattern of the theorems in this section, we stated Theorems 1.62, 1.65 and 1.68 as tautologies.
Finally, several tautologies are often used as steps in a proof.

Theorem 1.71 *The following statements are tautologies.*

1. $(p \wedge q) \Rightarrow q$ **(law of simplification)**

2. $p \Rightarrow (p \vee q)$ **(law of addition)**

3. $((p \Rightarrow q) \wedge (q \Rightarrow r)) \Rightarrow (p \Rightarrow r)$ **(law of syllogism**, *or,* **transitive law** *for implications)*

4. $(\neg p \wedge (p \vee q)) \Rightarrow q$ **(law of disjunction**, *or,* **modus tollendo ponens**, *"mode which, by denying, affirms")*

Proof. Left as Exercise 1-49.   ■

Standard Proof Technique 1.72 The tautologies in Theorem 1.71 are not so much proof techniques as they are steps that we frequently encounter in proofs. The law of simplification $(p \wedge q) \Rightarrow q$ states that if two statements are true, then, sensibly, each one of them is also true by itself. The law of addition $p \Rightarrow (p \vee q)$ says that if a statement is true, then it or anything else will still be true. The law of syllogism (transitive law for implications) $((p \Rightarrow q) \wedge (q \Rightarrow r)) \Rightarrow (p \Rightarrow r)$ simply states that if our conclusion is the hypothesis of another implication, then the implications can be chained together and the middle can be skipped. Most often the law of syllogism is used implicitly, when a proof establishes an intermediate result before arriving at the conclusion. The law of disjunction $(\neg p \wedge (p \vee q)) \Rightarrow q$ is the closest we can get to a law of simplification with the "or" connective instead of the "and" connective. In many situations, we know that one of two statements must be true. If we can exclude one of them, then the other one must be true. The law of disjunction formally expresses this insight. □

Aside from tautologies, we should also discuss how quantified statements are proved. The idea for either type of quantification is quite simple.

Standard Proof Technique 1.73 To prove the quantified statement $\forall x \in S : p(x)$, we must prove that $p(x)$ is true for all $x \in S$. Typically the proof starts by picking an arbitrary, but fixed, element $x \in S$ and the proof establishes $p(x)$ *for that element x.* Then, because x was *arbitrary* in S, the proof works indeed for all elements of S and the statement must therefore be true.
But we must be careful. We cannot use any property in the proof except that x is an element of S. □

Example 1.74 *Every even natural number is greater than or equal to* 2.

Proof. Let n be an *arbitrary, but fixed*, even natural number. (**Now we can work with n, because n is fixed.**) Then $n = 2k$ for some natural number k. Because all natural numbers are greater than or equal to 1, we have $k \geq 1$. Hence we infer that $n = 2k \geq 2$. Now, because n was an *arbitrary* even number, the result is proved for *all* even numbers.   ∎

Standard Proof Technique 1.75 To prove the quantified statement $\exists x \in S : p(x)$, we need to prove that an element x exists in S so that $p(x)$ holds. This is often done by explicitly constructing (or, in easy cases, just listing) such an element.   □

Example 1.76 *There exists an odd number that is not prime.*
   Proof. One such number is $9 = 3 \cdot 3$.   ∎

The simplicity of Example 1.76 should not lead to the conclusion that existence proofs typically are easy. The proofs of Theorems 5.44 and 6.78 show that existence proofs can be quite complicated.

As for implications, we should not forget that sometimes quantified statements are to be *dis*proved rather than proved. Because the negation of a universally quantified statement is an existentially quantified statement and vice versa, we do not need any new techniques. Regarding vocabulary, the example with which a universally quantified statement is disproved is also called a **counterexample**. So, the number 9 in Example 1.76 is a counterexample to the fallacious claim that all odd numbers are prime.

In general, counterexamples are used to dismiss incorrect statements. From our perspective, this will be useful when we encounter statements that are similar to statements we have proved, but which are incorrect. Whenever possible, exercises are included that require the reader to produce counterexamples to tempting, but incorrect statements. (See, for example, Exercise 1-50.) Because incorrect versions of proven results often have simple counterexamples, these exercises will (hopefully) be fairly simple. This fact should not lead to the conclusion that all counterexamples are easy to find. For example, all of Chapter 6 can be interpreted as the construction of *one* counterexample.

Although there are not that many possibilities, it is actually quite challenging to set up the right proof structure for a given claim. We have reached a stage that is similar to knowing the rules of algebra without having had much practice in working with them.[13] Moreover, mistakes can arise as we set up proofs, just as, in a computation, negative signs can get lost and other mistakes can sneak in. Exercise 1-51 shows a few common structural problems that can arise when a proof is constructed. The statements listed there are things to avoid. But just as on some bad days six times six gives sixteen in a computation[14], sometimes they will rear their ugly heads. The fact that such mistakes can happen is certainly not an excuse for making mistakes. It simply means that we

---

[13] In fact, with Boolean algebra, the reader is experiencing this situation, too.

[14] The author had that happen to him in a calculus class, much to the entertainment of all parties present, including the author. First of all, we caught it in time, which is the most important part of the story. But second, if you cannot laugh at yourself after detecting and fixing a silly mistake (and avoiding all negative consequences), you are missing part of life.

must carefully safeguard against mistakes. Knowing what mistakes can happen may help with that.

Along these lines, let us conclude with some general guiding rules for proof writing.

- *Only use what is given in the hypothesis, in axioms, in definitions and what has already been proved.*

  If you use more than that, even if everything else is correct, you will have proved less than what you set out to prove.

- *Restating the hypothesis at the start and the conclusion at the end can help.*

  By rewriting the hypothesis you familiarize yourself with the hypothesis once more, and you will be more likely to only use the hypothesis. By fully stating the conclusion, you are once more going over what you were supposed to prove, and you might be able to detect a gap as you mentally compare the last sentence with your thoughts in the proof.

- *Write complete sentences without abbreviations or unnecessary symbols.*

  Yes, the logical symbols are cute and they can be used to abbreviate statements. Plus, internet language seems to encourage the use of abbreviations and glyphs. But talking about "a number $\geq$ 3" rather than "a number greater than or equal to 3" makes you think less about what you write and you write it faster, too. Mathematics is about thinking deeply and working at a steady pace, which is often perceived as slow. So writing out all sentences can help produce better results.[15]

Exercises

1-39. Other proofs for Theorem 1.50.

  (a) Prove Theorem 1.50 by using a truth table.

  (b) Use $(p \Rightarrow q) = (\neg p \lor q)$ to prove algebraically that $\left[p \land (p \Rightarrow q) \right] \Rightarrow q$ is a tautology.

1-40. Finish the proof of Proposition 1.53.

1-41. Other proofs for Theorem 1.55.

  (a) Prove Theorem 1.55 verbally.

  Hint. Use the proof of Theorem 1.50 as guidance.

  (b) Prove Theorem 1.55 by using a truth table.

1-42. Another setup for **proofs by contradiction**.

  (a) Prove that $\left[(p \land q) \Rightarrow FALSE \right] \Rightarrow (\neg p \lor \neg q)$ is a tautology.

  (b) Explain why the tautology encodes a proof by contradiction. What do we prove?

1-43. Proving Theorem 1.59.

  (a) Prove Theorem 1.59 verbally.

  Hint. First consider what happens when q is true, then consider what happens when p is false, then start analyzing cases for A and B.

---

[15]To put it another way, how seriously would you take a love note that ends with "I will love you $\forall$ time"?

(b) Prove Theorem 1.59 algebraically.

Hint. Use that $(p \Rightarrow q) = (\neg p \vee q)$.

(c) Prove Theorem 1.59 using a truth table.

1-44. More on proofs by case distinction.

(a) State the tautology that governs a proof by case distinction with three cases A_1, A_2 and A_3.

(b) Can you state the tautology that governs a proof by case distinction with n cases?

1-45. Proving Theorem 1.62.

(a) Prove Theorem 1.62 verbally.

(b) Prove Theorem 1.62 algebraically.

Hint. Use that $(p \Rightarrow q) = (\neg p \vee q)$ and that $p \Leftrightarrow q = \big((p \wedge q) \vee (\neg p \wedge \neg q) \big)$.

(c) Prove Theorem 1.62 using a truth table.

1-46. Proving Theorem 1.65.

(a) Prove Theorem 1.65 verbally.

(b) Prove Theorem 1.65 algebraically.

(c) Prove Theorem 1.65 using a truth table.

1-47. Explain how the argument needed to prove the contrapositive of $p \Rightarrow q$ can be used to prove the implication itself by contradiction.

1-48. Proving Theorem 1.68.

(a) Prove Theorem 1.68 verbally.

(b) Prove Theorem 1.68 algebraically.

(c) Prove Theorem 1.68 using a truth table.

1-49. Proving Theorem 1.71.

(a) Prove part 1 of Theorem 1.71.

(b) Prove part 2 of Theorem 1.71.

(c) Prove part 3 of Theorem 1.71.

(d) Prove part 4 of Theorem 1.71.

1-50. For each of the following *false* universally quantified statements, provide a counterexample.

(a) All odd numbers are prime numbers.

(b) All odd numbers that are not divisible by 2, 3, 5, 7, 11, 13, 17, or 19 are prime numbers.

(c) All continuous functions are differentiable.

(d) All square roots of integers are rational numbers.

(e) All mammals have feet.

(f) All birds can fly.

(g) All integrable functions are continuous.

1-51. Let p, q be primitive propositions. Prove verbally and by constructing a truth table that the following are *not* tautologies.

(a) $\big[p \wedge (q \Rightarrow p) \big] \Rightarrow q$

Note. This statement encodes the mistake of using an implication "in the wrong direction," which is also called "affirming the consequent [the conclusion]," when we should actually "affirm the antecedent [the hypothesis]."

(b) $\big[\, p \wedge (\neg p \Rightarrow \neg q)\,\big] \Rightarrow q$

Note. This statement encodes the mistake of incorrectly setting up a contrapositive.

(c) $\big[\, p \wedge (A \vee B) \wedge (p \wedge A \Rightarrow q)\,\big] \Rightarrow q$

Note. This statement encodes the mistake of overlooking a case in a case distinction.

(d) $(p \vee q) \Rightarrow q$

Note. This statement encodes the mistake of misreading an "or."

(e) $p \Rightarrow (p \wedge q)$

Note. This statement encodes the mistake of concluding too much from a given statement.

(f) $\big[\, \neg p \wedge (p \Rightarrow q)\,\big] \Rightarrow q$

Note. This statement encodes the mistake of proving that the hypothesis is wrong rather than proving the hypothesis is right, which is also called "denying the antecedent [the hypothesis]," when we should actually "affirm the antecedent [the hypothesis]."

1-52. Use the law of disjunction and the fact that every natural number is even or odd to prove that every prime number greater than 2 is odd. Explicitly state where you use the law of disjunction.

1-53. Prove that $\sqrt{3}$ is not rational.

1-54. Prove that if p is a prime number, then $\sqrt{p}$ is not rational.

1-55. Prove that if the natural number k is not a perfect square, that is, if there is no natural number l so that $k = l^2$, then $\sqrt{k}$ is not rational.

1-56. Prove that the function of two variables $f(x, y) = \frac{8}{3}\left(\frac{1}{2} - x + y\right)$ cannot be factored into a product $f(x, y) = g(x)h(y)$ of functions of one variable.

Hint. Do a proof by contradiction using what would happen for $x = \frac{1}{2}$ and $y = 0$.

1-57. Prove that the sum of the angles in a quadrangle is $360°$.

1-58. Prove that the sum of the angles in a pentagon is $540°$.

1-59. Prove that the sum of the angles in a hexagon is $720°$.

1.7 Using Tautologies to Analyze Arguments

Tautologies and proof techniques can be used to analyze the validity of arguments made in real life. As soon as we can identify what is assumed to be true and what implications are given, that is, as soon as we can identify the logical structure, we can determine if an inference is valid or not. Being able to analyze rules of inference encoded in natural language is particularly important whenever legal issues are involved. Although such a schematic interpretation of language may feel artificial at first, with enough experience in mathematics, it does become automatic. Conversely, the use of tautologies in everyday language also helps train the skills that we want to acquire.

Example 1.77 *If I pass my analysis class, then I will graduate. If I do all my homework and do well on my tests, then I will pass my analysis class. I will do all the homework. Therefore I will graduate.*

Let's overlay a "mathematical template" over the sincere and laudable desire expressed above. Let $(G) :=$"I will graduate," let $(P) :=$"I pass my analysis class," let $(H) :=$"I do all my homework," let $(T) :=$"I do well on my tests." The first two sentences say $(H) \wedge (T) \Rightarrow (P)$ and $(P) \Rightarrow (G)$. The third sentence says that (H) will be true. From $(H) \wedge (T) \Rightarrow (P)$ and $(P) \Rightarrow (G)$ we can validly infer $(H) \wedge (T) \Rightarrow (G)$, but (H) alone does *not* give us $(H) \wedge (T)$. Hence the inference is not valid.

Claim.	**Proof attempt.**
(H)	$(H) \wedge (T) \Rightarrow (P)$
$(H) \wedge (T) \Rightarrow (P)$	$(P) \Rightarrow (G)$
$(P) \Rightarrow (G)$	$\therefore \quad (H) \wedge (T) \Rightarrow (G)$
$\therefore \quad (G)$	(H)
	?

Figure 1.8 Visualization of the claim and proof attempt in Example 1.77. The symbol ∴ reads "therefore." The first three lines of the proof attempt show that we can make some progress before getting stuck. But the remainder shows that with the third and fourth lines we are stuck.

Claim.	**Proof.**
(H)	$(H) \wedge (T) \Rightarrow (P)$
(T)	$(P) \Rightarrow (G)$
$(H) \wedge (T) \Rightarrow (P)$	$\therefore \quad (H) \wedge (T) \Rightarrow (G)$
$(P) \Rightarrow (G)$	
$\therefore \quad (G)$	(H)
	(T)
	$\therefore \quad (H) \wedge (T)$
	$(H) \wedge (T) \Rightarrow (G)$
	$(H) \wedge (T)$
	$\therefore \quad (G)$

Figure 1.9 Visualization of the claim and proof in Example 1.79. The symbol ∴ reads "therefore."

To explicitly show that an argument is invalid, we need to show that it is possible for all assumptions to be true and for the conclusion to still be false. In the above, if (H) is true and (T), (P) and (G) are false, then all assumptions, which were (H), $(H) \wedge (T) \Rightarrow (P)$ and $(P) \Rightarrow (G)$ are true, yet the conclusion (G) is false.

The real life interpretation is that even though homework is the best way to prepare for a test, doing well on the test itself is still necessary to succeed. A visualization of the argument is given in Figure 1.8. □

Remark 1.78 From a given argument, we may not be able to tell the truth values of certain statements. Therefore, we focus on the validity of the argument itself. In terms of language, we speak of *valid* or *invalid* arguments, not or true or false ones.

The inference in Example 1.77 is a lot more pleasant if we can say the following.

Example 1.79 *If I pass my analysis class (P), then I will graduate (G). If I do all the homework (H) and do well on my tests (T), then I will pass my analysis class. I did all the homework and I did well on my tests. Therefore I will graduate.*

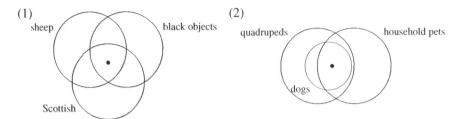

Figure 1.10 Visualization of the situations in Examples 1.81 (part 1) and 1.82 (part 2). We draw all possible sets under consideration in the most general configuration that satisfies the given constraints and we indicate entities that we know to exist with a dot. If this general configuration satisfies the claim, then the claim should be true. If not, we can use the figure to find a counterexample. Part (1) indicates that there may be Scottish sheep that are not black. Part (2) indicates that some household pets are quadrupeds.

This time the following are known to be true: (H), (T), $(H) \wedge (T) \Rightarrow (P)$ and $(P) \Rightarrow (G)$. Because (H) and (T) are true, $(H) \wedge (T)$ is true. Because $(H) \wedge (T)$ and $(H) \wedge (T) \Rightarrow (P)$ are true, (P) is true. Because (P) and $(P) \Rightarrow (G)$ are true, (G) is true. All right!

A visualization of the argument is given in Figure 1.9.   □

Remark 1.80 Visualizations as in Figures 1.8 and 1.9 can help put logical and verbal reasoning into a more schematic, in this case algebraic, context. But, similar to truth tables, they must not turn into a crutch. So we must ultimately be able to determine the validity of an argument straight from the argument, without detours through setups as shown in these figures.

Example 1.81 *A man is driving through Scotland when he sees a black sheep. He turns to his passenger and says: "Look, in Scotland the sheep are black."*

The driver uses the fact that he saw one black sheep to infer that all sheep in Scotland are black. Of course this inference is invalid. Although an example is good enough to prove an existential quantification, an example does not give a proof of a universal quantification.[16] So, even though we do not know from the given data if there are or are not sheep of different colors in Scotland, the argument itself is invalid.

A visualization of the argument is given in Figure 1.10, part (1).   □

Example 1.82 *All dogs are quadrupeds. Some household pets are dogs. Therefore some household pets are quadrupeds.*

This is a valid argument, because the existence of pets that are dogs implies the existence of pets that are quadrupeds.

---

[16]This example is based on an old joke in which the man has a scientist and a mathematician as passengers. The scientist replies "Actually you can only say that in Scotland some sheep are black." This reflects the justified scientific practice of inductive reasoning. If there is one black sheep, then it is likely that there are more. After that, the mathematician says "Actually, all we know at this stage is that there is one sheep in Scotland that is black on one side." This reflects the mathematical practice of only accepting that which is definitely proved. It's also why some people may find us a bit strange at times.

A visualization of the argument is given in Figure 1.10, part (2). ☐

Remark 1.83 Diagrams as in Figure 1.10 are called **Euler diagrams**. Similar to Remark 1.80, Euler diagrams are good visualizations, but we will be more efficient if we can identify valid and invalid arguments without resorting to such diagrams.

Exercises

1-60. Determine which of the following arguments are valid or invalid.

   (a) If I do well in the interview (I) and if my qualifications match the needs of the company (M), then I will get the job (J). If I sleep well the night before (S), I will do well in the interview. I slept well the night before and my qualifications match the needs of the company. Therefore I will get the job.

   (b) If string theory contradicts quantum mechanics or general relativity (C), then string theory is not valid (N). String theory does not contradict quantum mechanics or general relativity. Therefore string theory is valid.

   (c) If this car is made overseas, then parts are difficult to obtain. This car is expensive or it is not difficult to obtain parts. This car is not expensive. Hence it was not made overseas.

   (d) You should always do things that are good for the body. Drinking a half gallon of water per day is good for the body. Milk contains water. Therefore you should drink a half gallon of milk per day.

   (e) If Johnson wins the election, the aquifer will be saved. If we support Johnson, Johnson will win the election. We support Johnson. Therefore the aquifer will be saved.

   (f) If Johnson wins the election, the aquifer will be saved. If we don't support Johnson, Johnson will not win the election. We support Johnson. Therefore the aquifer will be saved.

1-61. Determine which of the following arguments are valid or invalid.

   (a) $p, (p \vee q) \Rightarrow r \qquad \therefore p$

   (b) $p, (p \wedge q) \Rightarrow r \qquad \therefore p$

   (c) $p \Rightarrow q, q \Rightarrow r, r \Rightarrow p \qquad \therefore p$

   (d) $p, p \Rightarrow q, q \vee r \Rightarrow s \qquad \therefore s$

   (e) $q, p \Rightarrow q, q \vee r \Rightarrow s \qquad \therefore s$

   (f) $q, p \Rightarrow q, p \Rightarrow r \qquad \therefore r$

   (g) $p \vee q, q \Rightarrow (r \wedge \neg r) \qquad \therefore p$

1-62. Each of the following arguments involve quantified statements. Determine which *arguments* are valid and which are not.

   (a) Some mathematicians are brilliant. Some brilliant people go crazy. Therefore some mathematicians go crazy.

   (b) All big houses are expensive. All expensive houses are on Chrysanthemum Lane. Therefore all houses on Chrysanthemum Lane are big.

   (c) I learned some things in Kindergarten that I needed. After Kindergarten I learned more. Some of what I learned after Kindergarten, I never needed. Therefore, everything I needed to know, I learned in Kindergarten.

   (d) Items weighing over 20 pounds are heavy. Humus bags weigh 40 points. Therefore, humus bags are heavy.

   (e) Some of my favorite games are video games. *Archon* is one of my favorite games. Therefore, *Archon* is a video game.

1-63. State Example 1.79 as a tautology similar to the tautologies for the proof methods in Section 1.6. Then prove that the statement really is a tautology.

1.8   Russell's Paradox

A strange thing about set theory is that the very entities with which set theory is concerned, sets and objects, actually remain undefined. The reason for this apparent gap is that any attempt to define sets and objects will ultimately lead to contradictions. One way to explain these contradictions is Russell's Paradox, which we discuss here.

Example 1.84 Russell's Paradox starts with three sensible-looking assumptions.

1. Assumption 1. *Sets and objects have been defined and can be identified using the definition.*

   Assumption 1 just states that "somehow" we have defined sets and objects, which means that we can identify them.

2. Assumption 2. *For any object and any set, we can determine if the object is in the set. For convenience of notation, we also write $x \in A$ iff the object x is in the set A.*

   Assumption 2 says that there is some "detection mechanism" that allows us to determine if the statement "$x \in A$" is true or false. This is a natural demand. If we are given a set and an object, then some type of "search through A" should allow us to detect if x is in A. If we could not do this, it would be rather pointless to work with sets, because we would never be sure what actually is in a set. Thus Assumption 2 is indispensable.

   Note that the second sentence merely introduces notation to make it easier to talk about Assumption 2. It does not introduce any new logical content.

3. Assumption 3. *Sets A can be formed by specifying their elements with open sentences $p(x)$, so that the element x is in the set A iff $p(x)$ is true. The set of elements for which $p(x)$ holds is also denoted $\{x : p(x)\}$.*

   Assumption 3 simply states that we want to be able to form sets by describing the objects in the set. We should note that the ability to define sets by specifying their objects is indispensable. Throughout mathematics, sets are defined by specifying the properties of their elements.

   Once more, the introduction of notation in the second sentence will make it easier to communicate, but it does not introduce new logical content.

Russell's Paradox, described below, shows that the above three assumptions lead to an absurdity, that is, to a self-contradictory statement. Because our reasoning will be sound, Standard Proof Technique 1.56 shows that there must be a problem with the assumptions. Because it was just announced that there will be a problem, we could start by saying "Suppose for a contradiction that Assumptions 1-3 can be satisfied." But note that if Assumptions 1-3 were used as axioms (as they were in some form in the late nineteenth century), then all the arguments below would be done assuming that we have a consistent system.

First note that, for given x, the sentence $p(x) = $ "x is a set" is an open sentence: Indeed, Assumption 1 guarantees that sets are defined. Thus we should be able to apply the definition of a set to any x and determine if it is a set or not.

Similarly, for given x, the sentence $q(x) =$ "$x \notin x$" is an open sentence. Assumption 2 guarantees that, for any given x, we can determine whether the statement is true or false. Granted, the statement looks a bit strange. But recall that the key to logic is the verifiability of truth or falsehood of a statement. Judgments of "strangeness" or not are not an issue in logic. Moreover, it should not bother us that sets can also act as elements of other sets. On one hand, we will see that it is quite common to talk about sets of subsets of another set (and naturally these sets have sets as their elements). On the other hand, from a logical point of view, if a set A could not act as an element of itself, then $q(A)$ would simply be false.

The above implies that $r(x) := p(x) \wedge q(x) = [x \text{ is a set }] \wedge [x \notin x]$ is an open sentence. Thus Assumption 3 guarantees that we can form the set $\mathcal{B} = \{A : r(A)\}$ of all A for which $r(A) = [A \text{ is a set }] \wedge [A \notin A]$ is a true statement.

Now we consider if the set $\mathcal{B}$ is an element of itself. The statement $\mathcal{B} \notin \mathcal{B}$ is either true or false. Suppose, for a contradiction, that $\mathcal{B} \notin \mathcal{B}$ is true. **Note that we are doing a proof by contradiction inside another proof by contradiction. That means we must carefully determine which contradiction is caused by which assumption.** Then $\mathcal{B}$ is not an element of itself. By definition of $\mathcal{B}$ being the set of all sets that do not contain themselves as an element, we must conclude that $\mathcal{B}$ is an element of itself, a contradiction. **This contradiction is caused by our assumption that $\mathcal{B} \notin \mathcal{B}$, not by any problem with Assumptions 1-3.** Hence $\mathcal{B} \notin \mathcal{B}$ must be false.

But then $\mathcal{B}$ is an element of itself, which means $\mathcal{B}$ does not satisfy the condition $A \notin A$ used to define $\mathcal{B}$. This in turn implies that $\mathcal{B}$ is not an element of itself, a contradiction. **This is the final contradiction, which shows that at least one of Assumptions 1-3 must be abandoned, because all three cannot hold at the same time. (See the discussion below, as well as Exercise 1-42, which displays the formal structure of the proof that one of Assumptions 1-3 must be abandoned.)** $\qquad \square$

Historically, this final contradiction did not arise from any setup for a proof by contradiction. It simply "materialized," which is nothing short of catastrophic. If self-contradictory statements could just "materialize" in a mathematical theory, then this theory would be worthless. After all, how could anything in this theory guarantee that a given statement is not an oxymoron? The very goal of mathematics is to exclude all oxymorons from discussion!

To determine what we can do about Russell's Paradox, we must first determine the root cause of the contradiction. The initial hypothesis of the argument was the conjunction [*Assumption 1* $\wedge$ *Assumption 2* $\wedge$ *Assumption 3*], and we have shown in correct logical fashion that "[*Assumption 1* $\wedge$ *Assumption 2* $\wedge$ *Assumption 3*] $\Rightarrow FALSE$" is true. The only way this implication can be true is if its hypothesis is false, which can only be if one of Assumptions 1, 2, or 3 is false.

As was stated after Assumption 2 was formulated, the ability to determine if an object is an element of a set is indispensable. Hence one of Assumptions 1 or 3 must be false. As was stated after Assumption 3 was formulated, the ability to define sets by specifying their objects using logic is indispensable, too. After all, with logical statements being the only unambiguous way to assign a truth value to anything, what else would we use to describe sets? But that means that Assumption 1 must be false! Therefore we cannot define sets and objects and then build mathematics from these definitions. This is the rather disappointing answer to question 5 on p. x.

The remainder of this text shows how a consistent set of axioms, called the Zermelo-Fraenkel axioms for set theory, can be used to put set theory (and thus all of mathematics) on a foundation that does not lead to paradoxes. Our exploration of these axioms should have two goals. First, we must avoid paradoxes, which means that the axioms must be more restrictive than Assumptions 1, 2, and 3 from Russell's Paradox. This point was made above. Mathematics is about eliminating logical paradoxes. Second, the axioms should be rich enough to allow the construction of objects that act like we expect the usual number systems (natural numbers, integers, rational numbers, real numbers and complex numbers) to act. After all, experience with algebra and calculus as well as with their applications in science and engineering shows that working with the number systems apparently does not produce paradoxes.[17] Moreover, because the applications of mathematics to the real world are the very reason why mathematics is important, a set theory that does not allow for the existence of these rather fundamental entities would not be useful.

The above two concerns, avoidance of paradoxes and the need for sufficient tools to construct the familiar number systems, should override any and all other personal perceptions as we explore the axioms for set theory. This situation is similar to what physicists faced at the end of the nineteenth and the beginning of the twentieth century. At that time, some people considered physics as a finished theory with only details left to be filled in.[18] Then, similar to our experience with Russell's Paradox, some experiments gave results that could not be explained with the established theories and viewpoints. This led to an exciting, but also to some a disconcerting, period during which established theories had to be questioned and revised. The end result was the rise of quantum mechanics and the theory of relativity. Both theories are challenging to the human mind. They could only be developed by adopting the point-of-view that any and all human preconceptions about what nature *should be* must be subordinate to what nature actually *is*.[19]

Of course, good scientists have an uncanny ability to use intuition (their preconception of what should be) to arrive at verifiable statements that actually turn out to be true. For example, Einstein's intuition (and a good measure of mathematics) told him that time should pass at different rates for objects that move at different velocities. This statement has since been verified experimentally with subatomic particles. On the other hand, in the early twentieth century there were attempts by highly qualified scientists to build theories that explain numerous experimental results without using the theory of relativity. None of these theories could ultimately describe reality as well as

---

[17]Indeed, several financial crises in the early twenty-first century have shown that the laws that govern the rational numbers cannot be violated: Companies that used "tricks" to spend more than they earned and yet still show a profit on their year-end ledger went bankrupt. Mortgages that were indiscriminately, or fraudulently, given to individuals who could not sustain the payments ended in default.

[18]One story tells that Max Planck's high school teacher advised him against studying physics, because the teacher believed that all physics had already been invented.

[19]Physics still is not a "finished" theory. In fact, with, for example, observations of phenomena that are explainable only with "dark matter" and "dark energy," physics may be more exciting than ever.

the theory of relativity, and hence these theories had to be abandoned. This was not done because one scientist had more clout or more followers than another. It was done because one theory gave verifiable results and useful models, while other theories did not (at all or at least not for as wide a scope before running into contradictions).

Similarly, in our exploration of set theory and mathematics, we are allowed (and indeed should be encouraged) to use visualizations (which are a reflection of what we *think* an entity should be). But the only validation of any claim will come through a correct *proof*. In particular, if we believe a certain result must be true and then we find a proof that the result is false, then our thinking must adjust, not the result.

Exercises

1-64. Consider a collection of databases that catalogue other databases. Prove that there cannot be a database that lists all databases that do not list themselves.

Chapter 2

Set Theory

As was mentioned at the end of Section 1.8, the axiomatic approach to set theory must provide a framework which, on one hand, does not lead to paradoxes, and which, on the other hand, is rich enough to allow the construction of entities that we are familiar with, such as real numbers. This chapter introduces the Zermelo-Frankel axioms as the foundation of set theory. In Chapters 2, 3, 4, and 5 we will construct the familiar number systems and familiar mathematical entities, such as functions and relations, as entities within Zermelo-Fraenkel set theory. These constructions will demonstrate that set theory is indeed rich enough to support the structures that were needed and used in mathematics, even before the invention of set theory. After all, set theory was invented around the end of the nineteenth and the beginning of the twentieth century. Calculus had been in use for over 200 years by then and significant strides, such as the formal definition of the limit, had already been made.

Every section of this chapter will introduce one or two axioms of set theory. Subsequently, abstract consequences of the axiom(s) will be discussed. At the end of each section, if appropriate, examples and visualizations will be provided. The precedence of abstract results over examples is deliberate. Mathematics at this level is all about abstract reasoning, and it leads to powerful results that can be applied to many examples. So, to practice appropriately, we must first consider the abstract setting to obtain insights. Then we can apply the results to examples.

Moreover, from here on, everything that we prove can be traced back to the axioms. So the insecurity we had in Chapter 1 about what we were allowed to use in a proof will no longer be an issue. Instead, we must make sure that any result we prove will be derived from an axiom or from another result that we have proved earlier, but after Chapter 1. The few exceptions to this approach will be marked in the text. If necessary, it will be explained why they are not a problem.

Fundamentals of Mathematics: An Introduction to Proofs, Logic, Sets, and Numbers.
By Bernd S. W. Schröder.
Copyright © 2010 John Wiley & Sons, Inc.

2.1   Sets and Objects

The axioms of set theory must start with sets and objects, because that is what set theory is all about. But Russell's Paradox shows that we cannot define sets. Consequently, we also cannot define objects. Indeed, if we could define what objects are, then we should be able to parlay this knowledge into a definition of sets. With objects and sets undefined, it is not surprising that the notion that an object belongs to a set is undefined, too. Terms that remain undefined in a theory are also called **primitive terms**. Primitive terms cannot be avoided. Any definition of a basic term would contain words that in turn must be defined, too. Thus any attempt to avoid primitive terms leads to an infinite sequence of more and more basic definitions that has no "most elementary" definition at its bottom. The terms "set," "object" and "belongs" (or, alternatively, "is an element of") are the primitive terms of set theory. Therefore, set theory must start with the humblest and weakest of axioms.

Axiom 2.1 *There is a set.*

We are not saying how many sets there are or what they must look like. The axiom simply says that there exists one of the entities that we are interested in. This is the second answer to question 5 on p. x. We don't know what a set is, but we do know that there is one.

Moreover, we must be able to determine if an object (note that we have not formally introduced objects) is an element of a set. Rather than defining what it means to be an element of a set, which would once more come dangerously close to defining sets, we demand that "somehow" we are able to check whether an object is an element of a set.

Axiom 2.2 *For every object x and every set S, we can determine whether x is an element of S or not.*

Note that it is not necessary to say anything more about objects: If there were no objects, the universally quantified statement above would remain valid, because it would be "vacuously true." To make the language simpler, we introduce the following notation.

Definition 2.3 *Let x be an object and let S be a set. Then we write $x \in S$ if x is an element of S and we write $x \notin S$ if x is not an element of S.*

Because we can determine for any object if it is an element of a set, we can define what it means for a set to be contained in another set.

Definition 2.4 *Let A, B be sets. Then we will say that A is **contained in** B iff every element of A is also an element of B. In this case, we will write $A \subseteq B$. If A is not contained in B we will write $A \nsubseteq B$. In general, if a symbol denotes a certain relation between two objects, then the same symbol with a strike through it will denote that the relation does not hold. We also sometimes write $B \supseteq A$ for $A \subseteq B$ and similarly $B \nsupseteq A$ for $A \nsubseteq B$. In general, if a symbol denotes a certain relation between two objects, then the mirror image of the same symbol denotes that the reversed relation holds.*

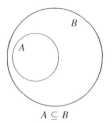
$A \subseteq B$

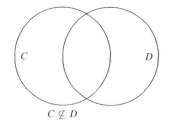
$C \not\subseteq D$

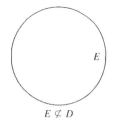
$E \not\subseteq D$

Figure 2.1 Visualization of set containment with Venn diagrams. The set A is contained in the set B, because the circle that represents A is contained in the circle that represents B. The set C is not contained in the set D, because, despite the overlap, the circle that represents C is not contained in the circle that represents D. Similarly, the set E is not contained in the set D, and this time, there is not even any overlap.

Standard Proof Technique 2.5 Definition 2.4 already states the standard way in which we prove that one set is contained in another. Namely, to prove that $A \subseteq B$, we will typically take an (arbitrary but fixed, see Standard Proof Technique 1.73) $x \in A$ and prove that it must also be in B. □

Note that for Definition 2.4 we are not concerned at all with *how* we would check that every element of A is an element of B. We simply say that if this can be done, then $A \subseteq B$. At this stage, even though we have a definition, we cannot give a sample result whose proof illustrates how the definition is used because, well, we have nothing to work with. The first proof in which Standard Proof Technique 2.5 will be used is the proof of Proposition 2.18. Until then, readers can try their hand at Exercise 2-1.

Visualization 2.6 Aside from concrete examples, visualizations that may or may not be accurate[1] are helpful in mathematics. For elementary set theory, **Venn diagrams** have turned out to be quite helpful. In a Venn diagram, every set under consideration is represented by a circle, an oval or some other region in the plane that is bounded by a closed curve that does not intersect itself. The area inside the curve contains the elements of the set. Figure 2.1 shows a visualization of set containment with Venn diagrams. For an idea how hard it is to draw simple symmetric Venn diagrams for more than 3 sets, consider [29]. □

Examples of situations in which Axioms 2.1 and 2.2 are satisfied should be easy to come by. After all, there is very little that must be satisfied. Such examples have a special place in set theory, and they are called models.

Definition 2.7 *A **model** for a set of axioms is a way to assign meanings to the primitive terms so that all the axioms become true statements.*

---

[1] And most visualizations of real abstract content are not accurate. For example, how could anyone *accurately* visualize five dimensions? Six? Seven? Eight? (Etc.) At some point the accuracy will break down. The amazing thing is that the visualizations can still help discover accurate abstract content.

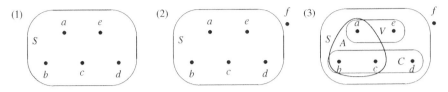

Figure 2.2 Venn diagrams representing Examples 2.8 (1), 2.10 (2) and 2.11 (3).

The above definition may not be quite as formal as what is done in model theory, but for our purposes, it shall suffice. The purpose of this text is to give a general overview of the fundamentals of mathematics for working mathematicians. Readers interested in becoming logicians or set theorists should consult specialized texts on these subjects.

Note that a model does not *define* the primitive terms. Essentially, a model provides *a possible way* in which the primitive terms can be interpreted without leading to a contradiction. Moreover, it should be noted that there can be many different models for the same set of axioms.

Example 2.8 Consider the following model.

1. The objects under consideration are the first five lowercase letters of the alphabet.

2. There is one set S, which has as its elements the first five lowercase letters of the alphabet. That is, $S = \{a, b, c, d, e\}$.

This model satisfies Axioms 2.1 and 2.2, because clearly, there is a set S and for every one of the objects a, b, c, d, e it can be verified that the object is in S. Although this example is simple, its very existence shows that Axioms 2.1 and 2.2 are not self-contradictory: If the axioms were self-contradictory, then there could be no example (model) in which they are valid.

So we know now that the two axioms are not self-contradictory despite the fact that Axiom 2.1 is *similar to* (but much weaker than) Assumption 1 in Russell's Paradox and despite the fact that Axiom 2.2 *is* Assumption 2 in Russell's Paradox.   ☐

In general, the existence of a model for a set of axioms proves that the set of axioms is not self-contradictory. So, for example, Assumptions 1-3 from Russell's Paradox cannot become the basis of any model. This is because we have shown in Section 1.8 that any model would contain the entity $B = \{x : x$ is a set $, x \notin x\}$, which cannot exist.

Visualization 2.9 Objects are represented as points in a Venn diagram. If a set is represented by a closed curve and the point representing an object is in the curve's interior, then the object is considered to be an element of the set. Otherwise, it is considered to not be an element of the set. Part (1) of Figure 2.2 shows the Venn diagram for the situation in Example 2.8.   ☐

Example 2.10 Consider the following model.

1. The objects under consideration are the first six lowercase letters of the alphabet.

2. There is one set S, which has as its elements the first five lowercase letters of the alphabet. That is, $S = \{a, b, c, d, e\}$.

The only difference between this example and Example 2.8 is that this time around there also is an object, the letter f, that is not in the set. For a Venn diagram, see part (2) of Figure 2.2. □

Example 2.11 Consider the following model.

1. The objects under consideration are the first six lowercase letters of the alphabet.

2. There are four sets. The set $S = \{a, b, c, d, e\}$, the set $V = \{a, e\}$ of all vowels in S, the set $C = \{b, c, d\}$ of all consonants in S and the set $A = \{a, b, c\}$.

The difference between this example and Example 2.10 is that this time we have more than one set. For a Venn diagram, see part (3) of Figure 2.2. □

Exercises

2-1. Let A be a set. Prove that $A \subseteq A$.

2-2. Consider the following model.

- The objects under consideration are the points in the plane.
- The sets under consideration are the straight lines in the plane.
- An object (point) belongs to a set (line) iff the point is on the line.

  (a) Explain why the model satisfies Axioms 2.1 and 2.2.

  (b) Can it be determined if two sets have elements in common?

  (c) If two sets have elements in common, must the collection of common elements be a set?

2-3. For each of the following situations, sketch a Venn diagram.

  (a) Three sets so that any two of them overlap and so that all three overlap. None of the overlaps of two sets is completely contained in another overlap of two or more sets.

  (b) Four sets so that any two of them overlap, any three of them overlap and so that all four overlap. None of the overlaps of two sets is completely contained in another overlap of two or more sets. None of the overlaps of three sets is completely contained in another overlap of three or more sets.

  (c) A model in which the objects are the numbers 1 through 10 and in which the sets are $\{1, \ldots, 5\}$, the even numbers between 1 and 10, the odd numbers between 1 and 10, and the prime numbers between 1 and 10.

2.2   The Axiom of Specification

Example 2.11 shows a key weakness of the first two axioms. Although we can detect that the set C of consonants and the set A of the first three lowercase letters have the elements b and c in common, the common elements do not form a set. The reason is simply that the set $\{b, c\}$ is missing from the model. This is possible because there is nothing in Axioms 2.1 and 2.2 that would force the existence of intersections. The next axiom fixes this problem by demanding that certain sets must always be present.

Axiom 2.12 Axiom of Specification. *If S is a set and $p(\cdot)$ is an open sentence for the elements of S, then the collection of all elements $x \in S$ that satisfy $p(x)$ is a set, too. It is denoted as $\{x \in S : p(x)\}$ or as $\{x \in S \mid p(x)\}$. This notation is also called* **set-builder notation**.

Note the similarity between the Axiom of Specification and Assumption 3 in Russell's Paradox. The only difference, and it is a crucial one, is that without a definition of sets and objects, the open sentence must always apply to the objects in a pre-specified set. That is, we cannot grab entities "out of thin air," which is what led to Russell's Paradox. Instead, we must make sure that everything we use is inside an already defined/known entity. The models at the end of this section will prove that Axioms 2.1 and 2.2 and the Axiom of Specification capture the indispensable parts of Russell's Paradox without leading to contradictions. So from here on, Russell's Paradox will not be an issue any more.

The Axiom of Specification allows us to make Axiom 2.1 a bit more specific, because it allows us to prove that a certain set must exist.

Proposition 2.13 *There is a set that contains no elements.*

Proof. To prove that there is a set without elements, we need to start with some set. This is possible, because by Axiom 2.1 there is a set, call it S. The open sentence $p(x) = \left[(x \notin S) \wedge (x \in S) \right]$ is false for all objects. By the Axiom of Specification the collection $A := \{x \in S : p(x)\} = \{x \in S : (x \notin S) \wedge (x \in S)\}$ of all elements x of S that satisfy $p(x)$ is a set. Because $p(x)$ is always false, the set A contains no elements. Therefore, we have constructed a set without elements, which completes the proof. ■

Definition 2.14 *The set that has no elements will be called the* **empty set**, *denoted $\emptyset$.*

The Axiom of Specification also allows us to define our first operation on sets. Formally, we should state Proposition 2.16 below for a *set $\mathcal{C}$ of sets C*. After all, sets and objects are all we have in set theory. To clarify that the elements of $\mathcal{C}$ are sets themselves, $\mathcal{C}$ is also called a family of sets. This terminology is often used for sets of sets in mathematics, simply to distinguish them from other sets under consideration.

Definition 2.15 *A set $\mathcal{S}$ whose elements are sets, too, is also called a* **family** *of sets.*

Proposition 2.16 *Let $\mathcal{C}$ be a nonempty family of sets. Then there is a set I so that an object x is an element of I iff for all $C \in \mathcal{C}$ we have $x \in C$.*

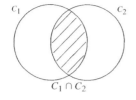
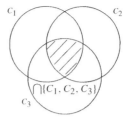

Figure 2.3 The intersection of two sets and the intersection of three sets.

Proof. To construct a set as specified, first note that $\forall C \in \mathcal{C} : x \in C$ is an open sentence. Now let $C_0 \in \mathcal{C}$ be any set in $\mathcal{C}$. Then $I := \left\{ x \in C_0 : [\forall C \in \mathcal{C} : x \in C] \right\}$ is a set by the Axiom of Specification. By definition of I, the elements of I are all the objects that are in all sets $C \in \mathcal{C}$, which completes the proof. ∎

The set I from Proposition 2.16 is called the intersection of the family $\mathcal{C}$.

Definition 2.17 *Let $\mathcal{C}$ be a family of sets. The set of elements x so that $x \in C$ for all $C \in \mathcal{C}$ is called the* **intersection** *of $\mathcal{C}$. The intersection is denoted $\bigcap \mathcal{C}$. The intersection of two sets A and B will also be denoted $A \cap B$.*

For a visualization of the intersection, see Figure 2.3. Clearly, the intersection should be a subset of every set in the family, but even that must be proved.

Proposition 2.18 *Let $\mathcal{C}$ be a family of sets and let $C \in \mathcal{C}$. Then $\bigcap \mathcal{C} \subseteq C$.*

Proof. To prove $\bigcap \mathcal{C} \subseteq C$, as stated in Standard Proof Technique 2.5, we must prove that every element of $\bigcap \mathcal{C}$ is an element of C, too. So let $x \in \bigcap \mathcal{C}$ be arbitrary but fixed (see Standard Proof Technique 1.73). By definition of the intersection, x is an element of every set that is in $\mathcal{C}$. So, in particular, $x \in C$. Because $x \in \bigcap \mathcal{C}$ was arbitrary, this means that every element of $\bigcap \mathcal{C}$ is in C. Thus we have proved that $\bigcap \mathcal{C} \subseteq C$. ∎

Of course, it can happen that an intersection is empty. In this case, we speak of disjoint sets.

Definition 2.19 *Let $\mathcal{C}$ be a family of sets. If the intersection $\bigcap \mathcal{C}$ has no elements, then the family $\mathcal{C}$ is called* **disjoint***.*

The Axiom of Specification also allows us to define complements.

Proposition 2.20 *Let A and B be sets. Then there is a set that contains all elements of A that are not in B.*

Proof. Because $x \notin B$ is an open sentence, the desired set is $\{x \in A : x \notin B\}$. ∎

Definition 2.21 *Let A and B be sets. The set of elements of A that are not in B is denoted $A \setminus B$, and it is called the* **(relative) complement** *of B in A.*

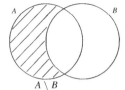 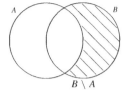

Figure 2.4 The complement of B in A and the complement of A in B.

Finally, regarding the feasibility of the Axiom of Specification, first note that none of the models in Examples 2.8, 2.10 and 2.11 satisfy the Axiom of Specification. For example, in Example 2.8 $\{x \in S : x = a\}$ is not a set. Nonetheless, the next two examples show that Axioms 2.1 and 2.2 and the Axiom of Specification can be satisfied simultaneously.

Example 2.22 Consider the following model.

1. There are no objects.

2. There is only one set, the empty set $\emptyset$.

This model satisfies Axioms 2.1 and 2.2 and the Axiom of Specification, which means that these three axioms are not self-contradictory.   □

Although it is a relief to know that our three axioms are not self-contradictory, Example 2.22 is not a very satisfying model. The next model gives us at least a few objects to work with.

Example 2.23 Consider the following model.

1. The objects are the first five letters of the alphabet.

2. The sets are all possible subsets of $\{a, b, c, d, e\}$ (see Exercise 2-4).

This model satisfies Axioms 2.1 and 2.2 and the Axiom of Specification, which means that these three axioms are not self-contradictory. We will not draw a Venn diagram here, because the picture would get a bit messy. Note, however, that Figure 2.10 on page 73 shows a similar model that uses three letters instead of five.   □

Example 2.23 indicates how models for set theory can be constructed, and it also indicates how a lot of good mathematics is done. In many situations there is, by default, some type of surrounding universal set, and we are only interested in what happens to entities inside this set. This surrounding set is also called the **universe**. In Example 2.23 the universe is the set $\{a, b, c, d, e\}$. In number theory, the universe often is the set of natural numbers. In calculus, the case could be made that the real numbers are the universe.

Returning to intersections once more, note that whenever a universe U is given, we can form the strange looking intersection $\bigcap \emptyset$, that is, the intersection of the empty family. Because for any $x \in U$, the statement $\forall C \in \emptyset : x \in C$ is (vacuously) true, we conclude that the intersection of the empty family is the universe, or, in symbols, $\bigcap \emptyset = U$.

Exercises

2-4. List all the sets in the model in Example 2.23.

2-5. Prove that there cannot be a set of all sets. That is, prove that the axioms of set theory we have introduced so far, together with the assumption that there is a set A that contains all existing sets as elements, lead to a contradiction.

   Hint. Russell's Paradox.

2-6. Let A, B and U be sets so that $A \subseteq U$ and $B \subseteq U$. Prove that $A \subseteq B$ iff $(U \setminus B) \subseteq (U \setminus A)$.

2-7. Let A, B and U be sets so that $A \subseteq U$ and $B \subseteq U$. Prove that $A \subseteq B$ iff $A \cap (U \setminus B) \subseteq \emptyset$.

   Note. We use "$\subseteq \emptyset$" instead of "$= \emptyset$" because the formal introduction of equality of sets is delayed to the next section.

2-8. Let A, B be sets.

   (a) Prove that $A \setminus B \subseteq A$.

   (b) Give a counterexample that shows that the containment $A \setminus B \subseteq B$ need not hold.

   (c) When does $A \setminus B \subseteq B$ hold?

2-9. For each of the following *false* statements, provide a counterexample.

   (a) Let A, B be sets. Then $A \subseteq A \cap B$.

   (b) Let A, B be sets so that $A \cap B \nsubseteq \emptyset$. Then $A \subseteq B$.

   (c) Let A, B, C be three sets so that the family $\{A, B, C\}$ is disjoint. Then at least one of $A \cap B$, $A \cap C$ and $B \cap C$ has no elements.

2.3 The Axiom of Extension

Now that we know that there are sets, now that we can determine if an object is in a set and now that we can form subsets, we can talk about what it means that two sets are equal. The Axiom of Extension tells when two sets are equal.

Axiom 2.24 Axiom of Extension. *Two sets are equal iff they have the same elements. That is, if A and B are sets, then $A = B$ iff for all objects x we have $x \in A$ iff $x \in B$.*

The fact that two sets are equal iff they have the same elements is quite unsurprising. What may be surprising is that we need to elevate this insight to the status of an axiom. The best way to understand this part is to realize that, using only the axioms that were introduced so far, we have no way of deriving or defining what equality of sets should mean. Hence we have no choice but to make an axiom out of the definition of equality of two sets.

Regarding the feasibility of the Axiom of Extension, note that the models in Examples 2.22 and 2.23 satisfy Axioms 2.1 and 2.2, the Axiom of Specification, and the Axiom of Extension. Therefore, these four axioms are not self-contradictory. On the other hand, the Axiom of Extension really cannot be derived from the other axioms: If we artificially add another set $\emptyset'$ to the model in Example 2.22, so that $\emptyset'$ has no elements and $\emptyset' \neq \emptyset$ (an artificial, but not illegal, demand), then the model satisfies Axioms 2.1 and 2.2, and the Axiom of Specification, but not the Axiom of Extension.

The Axiom of Extension immediately implies the following simple, but fundamental, theorem.

Theorem 2.25 *Two sets A and B are equal iff $A \subseteq B$ and $B \subseteq A$.*

Proof. Let A and B be sets. By the Axiom of Extension we have $A = B$ iff for every object x the statement $x \in A \Leftrightarrow x \in B$ is true. But by the biconditional law, this is the case iff for every object x the statement $(x \in A \Rightarrow x \in B) \wedge (x \in B \Rightarrow x \in A)$ is true. This statement, in turn, is true iff, for every object x, we have that $x \in A$ implies $x \in B$ and that $x \in B$ implies $x \in A$. Finally, by definition of set containment, this is true iff $A \subseteq B$ and $B \subseteq A$. ∎

Standard Proof Technique 2.26 To prove that two sets A and B are equal, instead of proving the equality directly, we typically use Theorem 2.25 and prove that $A \subseteq B$ and $B \subseteq A$. That is, we can prove equality by proving **mutual containment**. This approach to proving the equality of two sets should become fairly automatic. It is *the* standard approach to proving that two sets are equal. We will use this technique throughout this section, even if there are simpler ways to prove the equality of two sets. By internalizing certain standard approaches to mathematics, it will be easier to focus on the essentials of a new argument. For an example, see Proposition 2.29. □

To illustrate the technique of proving equality of two sets by proving mutual containment, we consider set intersection.

Theorem 2.27 *Set intersection is* **commutative**. *Let A and B be sets. Then the equality $A \cap B = B \cap A$ holds.*

Proof. Let A and B be sets. To prove that $A \cap B = B \cap A$, we must prove $A \cap B \subseteq B \cap A$ and $B \cap A \subseteq A \cap B$ (see Standard Proof Technique 2.26).

For the proof of $A \cap B \subseteq B \cap A$, let $x \in A \cap B$ be arbitrary, but fixed (see Standard Proof Techniques 2.5 and 1.73). Then $x \in A$ and $x \in B$, which implies that $x \in B$ and $x \in A$. Hence $x \in B \cap A$. Because x was arbitrary, every $x \in A \cap B$ must be in $B \cap A$, too. Therefore we have proved that $A \cap B \subseteq B \cap A$.

Conversely, for the proof of $B \cap A \subseteq A \cap B$, let $x \in B \cap A$. Then $x \in B$ and $x \in A$, which implies that $x \in A$ and $x \in B$. Hence $x \in A \cap B$. Because every $x \in B \cap A$ must be in $A \cap B$, too, we have proved that $B \cap A \subseteq A \cap B$.

Because $A \cap B \subseteq B \cap A$ and $B \cap A \subseteq A \cap B$, we conclude that $A \cap B = B \cap A$. ∎

Theorem 2.28 *Set intersection is* **associative** *(also see Figure 2.5). Let A, B and C be sets. Then $A \cap (B \cap C) = (A \cap B) \cap C$.*

Proof. Let A, B and C be sets. To prove that $A \cap (B \cap C) = (A \cap B) \cap C$ we must prove $A \cap (B \cap C) \subseteq (A \cap B) \cap C$ and $(A \cap B) \cap C \subseteq A \cap (B \cap C)$ (see Standard Proof Technique 2.26).

For the proof of $A \cap (B \cap C) \subseteq (A \cap B) \cap C$, let $x \in A \cap (B \cap C)$ be arbitrary, but fixed (see Standard Proof Techniques 2.5 and 1.73). Then $x \in A$ and $x \in B \cap C$,

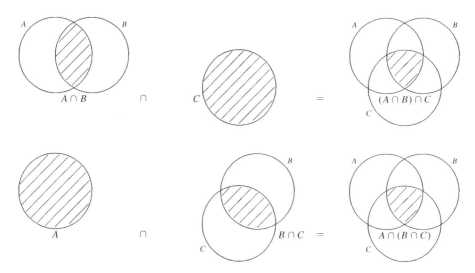

Figure 2.5 Visualization of the associativity of intersections.

which implies that $x \in A$ and $x \in B$ and $x \in C$. But then $x \in A \cap B$ and $x \in C$, and hence $x \in (A \cap B) \cap C$. Because x was arbitrary, every $x \in A \cap (B \cap C)$ must be in $(A \cap B) \cap C$, too. Therefore we have proved that $A \cap (B \cap C) \subseteq (A \cap B) \cap C$.

Conversely, for the proof of $(A \cap B) \cap C \subseteq A \cap (B \cap C)$, let $x \in (A \cap B) \cap C$. Then $x \in A \cap B$ and $x \in C$, which implies that $x \in A$ and $x \in B$ and $x \in C$. But then $x \in A$ and $x \in B \cap C$, and hence $x \in A \cap (B \cap C)$. Because every $x \in (A \cap B) \cap C$ must be in $A \cap (B \cap C)$, too, we have proved that $(A \cap B) \cap C \subseteq A \cap (B \cap C)$.

Because $A \cap (B \cap C) \subseteq (A \cap B) \cap C$ and $(A \cap B) \cap C \subseteq A \cap (B \cap C)$, we conclude that $(A \cap B) \cap C = A \cap (B \cap C)$. ∎

The proofs above are quite simple. They could probably be abbreviated, but presenting them in full detail emphasizes that *details matter*. The more attention is paid to possibly trivial details early on, the easier more complex tasks will be.

Let us now consider a surprisingly complex example that involves set containment and set intersection. We will first present the proof. After that, we will analyze the structure of the proof and how what we have done so far can help set up this structure. Mathematics texts present proofs, such as the one below, and rely on the readers to come up with their own analysis as done after the proof. In this fashion, mathematics texts can be kept at a reasonable length. Note, for example, that the similar first and second parts of the proofs of Theorems 2.27 and 2.28 are both intelligible, but that the second part is more efficient. If what is not said in a mathematical proof were to be spelled out every time, every proof could easily double in length. Such communication would not be efficient. In fact, it could keep the reader from making the very connections with the content that will ultimately allow the reader to go beyond what is presented. In lectures, the situation is similar. The proof that is written on the board will typically be correct and reasonably short, like the proof below. The teacher's verbal explanation, however, will typically involve many more words, as it fills in the parts that are **given in handwriting** in the analysis of the proof starting on page 63.

Proposition 2.29 *Let A and B be sets. Then $A \subseteq B$ iff $A \cap B = A$.*

Proof. Let A and B be sets. To prove the biconditional, we must prove that $A \subseteq B$ implies $A \cap B = A$ *(the direction "$\Rightarrow$" of the biconditional)* and that $A \cap B = A$ implies $A \subseteq B$ *(the direction "$\Leftarrow$" of the biconditional)*.

"$\Rightarrow$": We must prove that $A \subseteq B$ implies $A \cap B = A$. So let $A \subseteq B$. To prove $A \cap B = A$, we must prove that $A \cap B \subseteq A$ and $A \subseteq A \cap B$. To prove that $A \cap B \subseteq A$, let $x \in A \cap B$ be arbitrary, but fixed. Because $x \in A \cap B$, we know that $x \in A$ and $x \in B$, which, in particular, means that $x \in A$. Because $x \in A \cap B$ was arbitrary, every element of $A \cap B$ is in A, too. We conclude that $A \cap B \subseteq A$. To prove that $A \subseteq A \cap B$, let $x \in A$ be arbitrary, but fixed. Because $x \in A$ and $A \subseteq B$, we infer that $x \in B$. Therefore the statements $x \in A$ and $x \in B$ are true, which means that $x \in A \cap B = \{x \in A : x \in B\}$. Hence, because $x \in A$ was arbitrary, every element of A is in $A \cap B$, too. We conclude that $A \subseteq A \cap B$. We have proved that $A \subseteq A \cap B$ and that $A \cap B \subseteq A$, that is, $A \cap B = A$. This completes the proof that $A \subseteq B$ implies $A \cap B = A$.

"$\Leftarrow$": For the converse, we must prove that $A \cap B = A$ implies $A \subseteq B$. So let $A \cap B = A$. To prove $A \subseteq B$, let $x \in A$ be arbitrary, but fixed. Because $A = A \cap B$, we infer that $x \in A \cap B$. Thus $x \in A$ and $x \in B$ and, in particular, we obtain that $x \in B$. Hence, because $x \in A$ was arbitrary, every element of A is an element of B, too. We conclude that $A \subseteq B$. This completes the proof that $A \cap B = A$ implies $A \subseteq B$.

To summarize, because $A \cap B = A$ implies $A \subseteq B$, and because $A \subseteq B$ implies $A \cap B = A$, Proposition 2.29 is proved.   ∎

After reading a proof, it is natural, even useful, to try to find out how the writer could have come up with the idea. Sometimes the idea of a proof contains enough fertile substance for other specific results in the same area, or it may have the potential to be a standard technique. Especially when starting to read and write proofs, the question "How did they think of this?" can be quite frustrating. The difficulty with doing proofs for a beginner can be attributed to two factors. On one hand, the proof in question may simply be difficult. Nothing can be done about this possibility, because if a simpler proof existed, the writer would probably have presented it.[2] On the other hand, a beginner is not as accustomed to the language and the techniques of mathematics as an expert is. Thus certain approaches that are natural for an expert can seem challenging.

This text is supposed to acquaint the reader with mathematical communication and some standard techniques. So, to illustrate how an understanding of mathematical statements can structure the setup for a proof, let us analyze the proof of Proposition 2.29. Below, indentations are used to indicate how the proof splits into different "levels," as smaller results are proved to establish larger results and ultimately Proposition 2.29 itself. Notes in **handwriting** will be used to explain the structure and for reminders. An expert will *mentally* establish this structure *without giving it much*

---

[2]Sometimes it can be useful to check for the proof of a result in a different text, because some results have more than one nice proof. Overall, however, the author would assume that a writer will present the best possible proof at the time.

thought. For example, as the author wrote the proof of the "$\Rightarrow$" part, even though he did not yet write anything down for the "$\Leftarrow$" part, he had a little mental note that there was another, equally important, part that needed to be established. So, in a way, the outermost structure of the proof (the nonindented parts) existed before everything else. Similarly, within the proof of the "$\Rightarrow$" part, the author knew that the start had to be $A \subseteq B$ and the end had to be $A \cap B = A$. So, similar to the outermost "shell," the singly indented part existed before everything else. This nesting continued throughout the construction of the proof.

It should be said, though, that the reader should not read too much into the analysis below. It is purely an illustration of how mathematical language and standard proof techniques guide the construction of a proof. With enough practice, the reader will be very familiar with this approach to proofs. This familiarity then will allow the reader to focus on the truly hard part, which is what needs to be proved once the statement cannot be broken down any further. This part (the farthest indentation in the analysis below) typically is longer and more challenging than in this simple example.

Analysis of the proof of Proposition 2.29. By Standard Proof Technique 1.63 we know that we must prove the two separate statements "If A $\subseteq$ B, then A $\cap$ B=A." and "If A $\cap$ B=A, then A $\subseteq$ B." An introductory sentence to remind ourselves of this need (and of any other underlying hypotheses, such as that A and B are sets) can help start the proof. Mentally, we already know that the overall structure of the proof should be given by the parts of this narrative that are not indented and not italicized. That is, even though we have not written them down yet, we know the beginning, the middle and the end of the proof. Every further indentation will indicate that we are reaching another level of a "proof inside a proof."[3]

Let A and B be sets. To prove the biconditional, we must prove that $A \subseteq B$ implies $A \cap B = A$ *(the direction "$\Rightarrow$" of the biconditional)* and that $A \cap B = A$ implies $A \subseteq B$ *(the direction "$\Leftarrow$" of the biconditional).*

"$\Rightarrow$": We must prove that $A \subseteq B$ implies $A \cap B = A$.

> To start the proof of an implication, it is helpful to remind ourselves of the hypothesis. Re-stating the hypothesis accomplishes that.

So let $A \subseteq B$.

> > Once we have re-ascertained the hypothesis, we must prove the conclusion. Just as reminding ourselves of the hypothesis can be helpful, it can also be helpful to remind ourselves of the conclusion by starting out the proof by saying "To prove $\langle$conclusion$\rangle$, ..."
> >
> > In this case, we use Standard Proof Technique 2.26 to prove the conclusion: To prove that two sets are equal, we prove that they are mutually contained in each other.

---

[3]Using indentations to indicate different levels in a nested structure is common practice in programming. The program [35] uses indentations in this fashion to clarify the overall structure of a proof. With no current development on [35], the author of [35] recommended [5] as a program that has similar functionality and for which development is ongoing.

To prove $A \cap B = A$, we must prove that $A \cap B \subseteq A$ and $A \subseteq A \cap B$.

> Now we must prove two containments. We know that containments are proved using Standard Proof Technique 2.5: To prove one set is contained in another, we must show that every one of its elements is also an element of the larger set.

To prove that $A \cap B \subseteq A$, let $x \in A \cap B$ be arbitrary, but fixed.

>> This innermost part of the proof is the only one that is truly specific for Proposition 2.29. Up to this level of our nesting, we have only used standard techniques to decompose the proof into more manageable pieces.
>>
>> Because $x \in A \cap B$, we know that $x \in A$ and $x \in B$, which, in particular, means that $x \in A$.

> Now we have proved one of the two containments we need. In our hierarchy, we are back to the level of proving the equality A ∩ B=A by proving mutual containment.

Because $x \in A \cap B$ was arbitrary, every element of $A \cap B$ is in A, too. We conclude that $A \cap B \subseteq A$.

> Having proved one containment, we focus on the other one.

To prove that $A \subseteq A \cap B$, let $x \in A$ be arbitrary, but fixed.

>> Once again, this innermost part of the proof is the only one that is truly specific for Proposition 2.29. Up to this level of our nesting, we have only used standard techniques to decompose the proof into more manageable pieces.
>>
>> Because $x \in A$ and $A \subseteq B$ we infer that $x \in B$. Therefore the statements $x \in A$ and $x \in B$ are true, which means that $x \in A \cap B = \{x \in A : x \in B\}$.

> Now we have proved the second of the two containments we need.

Hence, because $x \in A$ was arbitrary, every element of A is in $A \cap B$, too. We conclude that $A \subseteq A \cap B$.

> With the two containments proved, we have established the desired equality. This equality is back at the level of proving the implication "A ⊆ B implies A ∩ B=A."

We have proved that $A \subseteq A \cap B$ and that $A \cap B \subseteq A$, that is, $A \cap B = A$.

> With the equality established, we have proved one direction of the biconditional. We are back at the level of proving the biconditional.

This completes the proof that $A \subseteq B$ implies $A \cap B = A$.

> Now we focus on the other direction. We are back at the outermost level of our mental nesting of proofs inside proofs.

"$\Leftarrow$": For the converse, we must prove that $A \cap B = A$ implies $A \subseteq B$.

Once again, we remind ourselves of the hypothesis.

So let $A \cap B = A$.

> To prove the conclusion, in this case we must prove that $A \subseteq B$. By Standard Proof Technique 2.5 we must prove that every element of A is an element of B.

To prove $A \subseteq B$, let $x \in A$ be arbitrary, but fixed.

>> Once again, this innermost part of the proof is the only one that is truly specific for Proposition 2.29. Up to this level of our nesting, we have only used standard techniques to decompose the proof into more manageable pieces.

Because $A = A \cap B$, we infer that $x \in A \cap B$. Thus $x \in A$ and $x \in B$ and, in particular, we obtain that $x \in B$.

> We have proved the required containment.

Hence, because $x \in A$ was arbitrary, every element of A is an element of B, too. We conclude that $A \subseteq B$.

> With the containment established, we have proved the other direction of the biconditional.

This completes the proof that $A \cap B = A$ implies $A \subseteq B$.

With both directions of the biconditional proved, we have proved Proposition 2.29. It can be a good idea to verbally express this fact and to check over everything one more time while writing the conclusion, just to make sure nothing was omitted.

To summarize, because $A \cap B = A$ implies $A \subseteq B$ and $A \subseteq B$ implies $A \cap B = A$, Proposition 2.29 is proved. ∎

Our analysis reveals that a substantial amount of thought is "between the lines" of even a simple mathematical proof. For certain proofs, we will use the above technique of indentations for proofs nested inside proofs to clarify the structure. Moreover, our analysis shows that certain sentences in a proof simply serve as a reminder of what we can assume and what we are supposed to prove. Hence, the proof of Proposition 2.29 could be condensed as follows.

"Expert proof" of Proposition 2.29. "$\Rightarrow$": Let $A \subseteq B$. First, let $x \in A \cap B$. Because $x \in A \cap B$, we know that $x \in A$ and $x \in B$, which, in particular, means that $x \in A$. Hence $A \cap B \subseteq A$. Conversely, let $x \in A$. Because $x \in A$ and $A \subseteq B$ we infer that $x \in B$. Therefore $x \in A \cap B$. Hence $A \subseteq A \cap B$ and thus $A \cap B = A$.

"$\Leftarrow$": For the converse, let $A \cap B = A$ and let $x \in A$. Because $A = A \cap B$, we infer that $x \in A \cap B$. Thus $x \in B$. Hence $A \subseteq B$. ∎

Note how the "expert proof" contains all the relevant information, but also note that it requires more concentration to read than the original proof. There are more gaps to be filled in by the reader, because when such a proof is written, it is expected that the reader can set up the relevant structures.

Throughout Chapter 2, the author has tried to be a bit more elaborate in proofs and to state what needs to be proved, how it will be proved and when the proof (or a part of it) is done. Later on, this practice will fade, as readers of mathematics are expected to automatically identify the correct structure for a proof, to recall hypotheses when proving a result and to keep track of the conclusion for which the argument is headed.

After all this analysis, readers may wish to refer back to their solutions of Exercises 2-6 and 2-7, which have similarly surprising complexity. **Do not be discouraged if your initial solutions for these exercises are not correct. You are currently learning about the structure of mathematical statements. It is, unfortunately, quite common to underestimate the difficulty of some statements.**

As was frequently the case so far, we conclude this section on a cautionary note. Although the Axiom of Extension sounds like an obvious demand, it is much more subtle to work with in practice than it seems at first. At this stage of development, we cannot give a natural example. But we can give an example that uses entities that we have not formally defined yet. Example 2.30 shows that even if we can state concise definitions for two sets, it may be impossible to determine if the sets are equal.

Example 2.30 A **twin prime** number is a prime number n that differs by 2 from another prime number. Examples of twin prime numbers are 3, 5, 7, 11, 13, 17, 19, 29, 31, 41, 43, etc. It is conjectured that there are infinitely many twin prime numbers, but the conjecture is unresolved at the time this example is written. That is, we do not know if the set of twin prime numbers is finite or infinite. Consider the set T of all twin prime numbers and consider the set K of all twin prime numbers that have been computed at the time the reader encounters these words. It is very hard to determine if the two sets are equal.[4]   □

Exercises

2-10. Let A, B be sets. Prove that $A \setminus (A \setminus B) = A \cap B$.

2-11. Let A, B be sets. Sketch a Venn diagram that illustrates the equation $A \setminus (A \setminus B) = A \cap B$.

2-12. Let A, B be sets. Prove that $A \setminus (A \cap B) = A \setminus B$.

2-13. Let A, B be sets. Prove that $A \setminus (B \setminus A) = A$.

2-14. Let A be a set. Prove that $A \cap A = A$.

2-15. Find the first 20 twin prime numbers.

2-16. For each of the following *false* statements, provide a counterexample.

    (a) Let A, B be sets so that $A \subseteq B$. Then $A = B$.

    (b) Let A, B be sets so that $A \subseteq B$. Then $A = B$ or $B \subseteq A$.

    (c) Let A, B be sets so that $A \subseteq B$. Then $A \cap B = B$.

    (d) Let A, B be sets so that $A \cap B = B$. Then $A \subseteq B$.

---

[4]Unless, by this time, it has been shown that there are infinitely many twin prime numbers, or, it has been shown that there are finitely many and they all have been found.

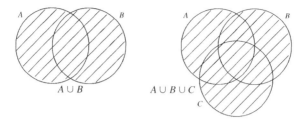

Figure 2.6 The union of two sets and the union of three sets.

2.4   The Axiom of Unions

Although intersection and union often go hand-in-hand, we did not introduce them together. The intersection is a *sub*set of one of the sets in the family that is intersected, which means that the Axiom of Specification guarantees its existence. But the union is a *super*set of the sets in the family that are unified. None of the axioms introduced so far allows us to prove that unions exists. In fact, if we remove the set $\{a, b, c, d, e\}$ from the model in Example 2.23, then we obtain a model that satisfies Axioms 2.1 and 2.2, the Axiom of Specification and the Axiom of Extension, but, for example, it does not contain the union of the sets $\{a, b, c\}$ and $\{c, d, e\}$. Hence, just like we needed the Axiom of Extension for equality of sets, we need an axiom to formally introduce unions.

Axiom 2.31   Axiom of Unions. *For every family C of sets, there exists a set whose elements are all the elements that belong to at least one element of the family. This set is denoted $\bigcup C$ and it is called the* **union** *of C. The union of two sets A and B is also denoted $A \cup B$.*

Of course the Axiom of Unions is feasible. The model in Example 2.23 satisfies Axioms 2.1 and 2.2, the Axiom of Specification, the Axiom of Extension and the Axiom of Unions.

For a visualization of unions, see Figure 2.6. Clearly, the union should be a superset of every set in the family.

Proposition 2.32 *Let C be a family of sets and let $C \in \mathcal{C}$. Then $C \subseteq \bigcup \mathcal{C}$.*

Proof. To prove $C \subseteq \bigcup \mathcal{C}$, as stated in Standard Proof Technique 2.5, we must prove that every element of C is an element of $\bigcup \mathcal{C}$, too. So let $x \in C$ be arbitrary but fixed (see Standard Proof Technique 1.73). By definition of the union $x \in \bigcup \mathcal{C}$. Because $x \in C$ was arbitrary, this means that every element of C is in $\bigcup \mathcal{C}$. Thus we have proved that $C \subseteq \bigcup \mathcal{C}$.   ■

Now that the three main operations for sets, intersection $\cap$, union $\cup$, and (relative) complement $\setminus$, have been introduced, we can investigate how these operations interact with each other. The similarity of the following results to Boolean algebra is not an accident. As we translate the sets in question into set notation, $\wedge$, $\vee$ and $\neg$ take the place of $\cap$, $\cup$ and $\setminus$, respectively (see the proofs of Theorems 2.35 and 2.36 below), and

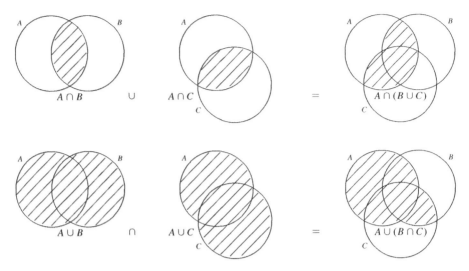

Figure 2.7 Visualization of the distributive laws.

then we can use the rules of Boolean algebra. Because of this correspondence, algebra with sets is called Boolean algebra, too. We leave some of the proofs as exercises.

Theorem 2.33 *Set union is* **commutative**. *Let* A, B *be sets. Then* $A \cup B = B \cup A$.

    Proof. Exercise 2-17.   ■

Theorem 2.34 *Set union is* **associative**. *Let* A, B, C *be sets. Then we have the equality* $A \cup (B \cup C) = (A \cup B) \cup C$.

    Proof. Exercise 2-18.   ■

Theorem 2.35 Distributivity *of intersections over unions and vice versa. Let* A, B, C *be sets. Then the following hold.*

    1. $A \cap (B \cup C) = (A \cap B) \cup (A \cap C)$.

    2. $A \cup (B \cap C) = (A \cup B) \cap (A \cup C)$.

For a visualization, consider Figure 2.7.

    Proof. We will prove part 1 here and leave part 2 to the reader. The key to an efficient proof is to reduce the expressions for the sets to logical expressions for which we know the laws of Boolean algebra.

$$A \cap (B \cup C) \quad = \quad \big\{ x \in A : x \in A \cap (B \cup C) \big\}$$

> By definition of intersections, $x \in A \cap (B \cup C)$ is true iff $x \in A \wedge (x \in B \cup C)$ is true.

$$= \quad \big\{ x \in A : x \in A \wedge (x \in B \cup C) \big\}$$

> By definition of unions, $x \in A \wedge (x \in B \cup C)$ is true iff $x \in A \wedge (x \in B \vee x \in C)$ is true.

$$= \{x \in A : x \in A \wedge (x \in B \vee x \in C)\}$$

> Now use distributivity of $\wedge$ over $\vee$ (see part 5 of Theorem 1.24).

$$= \{x \in A : (x \in A \wedge x \in B) \vee (x \in A \wedge x \in C)\}$$

> $(x \in A \wedge x \in B) \vee (x \in A \wedge x \in C)$ is, by definition of intersections, true iff $(x \in A \cap B) \vee (x \in A \cap C)$ is true.

$$= \{x \in A : (x \in A \cap B) \vee (x \in A \cap C)\}$$

> By definition of unions, $(x \in A \cap B) \vee (x \in A \cap C)$ is true iff $x \in (A \cap B) \cup (A \cap C)$ is true.

$$= \{x \in A : x \in (A \cap B) \cup (A \cap C)\}$$
$$= (A \cap B) \cup (A \cap C)$$

The redundant "$x \in A$:" must be carried through the computation above: The Axiom of Specification states that every set written in set builder notation must explicitly be declared to be a subset of another set. This is done by starting the expression in braces with "$x \in \langle set \rangle$."

The reader will prove part 2 in Exercise 2-19. ∎

The translation from algebra for sets to algebra for logical operators as seen in the proof of Theorem 2.35 (and also in the proof of Theorem 2.36 below) is very convenient and it makes for simple proofs. But we should not expect too much of this technique. It only works when there is an exact correspondence between logic operators and set operators. Thus, for proving the equality of sets in general, we will soon no longer think of this simple translation. Instead, we will come to cherish Standard Proof Technique 2.26, even though this technique looks a bit cumbersome right now.

Theorem 2.36 DeMorgan's Laws. *Let A, B, C be sets. Then the following hold.*

1. $A \setminus (B \cup C) = (A \setminus B) \cap (A \setminus C)$.

2. $A \setminus (B \cap C) = (A \setminus B) \cup (A \setminus C)$.

For a visualization, consider Figure 2.8.

Proof. Similar to the proof of Theorem 2.35, we only prove part 1 and we leave part 2 as an exercise. Similar to the boxes in the proof of Theorem 2.35, the reader should annotate each line to indicate how we went from one step to the next.

$$
\begin{aligned}
A \setminus (B \cup C) &= \{x \in A : x \in A \setminus (B \cup C)\} \\
&= \{x \in A : x \in A \wedge \neg(x \in B \cup C)\} \\
&= \{x \in A : x \in A \wedge \neg(x \in B \vee x \in C)\}
\end{aligned}
$$

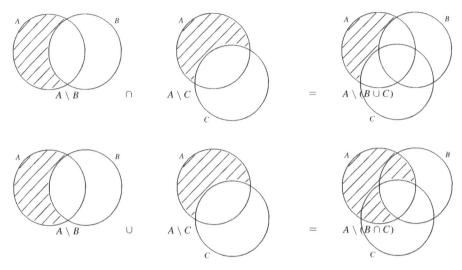

Figure 2.8 Visualization of DeMorgan's Laws.

$$= \ \big\{x \in A : x \in A \wedge \neg(x \in B) \wedge \neg(x \in C)\big\}$$

> Formally, we would need to write the condition for the set above as $x \in A \wedge \big(\neg(x \in B) \wedge \neg(x \in C)\big)$ and it would take 6 steps, mostly using associativity, to show that the set is the same as the set on the next line. We will ultimately prove in Theorem 3.59 that the parentheses that pair the primitive propositions in multiple AND operations can be omitted. To not be overwhelmed with unnecessary detail, we have omitted these parentheses in this proof. We will however pay careful attention to these parentheses after this proof and until we reach Theorem 3.59.

$$= \ \big\{x \in A : x \in A \wedge \neg(x \in B) \wedge x \in A \wedge \neg(x \in C)\big\}$$
$$= \ \big\{x \in A : x \in A \setminus B \wedge x \in A \setminus C\big\}$$
$$= \ \big\{x \in A : x \in (A \setminus B) \cap (A \setminus C)\big\}$$
$$= \ (A \setminus B) \cap (A \setminus C)$$

The reader will prove part 2 in Exercise 2-20.   ∎

Note that, although in DeMorgan's Laws intersections turn into unions and vice versa, the corresponding equations for excluding a single set from a union or an intersection are simpler.

Theorem 2.37 *Let A, B, C be sets. Then the following hold.*

1. $(A \cup B) \setminus C = (A \setminus C) \cup (B \setminus C)$.

2. $(A \cap B) \setminus C = (A \setminus C) \cap (B \setminus C)$.

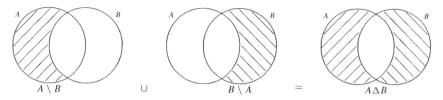

Figure 2.9 Visualization of the symmetric difference.

Proof. Exercise 2-21.   ∎

Theorems 2.27, 2.28, 2.33, 2.34, 2.35, 2.36, and 2.37 are the laws of **Boolean algebra** for sets. Similar to how Boolean algebra for logic can be used to prove certain logical statements, Boolean algebra for sets can be used to prove equality of certain sets.

Example 2.38 To prove that $(A \cup B) \cup (A \cup C) = (A \cup B) \cup C$ we could argue as follows.

$$
\begin{aligned}
(A \cup B) \cup (A \cup C) &= \big((A \cup B) \cup A\big) \cup C & &\text{Used associativity.} \\
&= \big((B \cup A) \cup A\big) \cup C & &\text{Used commutativity.} \\
&= \big(B \cup (A \cup A)\big) \cup C & &\text{Used associativity.} \\
&= (B \cup A) \cup C & &\text{Used } A \cup A = A \text{ (see Exercise 2-22).} \\
&= (A \cup B) \cup C & &\text{Used commutativity.}
\end{aligned}
$$

□

We conclude this section with the symmetric difference between two sets. This operation is not as widely used as the other operations, but we want to at least visualize it (see Figure 2.9) before it is used in the exercises.

Definition 2.39 *Let A, B be sets. Then the set $A \triangle B := (A \setminus B) \cup (B \setminus A)$ is called the* **symmetric difference** *of A and B.*

Exercises

2-17. Prove Theorem 2.33.

2-18. Prove Theorem 2.34.

2-19. Prove part 2 of Theorem 2.35.

2-20. Prove part 2 of Theorem 2.36.

2-21. Analyzing Theorem 2.37.

   (a) Prove part 1 of Theorem 2.37.

(b) Prove part 2 of Theorem 2.37.

(c) Sketch illustrations similar to Figure 2.8 for Theorem 2.37.

2-22. Let A be a set. Prove that $A \cup A = A$.

2-23. For the sets $A = \{a, b, c, f\}$, $B = \{c, d, e\}$ and $C = \{a, d\}$ find $A \cap B$, $A \cup C$, $A \cap (B \cup C)$ and $A \cup (B \cap C)$.

2-24. Alternative proofs for Theorems 2.35 and 2.36. For each of the following proofs, use Standard Proof Technique 2.26 (proving mutual containment).

(a) Prove part 1 of Theorem 2.35.

(b) Prove part 2 of Theorem 2.35.

(c) Prove part 1 of Theorem 2.36.

(d) Prove part 2 of Theorem 2.36.

2-25. Let A, B be sets. Prove that $A \subseteq B$ iff $A \cup B = B$.

Hint. See Proposition 2.29 and following discussion.

2-26. Let A, B be sets. Prove that $A \cap B = A$ iff $A \cup B = B$.

2-27. Let A, B and U be sets so that $A \subseteq U$ and $B \subseteq U$. Prove that $A \subseteq B$ iff $(U \setminus A) \cup B = U$.

2-28. Let A, B and U be sets so that $A \subseteq U$ and $B \subseteq U$. Prove that $A = \emptyset$ iff the equality $\big((U \setminus A) \cap B \big) \cup \big(A \cap (U \setminus B) \big) = B$ holds.

2-29. Let A, B, C be sets. Prove that if $B \subseteq A$ and $C \subseteq A$, then $B \cup C \subseteq A$.

2-30. For each of the following *false* statements, provide a counterexample.

(a) Let A, B be sets so that $A \cap B = \emptyset$. Then $A \cup B = \emptyset$.

(b) Let A, B, C be sets. Then $A \setminus (B \cup C) = (A \setminus B) \cup (A \setminus C)$.

(c) Let A, B, C be sets. Then $A \setminus (B \cap C) = (A \setminus B) \cap (A \setminus C)$.

(d) Let A, B, C be sets. Then $(A \cup B) \setminus C = (A \setminus C) \cap (B \setminus C)$.

(e) Let A, B, C be sets. Then $(A \cap B) \setminus C = (A \setminus C) \cup (B \setminus C)$.

2-31. Let A, B, C be sets. For each of the following equalities, sketch Venn diagrams that illustrate the equality.

(a) $(A \cap B) \cap (A \cap C) = (A \cap B) \cap C$

(b) $A \Delta B = (A \cup B) \setminus (A \cap B)$.

(c) $A \Delta (B \Delta C) = (A \Delta B) \Delta C$.

(d) $A \cap (B \Delta C) = (A \cap B) \Delta (A \cap C)$.

2-32. Let A, B, C be sets. Prove each of the following equalities using the Axiom of Extension. That is, prove each of the following equalities by proving that each element of the set on the left side is an element of the set on the right side and vice versa.

(a) $(A \cap B) \cap (A \cap C) = (A \cap B) \cap C$

(b) $A \Delta B = (A \cup B) \setminus (A \cap B)$.

(c) $A \Delta B = B \Delta A$.

(d) $A \Delta (B \Delta C) = (A \Delta B) \Delta C$.

(e) $A \cap (B \Delta C) = (A \cap B) \Delta (A \cap C)$.

2-33. Let A, B, C be sets. Prove each of the following equalities using Boolean algebra. That is, prove each of the following equalities using Theorems 2.27, 2.28, 2.33, 2.34, 2.35, 2.36, and 2.37.

(a) $(A \cap B) \cap (A \cap C) = (A \cap B) \cap C$

(b) $A \Delta B = (A \cup B) \setminus (A \cap B)$.

(c) $A \Delta B = B \Delta A$.

(d) $A \Delta (B \Delta C) = (A \Delta B) \Delta C$.

(e) $A \cap (B \Delta C) = (A \cap B) \Delta (A \cap C)$.

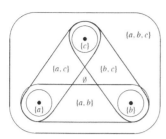
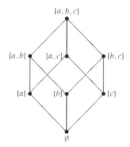

Figure 2.10 Visualization of the power set of a three element set. The left image shows a Venn diagram. The right image shows a hierarchical arrangement with sets represented by points, smaller sets placed lower than larger sets and containment indicated with connecting lines.

2.5   The Axiom of Powers, Relations, and Functions

The Axiom of Unions was our first step towards taking smaller sets and generating larger ones. To introduce more complex mathematical objects, such as functions, we need sets of sets. But this is exactly where, in Russell's Paradox, the bad entity B gave us trouble. So we must be careful at this juncture. It turns out that the set of all subsets of a set is exactly what we need.

Axiom 2.40 Axiom of Powers. *For every set S, there exists a set $\mathcal{P}(S)$, called the* **power set** *of S, whose elements are all the subsets of S.*

Example 2.41 Examples of power sets.

1. The power set of the empty set is $\mathcal{P}(\emptyset) = \{\emptyset\}$.

2. The power set of a one element set $\{a\}$ is $\mathcal{P}(\{a\}) = \{\emptyset, \{a\}\}$.

3. The power set of a two element set $\{a, b\}$ is $\mathcal{P}(\{a, b\}) = \{\emptyset, \{a\}, \{b\}, \{a, b\}\}$.

4. The power set of a three element set is given in Figure 2.10.   □

None of our models so far has included the set of subsets of its largest set *as another set*. For example, the set in Example 2.23 is itself a power set, but it does not *contain* any power sets. Thus the Axiom of Powers cannot be derived from the axioms we have so far. The superstructure that will be introduced in Definition 2.62 will allow us to present a model that satisfies all axioms, including the Axiom of Powers. But before we investigate this abstract structure, we discuss the more widely used entities that the Axiom of Power makes available.

A lot of mathematics is about functions. Intuitively, a function is a rule that assigns each input value a specific output value. But to put this idea on a logically firm footing, we must express functions as set theoretical entities. Functions are often represented via "assignment tables" that state which input value is mapped to which output value.

From a more abstract point-of-view, such assignment tables are ordered pairs of input and output values. So, to ultimately introduce functions, we must first introduce ordered pairs.

If a and b are distinct elements of a set, that is, if $a \neq b$, then the sets $\{a, b\}$ and $\{b, a\}$ have the same elements. Therefore, by the Axiom of Extension, they are the same, regardless of the order in which we list the elements. Our first goal is to obtain a mathematical construct in which the order of the "elements" matters. Theorem 2.42 provides the construct and Definition 2.43 states the formal definition.

In the following, we will assume that the objects we use are elements of some set. In this fashion, any set of sets exists as a subset of some power set. For example, if $\{a, b\}$ is a set, then $\{\{a\}, \{a, b\}\}$ is a subset of the power set $\mathcal{P}(\{a, b\})$.

Theorem 2.42 *Let $\{a, b\}$ and $\{c, d\}$ be sets with one or two elements. Then the equality $\{\{a\}, \{a, b\}\} = \{\{c\}, \{c, d\}\}$ holds iff $a = c$ and $b = d$.*

Proof. For "$\Leftarrow$," note that if $a = c$ and $b = d$, then $\{\{a\}, \{a, b\}\} = \{\{c\}, \{c, d\}\}$ clearly holds.

For the converse "$\Rightarrow$," let $\{\{a\}, \{a, b\}\} = \{\{c\}, \{c, d\}\}$. There are two cases to consider: $\{a\} = \{c, d\}$ and $\{a\} \neq \{c, d\}$ (see Standard Proof Technique 1.60).

Case 1: $\{a\} = \{c, d\}$. In this case, we have $a = c = d$. This equality implies $\{\{c\}\} = \{\{c\}, \{c, d\}\} = \{\{a\}, \{a, b\}\}$. Hence $\{a, b\} = \{c\}$ and thus $a = b = c$. But then $a = b = c = d$, which implies, in particular, that $a = c$ and $b = d$.

Case 2: $\{a\} \neq \{c, d\}$. In this case, we must have that $\{a\} = \{c\}$ and $\{a, b\} = \{c, d\}$. From $\{a\} = \{c\}$ we conclude $a = c$ and then $c \neq d$ (otherwise $\{a\} = \{c, d\}$). Then, because $c \neq d$ and $\{a, b\} = \{c, d\}$, both sets $\{a, b\}$ and $\{c, d\}$ have two elements. Because $a = c$ and $\{a, b\} = \{c, d\}$, this means we must have $b = d$. ∎

Definition 2.43 *Let A and B be sets and let $a \in A$ and $b \in B$ be objects. Then we define the **ordered pair** (a, b) to be $(a, b) := \{\{a\}, \{a, b\}\}$.*

The notation suggests that the ordered pair (a, b) is a set. We can verify this with the axioms, because both $\{a\}$ and $\{a, b\}$ are subsets of $A \cup B$, that is, they are elements of $\mathcal{P}(A \cup B)$. But then $\{\{a\}, \{a, b\}\}$ is a subset of $\mathcal{P}(A \cup B)$. Hence ordered pairs are part of set theory. In particular, an ordered pair of an element of A and an element of B is itself an element of $\mathcal{P}(\mathcal{P}(A \cup B))$.

More importantly, Theorem 2.42 shows that two ordered pairs (a, b) and (c, d) are equal iff $a = c$ and $b = d$, which is the property that an ordered pair should have. We can now define cartesian products as sets of ordered pairs.

Definition 2.44 *Let A and B be sets. Then we define the **cartesian product**, or **product**, to be the set $A \times B := \{(a, b) : a \in A, b \in B\}$*

With every ordered pair being an element of $\mathcal{P}(\mathcal{P}(A \cup B))$, the cartesian product is a subset of $\mathcal{P}(\mathcal{P}(A \cup B))$ and hence it is an element of $\mathcal{P}(\mathcal{P}(\mathcal{P}(A \cup B)))$. Cartesian products allow us to define relations and functions.

Definition 2.45 *Let A and B be sets. Then any subset $\sim \subseteq A \times B$ is also called a **relation** from A to B.*

With every ordered pair being an element of $\mathcal{P}(\mathcal{P}(A \cup B))$, any relation is a subset of $\mathcal{P}(\mathcal{P}(A \cup B))$ and hence any relation is an element of $\mathcal{P}\big(\mathcal{P}(\mathcal{P}(A \cup B))\big)$. Of course we would like to think of relations in simpler terms than as sets of ordered pairs, or, even worse, elements of $\mathcal{P}\big(\mathcal{P}(\mathcal{P}(A \cup B))\big)$.

Notation 2.46 *Let A and B be sets and let $\sim$ be a relation from A to B. To express what a relation does in more natural terms, whenever $(a, b) \in \sim$ we will write $a \sim b$ instead of $(a, b) \in \sim$ and we will say in this case that a is $\sim$ related to b, or, that a is related to b. We will write $a \nsim b$ instead of $(a, b) \notin \sim$. We also denote the fact that $\sim$ is a relation from A to B by writing $\sim: A \to B$.*

Definition 2.47 *Let A and B be sets and let $\sim: A \to B$ be a relation from A to B. Then we define the set $\operatorname{dom}(\sim) := \big\{ a \in A : (\exists b \in B : a \sim b) \big\}$ and call it the **domain** of $\sim$. We also define the set $\operatorname{rng}(\sim) := \big\{ b \in B : (\exists a \in A : a \sim b) \big\}$ and call it the **range** of $\sim$.*

Example 2.48 Examples of relations. (We refer to entities that we have not defined yet or that are outside mathematics to get better explanations.) Anything that matches objects into pairs of two is a relation.

1. If a social networking web site allows users to declare someone as a "friend," then friends are stored as pairs of two. The resulting set of pairs is the friend relation. Its domain is the set of all people who have a friend and its range is the set of all people who are someone's friend. Typically, because friendship is symmetric, the relation is symmetric, too. That is, if a is a friend of b, then b is a friend of a, too.

2. The strict inequality $<$ between numbers is a relation. Essentially, $a < b$ iff the pair (a, b) is part of the $<$-relation. Unlike the friend relation from part 1, this relation is not symmetric: If $a < b$, then definitely $b \nless a$.

   This example shows that we are already familiar with Notation 2.46. In fact, the actual definition of relations probably looks a bit strange compared to relations we know, such as $<$. Nonetheless, Definition 2.45 is useful. On one hand, Definition 2.45 allows us to put relations into the context of set theory (hooray for set theory). But, on the other hand, we must note that part 1 shows that Definition 2.45 is actually used in real-life software implementations. Relations that cannot be computed (and whereas computation can tell us whether $a < b$ or not, no computation can tell us who our friends are) are literally stored in the fashion of Definition 2.45 in the memory of servers. □

Another strange part of Definition 2.45 is that it talks about a relation *from* a set A *to* a set B, rather than about a relation *between* these sets. Part of the reason for this notation is that relations need not be symmetric and that A and B may be different sets. But it seems the notation is mostly chosen to be reminiscent of functions, because functions turn out to be special types of relations.

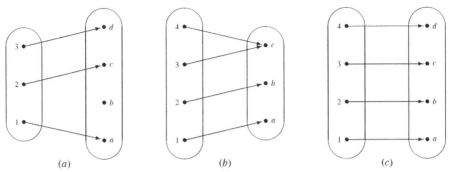

Figure 2.11 An injective function maps all elements of the domain to distinct images, but some elements of the range may not have a preimage (*a*). For a surjective function, every element of the range has a preimage, but some elements of the domain may be mapped to the same image (*b*). A bijective function maps all elements of the domain to distinct images and each element of the range has a preimage (*c*). This is why the existence of a bijection between two sets indicates that the two sets are "of the same size" (see Definition 2.59).

Definition 2.49 *Let A and B be sets and let* $f : A \to B$ *be a relation from A to B. Then f is called a* **function** *iff the following hold.*

1. *The relation f is* **well-defined**, *that is, for all* $a \in A$ *and* $b_1, b_2 \in B$ *we have that if* $(a, b_1) \in f$ *and* $(a, b_2) \in f$, *then* $b_1 = b_2$, *and*

2. *The relation f is* **totally defined**, *that is, for every* $a \in A$ *there is a* $b \in B$ *so that* $(a, b) \in f$.

Functions are also called **maps** *or* **mappings**.

Notation 2.50 *Let A and B be sets and let f be a function from A to B. Then, instead of* $(a, b) \in f$ *we will usually write* $f(a) = b$. *Moreover, we will often abbreviate the declaration "let f be a function from A to B" by "let* $f : A \to B$ *be a function."*

Example 2.51 Examples of functions. (Once more we refer to entities that we have not formally defined yet to clarify the definition.)

1. Any function *f* with a domain *D* and a range *R* that the reader has seen in classes such as algebra or calculus is a function in the sense of Definition 2.49. In fact, in algebra and calculus, the relation in Definition 2.49 is often called the **graph** $\{(x, f(x)) \in D \times R : x \in D\}$. So, once again we are considering familiar objects in the light of set theory.

2. As for relations, not every function is easily computable. For example, although the assignment of grades to students is ultimately computable from the rules on the syllabus and the student scores, it is reported as a discrete assignment, that

is, as a function $f : \langle$ set of students in the class $\rangle \to \{A, B, C, D, F\}$. Every student is assigned a grade and no student is assigned two grades.[5] □

Once functions are defined, we want to investigate their properties.

Definition 2.52 *Let A, B be sets and let $f : A \to B$ be a function. Then f is called* **injective** *or* **one-to-one** *iff for all $x, y \in A$ the inequality $x \neq y$ implies $f(x) \neq f(y)$. f is called* **surjective** *or* **onto** *iff for all $b \in B$ there is an $a \in A$ with $f(a) = b$. Finally, f is called* **bijective** *iff f is both injective and surjective. For a visualization, consider Figure 2.11.*

Because it is inconvenient to work with inequalities, we usually work with the contrapositive of the definition of injectivity.

Proposition 2.53 *Let A, B be sets and let $f : A \to B$ be a function. Then f is injective iff for all $x, y \in A$ we have that $f(x) = f(y)$ implies $x = y$.*

Proof.   This result is true, because the contrapositive of the implication "$x \neq y \Rightarrow f(x) \neq f(y)$" is "$f(x) = f(y) \Rightarrow x = y$."   ∎

Example 2.54 (Once more, we use entities that we have not formally introduced yet.) Consider the function $f : \mathbb{R} \setminus \{1\} \to \mathbb{R}$ defined by

$$f(x) = \frac{3x + 2}{x - 1}.$$

1. The function f is injective, because

$$f(x) = f(y)$$

$$\Rightarrow \quad \frac{3x + 2}{x - 1} = \frac{3y + 2}{y - 1}$$

$$\Rightarrow \quad (3x + 2)(y - 1) = (3y + 2)(x - 1)$$

$$\Rightarrow \quad 3xy - 3x + 2y - 2 = 3yx - 3y + 2x - 2$$

$$\Rightarrow \quad -3x + 2y = -3y + 2x$$

$$\Rightarrow \quad 5y = 5x$$

$$\Rightarrow \quad y = x$$

shows that $f(x) = f(y)$ implies $x = y$.

---

[5] We did not take incompletes into account. The author requests that readers strive to reduce the range to a smaller set, such as $\{A\}$ or $\{A, B\}$.

2. The function f is not surjective, because the value $3 \in \mathbb{R}$ does not have a preimage. Indeed, suppose for a contradiction that there is an $x \in \mathbb{R}$ so that $f(x) = 3$. Then

$$
\begin{aligned}
\frac{3x+2}{x-1} &= 3 \\
3x+2 &= 3x-3 \\
2 &= -3,
\end{aligned}
$$

a contradiction.   □

If the range of one function is in the domain of another, then we can apply the functions consecutively to obtain a new function.

Definition 2.55 *Let A, B, C be sets and let $g : A \to B$ and $f : B \to C$ be functions. Then the* **composition** *$f \circ g$ of the functions f and g is defined to be the function $f \circ g := \big\{(a, c) \in A \times C : [\exists b \in B : g(a) = b \wedge f(b) = c]\big\}$.*

The sign for composition looks similar to the familiar algebraic operation of multiplication. Just like for multiplication, there is a neutral element for composition.

Example 2.56 *Let A be a set. Then we define $\mathrm{id}_A := \big\{(a, a) \in A \times A : a \in A\big\}$ and we call it the* **identity function** *on A. When there is no doubt about the underlying set, the subscript is omitted.*

The identity function satisfies $\mathrm{id}_A(a) = a$ for all $a \in A$. Therefore for all functions $f : A \to B$ and all functions $g : C \to A$ we have $f \circ \mathrm{id}_A = f$ and $\mathrm{id}_A \circ g = g$.   □

Once a neutral element is available, the idea of an inverse can be discussed.

Definition 2.57 *Let A and B be sets and let $f : A \to B$ be a function. Then the function $g : B \to A$ is called the* **inverse** *of f iff $f \circ g = \mathrm{id}_B$ and $g \circ f = \mathrm{id}_A$.*

We speak of *the* inverse of a function, because if a function has an inverse, then the inverse is unique (see Exercise 2-35).

Example 2.58 The function $g : \mathbb{R} \setminus \{3\} \to \mathbb{R} \setminus \{1\}$ defined by

$$
g(x) = \frac{x+2}{x-3}
$$

is the inverse of the function $f : \mathbb{R} \setminus \{1\} \to \mathbb{R} \setminus \{3\}$ defined by

$$
f(x) = \frac{3x+2}{x-1},
$$

because for all $x \in \mathbb{R} \setminus \{3\}$ we have

$$
f \circ g(x) = f\left(\frac{x+2}{x-3}\right) = \frac{3\left(\frac{x+2}{x-3}\right)+2}{\left(\frac{x+2}{x-3}\right)-1} = \frac{\frac{3x+6+2x-6}{x-3}}{\frac{x+2-x+3}{x-3}} = \frac{\frac{5x}{x-3}}{\frac{5}{x-3}} = x
$$

and for all $x \in \mathbb{R} \setminus \{1\}$ we have

$$g \circ f(x) \;=\; g\left(\frac{3x+2}{x-1}\right) = \frac{\left(\frac{3x+2}{x-1}\right)+2}{\left(\frac{3x+2}{x-1}\right)-3} = \frac{\frac{3x+2+2x-2}{x-1}}{\frac{3x+2-3x+3}{x-1}} = \frac{\frac{5x}{x-1}}{\frac{5}{x-1}} = x.$$

$\square$

Bijective functions are also used to determine when two sets have the same size. This is the idea behind the definition below.

Definition 2.59 *Two sets A and B are called* **equivalent** *iff there is a bijective function* $f : A \to B$.

We will talk more about equivalence in Sections 4.6 and 5.6. For now, we can record that taking the power set always increases the size: Theorem 2.60 shows that the power set $\mathcal{P}(X)$ does not have the same size as the set X. Moreover, because the power set contains the set $\{\{x\} \in \mathcal{P}(X) : x \in X\}$, which clearly has the same size as X, it makes sense to say that the power set is "larger" than the original set. For more on sizes of sets, consider Section 7.3.

The proof of Theorem 2.60 is a bit reminiscent of Russell's Paradox.

Theorem 2.60 *If X is a set, then X is not equivalent to its power set* $\mathcal{P}(X)$.

Proof. Suppose for a contradiction that $f : X \to \mathcal{P}(X)$ is a bijection. Define $B := \{x \in X : x \notin f(x)\}$. Because f is surjective, there is a $b \in X$ with $B = f(b)$. Now $b \in B$ would imply $b \in B = f(b)$. By definition of B, this would mean that $b \notin f(b) = B$. Thus we infer $b \notin B$. But then $b \notin B = f(b)$, which by definition of B forces $b \in B$, a contradiction. $\blacksquare$

We conclude this section with the definition of a superstructure. First, we introduce indexed families of sets.

Definition 2.61 *Let X be a set. An* **indexed family** *of subsets* S_i *of X is denoted* $\{S_i\}_{i \in I}$, *where it is understood that there is a function from the* **index set** *I to* $\mathcal{P}(X)$ *that maps each* $i \in I$ *to* S_i. *If* $\{S_i\}_{i \in I}$ *is an indexed family, the union is denoted* $\bigcup_{i \in I} S_i := \{x \in X : (\exists i \in I : x \in S_i)\}$ *and the intersection of the indexed family is denoted* $\bigcap_{i \in I} S_i := \{x \in X : (\forall i \in I : x \in S_i)\}$.

There is not much of a difference between unions and intersections of families of sets and unions and intersections of indexed families of sets. In fact, every intersection or union of a family of sets $\mathcal{C}$ in the sense of Definition 2.17 and Axiom 2.31 is also an intersection or union of an indexed family: Simply use $\mathcal{C}$ as the index set and use $\mathrm{id}_{\mathcal{C}}$ as the indexing function. The converse is true, too, because the intersection and union of the indexed family $\{S_i\}_{i \in I}$ is simply the intersection or union of the range of the indexing function. The laws for intersections and unions of families of sets should be similar to those for indexed families of sets, and Exercise 2-47 confirms this conjecture.

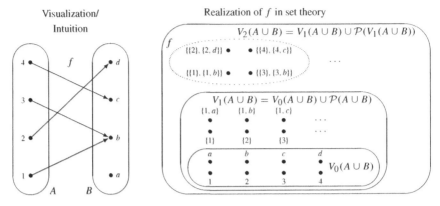

Figure 2.12 Visualization of a function and of its set theoretical definition. In set theory, f is a subset of $V_2(A \cup B)$, the second set in the iteration that produces the superstructure. Consequently, f is also an element of $V_3(A \cup B)$, the third set in the iteration that produces the superstructure.

Indexed families of subsets are useful, because sometimes sets are associated with certain elements or numbers. As these sets are given, some of them may be listed twice. Rather than painstakingly making sure doubles are eliminated, it is more efficient to work with an indexed family. For the definition of a superstructure below, there are no "doubles," but the sets are given as a sequence. The natural numbers with zero, which we use in the definition of superstructures, have not been formally introduced yet. But the natural numbers will be introduced in the next section without referring to superstructures, and an additional element 0 can formally be added. So there is no danger of circular reasoning here. Similarly, we will prove later, without referring to superstructures, that recursive definitions are allowed (see remarks before Definition 3.55).

Definition 2.62 *Let X be a set. Define $V_0(X) := X$ and for each $n \in \mathbb{N}$ define $V_n(X) := V_{n-1}(X) \cup \mathcal{P}(V_{n-1}(X))$. Then*

$$V(X) := \bigcup_{n \in \mathbb{N} \cup \{0\}} V_n(X)$$

is called the **superstructure** *over X.*[6]

The superstructure over $\emptyset$ (or, more precisely, the collection of all the sets in the superstructure) satisfies Axioms 2.1 and 2.2, the Axiom of Specification, the Axiom of Extension, the Axiom of Unions and the Axiom of Powers. Moreover, just about any set-theoretical object related to a set X is contained in the superstructure over X. For

---

[6] In the following models that involve superstructures, we will not consider $V(X)$ itself as a set. So the union in the definition is actually formal notation to indicate the "collection" of all the elements of the sets $V_n(X)$. Fine points about superstructures will be discussed in Exercise 7-15, when we have more formal set theoretical tools. Exercise 2-49 in this section shows that, as soon as we require the axiom of powers, our surrounding universe cannot be a set.

example, Figure 2.12 shows how functions $f : A \to B$ fit into the superstructure over $X = A \cup B$.

Exercises

2-34. Let A, B be sets and let $f : A \to B$ be a function.

    (a) Prove that f is injective iff there is a function $g : B \to A$ so that $g \circ f = \mathrm{id}_A$.

    (b) Prove that f is surjective iff there is a function $h : B \to A$ so that $f \circ h = \mathrm{id}_B$.

    (c) Prove that f is bijective iff there is a function $k : B \to A$ so that $k \circ f = \mathrm{id}_A$ and $f \circ k = \mathrm{id}_B$.

2-35. Let A and B be sets and let $f : A \to B$ be a function.

    (a) Prove that if f has an inverse, then f is bijective.

    (b) Prove that if $g : B \to A$ and $h : B \to A$ are both inverses of f, then $g = h$.

2-36. Power sets.

    (a) Compute the power set of the set $\{a, b, c, d\}$.

    (b) Compute the power set of the set $\{1, 2, 3, 4\}$.

    (c) Is there any significant difference between Exercises 2-36a and 2-36b?

2-37. List all two element subsets of the set $\{1, 2, 3, 4, 5\}$.

2-38. Sketch a visualization of the power set of $\{1, 2, 3, 4\}$ similar to that in Figure 2.10.

2-39. Let $A = \{1, 2, 3, 4\}$ and let $B = \{a, b, c, d, e\}$ Determine which of the following relations is a function. If it is a function, determine if it is injective, surjective, bijective or none of the above. Justify your answers using the definitions.

    (a) $\big\{ (1, a), (2, b), (3, c), (4, d) \big\}$ from A to B.

    (b) $\big\{ (1, a), (2, b), (3, c), (3, d) \big\}$ from $A \setminus \{4\}$ to B.

    (c) $\big\{ (1, a), (2, b), (3, c), (4, c) \big\}$ from A to B.

    (d) $\big\{ (1, a), (2, b), (3, c), (4, a) \big\}$ from A to $B \setminus \{d, e\}$.

2-40. For each pair of functions, check if they are inverses of each other. For this exercise, you may use what you know about real numbers and functions of real numbers.

    (a) $f, g : \mathbb{R} \to \mathbb{R}$, $f(x) = 3x + 6$, $g(x) = \dfrac{x}{3} - 2$

    (b) $f : \mathbb{R} \to \mathbb{R}$, $f(x) = x^2 + 1$, $g : [0, \infty) \to \mathbb{R}$, $g(x) = \sqrt{x} - 1$

    (c) $f, g : \mathbb{R} \setminus \{1\} \to \mathbb{R} \setminus \{1\}$, $f(x) = \dfrac{x + 1}{x - 1}$, $g(x) = \dfrac{x + 1}{x - 1}$

    (d) $f, g : \mathbb{R} \to \mathbb{R}$, $f(x) = \sqrt[5]{x^3 - 1}$, $g(x) = \sqrt[3]{x^5 + 1}$

2-41. Prove that the functions $f, g : \mathbb{R} \to \mathbb{R}$ defined by $f(x) = x^2$ and $g(x) = \sqrt{|x|}$ are not inverses of each other. Then determine if $f(x) = x^2$ and $h(x) = \sqrt{x}$ are inverses of each other. For this exercise, you may use what you know about real numbers and functions of real numbers.

2-42. Determine which of the following functions are one-to-one. For those that are one-to-one, compute the inverse from the range to the domain. For this exercise, you may use what you know about real numbers and functions of real numbers.

    (a) $f : \mathbb{R} \to \mathbb{R}$, $f(x) = 5x + 9$

    (b) $f : \mathbb{R} \to \mathbb{R}$, $f(x) = x^2 + 4$

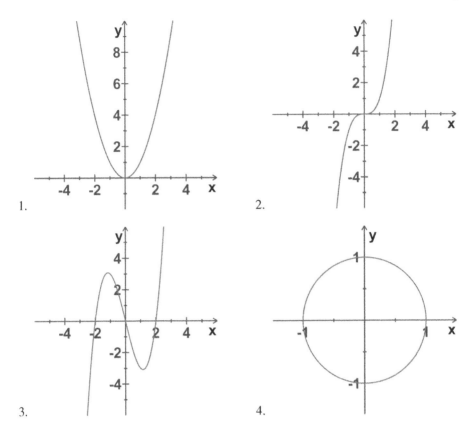

Figure 2.13 Graphs for Exercise 2-43. For each graph, assume that the graph continues just as a graph in calculus would. That is, if a graph looks as if it will continue to ∞ or $-\infty$, it does.

(c) $f : \mathbb{R} \setminus \left\{\dfrac{5}{2}\right\} \rightarrow \mathbb{R}, f(x) = \dfrac{5x + 4}{2x - 5}$

(d) $f : \mathbb{R} \rightarrow \mathbb{R}, f(x) = 2x^3 - 9$

(e) $f : \mathbb{R} \setminus \{1\} \rightarrow \mathbb{R}, f(x) = \dfrac{4}{x - 1}$

(f) $f : \mathbb{R} \rightarrow \mathbb{R}, f(x) = \sqrt{x + 3}$

(g) $f : \mathbb{R} \rightarrow \mathbb{R}, f(x) = \sqrt[3]{5 - x}$

2-43. Each of the graphs in Figure 2.13 depicts a relation from $\mathbb{R}$ to $\mathbb{R}$. For each graph, decide if the relation is an injective, surjective or bijective function.

2-44. For each of the following *false* statements, provide a counterexample.

  (a) Every injective function is surjective.

  (b) Every surjective function is injective.

  (c) Let $f : A \rightarrow B$ be a function so that there is a function $g : B \rightarrow A$ so that $g \circ f = \text{id}_A$. Then f is surjective.

(d) Let $f : A \to B$ be a function so that there is a function $g : B \to A$ so that $f \circ g = \mathrm{id}_B$. Then f is injective.

(e) Every relation is a function.

2-45. Let A, B be sets. Prove that $\{A, B\}$ is a set, too.

Note. The above result is also called the **Axiom of Pairing**. It is sometimes included in the Zermelo-Fraenkel axioms. As this exercise shows, it can actually be derived from the other axioms. Hence, if we are allowed to use the other axioms, then the Axiom of Pairing is redundant.

2-46. Let A, B, C be sets. Prove each of the following.

(a) $A \times (B \cup C) = A \times B \cup A \times C$

(b) $A \times (B \cap C) = A \times B \cap A \times C$

(c) $(A \cup B) \times C = A \times C \cup B \times C$

(d) $(A \cap B) \times C = A \times C \cap B \times C$

2-47. **DeMorgan's Laws** for families of sets. Let $\mathcal{C}$ be a family of sets, let $\{C_i\}_{i \in I}$ be an indexed family of sets and let X be another set. Prove each of the following.

(a) $X \setminus \bigcup_{i \in I} C_i = \bigcap_{i \in I} X \setminus C_i$

(b) $X \setminus \bigcup \mathcal{C} = \bigcap \{X \setminus C : C \in \mathcal{C}\}$.

(c) $X \setminus \bigcap_{i \in I} C_i = \bigcup_{i \in I} X \setminus C_i$

(d) $X \setminus \bigcap \mathcal{C} = \bigcup \{X \setminus C : C \in \mathcal{C}\}$.

2-48. Explain why no model for the axioms of set theory including the Axiom of Powers can be finite.

2-49. Explain why no model for the axioms of set theory including the Axiom of Powers can have a surrounding universe that is a set.

Hint. Proof of Theorem 2.60.

2.6 The Axiom of Infinity and the Natural Numbers

The Axiom of Powers has significantly expanded our horizon. By repeatedly forming power sets of power sets, it seems that we can form arbitrarily large sets. But, unfortunately, the axioms do not guarantee that we have very much to start with. In fact, the only entity that is guaranteed to exist is the empty set (see Proposition 2.13). If we start with the empty set and build the superstructure over the empty set, we can obtain finite sets of arbitrary size. But we will not obtain infinite sets at any stage, and the superstructure itself is not a set in the model of Definition 2.62. So, with the axioms at hand, we can have an infinite set theory in which all sets are finite. To get an infinite set, we need to somehow "cut across" all the stages of the superstructure and obtain another set. The Axiom of Infinity guarantees that this can be done.

Axiom 2.63 Axiom of Infinity. *There is a set I that contains the empty set $\emptyset$ as an element, and for each $a \in I$ the set $a \cup \{a\}$ is an element of I, too.*

The superstructure over $\emptyset$ shows that Axioms 2.1 and 2.2, the Axiom of Specification, the Axiom of Extension, the Axiom of Unions and the Axiom of Powers can all be satisfied and yet the Axiom of Infinity can still fail. The superstructure over the set I from the Axiom of Infinity shows that all axioms can be satisfied simultaneously.

With the Axiom of Infinity, we can construct the natural numbers in set theory. This is very important, because all our future investigations will be built upon the natural numbers as presented in Theorem 2.64 below. **The proof of Theorem 2.64 is the first really long proof that we encounter in this text. Do not get frustrated by the many details or by the fact that you might not see how anyone could have thought of it. As your level of experience increases, proofs like this one will feel more natural than they do upon first exposure. Eventually, you will simply read proofs like this one carefully and as often as necessary to understand it. Upon first reading of this text it may be permissible to skim it, though. We will not need the details of the proof of Theorem 2.64 until we reach Chapter 7.**

Theorem 2.64 *There is a set, denoted* $\mathbb{N}$ *and called the set of* **natural numbers***, so that the following hold.*

1. *There is a special element in* $\mathbb{N}$, *which we denote by* 1.[7]

2. *For each* $n \in \mathbb{N}$, *there is a corresponding element* $n' \in \mathbb{N}$, *called the* **successor** *of* n.

3. *The element* 1 *is not the successor of any natural number.*

4. **Principle of Induction.** *If* $S \subseteq \mathbb{N}$ *is such that* $1 \in S$ *and for each* $n \in S$ *we also have* $n' \in S$, *then* $S = \mathbb{N}$.

5. *For all* $m, n \in \mathbb{N}$ *if* $m' = n'$, *then* $m = n$.

The above properties are also called the **Peano Axioms** *for the natural numbers.*

Proof. To prove this result, we must construct a set $\mathbb{N}$ with properties 1-5. In particular, this set must be infinite. The only infinite set that is guaranteed by the axioms is the set I from the Axiom of Infinity, so we start with it. The special element 1 and the successor relation are constructed from the special element $\emptyset$ of I and from the fact that for each $a \in I$ the set $a \cup \{a\}$ is in I, too. We let 1 be the element $\{\emptyset\} = \emptyset \cup \{\emptyset\} \in I$,[8] and for each $n \in I$ we let $n' := n \cup \{n\}$.

The set I itself may have more elements than we need. (**It turns out that** $\mathbb{N}$ **is the smallest set with properties 1-5.**) Call a subset $S \subseteq I$ a **successor set** iff $\emptyset \notin S$, $1 \in S$ and for all $n \in S$ we have that $n' \in S$. Then $I \setminus \{\emptyset\}$ is a successor set. Moreover, by definition, all successor sets are subsets of I, which means that every

---

[7]The natural numbers can equally well be constructed with their first element being zero. In fact, starting with zero would simplify some proofs in Chapter 3. The reader may consider this an idiosyncrasy of this text, maybe even an idiotsyncrasy, but people count $1, 2, 3, \ldots$, not $0, 1, 2, 3, \ldots$. Hence the author believes that the natural numbers should start with 1.

[8]We will use $\emptyset$ as 0 later on.

successor set is an element of $\mathcal{P}(I)$. Hence, by the Axiom of Specification, there is a subset S of $\mathcal{P}(I)$ whose elements are all the successor sets. We now define $\mathbb{N} := \bigcap S$ to be the intersection of all successor sets (which exists, once more, by the Axiom of Specification).

To prove part 1, we must prove that $1 \in \mathbb{N}$. Note that every successor set contains 1. Therefore $1 \in \bigcap S = \mathbb{N}$, as was to be proved.

To prove part 2, we must prove that for every $n \in \mathbb{N}$ the successor n' is in $\mathbb{N}$, too. Let $n \in \mathbb{N}$ be arbitrary, but fixed (see Standard Proof Technique 1.73). We need to prove that $n' \in \mathbb{N}$. Because $n \in \mathbb{N} = \bigcap S$, we conclude that $n \in S$ for all $S \in S$. By definition of successor sets, $n' \in S$ for all $S \in S$. Hence $n' \in \bigcap S = \mathbb{N}$, as was to be proved. Because $n \in \mathbb{N}$ was arbitrary, we have proved part 2.

To prove part 3, we must prove that 1 is not the successor of any element of $\mathbb{N}$. Suppose for a contradiction that 1 was the successor $1 = x'$ of an $x \in \mathbb{N}$. Then $\{\emptyset\} = 1 = x' = x \cup \{x\}$. The only way this equation could hold is if $x = \emptyset$. But we have demanded that $\emptyset$ is not in any successor set, which means that $\emptyset \notin \mathbb{N}$. We have arrived at a contradiction (to $x \in \mathbb{N}$). Thus 1 cannot be the successor of any element of $\mathbb{N}$ and we have proved part 3.

To prove part 4, we must prove that if $S \subseteq \mathbb{N}$ contains 1 and the successor of each of its elements, then $S = \mathbb{N}$. Let $S \subseteq \mathbb{N}$ be an arbitrary, but fixed, subset so that $1 \in S$ and so that for every $n \in S$ we have that $n' \in S$ (see Standard Proof Technique 1.73). We will prove mutual containment for $\mathbb{N}$ and S (see Standard Proof Technique 2.26). Because S is a successor set, by definition of $\mathbb{N}$ and by Proposition 2.18 we conclude $\mathbb{N} \subseteq S$. But by definition of S, we also have $S \subseteq \mathbb{N}$, which means by Theorem 2.25 that $S = \mathbb{N}$. Because S was an arbitrary subset with the given properties, this shows that every $S \subseteq \mathbb{N}$ that contains 1 and the successor of each of its elements must be equal to $\mathbb{N}$, which proves part 4.

To prove part 5, we must prove that for all $m, n \in \mathbb{N}$ with $m' = n'$ we have that $m = n$. **This part of the proof is the longest, and we will use indentations to indicate the proof's structure.**

First we prove that, *for every $n \in \mathbb{N}$, every element of n is a subset of n.*

> To prove this claim, we will apply the Principle of Induction (part 4) to the set $S := \{n \in \mathbb{N} : [\forall m \in n : m \subseteq n]\}$. This will prove the claim, because the claim is equivalent to $S = \mathbb{N}$.
>
> **To apply the Principle of Induction, we must prove that $1 \in S$ and that for all $n \in S$ we have $n' \in S$.**
>
> > Trivially, $\{\emptyset\} \in S$, that is, $1 \in S$.
> >
> > Now let $n \in S$. To prove that $n' \in S$, first recall that $n' = n \cup \{n\}$. Hence, if $m \in n'$, then $m = n$ or $m \in n$. In the first case, $m = n \subseteq n'$. In case $m \in n$, because $n \in S$, we infer $m \subseteq n \subseteq n'$. Hence $n \in S$ implies that $n' \in S$.
>
> By the Principle of Induction (part 4), $S = \mathbb{N}$ and hence every element of every $n \in \mathbb{N}$ *(which is a set)* is a subset of *(that set)* n.

We can now use the above claim to prove part 5.

Let $m, n \in \mathbb{N}$ with $m' = n'$ be arbitrary but fixed (see Standard Proof Technique 1.73). Then, by definition of successors, $m \cup \{m\} = m' = n' = n \cup \{n\}$. Suppose for a contradiction that $m \neq n$. Then $\{n\} \neq \{m\}$, which implies $n \in m$ and $m \in n$. Therefore, by the above, $n \subseteq m$ and $m \subseteq n$. Hence $m = n$, a contradiction.

Because m, n were arbitrary elements of $\mathbb{N}$ with $m' = n'$ we have proved that $m' = n'$ implies $m = n$. Hence part 5 is proved.   ∎

Analysis of the proof of Theorem 2.64. Now that we are done with the proof, we can analyze how anyone may have come up with the requisite ideas. Once we remind ourselves that the natural numbers are used for counting, the Peano axioms themselves simply say that there is a first natural number (parts 1 and 3), that after every natural number, there is a "next natural number" and no two distinct natural numbers have the same "next natural number" (parts 2 and 5), and that all natural numbers are obtained by counting, starting at the first natural number (part 4). So, from this point-of-view, Theorem 2.64 is still abstract, but it encodes familiar ideas.

The proof itself must show that these ideas can be encoded in set theory. The first step towards the proof actually is the order in which we list the parts, because some parts, such as part 4, are used to prove other parts, such as part 5. The key concept is the idea of a successor. The Axiom of Infinity provides an infinite set I and it provides exactly one way to obtain another set from a given set $a \in I$, namely, by forming the new set $(a \cup \{a\}) \in I$. Therefore it is natural to define the successor as $n' := n \cup \{n\}$. But the set I need not be equal to the natural numbers. It turns out that I could be strictly larger than $\mathbb{N}$.[9] That's what spawns the idea of a successor set: Take all sets that contain the starting point $1 = \{\emptyset\}$ and which contain successors for all of their elements. Then find the smallest such set by forming their intersection $\mathbb{N}$. This approach of taking the intersection of a family of sets with certain properties is a standard technique for finding the smallest object with these properties.

Once we have the set $\mathbb{N}$ that should be the natural numbers, we still must prove that the set satisfies the Peano Axioms. Individually, the proofs of parts 1 to 4 are reasonably short, and they follow from the groundwork laid in the definition of our set $\mathbb{N}$. Part 5 is the one part whose proof is a bit harder. Its proof starts with a claim that, upon first reading, seems unrelated to the statement that we want to prove. This is a typical situation when reading certain proofs for the first time. Because information is best presented in logically linear fashion, it is not always clear why a particular claim will be useful. But reading the last paragraph of the proof of part 5 shows why the earlier claim was needed: The claim is crucial in coming up with a contradiction. More precisely, (supposing for a contradiction) if $m \neq n$ and $m' = n'$, then we would have $n \in m \in n$. If only we could replace the containment of an element with *set* containment, we would be done. And that is exactly what the claim provides.

Keep these ideas in mind as you re-read the proof of Theorem 2.64.

Comments on how the claim is proved are delayed to Standard Proof Technique 2.65 below.   □

---

[9]For example, the union of two disjoint copies $\mathbb{N}$ and $\mathbb{N}'$ of the natural numbers satisfies the Axiom of Infinity and all Peano Axioms, except the Principle of Induction.

Standard Proof Technique 2.65 The Principle of Induction leads to a standard proof technique, which is (sensibly) called **induction**. We have already seen this technique used in the proof of part 5 of Theorem 2.64, and we will see several proofs using the Principle of Induction in Section 3.1. The proofs may look a little strange to readers who are already familiar with induction. But note that until we have defined algebraic operations on the natural numbers, we cannot formulate induction in the form that readers may be more accustomed to. Once we have defined the algebraic operations, Section 3.4 will present induction in its more customary form. □

It may seem strange that a *theorem* like Theorem 2.64 is said to provide *axioms*, which, by definition, should be true statements that are not proved. This apparent contradiction of terms is easily resolved, though. Historically speaking, the Peano Axioms were discovered before Russell's Paradox and before Zermelo-Fraenkel set theory, and they have always been a solid foundation for mathematics. Practically speaking, the Peano Axioms are a higher level system of axioms that (together with some axioms on sets) can serve as the starting point into mathematics, just as well as the axioms of set theory do.

This situation is similar to different levels of programming languages being used for different purposes. Many game programmers use high level programming languages like C++ and Java, plus graphics packages to write their code. The functionality of these high-level programming languages basically has the status of an axiomatic system. Everything that can be done (and that is considerable) is determined by the functionality of the languages and packages. On the other hand, the engineers who design CPUs will work at the assembly language level. That is, everything is not just decomposed into bytes, but sometimes even into bits. This is a more fundamental level, or, in mathematical language, a more fundamental axiomatic system. But at the same time, CPU design is set up so that the fundamental functions will ultimately allow the creation and use of higher level programming languages.

Note that people can work successfully at either level. Engineers continue to work on improving CPUs that can deliver the functionality needed for higher levels of programming. Game programmers, in turn, use fundamental operations, like bit manipulations or arithmetic, in the context of higher level work, but without much thought about "what lies beneath."

In similar fashion, we will now leave fundamental set theory behind.[10] We will still use sets, functions, relations, etc., but we will increasingly do so without worrying about set theoretical details. As we do this, the Peano Axioms provide a higher level axiomatic system that simplifies the constructions, because we will not need to go back to the fundamental set theoretical level. (Imagine a game programmer having to go back to assembly language every time floating point numbers are added. This approach would not be very efficient.) At the same time, the Peano Axioms are guaranteed to be a rock solid foundation, because they are derived from set theory. In fact, *in principle* we could still go back to set theory. We would just need to replace each Peano Axiom with its underlying set theoretical constructions. This sounds cumbersome, and it is, so we will not do it. But, in the programming analogy, this is what

---

[10]We will eventually return to set theory in Chapter 7.

higher level programming languages do. They provide a convenient way to work with, say, floating point numbers, and the connection to assembly language (and, ultimately, machine code execution) is achieved by the compiler replacing symbols with the requisite assembly code.

Exercises

2-50. More on natural numbers acting as elements of natural numbers.

   (a) Let $m, n \in \mathbb{N}$ be so that $n \subseteq m$. Prove that $m \notin n$.

   Hint. Induction. Use $S := \{ n \in \mathbb{N} : [\forall m \supseteq n : m \notin n] \}$.

   (b) Prove that no natural number is an element of itself.

   (c) Explain why it is easier to prove that no superset m of a natural number n is an element of n than it is to prove directly that no natural number is an element of itself.

   Hint. Try to prove with the Principle of Induction that no natural number is an element of itself and determine which parts of this proof become harder than the corresponding parts in the proof of part 2-50a.

2-51. Prove that the function $f : \mathbb{N} \to \mathbb{N}$ defined by $f(n) = n'$, which maps every natural number to its successor, is injective, but not surjective.

2-52. Prove that no natural number is equal to its successor.

   Hint. Do a proof by contradiction or a proof by induction on the set of all $n \neq n'$.

2-53. Let $m \in \mathbb{N}$. Prove that either $m = 1$ or there is an $n \in \mathbb{N}$ so that $n' = m$.

   Hint. Induction.

Chapter 3

Number Systems I: Natural Numbers

Although mathematics formally is founded upon logic and set theory, when people think of mathematics, they typically think of numbers. In fact, the mathematician Leopold Kronecker once said that "God made the integers, all else is man's work." He may well have been right: Even anything beyond $\mathbb{N}$ may be man's invention. After all, the world is quantized. That is, every quantity is a natural number multiplied by some base quantity. Of course, number systems like $\mathbb{R}$ and $\mathbb{C}$ remain extremely useful, but nonetheless.

Historically, numbers most likely were the first mathematical entities known to mankind. For example, when determining if stocks were sufficient to survive the winter, people had to count animals, loaves of bread, days, etc. The natural numbers are inspired exactly by the counting of whole objects. This chapter first introduces some properties of the natural numbers in Sections 3.1-3.6. Equivalence relations, which are needed to construct the integers, are introduced in Section 3.7, the "wraparound" arithmetic of computer processors is discussed in Section 3.8, and Section 3.9 presents public key encryption as an important application.

3.1   Arithmetic With Natural Numbers

The natural numbers were introduced in Theorem 2.64. But the Peano Axioms in Theorem 2.64 look a bit bare compared to what we "know" about the natural numbers. For example, the Peano Axioms do not mention anything about arithmetic. This section introduces the arithmetic for natural numbers and some of the abstract properties of addition and multiplication. In later sections, we will see that working with these

Fundamentals of Mathematics: An Introduction to Proofs, Logic, Sets, and Numbers.
By Bernd S. W. Schröder.

properties, rather than with the objects that have the properties, will allow us to proceed more efficiently as we construct the familiar number systems.

Because the abstract viewpoint will turn out to be most efficient one, we start on an abstract note: To do arithmetic, we need operations that take two numbers and produce a third number. Because these operations must exist within our framework of set theory, we define them as functions.

Definition 3.1 *A* **binary operation** *on a set S is a function $\circ : S \times S \to S$.*

Definition 3.1 captures exactly what a binary operation, like an operation in arithmetic, should do: We enter two numbers into $\circ$ and the operation $\circ$ "somehow" (and in the abstract realm we don't care how) produces another number as its output. The familiar algebraic operations all fit this pattern. Hence there are no set theoretical problems as we explore the arithmetic of the natural numbers. Just as we did for relations and for functions, we adopt more familiar notation for binary operations.

Notation 3.2 *Let S be a set and let $\circ : S \times S \to S$ be a binary operation. For all $a, b \in S$ we set $a \circ b := \circ(a, b)$.*

Now that we can encode binary operations in set theory, we can define the usual binary operations for the natural numbers. We start with addition.

Definition 3.3 *For all $m, n \in \mathbb{N}$ the relation $+ : \mathbb{N} \times \mathbb{N} \to \mathbb{N}$ is defined by*

$$
\begin{aligned}
n + 1 &:= n', \qquad \text{and} \\
n + m' &:= (n + m)'.
\end{aligned}
$$

Or, in relation notation,

1. *For all $n, m \in \mathbb{N}$ the pair $\big((n, m), 1\big)$ is not in $+$.*

2. *For all $n \in \mathbb{N}$, the relation $+ \subseteq (\mathbb{N} \times \mathbb{N}) \times \mathbb{N}$ contains the pair $\big((n, 1), n'\big)$ and it contains no other pairs for which the second component of the first pair is a 1.*

3. *For all $n, m, k \in \mathbb{N}$ we have $\big((n, m'), k'\big) \in +$ iff $\big((n, m), k\big) \in +$.*

The relation (which turns out to be a binary operation) is called **addition***.*

Although it seems obvious, at this fundamental stage we do not know if the relation $+ : \mathbb{N} \times \mathbb{N} \to \mathbb{N}$ is a function (a binary operation) or merely a relation. Thus we must formally establish that $+$ really is a binary operation. The proof of Proposition 3.4 is a bit on the abstract side. On first reading of this section, the reader may want to skim it and continue with Notation 3.5. In particular, if the property of well-definedness looks foreign, rest assured that this impression is quite normal. We typically expect entities that are meant to be functions to be well-defined. The discussion before Definition 3.106 will show a familiar looking and less abstract operation that should be (and is) well-defined, but for which well-definedness is not obvious.

Proposition 3.4 *The relation $+$ is a binary operation on $\mathbb{N}$.*

Proof. Definition 3.3 gives us a subset $+$ of $(\mathbb{N} \times \mathbb{N}) \times \mathbb{N}$ that contains pairs $((n, m), k)$ of an ordered pair (n, m) and a number k. That is, *a priori*, $+$ really is nothing but a relation. To emphasize this fact, we use relation notation in this proof. Once we have established that $+$ is a binary operation, we will go back to the usual notation. We use the Principle of Induction to prove that the relation $+ : \mathbb{N} \times \mathbb{N} \to \mathbb{N}$ is a binary operation, that is, to prove that $+$ is totally defined and well-defined.

To prove that the relation $+$ is totally defined, we must prove that the sum $n + m$ is defined for all $n, m \in \mathbb{N}$. Let $n \in \mathbb{N}$ be arbitrary, but fixed. We now consider the set $S := \{m \in \mathbb{N} : [\exists k \in \mathbb{N} : ((n, m), k) \in +]\}$. **(Note how the condition that defines S is exactly the condition that n+m is defined.)** We will use the Principle of Induction (part 4 of Theorem 2.64) to prove that $S = \mathbb{N}$. **To do this, we must prove that $1 \in S$ and that if $m \in S$, then $m' \in S$.** By definition of $+$, $((n, 1), n') \in +$, so $1 \in S$. Moreover, for all $m \in S$ there is a $k \in \mathbb{N}$ so that $((n, m), k) \in +$. By definition of $+$, we conclude that $((n, m'), k') \in +$. Hence $m' \in S$. By the Principle of Induction, we conclude that $S = \mathbb{N}$. Thus $n + m$ is defined for all $m \in \mathbb{N}$. Moreover, because $n \in \mathbb{N}$ was arbitrary, $n + m$ is defined for all $n, m \in \mathbb{N}$. Hence $+$ is totally defined.

To prove that the relation $+$ is well-defined, we must prove that the sum $n + m$ is unique for all $n, m \in \mathbb{N}$. Let $n \in \mathbb{N}$ be arbitrary, but fixed. We now consider the set $S := \{m \in \mathbb{N} : [\forall k, l \in \mathbb{N} : ((n, m), k), ((n, m), l) \in + \Rightarrow k = l]\}$. **(Note how the condition that defines S is exactly the condition that n+m is unique.)** We will use the Principle of Induction to prove that $S = \mathbb{N}$. **To do this, we must prove that $1 \in S$ and that if $m \in S$, then $m' \in S$.**

For $m = 1$, note that if $((n, 1), k), ((n, 1), l) \in +$, then, by part 2 of Definition 3.3 (and because 1 is not the successor of any natural number), we obtain $k = n' = l$ and we conclude that $1 \in S$.

Now let $m \in S$ and let $((n, m'), k), ((n, m'), l) \in +$. By part 1 of Definition 3.3, there are $j_k, j_l \in \mathbb{N}$ so that $k = j_k'$ and $l = j_l'$. By part 3 of Definition 3.3 and by $m \in S$, we conclude that $j_k = j_l$. Hence $k = j_k' = j_l' = l$, and therefore $m' \in S$.

Hence, by the Principle of Induction, $S = \mathbb{N}$, and $n + m$ is unique for all $m \in \mathbb{N}$. Because $n \in \mathbb{N}$ was arbitrary, this means that $+$ is well-defined. ∎

The proof of Proposition 3.4 exhibits a key feature of proofs using the Principle of Induction from Theorem 2.64: If we choose the set S to be the set of all elements that are supposed to have a certain property, and we did that twice in the proof of Proposition 3.4, then the Principle of Induction will allow us to establish this property. All proofs with the Principle of Induction in this section will follow this pattern.

Notation 3.5 Now that we have established that addition is a binary operation, we will use the more typical notation $n + m = k$ instead of $((n, m), k) \in +$. (See Notation 3.2.)

We can now establish some properties of addition.

Proposition 3.6 *Properties of the addition operation on* $\mathbb{N}$.

1. *Addition is* **associative**, *that is, for all* $n, m, k \in \mathbb{N}$ *we have that*
   $(n + m) + k = n + (m + k)$.

2. *Addition is* **commutative**, *that is, for all* $n, m \in \mathbb{N}$ *we have that* $n + m = m + n$.

Proof. Both parts are proved with the Principle of Induction (see part 4 of Theorem 2.64).

To prove part 1, let $n, m \in \mathbb{N}$ be arbitrary, but fixed, natural numbers and consider the set $S := \{ k \in \mathbb{N} : (n + m) + k = n + (m + k) \}$. We will use the Principle of Induction to prove that $S = \mathbb{N}$. **To do this, we must prove that** $1 \in S$ **and that if** $k \in S$**, then** $k' \in S$.

First note that $1 \in S$, because

$$(n + m) + 1 \quad = \quad (n + m)' \qquad \boxed{\text{Used definition of } +1 \text{ for } (n + m).}$$

$$= \quad n + m' \qquad \boxed{\text{Used definition of } n + m'.}$$

$$= \quad n + (m + 1) \qquad \boxed{\text{Used definition of } +1 \text{ for } m.}$$

Note that in the argument above, we did not distinguish between using the equations from Definition 3.3 read left-to-right, as in x+1=x' (first step), or right-to-left, as in x'=x+1 (last step). This approach is common when working with equations and we must become very comfortable with it.

Now let $k \in S$. Then $k' \in S$, because

$$(n + m) + k' \quad = \quad \big((n + m) + k \big)' \qquad \boxed{\text{Used definition of } + \text{ for } (n + m) \text{ and } k'.}$$

$$= \quad \big(n + (m + k) \big)' \qquad \boxed{\text{Used that } k \in S.}$$

$$= \quad n + (m + k)' \qquad \boxed{\text{Used definition of } + \text{ for } n \text{ and } (m + k)'.}$$

$$= \quad n + \big(m + k' \big) \qquad \boxed{\text{Used definition of } + \text{ for } m \text{ and } k'.}$$

By the Principle of Induction, we conclude that $S = \mathbb{N}$, that is, $(n+m)+k = n+(m+k)$ for all $k \in \mathbb{N}$. Because $n, m \in \mathbb{N}$ were arbitrary, we conclude that $+$ is associative.

To prove part 2, let $n \in \mathbb{N}$ be an arbitrary, but fixed, natural number and consider the set $S := \{ m \in \mathbb{N} : n + m = m + n \}$. We will use the Principle of Induction to prove that $S = \mathbb{N}$. **To do this, we must prove that** $1 \in S$ **and that if** $m \in S$**, then** $m' \in S$.

To prove that $1 \in S$, we will need to use the Principle of Induction. **That is, we will perform an induction inside an induction.**

Let $T := \{ k \in \mathbb{N} : k + 1 = 1 + k \}$. Then, trivially, $1 \in T$. Now let $k \in T$. Then

$$k' + 1 = (k + 1) + 1 = (1 + k) + 1 = 1 + (k + 1) = 1 + k',$$

so $k' \in T$. **In the above computation, you should justify every step, similar to what was done twice in the proof of associativity.** By the Principle of Induction, $T = \mathbb{N}$, which shows that $k + 1 = 1 + k$ for all $k \in \mathbb{N}$.

Now we can get back to proving $1 \in S$ **in the "main induction."**
The above shows that for $k = n$ we obtain $n + 1 = 1 + n$. Hence $1 \in S$.
Now let $m \in S$. Then

$$
\begin{aligned}
n + m' &= n + (m + 1) = (n + m) + 1 = (m + n) + 1 = m + (n + 1) \\
&= m + (1 + n) = (m + 1) + n = m' + n,
\end{aligned}
$$

that is, $m' \in S$. By the Principle of Induction, $S = \mathbb{N}$, which means that for all $m \in \mathbb{N}$ we have that $n + m = m + n$. Because $n \in \mathbb{N}$ was arbitrary, $+$ is commutative. ∎

Commutativity and associativity can be defined for any binary operation. The reader will translate the abstract properties of the binary operations in this section to properties for arbitrary binary relations in Exercise 3-1.

When we were introduced to addition in elementary school, there was no mention of binary relations, there was no mention of commutativity, and there was no mention of associativity.[1] What we had were numbers, and addition was an abbreviation for counting. The definition of $+$ in Definition 3.3 reflects this idea. Adding 1 to a number n leads to the successor of the number: $n + 1 = n'$. Adding the successor m' of a number m to a number n adds 1 to the original sum: $n + m' = (n + m)' = (n + m) + 1$. In terms of the familiar digits, the connection to this fundamental level of mathematics is rather anticlimactic: The digits are simply symbols for numbers that are obtained by adding 1 a certain number of times. (Also see Theorem 3.89 for the idea of place value.)

Definition 3.7 *The* **usual representation of natural numbers**. *We define*

$$
\begin{aligned}
2 &:= 1 + 1 \\
3 &:= 2 + 1 = (1 + 1) + 1 \\
4 &:= 3 + 1 = \big((1 + 1) + 1\big) + 1 \\
5 &:= 4 + 1 = \Big(\big((1 + 1) + 1\big) + 1\Big) + 1
\end{aligned}
$$

and so on, in the usual fashion.

So the answer to question 1 on page x is very simple: 2 is a symbol and the definition of the symbol is that 2 equals $1 + 1$. In particular, there is nothing to prove. This view of numerals may seem strange, but all numerals are symbols. We are just very much accustomed to the arabic number system, and we instinctively attach meaning to the numbers we see. For comparison, consider roman numerals. The fact that we are not as familiar with roman numerals as well as the fact that roman numerals are more "concrete" (though less convenient) make the idea that numerals are symbols more obvious. The roman numerals I, II and III are simply tallies (which are probably the most primitive numerical symbols). After that, we use other symbols to avoid putting down too many tallies. But every new symbol simply represents a certain number of tallies, that is, $IV = IIII$, $V = IIIII$, etc.

---

[1] At least the author hopes so, because he certainly does not remember anything like that from elementary school.

So even though non-mathematicians sometimes ask if mathematicians can prove that $1 + 1 = 2$, we can't. It's a definition. But once we have defined the numerals, we can prove other familiar equations, such as $4 + 3 = 7$.

Proposition 3.8 $4 + 3 = 7$.

Proof. The challenge for the proof is that we must accurately use the definitions and previously proved results to turn $4 + 3$ into $6 + 1 = 7$.

$$
\begin{aligned}
4 + 3 &= 4 + (2 + 1) &&\text{Used the definition of 3.} \\
&= (4 + 2) + 1 &&\text{Used associativity.} \\
&= \big(4 + (1 + 1)\big) + 1 &&\text{Used the definition of 2.} \\
&= \big((4 + 1) + 1\big) + 1 &&\text{Used associativity.} \\
&= (5 + 1) + 1 &&\text{Used the definition of 5.} \\
&= 6 + 1 &&\text{Used the definition of 6.} \\
&= 7 &&\text{Used the definition of 7.}
\end{aligned}
$$

∎

Of course, it is not the purpose of a mathematics class to prove the obvious. But at the same time, results like Proposition 3.8 are a perfect opportunity to practice the formal justification of each step. Exercise 3-2 allows the reader more practice of this skill.

We now turn our attention to multiplication.

Definition 3.9 *For all $m, n \in \mathbb{N}$ the relation $\cdot : \mathbb{N} \times \mathbb{N} \to \mathbb{N}$ is defined by*

$$
\begin{aligned}
n \cdot 1 &:= n, &&\text{and} \\
n \cdot m' &:= n \cdot m + n.
\end{aligned}
$$

Or, in relation notation,

1. *For all $n, m \in \mathbb{N}$ the pair $\big((n, m), 1\big)$ is in $\cdot$ iff $n = m = 1$.*

2. *For all $n \in \mathbb{N}$, the relation $\cdot \subseteq (\mathbb{N} \times \mathbb{N}) \times \mathbb{N}$ contains the pair $\big((n, 1), n\big)$ and it contains no other pairs for which the second component of the first pair is a 1.*

3. *For all $n, m, k \in \mathbb{N}$ we have $\big((n, m'), k + n\big) \in \cdot$ iff $\big((n, m), k\big) \in \cdot$.*

The relation (which turns out to be a binary operation) is called **multiplication**.

Of course we must establish *that* multiplication really is a binary operation.

Proposition 3.10 *The relation · is a binary operation on* $\mathbb{N}$.

    Proof. Exercise 3-3. ■

Just as addition was a way to abbreviate counting (counting amounts to repeated additions of 1), multiplication is a way to abbreviate repeated additions of the same number. Just as for addition, the connection is encoded in Definition 3.9: n by itself (multiplied by 1) gives n and multiplication of n with the successor of m adds another n to the product $n \cdot m$. The reader can practice with the connection between numerals and the definition of multiplication in Exercise 3-4.

Notation 3.11 Of course, we will use the more typical notation $n \cdot m = k$ instead of $\big((n, m), k\big) \in \cdot$. (See Notation 3.2.) Moreover, for any type of multiplication operation, it is customary to leave out the multiplication sign · and write nm instead of $n \cdot m$. From now on, we will no longer mention obvious shifts to more common notation.

Proposition 3.12 *Properties of the multiplication operation on* $\mathbb{N}$.

    1. Multiplication is **associative***, that is, for all* $n, m, k \in \mathbb{N}$ *we have that* $(n \cdot m) \cdot k = n \cdot (m \cdot k)$.

    2. Multiplication is **commutative***, that is, for all* $n, m \in \mathbb{N}$ *we have that* $n \cdot m = m \cdot n$.

    3. The number 1 *is a* **neutral element** *or* **identity element** *with respect to multiplication, that is, for all* $n \in \mathbb{N}$ *we have that* $n \cdot 1 = 1 \cdot n = n$.

    Proof. Exercise 3-5. The proof is actually quite tricky in terms of what is proved first. The author recommends that readers follow the suggested order in the exercise. Each of the later parts uses one or more of the earlier parts, and all but part 3-5d are proved with the Principle of Induction. ■

Addition and multiplication do not exist independent of each other. The next result shows how addition and multiplication interact.

Proposition 3.13 *Multiplication is* **right distributive** *over addition. That is, for all* $n, m, k \in \mathbb{N}$ *we have* $(n + m) \cdot k = n \cdot k + m \cdot k$.

    Proof. Exercise 3-5b. (So, technically, we already "knew" this result, because Exercise 3-5b was needed to prove Proposition 3.12. It's just presented once more to emphasize its importance.) ■

Of course, addition is distributive "from the other side," too. Thankfully, we will not need to go through the same proof as for Proposition 3.13. As mentioned earlier, once we know enough about the abstract properties of operations, we can leave the definitions of the operations behind and just use the properties. Similar to how the Peano Axioms make us more efficient because we don't need to go back to the fundamental set theoretical level, this approach makes us more efficient, too: Properties are ultimately easier to work with than the definitions of addition and multiplication presented here. The final two results in this section show the efficiency of this approach. Just think of how convoluted induction proofs of these results would have been!

Proposition 3.14 *Multiplication is* **left distributive** *over addition. That is, for all* $n, m, k \in \mathbb{N}$ *we have* $n \cdot (m + k) = n \cdot m + n \cdot k$.

Proof. Let $m, n, k \in \mathbb{N}$ be arbitrary, but fixed. Then

$$
\begin{aligned}
n \cdot (m + k) &= (m + k) \cdot n & \boxed{\text{Used commutativity of } \cdot.} \\
&= m \cdot n + k \cdot n & \boxed{\text{Used right distributivity.}} \\
&= n \cdot m + n \cdot k & \boxed{\text{Used commutativity of } \cdot.}
\end{aligned}
$$

Because n, m, k were arbitrary elements of $\mathbb{N}$, the result is proved. (Technically, we already "knew" that Proposition 3.14 is true, because Exercise 3-5d was needed to prove Proposition 3.12. So the reader may consider this proof a "really good hint" for Exercise 3-5d.)   ∎

Once distributivity is established, we can talk about multiplying out parentheses. This is the "**FOIL**: First-Outer-Inner-Last" from bygone days. Also see Exercise 3-7 for a situation in which a too rigid use of the acronym fails.

Proposition 3.15 *Let* $a, b, c, d \in \mathbb{N}$. *Then* $(a + b)(c + d) = (ac + ad) + (bc + bd)$.

Proof. Let $a, b, c, d \in \mathbb{N}$ be arbitrary, but fixed. Then

$$
\begin{aligned}
(a + b)(c + d) & \\
&= (a + b)c + (a + b)d & \boxed{\begin{array}{l}\text{Left distributivity with } n = (a + b), m = c, \\ k = d.\end{array}} \\
&= (ac + bc) + (ad + bd) & \boxed{\begin{array}{l}\text{Right distributivity with } n = a, m = b, k = c \\ \text{and } n = a, m = b, k = d, \text{respectively.}\end{array}} \\
&= \big((ac + bc) + ad\big) + bd & \boxed{\text{Associativity of } +.} \\
&= \big(ac + (bc + ad)\big) + bd & \boxed{\text{Associativity of } +.} \\
&= \big(ac + (ad + bc)\big) + bd & \boxed{\text{Commutativity of } +.} \\
&= \big((ac + ad) + bc\big) + bd & \boxed{\text{Associativity of } +.} \\
&= (ac + ad) + (bc + bd) & \boxed{\text{Associativity of } +.}
\end{aligned}
$$

Because a, b, c, d were arbitrary elements of $\mathbb{N}$, the result is proved. *This result can be proved more efficiently, and the reader is encouraged to do so in Exercise 3-6.*   ∎

We will show soon (see Theorem 3.58) that parentheses that group summands in groups of two can be dropped in additions. After that, we will no longer need to include them. But until then, we formally must include enough parentheses so that every individual addition has exactly two summands.

Exercises

3-1. Let S be a set and let $\circ : S \times S \to S$ and $* : S \times S \to S$ be binary operations on S.

    (a) State the definition of "$\circ$ is associative."

    (b) State the definition of "$\circ$ is commutative."

    (c) State the definition of "$e \in S$ is a neutral element for the operation $\circ$."

    (d) State the definition of "$\circ$ is left distributive over $*$."

    (e) State the definition of "$\circ$ is right distributive over $*$."

3-2. Use Definition 3.7 and the properties of addition of the natural numbers to prove each of the following.

    (a) $3 + 4 = 7$

    (b) $2 + 2 = 4$

    (c) $2 + 5 = 7$

    (d) $4 + 4 = 8$

3-3. Prove Proposition 3.10.

3-4. Use Definition 3.7 and the properties of addition and multiplication of the natural numbers to prove each of the following.

    (a) $2 \cdot 3 = 6$

    (b) $4 \cdot 2 = 8$

    (c) $3 \cdot 4 = 12$

3-5. Prove Proposition 3.12 as follows.

    (a) Prove that for all $n \in \mathbb{N}$ we have that $1 \cdot n = n$.

    (b) Prove that for all $n, m, k \in \mathbb{N}$ we have that $(n + m) \cdot k = n \cdot k + m \cdot k$.

    (c) Prove that for all $n, m \in \mathbb{N}$ we have that $n \cdot m = m \cdot n$.

    (d) Prove that for all $n, m, k \in \mathbb{N}$ we have that $n \cdot (m + k) = n \cdot m + n \cdot k$.

    (e) Prove that for all $n, m, k \in \mathbb{N}$ we have that $(n \cdot m) \cdot k = n \cdot (m \cdot k)$.

3-6. Find a faster proof for Proposition 3.15.

    Hint. The fastest proof has two steps.

3-7. The FOIL acronym is limited to sets of parentheses with two summands. It must not become a crutch for middle school students.

    (a) Let $a, b, c, d, e \in \mathbb{N}$. Prove that $(a + b)((c + d) + e) = ((((ac + ad) + ae) + bc) + bd) + be$.

    (b) Explain why a mechanical application of "FOIL" fails in part 3-7a.

    (c) State a verbal explanation how to multiply out parentheses with arbitrarily many summands. (You may ignore the extra parentheses from the associative law.)

3-8. Solving equations in the natural numbers. That is, finding solutions for the equations *that are in* $\mathbb{N}$.

    (a) Solve the equation $x + 2 = 6$ in $\mathbb{N}$.

    (b) Solve the equation $x + 6 = 2$ in $\mathbb{N}$.

    (c) Solve the equation $2x + 5 = 9$ in $\mathbb{N}$.

    (d) Solve the equation $3x + 2 = 4$ in $\mathbb{N}$.

    (e) Solve the equation $x^2 - 1 = 0$ in $\mathbb{N}$.

    (f) Solve the equation $x^2 - 2 = 0$ in $\mathbb{N}$.

    (g) Solve the equation $x^2 + 1 = 0$ in $\mathbb{N}$.

3.2   Ordering the Natural Numbers

We now turn to the comparison of natural numbers. The answer to question 2 from page x seems obvious: 3 is smaller than 9, because $9 - 3$ is positive. But what does it mean to be positive? For that matter, what is subtraction? Positivity will not need to be defined for $\mathbb{N}$, because all natural numbers are positive. But we must replace subtraction, which is as yet undefined, with a construction that only involves addition. The next result allows us to do just that.

Theorem 3.16 *The natural numbers satisfy the following* **trichotomy condition.** *For all $n, m \in \mathbb{N}$, exactly* one *of the following three statements is true.*

1. *Either $n = m$, or*

2. *There is a $d \in \mathbb{N}$ so that $n + d = m$, or*

3. *There is a $d \in \mathbb{N}$ so that $m + d = n$.*

Proof. To better emphasize the structure of this proof, we use indentations once more. We first prove that the three conditions are mutually exclusive, that is, that no two of 1, 2 and 3 can be satisfied at the same time.

> To prove this claim we first prove that for all $n, d \in \mathbb{N}$ we have the inequality $n + d \neq n$. Let $d \in \mathbb{N}$ be arbitrary but fixed.
>
> > Let $S := \{n \in \mathbb{N} : n + d \neq n\}$. We use the Principle of Induction to prove that $S = \mathbb{N}$.
> >
> > > To prove that $1 \in S$, note that $1 + d = d + 1 = d'$, and that 1 is not the successor of any natural number. Hence $1 + d \neq 1$ and $1 \in S$.
> > >
> > > Now let $n \in S$ and suppose for a contradiction that $n' + d = n'$. Then $n' = n' + d = d + n' = (d + n)' = (n + d)'$ would imply, by part 5 of Theorem 2.64, that $n = n + d$, contradicting $n \in S$. Hence $n' + d \neq n'$ and $n' \in S$.
> >
> > Thus by the Principle of Induction we conclude that $S = \mathbb{N}$.
>
> Because d was arbitrary, we conclude that $n + d \neq n$ for all $n, d \in \mathbb{N}$.

Now let $n, m \in \mathbb{N}$ and suppose for a contradiction they satisfy any two of 1-3. In case 1 and one of 2 or 3 are satisfied we infer $n + d = n$ or $m + d = m$ for some $d \in \mathbb{N}$, a contradiction. In case 2 and 3 are satisfied, we infer that there are $d, \tilde{d} \in \mathbb{N}$ so that $m = n + d$ and $n = m + \tilde{d}$. But then $m = n + d = \left(m + \tilde{d} \right) + d = m + \left(d + \tilde{d} \right)$, a contradiction. Thus any two $n, m \in \mathbb{N}$ cannot satisfy any more than one of the three conditions 1-3.

To prove that any $n, m \in \mathbb{N}$ satisfy at least one of the conditions 1-3, we proceed as follows. For each $n \in \mathbb{N}$ let $S(n)$ be the set of all $m \in \mathbb{N}$ so that one of 1-3 holds for n and m. Moreover, let $S := \left\{ n \in \mathbb{N} : S(n) = \mathbb{N} \right\}$. The result will be proved if we can show that $S = \mathbb{N}$. We use the Principle of Induction to prove $S = \mathbb{N}$. So for this part of the proof the challenge is that the set S is defined in terms of other sets S(n), which will lead to an induction inside an induction.

To prove that $1 \in S$ we must show that $S(1) = \mathbb{N}$. We will prove this equality by proving mutual containment (see Standard Proof Technique 2.26).

Clearly $S(1) \subseteq \mathbb{N}$.

To prove that $\mathbb{N} \subseteq S(1)$, let $m \in \mathbb{N}$. Then by Exercise 2-53 either we have that $m = 1$, in which case condition 1 is satisfied, or there is a $d \in \mathbb{N}$ so that $m = d' = d + 1 = 1 + d$, in which case condition 2 holds. Either way, $m \in S(1)$ and hence $\mathbb{N} \subseteq S(1)$.

Therefore $S(1) = \mathbb{N}$ and $1 \in S$.

Let $n \in S$. Then, assuming $S(n) = \mathbb{N}$, we must prove $S(n') = \mathbb{N}$. **We do this with an induction inside an induction. That is, we must prove that $1 \in S(n')$ and that for all $m \in S(n')$ we have $m' \in S(n')$.**

Because $S(1) = \mathbb{N}$ we conclude that $1 \in S(n')$.

Now let $m \in S(n')$. We need to prove that $m' \in S(n')$. Because $m \in S(n')$, we have that $m = n'$ or there is a $d \in \mathbb{N}$ so that $m + d = n'$ or there is a $\tilde{d} \in \mathbb{N}$ so that $n' + \tilde{d} = m$. We will prove that $m' \in S(n')$ by distinguishing these three cases.

Case 1: $m = n'$. If $m = n'$, then $m' = m + 1 = n' + 1$ and hence $m' \in S(n')$.

Case 2: There is a $d \in \mathbb{N}$ so that $m + d = n'$. If there is a $d \in \mathbb{N}$ so that $m + d = n'$, then by Exercise 2-53 either $d = 1$ or $d = \hat{d}'$ for some $\hat{d} \in \mathbb{N}$. **(So now we are doing a case distinction inside a case distinction.)**

If $d = 1$, then $m' = m + 1 = m + d = n'$, which means $m' \in S(n')$.

If $d = \hat{d}'$ for some $\hat{d} \in \mathbb{N}$, then

$$
\begin{aligned}
n' &= m + d = m + \hat{d}' = m + (\hat{d} + 1) = m + (1 + \hat{d}) \\
&= (m + 1) + \hat{d} = m' + \hat{d},
\end{aligned}
$$

which means $m' \in S(n')$.

Case 3: There is a $\tilde{d} \in \mathbb{N}$ so that $n' + \tilde{d} = m$. Finally, if there is a $\tilde{d} \in \mathbb{N}$ so that $n' + \tilde{d} = m$, then

$$
n' + \tilde{d}' = n' + (\tilde{d} + 1) = (n' + \tilde{d}) + 1 = (n' + \tilde{d})' = m',
$$

which means $m' \in S(n')$.

Hence, in any case, $m' \in S(n')$. This completes the case distinction.

The induction inside an induction is completed. We conclude that $S(n') = \mathbb{N}$.

Therefore the Principle of Induction for S now implies that $S = \mathbb{N}$. This completes the proof of the result.   ∎

Before we continue with the ordering of the natural numbers, we briefly show that the trichotomy condition allows us to cancel terms on both sides of an equation, just like we would if subtraction was available. This result indicates that the trichotomy condition will be a good substitute for subtraction.

Proposition 3.17 *Let $a, b, c \in \mathbb{N}$ be so that $c + a = c + b$. Then $a = b$.*

Proof. Suppose for a contradiction that $a \neq b$. Then by the trichotomy condition (see Theorem 3.16) there is a $d \in \mathbb{N}$ so that $a+d = b$ or so that $b+d = a$. Without loss of generality assume that $a+d = b$. Then $c+b = c+(a+d) = (c+a)+d = (c+b)+d$. Moreover, trivially, $c + b = c + b$. We have obtained a contradiction to Theorem 3.16 with $n = m = c + b$.   ∎

Standard Proof Technique 3.18 Sometimes two parts of a proof are very similar, such as the part for $a + d = b$ and the part for $b + d = a$ in the proof of Proposition 3.17. In such cases, in a proof it is often assumed **without loss of generality** that one of the conditions is true. When doing so, the writer assumes that the reader can fill in the argument in case the other condition is true. The reader will show in Exercise 3-9 that the proof for $b + d = a$ is indeed very similar to the proof for $a + d = b$. ☐

Of course, by commutativity, if we can cancel on the left, then we can cancel on the right, too.

Proposition 3.19 *Let $a, b, c \in \mathbb{N}$ be so that $a + c = b + c$. Then $a = b$.*

Proof. Because $a + c = b + c$ we obtain $c + a = a + c = b + c = c + b$. Now by Proposition 3.17 we infer $a = b$.   ∎

Cancelation laws for multiplication are proved similarly (see Exercise 3-10). Aside from cancelation results, the trichotomy condition allows us to define the comparability of natural numbers.

Definition 3.20 *For $m, n \in \mathbb{N}$, we define $m < n$ iff there is a $d \in \mathbb{N}$ so that $m+d = n$. Moreover, we define $m \leq n$ iff $m = n$ or $m < n$. The symbol "$<$" reads "**less than**" or "**strictly less than**" and the symbol "$\leq$" reads "**less than or equal**."*

So now we know the answer to question 2 on page x: 3 is smaller than 9, because there is a natural number d, namely $d = 6$, so that $3 + d = 9$, and Theorem 3.16 guarantees that there is no other natural number $\tilde{d}$ so that $9 + \tilde{d} = 3$.

Of course we should establish that the relation $\leq$ has all the properties that an order relation should have. In a (non-strict) order relation, every object should be related to itself; ties should not be possible, that is, the only way the relation could work both ways is if an object is compared to itself; and the relation should be preserved "through any middleman," that is, if one object is less than or equal to another, which is yet less than or equal to a third, then the original object is also less than or equal to the third. These ideas, plus the desire that any two natural numbers should be comparable, lead to the following definition.

Definition 3.21 *Let P be a set and let $\leq \subseteq P \times P$ be a relation on P. Then $\leq$ is called an* **order** *relation iff*

1. *$\leq$ is* **reflexive***. That is, for all $x \in P$ we have $x \leq x$,*

2. *$\leq$ is* **antisymmetric***. That is, for all $x, y \in P$ we have that $x \leq y$ and $y \leq x$ implies $x = y$,*

3. *$\leq$ is* **transitive***. That is, for all $x, y, z \in P$, we have that $x \leq y$ and $y \leq z$ implies $x \leq z$.*

The relation $\leq$ is called a **total order** *relation iff for any two $x, y \in P$ we have that $x \leq y$ or $y \leq x$.*

Example 3.22 Before we continue with natural numbers, we give two abstract examples that show that general order relations can be quite different from the total order relations that we are accustomed to from working with numbers.

1. Let X be a set and let $\leq := \big\{ (x, x) \in X \times X : x \in X \big\}$. Then $\leq$ is an order relation, albeit a rather uninteresting one. Elements are only related to themselves and that's it.

2. Let X be a set of sets and let $\leq := \subseteq$ be set containment. Then $\leq$ is an order relation, and a very common one at that. Moreover, this example shows that in a general order relation there can be elements that are not comparable to each other. For example, if $\{a\}, \{b\} \in X$ are distinct sets, then neither $\{a\} \subseteq \{b\}$ nor $\{b\} \subseteq \{a\}$. $\qquad\square$

Now we continue with our discussion of the order relation for natural numbers.

Theorem 3.23 *The relation $\leq$ on $\mathbb{N}$ is a total order relation on $\mathbb{N}$.*

Proof. For reflexivity, let $x \in \mathbb{N}$. Then $x = x$, and hence $x \leq x$.

For antisymmetry, let $x, y \in \mathbb{N}$ be so that $x \leq y$ and $y \leq x$. Suppose for a contradiction that $x \neq y$. Then $x < y$ and $y < x$. Therefore there are $d, \tilde{d} \in \mathbb{N}$ so that $x + d = y$ and $y + \tilde{d} = x$. But then $y + (\tilde{d} + d) = (y + \tilde{d}) + d = x + d = y$. By Theorem 3.16 this is not possible. Hence we must have $x = y$.

For transitivity, let $x \leq y$ and $y \leq z$. If either inequality is an equality, then there is nothing to prove. Hence we can assume that $x < y$ and $y < z$. Then there is a $d \in \mathbb{N}$ so that $x + d = y$ and there is a $\tilde{d} \in \mathbb{N}$ so that $y + \tilde{d} = z$. But then $z = y + \tilde{d} = (x + d) + \tilde{d} = x + (d + \tilde{d})$, which means that $x < z$, and in particular $x \leq z$.

To prove that the order is a total order, note that by the trichotomy condition we must have $x \leq y$ or $y \leq x$. $\qquad\blacksquare$

Theorem 3.24 *Properties of the order relation. Let $x, y, z \in \mathbb{N}$.*

1. *If $x \leq y$, then $x + z \leq y + z$.*

2. *If $x \leq y$, then $xz \leq yz$.*

Similar results can be proved for strict inequalities. We will not state these here. Instead, we trust that the reader can make the requisite translation from the statements in this theorem.

Proof. To prove part 1, let $x \leq y$ and let $z \in \mathbb{N}$. In case $x = y$, we obtain $x + z = y + z$, which completes the proof. In case $x \neq y$, there is a $d \in \mathbb{N}$ so that $x + d = y$. But then $(x + z) + d = x + (z + d) = x + (d + z) = (x + d) + z = y + z$, which implies $x + z \leq y + z$.

The proof of part 2 is left to Exercise 3-11. ∎

Exercises

3-9. Provide the part of the proof of Proposition 3.17 in which $b + d = a$.

3-10. Cancelation of common factors. Let $a, b, c \in \mathbb{N}$.

    (a) Prove that if $ca = cb$, then $a = b$.

       Hint. Induction on a.

    (b) Prove that if $ac = bc$, then $a = b$.

3-11. Prove part 2 of Theorem 3.24.

3-12. For each of the following relations, determine if they are reflexive, antisymmetric or transitive. Prove your claims for the properties that hold, give a counterexample for the properties that don't.

    (a) The relation $\sim := \{ (b, a), (a, b) \}$ on the set $\{a, b\}$.

       Hint. First translate the definitions of reflexivity, antisymmetry and transitivity (see Definition 3.21) to relations that are given by ordered pairs. For example, a relation $\sim \subseteq X \times X$ is reflexive iff for all $x \in X$ we have that $(x, x) \in \sim$.

    (b) The relation $\sim := \{ (a, a), (a, b), (b, b) \}$ on the set $\{a, b\}$.

    (c) The relation $\sim := \{ (a, b), (b, c), (a, c) \}$ on the set $\{a, b, c\}$.

    (d) The relation $\sim := \{ (a, a), (b, b), (c, c), (a, b), (a, c) \}$ on the set $\{a, b, c\}$.

    (e) The relation $\sim := \{ (a, a), (b, b), (c, c), (a, b), (b, c) \}$ on the set $\{a, b, c\}$.

    (f) The relation $\sim := \{ (a, a), (b, b), (c, c), (a, b), (b, c), (a, c) \}$ on the set $\{a, b, c\}$.

    (g) The superset relation $\supseteq$ on a set of sets.

    (h) The relation $(i, j) \leq (m, n)$ iff $i < m$ or $i = m$ and $j \leq n$ on the set $\mathbb{N} \times \mathbb{N}$. Comparabilities between numbers are taken in the usual sense.

    (i) The relation $(i, j) \leq (m, n)$ iff $i < m$ or $i = m$ and $j < n$ on the set $\mathbb{N} \times \mathbb{N}$. Comparabilities between numbers are taken in the usual sense.

    (j) The relation $(i, j) \leq (m, n)$ iff $i \leq m$ and $j \leq n$ on the set $\mathbb{N} \times \mathbb{N}$. Comparabilities between numbers are taken in the usual sense.

    (k) The relation $(i, j) \leq (m, n)$ iff $ij \leq mn$ on the set $\mathbb{N} \times \mathbb{N}$. Comparabilities between numbers are taken in the usual sense.

    (l) The relation $(i, j) \leq (m, n)$ iff $ij \leq m + n$ on the set $\mathbb{N} \times \mathbb{N}$. Comparabilities between numbers are taken in the usual sense.

3-13. The **pointwise comparability** of functions $f : \mathbb{N} \to \mathbb{N}$ is defined by $f \leq g$ iff for all $n \in \mathbb{N}$ we have $f(n) \leq g(n)$ (in the usual sense).

    (a) Prove that pointwise comparability is an order relation on the set of functions from $\mathbb{N}$ to $\mathbb{N}$.

    (b) Is the pointwise comparability of functions $f : \mathbb{N} \to \mathbb{N}$ a total order?

3-14. Consider the relation $\sqsubseteq$ on $\mathbb{N}$ defined by $n \sqsubseteq m$ iff $n = m$ or $n + 5 \leq m$, where $\leq$ denotes the usual order of natural numbers.

    (a) Prove that $\sqsubseteq$ is an order relation on $\mathbb{N}$.

    (b) Is $\sqsubseteq$ a total order relation?

3-15. Let $m, n \in \mathbb{N}$. Prove that, with the natural numbers constructed as in the proof of Theorem 2.64, we have $n \leq m$ iff $n \subseteq m$.

3.3   A More Abstract Viewpoint: Binary Operations

Now that we have introduced the natural numbers, we can, on one hand, investigate the natural numbers themselves, and, on the other hand, we can investigate their abstract properties. Investigation of the numbers would be more familiar/comfortable to us. But, for the development of a theory, working with abstract properties is more efficient. For example, the proofs of properties of the natural numbers that were based on the Peano Axioms were more complicated than proofs, such as the ones for Propositions 3.14 or 3.15, that derived new properties from established properties. Now note that just about all the properties that we have established for the natural numbers so far must be established for other number systems, such as the integers, the rational numbers, the real numbers and the complex numbers, too. So, on one hand, it will be unavoidable to establish certain properties from the definitions of these number systems. But, on the other hand, any property that we know can be *derived from other properties* does not need to be proved all over again. After all, if property B follows from property A and our number system has property A, then it must have property B, too.

This idea turns out to be very powerful. Aside from making our construction of the number systems easier, the work with abstract properties has allowed mathematicians to comfortably work with less intuitive entities, such as number systems that "wrap around" (see Section 3.8) or infinite dimensional spaces (about which we will not talk in this text). The trick is that once we know which properties hold, we can use all the consequences of these properties, too, *and* we can use a simpler visualization of the entity in question.

So, in this section, rather than considering natural numbers and addition, multiplication and comparability, we consider sets and operations. Binary operations were already defined in Definition 3.1, so we can start with a property.

Definition 3.25 *Let S be a set and let $\circ : S \times S \to S$ be a binary operation on S. Then $\circ$ is called* **associative** *iff for all $a, b, c \in S$ we have that $(a \circ b) \circ c = a \circ (b \circ c)$.*

This definition should not be a surprise, because it follows the pattern set in Theorem 1.24 (parts 3 and 4), Theorem 2.28, Theorem 2.34, Proposition 3.6 (part 1), and Proposition 3.12 (part 1). The only difference is that this time the operation is called $\circ$ instead of $\wedge, \vee, \cap, \cup, +,$ or $\cdot$, respectively. Similarly, the other properties introduced in this section should be familiar.

Although an operation that is only associative is very weak indeed, sometimes associativity is all that is available. Associative operations give rise to the notion of a semigroup.

Definition 3.26 *Let* S *be a set and let* $\circ : S \times S \to S$ *be a binary operation on* S. *Then* $(S, \circ)$ *is called a* **semigroup** *iff the operation* $\circ$ *is associative, that is, iff for all* $x, y, z \in S$ *we have* $(x \circ y) \circ z = x \circ (y \circ z)$.

Example 3.27 Associativity is such a common property that we rarely even think about it in algebra. *Of course* we should be able to put in parentheses in any order we want to.

1. Addition of natural numbers is an associative binary operation. That is, $(\mathbb{N}, +)$ is a semigroup.

2. Multiplication of natural numbers is an associative binary operation. That is, $(\mathbb{N}, \cdot)$ is a semigroup.

3. Let S be a set and let $\mathcal{F}(S, S)$ be the set of all functions $f : S \to S$ from S to itself. Then composition $\circ$ is an associative binary operation on $\mathcal{F}(S, S)$. That is, $\big(\mathcal{F}(S, S), \circ\big)$ is a semigroup (see Exercise 3-16b). □

It is not an accident that the composition symbol $\circ$ is typically used for abstract binary operations. On one hand, sets of functions are often the target of more abstract investigations. On the other hand, we know that composition has many nice properties, but that it is not totally well-behaved (see part 3 of Example 3.30 below). So using the composition symbol for an abstract binary operation reminds us to think twice before using a certain property (namely, commutativity, see Definition 3.29 below).

Example 3.28 Associativity cannot be taken for granted.

1. Subtraction of integers, considered as a binary operation, is not associative. *(We will give the formal definition of integers and of subtraction later. Readers should rely on existing knowledge about integers here.)*

2. Although we usually do not use parentheses in natural language, we often group words in a way that is similar to how we group symbols in a computation. Consider the term "child tax credit." For a parent who paid taxes in the United States in the early twenty-first century, these words read as "child (tax credit)," because the tax code allowed a tax credit for every qualifying child. But the words could also be read as "(child tax) credit," which would grammatically be a credit on a tax that was paid for a child. (Strange as that may sound.) □

There are a few other abstract properties that we should introduce at this stage.

Definition 3.29 *Let* S *be a set and let* $\circ : S \times S \to S$ *be a binary operation on* S. *Then* $\circ$ *is called* **commutative** *iff for all* $a, b \in S$ *we have that* $a \circ b = b \circ a$. *A semigroup* $(S, \circ)$ *with commutative operation* $\circ$ *is also called a* **commutative semigroup**.

Example 3.30 Commutativity is fairly common, but is is not as common as associativity.

1. Addition of natural numbers is a commutative binary operation. That is, $(\mathbb{N}, +)$ is a commutative semigroup.

2. Multiplication of natural numbers is a commutative binary operation. That is, $(\mathbb{N}, \cdot)$ is a commutative semigroup.

3. Composition of functions is associative, but not commutative (see Exercise 3-16c). That is, if S is a set, then the semigroup $\big(\mathcal{F}(S, S), \circ\big)$ is a non-commutative semigroup. □

Definition 3.31 *Let S be a set and let $\circ : S \times S \to S$ be a binary operation on S. An element $e \in S$ is called a* **neutral element** *iff for all $a \in S$ we have $e \circ a = a = a \circ e$. A semigroup that contains a neutral element is also called a* **semigroup with a neutral element** *or a* **monoid**.

Example 3.32 Neutral elements may or may not exist for an operation.

1. The number 1 is a neutral element for multiplication of natural numbers. That is, $(\mathbb{N}, \cdot)$ is a semigroup with neutral element 1.

2. There is no neutral element (in $\mathbb{N}$ as we defined it here) for addition of natural numbers.

3. The semigroup $\big(\mathcal{F}(S, S), \circ\big)$ has a neutral element (see Exercise 3-16d). □

Finally, we often have more than one binary operation on a set. When that is the case, we need to know how the operations interact with each other.

Definition 3.33 *Let S be a set and let $\circ : S \times S \to S$ and $* : S \times S \to S$ be binary operations on S. The operation $\circ$ called* **left distributive** *over $*$ iff for all $a, b, c \in S$ we have that $a \circ (b * c) = a \circ b * a \circ c$. The operation $\circ$ called* **right distributive** *over $*$ iff for all $a, b, c \in S$ we have that $(a * b) \circ c = a \circ c * b \circ c$. Finally, $\circ$ is called* **distributive** *over $*$ iff $\circ$ is left distributive and right distributive over $*$.*

We can summarize what we have proved so far as follows.

Theorem 3.34 $(\mathbb{N}, +)$ *is a commutative semigroup, $(\mathbb{N}, \cdot)$ is a commutative semigroup with neutral element 1, and $\cdot$ is distributive over $+$.* ∎

It may look as if we are creating notation for the sake of creating notation. But the fact that functions form semigroups, too, is a first indication that the abstract viewpoint can be helpful. After all, we know now that, as long as we do not use commutativity, we can treat functions like numbers. The following results further explain why an abstract approach can be very powerful.

We know from experience that the number 1 is the *only* neutral element for multiplication. In our construction of the number systems, we will revisit the number 1 as an element of the integers, as an element of the rational numbers, as an element of the real numbers and as an element of the complex numbers. In each of these contexts, the number 1 will be constructed with a different method. Hence, formally we would need to prove all its properties from scratch. That would mean we would need to prove five times (for $\mathbb{N}$, for which we have not proved it yet, as well as for $\mathbb{Z}$, $\mathbb{Q}$, $\mathbb{R}$ and $\mathbb{C}$) that 1 is the only neutral element with respect to multiplication. Moreover, we would still not know if neutral elements would be unique in other contexts, such as the composition of functions. Proposition 3.35 below takes care of all these concerns at once, by showing that if an operation is associative, then neutral elements will be unique, if they exist.

Proposition 3.35 *Let* $(S, \circ)$ *be a semigroup. Then S has at most one neutral element. That is, if e, e' are both elements so that for all $x \in S$ we have $e \circ x = x = x \circ e$ and $e' \circ x = x = x \circ e'$, then $e = e'$.*

Proof. Let $e, e' \in S$ be so that for all $x \in S$ we have the equalities $e \circ x = x = x \circ e$ and $e' \circ x = x = x \circ e'$. Then, by first using that e' is neutral and then that e is neutral, we infer $e = e \circ e' = e'$. ∎

Standard Proof Technique 3.36 For many mathematical objects it is important to assure that they are the *only* object that has certain properties. That is, we want to assure that the object is unique. In a typical **uniqueness proof**, such as the proof of Proposition 3.35, we assume that there is more than one object with the properties under investigation and we prove that any two of these objects must be equal. □

Standard Proof Technique 3.37 In mathematics texts, it is not common to put boxes into computations that explain why certain steps are true.[2] Instead, just as in the proof of Proposition 3.35, it is often customary to say before or after a computation which facts have been used at what place. It is the reader's job to make the connections. □

Proposition 3.35 does not look very special, and it certainly should not be overrated. But we can use it to exhibit the power of the abstract approach. Figure 3.1 shows a Venn diagram of all the abstract structures that will be investigated in this text. Readers should not be discouraged by the fact the figure contains new terminology. Note instead that all the structures are semigroups. Consequently, in *all* these structures neutral elements are unique. Moreover, any other results that are proved in more general contexts (larger sets in Figure 3.1) will hold in more specialized contexts, too. This fact is independent of whether we are familiar with a context, such as for example the real numbers $\mathbb{R}$, or whether the context is new, such as for example the finite fields $\mathbb{Z}_p$. In particular, we will not need to re-state proofs, which will make our construction of the number systems a lot more efficient. Proposition 3.38 below translates Proposition 3.15 to a more general setting. Consequently, it shows that, as soon as we have commutativity and distributivity, parentheses are multiplied out in the usual fashion. Note that, despite the familiar notation, the operations $+$ and $\cdot$ in Proposition 3.38 are nothing more than abstract binary operations with certain properties.

Proposition 3.38 *Let* $(S, +)$ *be a commutative semigroup and let $\cdot$ be an associative binary operation that is distributive over $+$. Then for all elements $x, y, z, u \in S$ we have* $(x + y)(z + u) = (xz + xu) + (yz + yu)$.

Proof. Exercise 3-17. ∎

---

[2]The author really has started to like this idea, but it is mainly an educational tool.

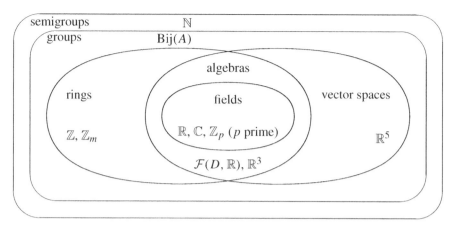

Figure 3.1 Hierarchy of the abstract algebraic structures (names as well as examples) that will be introduced in this text.

The above results indicate that the familiar laws of algebra should hold in all structures that we will investigate, as long as fundamental laws such as associativity, distributivity and commutativity hold. This is what makes the abstract approach so powerful. Independent of what the objects under investigation *are*, as long as they have the right *properties*, they will behave like, say, numbers. In particular, in a lot of abstract settings we can visualize a familiar number system and still obtain general results.

At this stage it would be natural to also investigate order relations in semigroups. But it turns out that the hypotheses we need to order semigroups are a little more restrictive than we would like them to be. Hence, order relations in semigroups have been relegated to Exercise 3-28 and we will investigate order relations in the context of rings in Section 4.4.

We conclude with a type of operation that is used less often than binary operations. Although we cannot subtract any two natural numbers from each other, subtraction should be possible if the first number is larger than the second number. That means, on $\mathbb{N}$, subtraction is an operation that is defined for some, but not all pairs in $\mathbb{N} \times \mathbb{N}$. Such an operation is called a partial operation.

Definition 3.39 *Let S be a set. A* **partial (binary) operation** *on the set S is a function* $\circ : A \to S$, *where A is a subset of* $S \times S$.

To define subtraction, we must reduce the idea of subtraction to the idea of adding the difference. This is exactly how subtraction is often introduced in elementary school: The difference between two numbers is the number that must be added to the smaller number to get the larger number. Of course we must make sure that this can only be accomplished in one way.

Proposition 3.40 *Let* $n, m \in \mathbb{N}$ *be so that* $n < m$. *Then the number d so that* $n + d = m$ *is unique.*

Proof. To prove the uniqueness of d, we use Standard Proof Technique 3.36. So let $n, m \in \mathbb{N}$ be so that $n < m$ and let $d, \tilde{d}$ be so that $n + d = m$ and $n + \tilde{d} = m$. Then $n + \tilde{d} = m = n + d$ and by Proposition 3.17 we infer that $d = \tilde{d}$. Hence the number d is unique. ∎

Definition 3.41 *Let $n, m \in \mathbb{N}$ be so that $n < m$. Then we set $m - n := d$, where d is the unique number so that $n + d = m$. The number d is also called the* **difference** *between m and n, and the partial operation $-$ on $\mathbb{N}$ is called* **subtraction**.

Definition 3.41 claims that $-$ is a binary operation, which can be seen as follows: The set $A \subseteq \mathbb{N} \times \mathbb{N}$ on which $-$ is defined is $A := \{(m, n) : n < m\}$ and $-$ is the set $\{((m, n), d) \in A \times \mathbb{N} : n + d = m\}$. Proposition 3.40 implies that $-$ is a function, and hence a partial binary operation.

Once we have defined differences, we want to know about their properties.

Proposition 3.42 *Let $m, n, x, y \in \mathbb{N}$ be so that $n < m$ and $y < x$. Then the following hold.*

1. *$n + y < m + x$ and $(m + x) - (n + y) = (m - n) + (x - y)$.*

2. *$nx < mx$ and $mx - nx = (m - n)x$.*

3. *If $n + x = m + y$, then $m - n = x - y$.*

Proof. We will prove part 1, leaving the proofs of parts 2 and 3 to Exercise 3-18.

This proof is a bit tedious, but it is a good exercise exactly because of that tedium: We simply want to distribute the negative sign, and yet we cannot do that, because we can only use the definition. So the very laws of arithmetic that we are tempted to use are not available for the formal differences in this result. Instead, we argue as follows.

Let d_{mn} and d_{xy} be the numbers so that $n + d_{mn} = m$ and $y + d_{xy} = x$, which are guaranteed to exist by Definition 3.41. Then by Definition 3.41 we have the equality $(m - n) + (x - y) = d_{mn} + d_{xy}$. To prove that this number is the difference between $(m + x)$ and $(n + y)$ (which will also prove that $n + y < m + x$), we must show that $(n + y) + (d_{mn} + d_{xy}) = m + x$. We compute

$$
\begin{aligned}
(n + y) + (d_{mn} + d_{xy}) &= \big((n + y) + d_{mn}\big) + d_{xy} \\
&= \big(n + (y + d_{mn})\big) + d_{xy} \\
&= \big(n + (d_{mn} + y)\big) + d_{xy} \\
&= \big((n + d_{mn}) + y\big) + d_{xy} \\
&= (m + y) + d_{xy} \\
&= m + (y + d_{xy}) \\
&= m + x,
\end{aligned}
$$

which proves part 1. ∎

Division can be defined as a partial operation, too. (See Exercise 3-19 for details.) We introduce the main properties that we need for the introduction of binomial coefficients in Section 3.5, and we leave their proofs as good exercises for the reader (see the comment in handwriting at the start of the proof of Proposition 3.42).

Definition 3.43 *Let* $n, d \in \mathbb{N}$ *be so that* $n > d$ *and so that there is a* $q \in \mathbb{N}$ *so that* $n = dq$. *Then we set* $\dfrac{n}{d} := q$, *and call it the* **quotient** *of* n *and* d. *The number* n *is also called the* **numerator** *and the number* d *is called the* **denominator**.

Proposition 3.44 *Let* $m, n, d, e \in \mathbb{N}$.

1. *If* $\dfrac{n}{d}$ *and* $\dfrac{m}{d}$ *both exist, then so does* $\dfrac{m+n}{d}$ *and* $\dfrac{m+n}{d} = \dfrac{m}{d} + \dfrac{n}{d}$.

2. *If* $\dfrac{n}{d}$ *and* $\dfrac{m}{e}$ *both exist, then so does* $\dfrac{mn}{de}$ *and* $\dfrac{mn}{de} = \dfrac{m}{e} \cdot \dfrac{n}{d}$.

3. *If* $\dfrac{n}{d}$ *and* $\dfrac{m}{d}$ *both exist and* $n < m$, *then so does* $\dfrac{m-n}{d}$ *and* $\dfrac{m-n}{d} = \dfrac{m}{d} - \dfrac{n}{d}$.

4. *If* $\dfrac{n}{d}$ *and* $\dfrac{m}{e}$ *both exist and* $ne = md$, *then* $\dfrac{n}{d} = \dfrac{m}{e}$.

Proof. Exercise 3-20 ∎

Exercises

3-16. Composition of functions.

    (a) Let S be a set. Prove that composition $\circ$ of functions is a binary operation on $\mathcal{F}(S, S)$.

    (b) Let A, B, C, D be sets. Prove that if $h : A \to B$, $g : B \to C$ and $f : C \to D$ are functions, then $f \circ (g \circ h) = (f \circ g) \circ h$. Conclude that composition is an associative binary operation on $\mathcal{F}(S, S)$.

    Hint. Argue pointwise.

    (c) Give an example that shows that composition of functions is not commutative. That is, find a set S and two functions $f, g : S \to S$ so that $f \circ g$ is not equal to $g \circ f$.

    (d) The semigroup $\left(\mathcal{F}(S, S), \circ \right)$ has a neutral element. Find it.

3-17. Prove Proposition 3.38.

    Hint. Proof of Proposition 3.15.

3-18. Finish the proof of Proposition 3.42.

    (a) Prove part 2 of Proposition 3.42.

    (b) Prove part 3 of Proposition 3.42.

3-19. Division is a partial binary operation.

    (a) Let $n, d, q, \tilde{q} \in \mathbb{N}$ be so that $n = dq$ and $n = d\tilde{q}$. Prove that $q = \tilde{q}$.

    Hint. Exercise 3-10.

    (b) Explicitly state the definition of the set A on which division is defined.

    Hint. See the remark after Definition 3.41.

    (c) Explicitly state the definition of division as a subset of $(\mathbb{N} \times \mathbb{N}) \times \mathbb{N}$ that is a partial binary operation.

3-20. Prove Proposition 3.44.

  (a) Prove part 1 of Proposition 3.44.

  (b) Prove part 2 of Proposition 3.44.

  (c) Prove part 3 of Proposition 3.44.

  (d) Prove part 4 of Proposition 3.44.

3-21. Prove that there is no $n \in \mathbb{N} \setminus \{1\}$ for which there is another $m \in \mathbb{N}$ so that $nm = 1$.

3-22. A **noncommutative** operation. Consider a six-sided die whose center is at the origin of a three dimensional coordinate system and which is oriented so that each of its faces is perpendicular to a coordinate axis. The set under consideration shall be the set R of all possible rotations of the die around its center so that at the end of the rotation each face is again perpendicular to a coordinate axis.

  (a) Prove that if one such rotation is performed after another, then the resulting operation is also a rotation as indicated above.

  This means that "performing one rotation r_2 after another rotation r_1" is a binary operation on the set R. Call this composition of rotations $\circ$, and let the result of performing a rotation r_2 after another rotation r_1 be denoted $r_2 \circ r_1$.

  (b) Prove that $\circ$ is not commutative.

  Hint. Envision the die initially so that the faces with the 1 and the 6 are perpendicular to the x-axis, the faces with the 2 and the 5 are perpendicular to the y-axis, and the faces with the 3 and the 4 are perpendicular to the z-axis. Now consider $90°$ rotations about two different axes.

3-23. A commutative, **nonassociative** operation. For this exercise, you may use what you know about real numbers. For $x, y \in [-1, 1]$ define $x \oplus y := \begin{cases} \frac{a+b}{2}; & \text{if } a, b \neq 0, \\ a; & \text{if } b = 0, \\ b; & \text{if } a = 0, \end{cases}$ and let $\cdot$ stand for regular multiplication. Prove that $\oplus$ is commutative, that 0 is a neutral element for addition, that every element in $[-1, 1]$ has an additive inverse (that is, for every x there is a y so that $x + y = 0$), that multiplication is associative and commutative with neutral element 1, but that $\oplus$ is not associative.

3-24. For each of the following binary operations, determine if they are associative, commutative, both or neither. In case the operation is not associative or not commutative, provide a counterexample that shows that the property does not hold.

  (a) Set union as a binary operation on a power set $\mathcal{P}(A)$.

  (b) Set intersection as a binary operation on a power set $\mathcal{P}(A)$.

  (c) The AND-operation $\wedge$ on a set of primitive propositions.

  (d) The OR-operation $\vee$ on a set of primitive propositions.

  (e) Subtraction of integers.

  (f) Division of nonzero rational numbers.

  (g) Composition of functions on a set of injective functions from a set S to itself.

  (h) Composition of functions on a set of surjective functions from a set S to itself.

  (i) Composition of functions on a set of injective or surjective functions from a set S to itself.

  Note. This last part is a trap. Be very careful.

3-25. For each of the following pairs of binary operations, determine which binary operation is distributive over the other one. Quote the requisite results in case distributivity holds, give a counterexample otherwise.

  (a) Addition of natural numbers and multiplication of natural numbers.

  (b) On a power set $\mathcal{P}(A)$ of a set, set union and set intersection.

(c) On a set of primitive propositions, $\wedge$ and $\vee$.

3-26. Prove that addition as a function from $\mathbb{N} \times \mathbb{N}$ to $\mathbb{N}$ is neither injective nor surjective.

3-27. Prove that multiplication as a function from $\mathbb{N} \times \mathbb{N}$ to $\mathbb{N}$ is surjective, but not injective.

3-28. Let $(S, +)$ be a commutative semigroup and let $S^+ \subseteq S$ be so that for any two $x, y \in S$ we have that $x + y \in S$ and that exactly one of the following holds.

   (I) $x = y$,

   (II) There is a $d \in S^+$ so that $x + d = y$,

   (III) There is a $d \in S^+$ so that $y + d = x$.

Prove each of the following.

   (a) For all $x, y, z \in S$ we have that $z + x = z + y$ implies $x = y$.

   (b) For all $x, y, z \in S$ we have that $x + z = y + z$ implies $x = y$.

   (c) The relation $\leq \subseteq S \times S$ defined by $x \leq y$ iff $x = y$ or there is a $d \in S^+$ so that $x + d = y$ is a total order relation on S.

   (d) If $x, y, z \in S$ and $x \leq y$, then $x + z \leq y + z$ and $z + x \leq z + y$.

   (e) If $\cdot$ is an associative operation that is distributive over $+$ and if for all $x, y \in S^+$ we have $xy \in S^+$, then for all $x, y, z \in S$ we have that if $x \leq y$ and $z \in S^+$, then $xz \leq yz$ and $zx \leq zy$.

   (f) Explain why we must demand that $z \in S^+$ in part 3-28e.

3-29. For each of the terms below, give two interpretations that depend on which two words are grouped together.

   (a) Artificial fruit flavor

   (b) Old school approach

   (c) New mathematics teacher

   (d) Tall book box

   (e) Canned food drive.

   (f) Good test problem.

3.4 Induction

Induction as stated in Standard Proof Technique 2.65 is not the typical way in which induction is used. Usually, induction is about proving statements $P(n)$ about a natural number n. In this section, we present the Principle of Induction in its more common form and we give examples of its use.

Theorem 3.45 Principle of Induction. *Let $P(n)$ be a statement about the natural number n. If $P(1)$ is true and if, for all $n \in \mathbb{N}$, truth of $P(n)$ implies truth of $P(n + 1)$, then $P(n)$ holds for all natural numbers.*

   Proof. Let $P(n)$ be as indicated and let $S := \{n \in \mathbb{N} : P(n) \text{ is true }\}$. Then $1 \in S$. Moreover, if $n \in S$, then the statement $P(n)$ is true. By assumption, this implies that $P(n + 1)$ is true, that is, $n + 1 \in S$. By the Principle of Induction as in Theorem 2.64 we conclude $S = \mathbb{N}$. Therefore for all $n \in \mathbb{N}$ the statement $P(n)$ is true. ∎

Standard Proof Technique 3.46 As noted before, induction is a standard proof technique. When used in the form of Theorem 3.45, the proof of $P(1)$ is usually called the **base step** and the proof that $P(n)$ implies $P(n+1)$ is usually called the **induction step**. In the induction step, the hypothesis $P(n)$ is also called the **induction hypothesis**. ☐

The way induction works is nicely demonstrated in the proofs of summation formulas. For the next example, we omit the pesky parentheses that are formally required because $+$ and $\cdot$ are binary operations, that is, because each has exactly two inputs, not n. We will prove that we can do so in Theorem 3.58, but we want to have a readable and fairly simple first example.

Proofs of summation formulas are fairly canonical examples of how induction proofs work. You should make sure that you understand the proof in Example 3.47, and maybe do some parts of Exercise 3-32, before you move on to the rest of this section. In the remainder of this section we will explore the wide variety of scenarios in which induction is useful.

Example 3.47 *For each $k \in \mathbb{N}$ let $k^2 := k \cdot k$. Then for all $n \in \mathbb{N}$ we have*

$$1^2 + 2^2 + 3^2 + \cdots + n^2 = \frac{1}{6}n(n+1)(2n+1).$$

Proof. *Base step, $n = 1$.* For $n = 1$ note that $1^2 = \frac{1}{6}1(1+1)(2 \cdot 1 + 1)$, which proves the base step.

Induction hypothesis. We can assume that for n the formula

$$1^2 + 2^2 + 3^2 + \cdots + n^2 = \frac{1}{6}n(n+1)(2n+1)$$

holds. We must now prove that the above formula holds when n is replaced with $(n+1)$, and we are allowed to use the induction hypothesis.

Induction step. We must prove that

$$1^2 + 2^2 + 3^2 + \cdots + (n+1)^2 = \frac{1}{6}(n+1)\big((n+1)+1\big)\big(2(n+1)+1\big)$$

and we are free to use the induction hypothesis.

$$
\begin{aligned}
1^2 + 2^2 + 3^2 + \cdots + n^2 + (n+1)^2 &= \left(1^2 + 2^2 + 3^2 + \cdots + n^2\right) + (n+1)^2 \\
&= \frac{1}{6}n(n+1)(2n+1) + (n+1)^2 \\
&= \frac{1}{6}(n+1)\big[n(2n+1) + 6(n+1)\big] \\
&= \frac{1}{6}(n+1)\left[2n^2 + 7n + 6\right] \\
&= \frac{1}{6}(n+1)\big[(n+2)(2n+3)\big] \\
&= \frac{1}{6}(n+1)\big((n+1)+1\big)\big(2(n+1)+1\big),
\end{aligned}
$$

which was to be proved.   ■

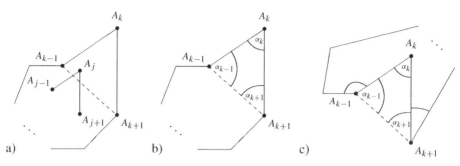

Figure 3.2 In the proof of Example 3.50, we first ascertain there is a vertex that is not in a configuration as in part a) in the above figure. Parts b) and c) show how an added vertex adds π to the sum of the interior angles.

Remark 3.48 Note how similar this approach is to the induction proofs we have done so far. The proof of $P(1)$ (the base step) takes the place of proving that $1 \in S$ and the proof of $P(n) \Rightarrow P(n+1)$ (the induction step) takes the place of proving that $n \in S$ implies $n' \in S$. Induction in the form presented in this section is more common and simpler than the Principle of Induction from the Peano Axioms, because we can talk directly about statements and we do not need sets to encode statements. □

Remark 3.49 An easier way to prove the equation

$$1^2 + 2^2 + 3^2 + \cdots + (n+1)^2 = \frac{1}{6}(n+1)\big((n+1)+1\big)\big(2(n+1)+1\big)$$

in the induction step would be to replace the sum $1^2 + 2^2 + 3^2 + \cdots + n^2$ on the left side with $\frac{1}{6}n(n+1)(2n+1)$, and to replace the n on the right side with $(n+1)$, and to then show that both sides are equal. The resulting presentation is not as elegant as the proof in Example 3.47, but, if desired, it can be translated into a "left to right" argument as in Example 3.47: Simply go down the left column of the simplification and present the terms in left to right fashion, then go up the right column and present those terms in left to right fashion. □

Of course, induction is not solely tied to summation formulas. We briefly leave our set theoretical framework, to present an example from geometry.

Example 3.50 *The sum of the interior angles of a plane n-gon with $n \geq 3$ vertices is $(n-2)\pi$.*

Proof. The *base step* for $n = 3$ was proved in Example 1.52. *Starting an induction with a number other than 1 is not a problem. The result will simply be valid from that point on, rather than for all $n \in \mathbb{N}$ (see Exercise 3-37).*

So, as the *induction hypothesis*, we can assume that the sum of the interior angles of a plane n-gon with $n \geq 3$ vertices is $(n-2)\pi$. Moreover, to avoid technical complications, we will allow an angle at a vertex to be straight, that is, it will be allowed for a vertex to lie on a straight line segment of the boundary of the n-gon and the angle at that vertex will count as π. We must now prove that the sum of the interior angles of a plane $(n + 1)$-gon with $n \geq 3$ vertices is $\big((n + 1) - 2\big)\pi$.

For the *induction step*, consider a plane $(n+1)$-gon P with vertices $A_1, \ldots, A_{n+1}$. Let A_k be a vertex of P. For convenience of notation, we will assume throughout the proof that the index of each vertex in question is in the set $\{2, \ldots, n\}$. We want A_k to be so that the interior of the triangle $A_{k-1}A_kA_{k+1}$ and the line segment $\overline{A_{k-1}A_{k+1}}$ (except for the end points) lie entirely in the interior or the exterior of the n-gon. To do this, we may need to change the vertex we originally selected. The next paragraph explains how this can be done.

If the interior of the triangle $A_{k-1}A_kA_{k+1}$ and the line segment $\overline{A_{k-1}A_{k+1}}$ (except for the end points) lie entirely in the interior or the exterior of the n-gon, we continue with the vertex A_k in the next paragraph. If not, then there is a vertex A_j of the n-gon so that $j \notin \{k - 1, k, k + 1\}$ and A_j is in the interior of the triangle $A_{k-1}A_kA_{k+1}$ or on the line segment $\overline{A_{k-1}A_{k+1}}$ [see Figure 3.2, part a) for a possible placement of the triangle $A_{j-1}A_jA_{j+1}$]. Restart the argument in this paragraph with A_j in place of A_k. This repeatedly restarted argument never visits the same vertex twice. Because P has only finitely many vertices, the argument must ultimately stop with a vertex A_k so that the interior of the triangle $A_{k-1}A_kA_{k+1}$ and the line segment $\overline{A_{k-1}A_{k+1}}$ (except for the end points) lie entirely in the interior or the exterior of the n-gon [see parts b) and c) of Figure 3.2].

Now consider the n-gon Q obtained from P by removing $\overline{A_{k-1}A_k}$ and $\overline{A_kA_{k+1}}$ and replacing them with $\overline{A_{k-1}A_{k+1}}$. By induction hypothesis, the sum of the interior angles of Q is $(n - 2)\pi$. Now we must consider two cases: The triangle $A_{k-1}A_kA_{k+1}$ can be contained in the interior or in the exterior of the $(n + 1)$-gon P.

First consider the case that the triangle $A_{k-1}A_kA_{k+1}$ is contained in P [see part b) of Figure 3.2]. In this case, the sum of the interior angles of P is the sum of the interior angles of Q plus the interior angles of $A_{k-1}A_kA_{k+1}$. As was to be proved, this sum is $(n - 2)\pi + \pi = \big((n + 1) - 2\big)\pi,$.

In case the triangle $A_{k-1}A_kA_{k+1}$ is not contained in P [see part c) of Figure 3.2], let α_i be the interior angle of $A_{k-1}A_kA_{k+1}$ at the vertex A_i ($i = k - 1, k, k + 1$). The interior angle of P at A_{k-1} (marked) is the interior angle of Q at A_{k-1} minus α_{k-1}. The interior angle of P at A_{k+1} (marked) is the interior angle of Q at A_{k+1} minus α_{k+1}. Finally, the interior angle of P at A_k is $2\pi - \alpha_k = \pi + \alpha_{k-1} + \alpha_{k+1}$. Hence the sum of the interior angles is $(n - 2)\pi - \alpha_{k-1} - \alpha_{k+1} + \pi + \alpha_{k-1} + \alpha_{k+1} = \big((n + 1) - 2\big)\pi,$ as was to be proved.   ∎

After Examples 3.47 and 3.50 briefly interrupted our building mathematics from the axioms of set theory, we now resume with this main thrust of our presentation. The Principle of Induction can be used to prove that every nonempty set A of natural numbers has a **smallest element**, that is, an element $s \in A$ so that for all $a \in A$ we have $s \leq a$. From here on, we will not explicitly re-state the induction hypothesis.

Theorem 3.51 Principle of Induction. *Every nonempty subset A of $\mathbb{N}$ has a smallest element.*

Proof. Let $A \subseteq \mathbb{N}$ be a nonempty set, and suppose for a contradiction that A does not have a smallest element. Consider the statement $P(n) =$ "$\{1, \ldots, n\} \cap A = \emptyset$." We will prove by induction that $P(n)$ holds for all $n \in \mathbb{N}$.

Base Step. Because A does not have a smallest element and 1 is the smallest element of $\mathbb{N}$, we must have that $1 \notin A$. (Otherwise 1 would be the smallest element of A.) Hence $P(1)$ is true.

Induction step. We may assume that $\{1, \ldots, n\} \cap A = \emptyset$. Consider the element $n+1$. Because $n + 1$ is the smallest element of $\mathbb{N} \setminus \{1, \ldots, n\}$, we must have that $n + 1 \notin A$. (Otherwise $n + 1$ would be the smallest element of A.) But then $\{1, \ldots, n+1\} \cap A = \emptyset$ and $P(n + 1)$ is true.

By induction, $P(n)$ holds for all $n \in \mathbb{N}$ and we must conclude that A is empty. This is a contradiction to the assumption that A is not empty. Therefore A must have a smallest element. ∎

Standard Proof Technique 3.52 Theorem 3.51 is called "Principle of Induction," too, because it is another standard way to prove results about natural numbers. To use Theorem 3.51, we assume for a contradiction that there is a counterexample to our result. Theorem 3.51 then guarantees that there is a *smallest* counterexample. We then use this smallest counterexample to obtain a contradiction. □

The approach of Standard Proof Technique 3.52 can be demonstrated with a result that we will need in the proof of Proposition 3.107. Definition 2.59 says that two sets are equivalent iff they have the same "size," because the elements of equivalent sets are matched in a one-to-one and onto fashion. But when we think of size, we typically think of a number. The next definition assigns a numerical size to sets, when possible.

Definition 3.53 *A set F is called* **finite** *iff F is empty or there is an $n \in \mathbb{N}$ and a bijective function $f : \{1, \ldots, n\} \to F$. Sets that are not finite are called* **infinite**. *For finite sets $F \neq \emptyset$ we set $|F| := n$ with n as above and we set $|\emptyset| := 0$, where 0 is an element that is not in $\mathbb{N}$.[3] For infinite sets I we set $|I| := \infty$, where ∞ is an element that is not in $\mathbb{N} \cup \{0\}$. The number $|F|$ is also called the* **cardinality** *of the set F. We will refine the notion of cardinality for infinite sets in Section 7.3.*

It is now natural to assume that if one set is contained in another and both sets have the same size, then both sets should be equal. The next result shows that this is indeed the case for finite sets. For infinite sets, the situation is more complicated, as Theorem 4.60 will show.

---

[3]At this stage, 0 is basically a symbol. We will introduce the arithmetic with this symbol in Definition 3.72. If Theorem 2.64 had constructed natural numbers that included 0, we would have started $\mathbb{N}$ with $0 := \emptyset$. Everything would be consistent with the presentation here, because $1 = 0' = \{\emptyset\}$ would be as in the construction in Theorem 2.64.

Theorem 3.54 *Let A and B be finite sets so that $A \subseteq B$ and so that $|A| = |B|$. Then $A = B$.*

Proof. Clearly, if $|A| = 0$, then $|B| = |A| = 0$. Therefore, in this case both sets A and B are empty and hence they are equal. Thus, for the remainder of the proof, we can assume that $|A| \in \mathbb{N}$.

Now we apply the Principle of Induction as in Theorem 3.51 to the cardinality $|A| = |B| \in \mathbb{N}$. So suppose for a contradiction that there is a counterexample, that is, that there are finite sets A and B so that $A \subseteq B$, $|A| = |B| \in \mathbb{N}$ and $A \neq B$. Let $M \subseteq \mathbb{N}$ be the set of all sizes $|A| = |B| \in \mathbb{N}$ of such sets. Then by Theorem 3.51 the set M has a smallest element m. So, we assume there is a smallest counterexample. We will now prove that, if m is not 1, then the size of this smallest counterexample can be reduced, which is a contradiction.

Suppose for a contradiction that m was greater than 1 and let A and B be so that $A \subseteq B$, $|A| = |B| = m$ and $A \neq B$. Choose an element $a \in A$, let $\tilde{A} := A \setminus \{a\}$ and let $\tilde{B} := B \setminus \{a\}$. Then $|\tilde{A}| = |\tilde{B}| = m - 1$ (see Exercise 3-30), $\tilde{A} = A \setminus \{a\} \subseteq B \setminus \{a\} = \tilde{B}$ and $\tilde{A} \neq \tilde{B}$ (because otherwise $A = \tilde{A} \cup \{a\} = \tilde{B} \cup \{a\} = B$, which was not true). But then $\tilde{A}$ and $\tilde{B}$ are a counterexample of size $m - 1 < m$, contradicting the choice of m. Therefore m is not greater than 1.

But that means that $m = 1$. Hence $A = \{a\}$ and $B = \{b\}$. Therefore, by the definition of containment we have $a \in B = \{b\}$, which means that $a = b$ and then $A = B$, contradiction. Thus the theorem must hold. ■

We will continue our investigation of sizes of sets in Section 4.6. To conclude this section, we tackle all these extra parentheses in long sums, which we know should be superfluous. This proof is rather technical, and we should see it for what it is: A necessary evil to overcome a serious technical nuisance. But at the same time, the argument demonstrates beautifully how technical features can be encoded correctly and then be treated rigorously.

We start with a recursive definition, that is, with a definition that defines something for $n = 1$ and then uses the definition for n to give the definition for $n + 1$. Note that induction can be used to prove that recursive definitions are allowed: Let $q(n)$ be a quantity that is to be defined for every $n \in \mathbb{N}$. If we can define $q(1)$ and if, whenever $q(n)$ is defined, we can define $q(n + 1)$, then $q(n)$ is defined for all $n \in \mathbb{N}$: Simply apply the Principle of Induction from Standard Proof Technique 2.65 to the set $S := \{n \in \mathbb{N} : q(n) \text{ is defined}\}$.

Because quantities can be defined recursively, we can define a canonical way to bracket sums.

Definition 3.55 *For each $j \in \mathbb{N}$ let $a_j \in \mathbb{N}$. Recursively define $C(a_1, \ldots, a_n)$ by $C(a_1) := a_1$ and $C(a_1, \ldots, a_n, a_{n+1}) := \big(C(a_1, \ldots, a_n)\big) + a_{n+1}$. We will call $C(a_1, \ldots, a_n)$ the* **canonical bracketing** *of the sum of $a_1, \ldots, a_n$.*

This canonical bracketing of a sum with n terms is just the sequential addition from left to right: $C(a_1, \ldots, a_n) = \big(\cdots (((a_1) + a_2) + a_3) + \cdots \big) + a_n$. We can define arbitrarily bracketed sums in the same fashion. But because we do not know where an "arbitrarily bracketed sum" breaks into two shorter sums, we first need another Principle of Induction.

Theorem 3.56 Principle of Induction ("strong induction"). *Let $P(n)$ be a statement about the natural number n. If $P(1)$ is true and if for all $n \in \mathbb{N}$ truth of $P(1) \wedge \cdots \wedge P(n)$ implies truth of $P(n+1)$, then $P(n)$ holds for all natural numbers.*[4]

Proof. Let P be as indicated and consider the statement $Q(n) := P(1) \wedge \cdots \wedge P(n)$. We will prove by induction that $Q(n)$ is true for all $n \in \mathbb{N}$.

Base step. By assumption, $Q(1) = P(1)$ is true.

Induction step. If $Q(n)$ is true, then $P(1) \wedge \cdots \wedge P(n)$ is true. By assumption this implies that $P(n+1)$ is true, and thus $Q(n+1) = \big(P(1) \wedge \cdots \wedge P(n)\big) \wedge P(n+1)$ is true.

Thus $Q(n)$ holds for all $n \in \mathbb{N}$, and in particular, $P(n)$ holds for all $n \in \mathbb{N}$. ∎

We will see how to use strong induction as we consider "arbitrarily bracketed sums." Note that the next recursive definition is possible because we have strong induction.

Definition 3.57 *For each $j \in \mathbb{N}$ let $a_j \in \mathbb{N}$. Define the term $S(a_j) := a_j$ and for any $m - k + 1$ numbers $a_k, \ldots, a_m$, call the term $S(a_k, \ldots, a_m)$ a* **sum** *of these numbers iff there is an $l \in \{k, \ldots, m-1\}$ so that*

$$S(a_k, \ldots, a_m) = \big(S(a_k, \ldots, a_l)\big) + \big(S(a_{l+1}, \ldots, a_m)\big),$$

where $S(a_k, \ldots, a_l)$ and $S(a_{l+1}, \ldots, a_m)$ are shorter sums.

Definition 3.57 expresses the idea of an "arbitrarily bracketed sum:" The order of the terms is preserved[5], and every summation has two summands, as dictated by the definition of addition. Each summand is a shorter sum, and we do not specify how long either of the shorter sums needs to be.

We can now show that all "arbitrarily bracketed sums" yield the same result by proving that any of them is equal to the sum in canonical form.

Theorem 3.58 *For each $j \in \mathbb{N}$ let $a_j \in \mathbb{N}$ and for each $n \in \mathbb{N}$ let $S(a_1, \ldots, a_n)$ be a sum of $a_1, \ldots, a_n$. Then for all $n \in \mathbb{N}$ we have that $S(a_1, \ldots, a_n) = C(a_1, \ldots, a_n)$.*

Proof. The proof is a strong induction on n. The base step $n = 1$ is trivial, because $S(a_1) = C(a_1)$.

For the induction step $\{1, \ldots, n\} \to n + 1$ note that by Definition 3.57 there is an $m \in \{1, \ldots, n\}$ so that $S(a_1, \ldots, a_{n+1}) = \big(S(a_1, \ldots, a_m)\big) + \big(S(a_{m+1}, \ldots, a_{n+1})\big)$. By induction hypothesis for m summands, $S(a_1, \ldots, a_m) = C(a_1, \ldots, a_m)$ and, if $n > m$, by induction hypothesis for $n - m$ summands, because any two arbitrarily bracketed sums are equal, there is a sum $T(a_{m+2}, \ldots, a_{n+1})$ so that we have the equality $S(a_{m+1}, \ldots, a_{n+1}) = (a_{m+1}) + \big(T(a_{m+2}, \ldots, a_{n+1})\big)$. But then we can argue as

---

[4]Note that formally we would need to bracket the statement $P(1) \wedge \cdots \wedge P(n)$ in a canonical fashion, too. Then we would need to prove that brackets do not matter in ANDs either. We forego this rather pedantic detail. The argument for brackets in sums will be rough enough.

[5]We are not taking on commutativity yet. That will have to wait until Theorem 3.65.

follows (in case $n = m$ we would leave out T).

$$
\begin{aligned}
S(a_1, \ldots, a_{n+1}) &= \big(S(a_1, \ldots, a_m)\big) + \big(S(a_{m+1}, \ldots, a_{n+1})\big) \\
&= \big(C(a_1, \ldots, a_m)\big) + \Big((a_{m+1}) + \big(T(a_{m+2}, \ldots, a_{n+1})\big)\Big) \\
&= \big(C(a_1, \ldots, a_m)\big) + \Big(a_{m+1} + \big(T(a_{m+2}, \ldots, a_{n+1})\big)\Big)
\end{aligned}
$$

> Now use regular associativity for these three terms.

$$
\begin{aligned}
&= \Big(\big(C(a_1, \ldots, a_m)\big) + a_{m+1}\Big) + \big(T(a_{m+2}, \ldots, a_{n+1})\big) \\
&= \big(C(a_1, \ldots, a_{m+1})\big) + \big(T(a_{m+2}, \ldots, a_{n+1})\big)
\end{aligned}
$$

> Now treat the first sum as *one* term and use the induction hypothesis on the thus resulting $n + 1 - m \leq n$ terms.

$$
\begin{aligned}
&= C\Big(\big(C(a_1, \ldots, a_{m+1})\big), a_{m+2}, \ldots, a_{n+1}\Big) \\
&= C(a_1, \ldots, a_{n+1})
\end{aligned}
$$

∎

That proof certainly was ugly enough that we don't want to do it again. But addition of natural numbers is not the only associative operation. Multiplication is associative, too, and we will consider addition and multiplication in the context of integers, rational numbers, real numbers and complex numbers, too. Plus we have other associative operations, like ANDs and ORs in logic. Being required to do a proof as above at every stage would certainly be **cruel and unusual punishment**. Here is where the more abstract viewpoint of Section 3.5 can come to the rescue.

In summary, whereas Example 3.47 indicates that the standard way to prove summation formulas is to split off the last term and apply the induction hypothesis, the rest of the section shows that the idea for induction is similar, but more general. Any time we have a statement that involves a parameter $n \in \mathbb{N}$ and there is a way to reduce a statement for n to a statement for a smaller natural number, induction is a natural proof technique to use. The reduction can be done by splitting off a single entity (a term in a sum, a corner of an n-gon, an element of a smallest counterexample) by or by "splitting in the middle" (as done in the proof of Theorem 3.58).

The exercises in this section are primarily geared towards practicing proofs by induction. So in an exercise that uses operations that we have not formally defined yet, but which should be known, readers should use what they know to do the algebra.

Exercises

3-30. Let A be a finite set of size $m \in \mathbb{N} \setminus \{1\}$ and let $a \in A$. Prove that $\big| A \setminus \{a\} \big| = m - 1$.

   Hint. This the only exercise for this section that does not use induction. Use a bijective function $f : \{1, \ldots, m\} \to A$.

3-31. **Sums of powers**, special cases. Prove each of the following by induction.

(a) $1 + 2 + \cdots + n = \dfrac{n(n+1)}{2}$

(b) $1^3 + 2^3 + \cdots + n^3 = \dfrac{n^2(n+1)^2}{4}$

(c) $1^4 + 2^4 + \cdots + n^4 = \dfrac{n(n+1)(2n+1)\left(3n^2 + 3n - 1\right)}{30}$

Note. For an indication how formulas for the sum of the powers of the first n natural numbers can be derived, consider [30], Exercise 12-31.

3-32. Prove each of the following by induction.

(a) $1 + 3 + \cdots + (2n - 1) = n^2$

(b) $2^0 + 2^1 + \cdots + 2^n = 2^{n+1} - 1$

(c) $3 + 3^2 + \cdots + 3^n = \dfrac{3^{n+1} - 3}{2}$

(d) For all $q \in \mathbb{R} \setminus \{1\}$ we have $1 + q + q^2 + \cdots + q^n = \dfrac{1 - q^{n+1}}{1 - q}$

Note. This is the formula for **geometric sums**.

(e) $2^n > n^2$ for all $n \geq 5$

(f) $k < 3^k$ for all $k > 0$

3-33. The **Compound Interest Formula**. For this exercise, it is permissible to start induction at $n = 0$.

(a) Let $r, B \in \mathbb{R}$, $k \in \mathbb{N}$ and let $\{a_n\}_{n=0}^{\infty}$ be so that $a_0 = B$ and $a_{n+1} = \left(1 + \frac{r}{k}\right) a_n$. Prove that for all $n \in \mathbb{N}$ we have $a_n = \left(1 + \frac{r}{k}\right)^n B$. You may assume that the customary properties of the real numbers hold.

(b) Explain why the formula in part 3-33a is called the Compound Interest Formula.

3-34. **Mortgage or Loan Formulas**. Let $r, B, P \in \mathbb{R}$, $k \in \mathbb{N}$, and let $\{a_n\}_{n=0}^{\infty}$ be so that $a_0 = B$ and $a_{n+1} = \left(1 + \dfrac{r}{k}\right) a_n - P$. In financial terms, we consider the payoff of a loan (such as, for example, a mortgage or a car loan) of an initial balance of B dollars, with a fixed yearly interest rate r, and k payments per year of P dollars per payment. The number a_n is the amount left to be paid off after the nth payment has just been made. For this exercise, it is permissible to start induction at $n = 0$.

(a) Explain (in financial terms) why $a_0 = B$ and $a_{n+1} = \left(1 + \dfrac{r}{k}\right) a_n - P$.

(b) Prove that for all $n \in \mathbb{N}$ we have $a_n = \left(1 + \dfrac{r}{k}\right)^n B - P \dfrac{\left(1 + \frac{r}{k}\right)^n - 1}{\frac{r}{k}}$.

(c) Prove that for a 30 year fixed rate mortgage for an initial balance of B dollars, at a yearly interest rate r with monthly payments, the monthly payment amount is

$$P = \frac{r}{12} \frac{\left(1 + \frac{r}{12}\right)^{360} B}{\left(1 + \frac{r}{12}\right)^{360} - 1}.$$

(d) Compute the monthly payment and the total payment required to pay off a 30-year fixed rate mortgage with an initial balance of $150,000 and an interest rate of 6%.

(e) Compute the monthly payment and the total payment required to pay off a 15-year fixed rate mortgage with an initial balance of $150,000 and an interest rate of 6%.

(f) For each of the mortgages in parts 3-34d and 3-34e suppose you accelerate the payoff by making "one extra payment per year." To simplify the computation, add $\frac{1}{12}$ of a monthly payment to each of the monthly payments. Determine in each case how long it takes to pay off the mortgage with the new monthly payment. Then determine the total payment for each accelerated mortgage (that is, the sum of all monthly payments over the lifetime of the loan) and compare it to the total payment for the regularly paid mortgages.

(g) Part 3-34b shows a key strength and a key weakness of induction. Induction can show in a fairly concise manner *that* a formula is true, but it does not show how to *get* the formula. Explain how one can obtain the formula in part 3-34b by expressing a_n first in terms of a_{n-1} then in terms of a_{n-2}, and so on until you have a long sum in which the only a_j that occurs is a_0. (You'll need **"magic dots"** $\cdots$ for this part.) Collect terms and use pattern recognition and the summation formula from Exercise 3-32d to obtain the formula in part 3-34b.

3-35. **Pigeonhole Principle.** Let $m, n \in \mathbb{N}$ be so that $n > m$. Prove that there is no injective function f from $\{1, \ldots, n\}$ to $\{1, \ldots, m\}$.

Note. The inspiration for the name is the idea of having n pigeons in a roost with m holes. If $n > m$, then some holes must be occupied by more than one pigeon.

3-36. Although the base step often seems like a mere formality, it is important to prove the base step. To see this, show that for the <u>incorrect</u> formula $1 + 2 + \cdots + n = \frac{n}{2}(n+1) + \pi$ the induction step works, but (naturally) no base step can ever be proved.

3-37. Let $n_0 \in \mathbb{N}$ and let $P(n)$ be a statement for $n \in \mathbb{N}$. Prove that if $P(n_0)$ is true and truth of $P(n)$ implies truth of $P(n+1)$, then $P(n)$ is true for all $n \geq n_0$.

Hint. Consider $Q(n) := (n = 1) \vee (n = 2) \vee \cdots \vee (n = n_0 - 1) \vee P(n)$.

3-38. Let $n_0 \in \mathbb{N}$ and let $P(n)$ be a statement for $n \in \mathbb{N}$. Prove that if $P(1), \ldots, P(n_0)$ are true and, for $n > n_0$, truth of $P(1), \ldots, P(n-1)$ implies truth of $P(n)$, then $P(n)$ is true for all $n \in \mathbb{N}$.

3-39. *The various principles of induction in this section are actually equivalent.* We have already used Theorem 3.45 to prove Theorems 3.51 and 3.56. In this exercise we prove the converse directions.

(a) Use Theorem 3.51 to prove Theorem 3.45.

(b) Use Theorem 3.56 to prove Theorem 3.45.

3-40. Let $o(p_1, \ldots, p_n)$ be a compound statement that depends on the primitive propositions $p_1, \ldots, p_n$. Prove that there is a formula that represents o, which consists solely of $\wedge, \vee, \neg, TRUE, FALSE$ and $p_1, \ldots, p_n$.

Hint. In the induction step, first hold p_{n+1} constant at $TRUE$, then hold it constant at $FALSE$.

3.5   Sums and Products

Section 3.3 introduced binary operations and semigroups. The proof of Theorem 3.58 and the remarks afterwards provide good further motivation to work with abstract operations and structures: Once a result is proved for an abstract structure, such as a semigroup, it can easily be quoted and used in more concrete settings, such as natural numbers, integers, logic, etc. To do this efficiently, we must learn to think more abstractly. Ultimately it is best to just think of the abstract structures as entities in their own right. But at the beginning[6] it can be helpful to think of the results and proofs in a concrete setting, such as natural numbers. The result and proof then become abstract when we realize that we only needed to use a few properties of the concrete structure that we actually had in mind.

---

[6]And that's where we still are, even at this stage.

So, if a proof in this section feels really tough, first pretend you work with natural numbers. Then re-read the proof to ascertain that we only used the properties mentioned in the statement of the result.

Consider an arbitrary semigroup $(S, \circ)$. The whole argument from Definition 3.55 through Theorem 3.58 and its proof can be written for the semigroup S and the operation $\circ$. As such, the argument then proves that in any semigroup, the "multiplication" via the semigroup operation $\circ$ can be bracketed in any way, and the result will always be the same. In fact, we don't even need to re-state the argument.[7] We simply need to realize that, except for associativity, we did not use any special property of the natural numbers for the argument from Definition 3.55 through the proof of Theorem 3.58. This means the argument must work in an arbitrary semigroup with operation $\circ$, and we have the following theorem.

Theorem 3.59 *Let $(S, \circ)$ be a semigroup and for each $j \in \mathbb{N}$ let $a_j \in S$. Then for any multiplication $M(a_1, \ldots, a_n)$, defined similar to Definition 3.57 (see Exercise 3-41b), we have that $M(a_1, \ldots, a_n) = C(a_1, \ldots, a_n)$, the "canonical product," defined similar to Definition 3.55 (see Exercise 3-41a). Consequently, for products in semigroups, parentheses can be omitted.* ∎

It is tempting to think that, rather than formulating results for abstract objects like semigroups, it might be easier to remember the argument and, any time the result is needed, argue as above that "a similar proof will work." But that would require detailed memory of the argument, so that we can be *absolutely sure* that we did not miss anything. It is much easier to state an abstract theorem and then apply it when needed. In this fashion our brains are relieved from remembering detailed, possibly technical, arguments. Instead, we remember shorter, easier to apply, theorems.

Notation 3.60 Because the bracketing of terms in a semigroup $(S, \circ)$ does not matter, in the future we will omit parentheses when only one associative operation is involved. That is, we will write products such as $a_1 \circ a_2 \circ a_3$, etc., without parentheses.

Next, we consider a formal shortcoming of Example 3.47: We do not have a precise definition of a sum with n terms. But we need such a definition to give precise meaning to the **"magic three dots"** $\cdots$ in Example 3.47. We could argue that we can "see" what the sum should be. But, on one hand, in mathematics nothing is left between the lines.[8] On the other hand, precision is needed in applications, too. To compute a sum with a computer, we cannot use "magic three dots." Instead, we must exactly encode what terms to sum from where to where. This is where the formal definition of sums comes in.

Below, we state the definition of a sum in a semigroup $(S, +)$. In this fashion, we will not need to re-state the definition of sums when we introduce other number systems. Readers who feel more comfortable with $\mathbb{N}$ may substitute $\mathbb{N}$ for the semigroups in the following results. But they should remember that the results work in general semigroups. With regard to the notation, note that it is quite common to denote the operation in a commutative semigroup as $+$. Moreover, any time the operation is denoted

---

[7]This is good, because the author certainly does not want to, and likely neither does the reader.

[8]Where else would anyone prove that obviously superfluous parentheses can be omitted?

$+$, we will use summation notation for longer sums, independent of what we sum, be it numbers, matrices, functions or something else.

Definition 3.61 *Let $(S, +)$ be a semigroup and for each $j \in \mathbb{N}$ let $a_j \in S$. Define the* **sum**

$$\sum_{j=1}^{1} a_j := a_1$$

and for $n \in \mathbb{N}$ define the **sum**

$$\sum_{j=1}^{n+1} a_j := \sum_{j=1}^{n} a_j + a_{n+1}.$$

The parameter j is also called the **summation index**. *Sums $\sum_{j=k}^{m} a_j$ with a higher starting index $k > 1$ are defined similarly.*

Example 3.62 Evaluating sums is quite simple. The a_j typically are given by some expression. To compute the sum, we compute a_j for j going from the starting index to the finishing index and then we add the terms. For example, the sum $\sum_{j=1}^{5} j^2$ equals $\sum_{j=1}^{5} j^2 = 1^2 + 2^2 + 3^2 + 4^2 + 5^2 = 55$. □

With summation notation, we can give a "cleaner" proof of Example 3.47. Also recall that by Definition 3.43 and Proposition 3.44 we can work with fractions as long as the results are natural numbers (they are) and that, because multiplication is associative, we can leave out superfluous parentheses in products. So at this stage we can present a proof of the summation formula from Example 3.47 that is "watertight" within the framework of set theory that we are building, as well as readable from the point-of-view of calculus. The following induction proofs for sums re-emphasize the idea that, for induction with sums, we split off a term, usually the last one, and then we apply the induction hypothesis.

Example 3.63 *(Example 3.47 revisited.)* $\displaystyle\sum_{j=1}^{n} j^2 = \frac{n(n+1)(2n+1)}{6}.$

Proof. Induction on n.

Base step. For $n = 1$ note that $\displaystyle\sum_{j=1}^{1} j^2 = 1 = \frac{1(1+1)(2 \cdot 1 + 1)}{6}.$

Induction step $n \to n + 1$.

$$\begin{aligned}
\sum_{j=1}^{n+1} j^2 &= \sum_{j=1}^{n} j^2 + (n+1)^2 \\
&= \frac{n(n+1)(2n+1)}{6} + (n+1)^2 \\
&= \frac{(n+1)\big[n(2n+1) + 6(n+1)\big]}{6}
\end{aligned}$$

$$= \frac{(n+1)\left[2n^2 + 7n + 6\right]}{6}$$

$$= \frac{(n+1)\left[(n+2)(2n+3)\right]}{6}$$

$$= \frac{(n+1)\big((n+1)+1\big)\big(2(n+1)+1\big)}{6}$$

$\square$

We start our investigation of sums in general with some properties that hold in any semigroup. You may have seen the index shifting and merging of sums in Theorem 3.64 when you were computing series solutions for differential equations.

Theorem 3.64 Properties of sums. *Let $(S, +)$ be a semigroup and for all $j \in \mathbb{N}$ let $a_j \in \mathbb{N}$. Then the following hold.*

1. For all $k \in \mathbb{N}$ we have $\displaystyle\sum_{j=1}^{n} a_{j+k} = \sum_{i=k+1}^{k+n} a_i$

2. For all $m, n \in \mathbb{N}$ with $m < n$ we have $\displaystyle\sum_{j=1}^{m} a_j + \sum_{j=m+1}^{n} a_j = \sum_{j=1}^{n} a_j$

Proof. The proof is an induction on n, similar to Example 3.47. In each case, the idea is to split off the last term. For part 1, we argue as follows.

The *base step* is trivial, because $\sum_{j=1}^{1} a_{j+k} = a_{1+k} = a_{k+1} = \sum_{i=k+1}^{k+1} a_i$.

For the *induction step $n \to n + 1$*, note that by induction hypothesis we have the equality $\sum_{j=1}^{n} a_{j+k} = \sum_{i=k+1}^{k+n} a_i$. Now

$$\sum_{j=1}^{n+1} a_{j+k} = \sum_{j=1}^{n} a_{j+k} + a_{(n+1)+k} = \sum_{i=k+1}^{k+n} a_i + a_{k+(n+1)} = \sum_{i=k+1}^{k+(n+1)} a_i,$$

which proves part 1.

Part 2 is left to the reader as Exercise 3-42. ∎

By Theorem 3.59, in a semigroup we do not need to worry about parentheses that make sure each binary operation really acts on exactly two objects. We will now prove that for sums in a *commutative* semigroup, the order of the terms does not matter either and that we can combine sums. But how do we encode that the order of the terms in a sum does not matter? Reordering terms basically means that the original order (first term, second term, etc.) is changed, but that no term is omitted and no term is added or doubly listed. Formally, this means that the numbering $1, 2, \ldots$ of the terms was put into a different order, which is done with a bijective function, say, σ. The new first term is numbered $\sigma(1)$, the new second term is numbered $\sigma(2)$, etc. Bijectivity of σ assures that no terms are omitted, added or doubly listed in the reordered sum. This argument explains why we need a bijective function in part 1 of Theorem 3.65 below.

Theorem 3.65 *Let $(S, +)$ be a commutative semigroup, let $n \in \mathbb{N}$, and for each $j \in \mathbb{N}$ let $a_j, b_j \in S$.*

1. *If $\sigma : \{1, \ldots, n\} \to \{1, \ldots, n\}$ is a bijective function, then $\sum_{j=1}^{n} a_j = \sum_{j=1}^{n} a_{\sigma(j)}$.*

2. $\sum_{j=1}^{n} (a_j + b_j) = \sum_{j=1}^{n} a_j + \sum_{j=1}^{n} b_j$.

Proof. This proof of part 1 is an induction on the length n of the sum.

The *base step* for $n = 1$ is trivial, because there is only one function $\sigma : \{1\} \to \{1\}$.

For the *induction step* $n \to n + 1$, note that, by induction hypothesis, for arbitrary bijective functions $\mu : \{1, \ldots, n\} \to \{1, \ldots, n\}$ we have that $\sum_{j=1}^{n} a_j = \sum_{j=1}^{n} a_{\mu(j)}$. Now let $\sigma : \{1, \ldots, n, n + 1\} \to \{1, \ldots, n, n + 1\}$ be an arbitrary bijective function. If $\sigma(n + 1) = n + 1$, then σ maps $\{1, \ldots, n\}$ to itself. That means we can apply the induction hypothesis and obtain

$$\sum_{j=1}^{n+1} a_{\sigma(j)} = \sum_{j=1}^{n} a_{\sigma(j)} + a_{n+1} = \sum_{j=1}^{n} a_j + a_{n+1} = \sum_{j=1}^{n+1} a_j.$$

Unfortunately, $\sigma(n+1)$ need not be equal to $n+1$. In this case, splitting off the last term will not work. Instead, we split off the term for which the index is equal to $n+1$.

If $\sigma(n + 1) \leq n$, let $i \in \{1, \ldots, n\}$ be so that $\sigma(i) = n + 1$. Then, repeatedly using Theorem 3.64 and assuming that the sum that ends at $i - 1$ is not present in case $i = 1$, we obtain the following.

$$
\begin{aligned}
\sum_{j=1}^{n+1} a_{\sigma(j)} &= \sum_{j=1}^{i-1} a_{\sigma(j)} + \sum_{j=i}^{n+1} a_{\sigma(j)} \\
&= \sum_{j=1}^{i-1} a_{\sigma(j)} + \left(a_{\sigma(i)} + \sum_{j=i+1}^{n+1} a_{\sigma(j)} \right) \\
&= \sum_{j=1}^{i-1} a_{\sigma(j)} + \left(\sum_{j=i+1}^{n+1} a_{\sigma(j)} + a_{\sigma(i)} \right) \\
&= \sum_{j=1}^{i-1} a_{\sigma(j)} + \sum_{j=i}^{n} a_{\sigma(j+1)} + a_{n+1}
\end{aligned}
$$

Now let $\tau(j) := \begin{cases} \sigma(j); & \text{for } j \leq i - 1, \\ \sigma(j + 1); & \text{for } j \geq i, \end{cases}$. Then, because $\sigma|_{\{1,\ldots,i-1,i+1,\ldots,n+1\}}$ is injective and surjective onto $\{1, \ldots, n\}$, τ is a bijective function from $\{1, \ldots, n\}$ to $\{1, \ldots, n\}$.

$$= \sum_{j=1}^{i-1} a_{\tau(j)} + \sum_{j=i}^{n} a_{\tau(j)} + a_{n+1}$$

$$= \sum_{j=1}^{n} a_{\tau(j)} + a_{n+1}$$

$$= \sum_{j=1}^{n} a_j + a_{n+1}$$

$$= \sum_{j=1}^{n+1} a_j$$

Part 2 is left as Exercise 3-43. The idea is to split off the last term of the sum, as in the proof of part 1 of Theorem 3.64. ∎

Finally, note that if there is another operation that is distributive over the addition, then we can "multiply through sums."

Theorem 3.66 *Let $(S, +)$ be a semigroup and for all $j \in \mathbb{N}$ let $a_j, b_j \in \mathbb{N}$. If the binary operation $\cdot : S \times S \to S$ is distributive over $+$ and $c \in S$, then for all $n \in \mathbb{N}$ we have*

$$\sum_{j=1}^{n} c a_j = c \sum_{j=1}^{n} a_j.$$

Proof. The proof is an induction on n, similar to the proof of part 1 of Theorem 3.64. (See Exercise 3-44.) ∎

Summations allow us to define multiples of an element, too. Note that the "multiplication" in the definition below is formal notation, not actual multiplication: The natural number n need not be an element of the semigroup in question.

Definition 3.67 *Let $(S, +)$ be a semigroup, let $a \in S$ and let $n \in \mathbb{N}$. Then we define*

$$na := \sum_{j=1}^{n} a.$$

Independent of the formal problem that the semigroup may not contain $\mathbb{N}$, multiples of elements have the properties we expect them to have. Aside from establishing these properties, the proof of Theorem 3.68 is a good first example of how to use the properties of summations.

Theorem 3.68 *Let $(S, +)$ be a semigroup, let $a, b \in S$ and let $m, n \in \mathbb{N}$. Then the following hold.*

1. *$(m + n)a = ma + na$*

2. *If $+$ is commutative, then $n(a + b) = na + nb$*

Proof. Part 1 follows from Theorem 3.64 via breaking up and re-indexing the sum $\sum_{j=1}^{m+n} a$. (Specifically, because the semigroup need not contain $\mathbb{N}$, Theorem 3.66 cannot be used.)

$$\sum_{j=1}^{m+n} a = \sum_{j=1}^{m} a + \sum_{j=m+1}^{m+n} a = \sum_{j=1}^{m} a + \sum_{j=1}^{n} a = ma + na.$$

Part 2 follows from part 2 of Theorem 3.65.

$$n(a+b) = \sum_{j=1}^{n}(a+b) = \sum_{j=1}^{n} a + \sum_{j=1}^{n} b = na + nb.$$

■

Similar to sums, we can define products.

Definition 3.69 *Let $(S, \cdot)$ be a semigroup and for each $j \in \mathbb{N}$ let $a_j \in S$. (For products, it is customary to not demand commutativity. There are product-like operations, such as composition of functions, that are not commutative.) Define the* **product**

$$\prod_{j=1}^{1} a_j := a_1$$

and for $n \in \mathbb{N}$ define the **product**

$$\prod_{j=1}^{n+1} a_j := \prod_{j=1}^{n} a_j \cdot a_{n+1}.$$

The parameter j is also called the **product index**.

Of course, at this stage, the only difference between the definitions of sums and products is the notation. But we have seen that there are structures, such as the natural numbers, that are endowed with an addition operation *and* a multiplication operation. Hence it makes sense to have separate notations. But the fact that, for abstract binary operations, sums and products are "the same idea" implies that similar results must hold. For example, Theorems 3.64 and 3.65 can be proved for multiplicative notation in Exercise 3-46. Moreover, just as repeated additions are abbreviated as positive integer multiples, repeated multiplications are abbreviated as positive integer powers.

Definition 3.70 *Let $(S, \cdot)$ be a semigroup, let $a \in S$ and let $n \in \mathbb{N}$. Then we define*

$$a^n := \prod_{j=1}^{n} a.$$

Theorem 3.71 *Let $(S, \cdot)$ be a semigroup, let $a, b \in S$ and let $m, n \in \mathbb{N}$. Then the following hold.*

1. *$a^{m+n} = a^m \cdot a^n$*

2. *If $a \cdot b = b \cdot a$, then $(a \cdot b)^n = a^n \cdot b^n$*

3. *$\left(a^m\right)^n = a^{mn}$*

Proof. See Exercise 3-47. ∎

Viewed as formal entities, sums and products are very similar. But they are quite different in practice. When two operations like addition and multiplication are available, then the multiplication, typically denoted $\cdot$, is usually distributive over the addition, typically denoted $+$.[9] This means that powers of the form $(a + b)^n$ can be "multiplied out." We will devote the rest of this section to this process, culminating in the proof of the Binomial Theorem. Note that because we prove the result for semigroups with the right properties, we will not need to re-prove the Binomial Theorem for the other number systems that we will introduce.

In some computations for natural numbers, we need an element 0, which is so that $0 + n = n + 0 = n$ for all $n \in \mathbb{N} \cup \{0\}$. Philosophically as well as historically, it can be argued that zero is not a natural number. After all, the philosophy can be that people typically count "1, 2, 3, . . ." and not "0, 1, 2, 3, . . .," whereas the history is that it took around a thousand years from the times of the ancient Greeks until the idea of a number zero made it from India through Arabia to Europe. Nonetheless, at times it is convenient to have such a null element.

There are two ways to introduce zero. One way would be to go back to the axioms of the natural numbers and call the first element "0" instead of "1." We could define the arithmetic so that $n + 0 = n, n + m' = (m + n)'$ and $n \cdot 0 = 0, n \cdot m' = n \cdot m + n$. This arithmetic works just like the one we have constructed. Moreover, $1 := 0'$ has the same properties as the element 1 that starts the natural numbers as constructed in Sections 2.6 and 3.1. Of course, this would require that we go through all requisite earlier proofs once more. It is therefore easier and more appropriate to do the following.

Definition 3.72 *Let 0 be the element from Definition 3.53[10] and consider the set $\mathbb{N}_0 := \mathbb{N} \cup \{0\}$. For all $n \in \mathbb{N}_0$ we define $0 + n := n + 0 := n$, $0 \cdot n := n \cdot 0 := 0$ and for $n, m \in \mathbb{N}$ we define the operations $+$ and $\cdot$ as before. Moreover, we set $n - 0 := n$, $n - n := 0$ and $n^0 := 1$. A sum $\sum_{j=n}^{m} a_j$ of natural numbers so that $m < n$ is set equal to 0 and it is called an **empty sum**. Finally, in a semigroup $(S, \cdot)$ with neutral element 1 we set $a^0 := 1$ for all $a \in S$.*

Definition 3.72 provides us with a new number system $\mathbb{N}_0$ with commutative and associative operations $+$ and $\cdot$ (see Exercise 3-49). More importantly, the Principle of Induction holds in the following form.

---

[9]In fact, we are so accustomed to this notation that switching the roles of addition and multiplication looks downright strange (see Exercise 3-48).

[10]In this fashion everything is consistent with how we expect it to be.

Theorem 3.73 Principle of Induction *for* $\mathbb{N}_0$. *Let* $P(n)$ *be a statement about* $n \in \mathbb{N}_0$. *If* $P(0)$ *is true and if for all* $n \in \mathbb{N}_0$ *truth of* $P(n)$ *implies truth of* $P(n+1)$, *then* $P(n)$ *holds for all* $n \in \mathbb{N}_0$.

Proof. By assumption, $P(0)$ is true and hence $P(1) = P(0+1)$ is true. Thus, by Theorem 3.45 $P(n)$ is true for all $n \in \mathbb{N}$. Because $P(n)$ holds for 0, too, $P(n)$ is true for all $n \in \mathbb{N}_0$. ∎

The above properties show that $\mathbb{N}_0$ acts exactly as we expect it to act. Next, we need binomial coefficients.

Definition 3.74 *For all* $n \in \mathbb{N}$, *we define*

$$n! := \prod_{j=1}^{n} j$$

and call it the **factorial** *of* n. *We also set* $0! := 1$. *For all* $n, k \in \mathbb{N}_0$ *with* $k \leq n$, *we define the* **binomial coefficient** *as*

$$\binom{n}{k} := \frac{n!}{k!(n-k)!}.$$

Note that there may be a problem with the definition of binomial coefficients: *A priori*, we do not know if the quotients in the definition are natural numbers or not. Of course it turns out that they are, as the following lemma shows. We should not be bothered that we justify the definition "after the fact." It is not uncommon to define a quantity and explain afterwards why the definition is allowed.

Lemma 3.75 *For all* $n, k \in \mathbb{N}_0$ *the quotient* $\binom{n}{k}$ *exists in* $\mathbb{N}$. *That is, in the terms of Definition 3.43, there is a number* q *so that* $n! = qk!(n-k)!$.

Proof. First note that for all $n \in \mathbb{N}$ we have that $\binom{n}{0} = \binom{n}{n} = \frac{n!}{0!n!} = 1$. Thus we only need to consider $\binom{n}{k}$ for $k \in \{1, \ldots, n-1\}$. For these values of k, the result is proved by induction on n.

The *base step* for $n = 1$ is simply $\binom{1}{0} = \binom{1}{1} = \frac{1!}{0!1!} = 1$.

For the *induction step* $n \to n+1$ we will use part 1 of Proposition 3.44. Let the number $k \in \{1, \ldots, n\}$ be arbitrary but fixed. By induction hypothesis, both $\binom{n}{k-1}$ and $\binom{n}{k}$ are natural numbers. But then

$$
\begin{aligned}
\binom{n}{k-1} + \binom{n}{k} &= \frac{n!}{(k-1)!\bigl(n-(k-1)\bigr)!} + \frac{n!}{k!(n-k)!} \\
&= \frac{n!k}{k!(n+1-k)!} + \frac{n!(n+1-k)}{k!(n+1-k)!} = \frac{n!(k+n+1-k)}{k!(n+1-k)!} \\
&= \frac{(n+1)!}{k!(n+1-k)!} = \binom{n+1}{k},
\end{aligned}
$$

Figure 3.3 Illustration of Theorem 3.76. The formula in Theorem 3.76 says that, except for the ones, binomial coefficients in a row of the triangular setup in the figure are obtained by adding the adjacent binomial coefficients from the row above.

and hence $\binom{n+1}{k}$ must be a natural number, too. Because k was an arbitrary element of $\{1, \ldots, n\}$, this completes the induction step. Hence the result is proved. ∎

Lemma 3.75 does more than just assure that binomial coefficients are indeed natural numbers. The proof of Lemma 3.75 contains a way to compute binomial coefficients for $n + 1$ from binomial coefficients for n. This process is known by a more familiar name.

Theorem 3.76 Pascal's Triangle. *The equation*

$$\binom{n}{k-1} + \binom{n}{k} = \binom{n+1}{k}$$

holds for all $n, k \in \mathbb{N}$ with $k \leq n$. (For a visualization, see Figure 3.3.)

Proof. This result is proved by the computation in the proof of Lemma 3.75. ∎

Now we are ready to prove the Binomial Theorem. Visualization of numbers rather than an abstract semigroup with another operation is admissible as long as we make sure that we only use what is given in the hypothesis.

Theorem 3.77 *The **Binomial Theorem**. Let $(S, +)$ be a commutative semigroup and let $\cdot$ be an associative, commutative operation that is distributive over $+$ and that has a neutral element. Then for all elements $a, b \in S$, and all natural numbers $n \in \mathbb{N}$, we have*

$$(a + b)^n = \sum_{k=0}^{n} \binom{n}{k} a^k b^{n-k}.$$

Proof. The proof is by induction on n, with $P(n)$ being the statement about $(a+b)^n$. *Base step.* For $n = 1$, note that

$$(a + b)^1 = a + b = \binom{1}{0} a^0 b^{1-0} + \binom{1}{1} a^1 b^{1-1} = \sum_{k=0}^{1} \binom{1}{k} a^k b^{1-k},$$

which proves the base step.

Induction step. Assuming that the result holds for n, we must prove it for $n + 1$.

$(a + b)^{n+1}$

The first step is to split off the last term of the power.

$= (a + b)(a + b)^n$

Now we can apply the induction hypothesis to $(a + b)^n$.

$$= (a + b) \sum_{k=0}^{n} \binom{n}{k} a^k b^{n-k}$$

Multiply out the parentheses and absorb a and b into the sums.

$$= \sum_{k=0}^{n} \binom{n}{k} a^{k+1} b^{n-k} + \sum_{k=0}^{n} \binom{n}{k} a^k b^{n+1-k}$$

After multiplying out parentheses, we want to combine the sums. To do this, we shift the indices to obtain similar terms in both sums. In the first sum we set $j := k + 1$ and in the second sum we set $j := k$.

$$= \sum_{j=1}^{n+1} \binom{n}{j-1} a^j b^{n+1-j} + \sum_{j=0}^{n} \binom{n}{j} a^j b^{n+1-j}$$

To combine the sums, the indices must start and end at the same numbers. Thus we split off the last term of the first sum and the first term of the second sum. Then we can combine the sums.

$$= \binom{n}{n} a^{n+1} b^{n+1-(n+1)} + \binom{n}{0} a^0 b^{n+1-0} + \sum_{j=1}^{n} \left[\binom{n}{j-1} + \binom{n}{j} \right] a^j b^{n+1-j}$$

Now we can apply Theorem 3.76. Moreover, by rewriting the terms outside the sum, we see that they fit the requisite pattern and can be absorbed into the sum.

$$= \binom{n+1}{n+1} a^{n+1} b^{n+1-(n+1)} + \binom{n+1}{0} a^0 b^{n+1-0} + \sum_{j=1}^{n} \binom{n+1}{j} a^j b^{n+1-j}$$

$$= \sum_{j=0}^{n+1} \binom{n+1}{j} a^j b^{n+1-j}. \qquad \blacksquare$$

Exercises

3-41. **Products in semigroups.** Let $(S, \circ)$ be a semigroup.

   (a) Translate Definition 3.55 into the definition of a "canonical product" in a semigroup.

(b) Translate Definition 3.57 into the definition of an "arbitrarily bracketed product" in a semi-group.

3-42. Prove part 2 of Theorem 3.64.

3-43. Prove part 2 of Theorem 3.65.

3-44. Prove Theorem 3.66.

3-45. Prove that for all $n \in \mathbb{N}$ we have $\sum_{j=1}^{n} 1 = n$.

3-46. Let $(S, \cdot)$ be a semigroup, let $a, b \in S$, $m, n \in \mathbb{N}$ and for all $j \in \mathbb{N}$ let $a_j, b_j \in S$

(a) Prove that $\prod_{j=1}^{n} a_{j+k} = \prod_{i=k+1}^{k+n} a_i$

(b) Prove that for all $m < n$ we have $\prod_{j=1}^{m} a_j \cdot \prod_{j=m+1}^{n} a_j = \prod_{j=1}^{n} a_j$.

(c) Prove that if $\cdot$ is commutative and $\sigma : \{1, \ldots, n\} \to \{1, \ldots, n\}$ is a bijective function, then
$$\prod_{j=1}^{n} a_j = \prod_{j=1}^{n} a_{\sigma(j)}.$$

(d) Prove that if $\cdot$ is commutative, then $\prod_{j=1}^{n} (a_j \cdot b_j) = \left(\prod_{j=1}^{n} a_j \right) \cdot \left(\prod_{j=1}^{n} b_j \right)$.

3-47. Proving Theorem 3.71. In each of the parts below, feel free to use the results from Exercise 3-46.

(a) Prove part 1 of Theorem 3.71.

(b) Prove part 2 of Theorem 3.71.

(c) Prove part 3 of Theorem 3.71.

3-48. State the equations that encode the phrase "The operation $+$ is distributive over the operation $\cdot$." Do these equations look natural?

3-49. Properties of $+$ and $\cdot$ on $\mathbb{N}_0$.

(a) Prove that $+$ is associative on $\mathbb{N}_0$.

(b) Prove that $+$ is commutative on $\mathbb{N}_0$.

(c) Prove that $\cdot$ is associative on $\mathbb{N}_0$.

(d) Prove that $\cdot$ is commutative on $\mathbb{N}_0$.

(e) Prove that $+$ is distributive over $\cdot$ on $\mathbb{N}_0$.

3-50. For $a, b \in \mathbb{N}$ define $a * b := a^b$.

(a) Prove that $*$ is not associative.

(b) Prove that $*$ is not commutative.

(c) Prove that $*$ is not distributive over $+$.

(d) Prove that $+$ is not distributive over $*$.

3-51. Prove each of the following by induction.

(a) $\sum_{j=1}^{n} (2j - 1) = n^2$

(b) $\displaystyle\sum_{j=0}^{n} 2^j = 2^{n+1} - 1$

(c) $\displaystyle\sum_{j=1}^{n} 3^j = \frac{3^{n+1} - 3}{2}$

3-52. **Sums of powers**, special cases. Prove each of the following by induction.

(a) $\displaystyle\sum_{j=1}^{n} j = \frac{n(n + 1)}{2}$

(b) $\displaystyle\sum_{j=1}^{n} j^3 = \frac{n^2(n + 1)^2}{4}$

(c) $\displaystyle\sum_{j=1}^{n} j^4 = \frac{n(n + 1)(2n + 1)\left(3n^2 + 3n - 1\right)}{30}$

Note. For an indication how formulas for the sum of the powers of the first n natural numbers can be derived, consider [30], Exercise 12-31.

3-53. **Sums of powers**, general case. For $p, n \in \mathbb{N}$ let $\displaystyle S_p(n) := \sum_{j=1}^{n} j^p$.

(a) Prove by induction on n that $\displaystyle S_p(n) = \frac{1}{p+1}\left(n(n+1)^p - \sum_{j=1}^{p-1}\binom{p}{j-1} S_j(n)\right)$.

(b) Derive a formula that does not involve summations for $\displaystyle S_5(n) = \sum_{j=1}^{n} j^5$.

3-54. Let A be a finite set. Prove that for every $k \in \{0, \ldots, |A|\}$, A has $\displaystyle\binom{|A|}{k}$ distinct subsets with k elements.

3-55. Prove each of the following. You may assume that the customary properties of the real numbers hold.

(a) For all $q \in \mathbb{R} \setminus \{1\}$ we have $\displaystyle\sum_{j=1}^{n} q^j = \frac{q - q^{n+1}}{1 - q}$.

Note. This is the formula for **geometric sums**.

(b) $\displaystyle\sum_{j=1}^{n} \frac{1}{j} - \frac{1}{j+1} = 1 - \frac{1}{n+1}$,

(c) For $x, y \in \mathbb{R}$ with $x \neq y$ we have $\displaystyle\sum_{j=1}^{n} x^{n-j} y^{j-1} = \frac{x^n - y^n}{x - y}$.

(d) For all $x \in \mathbb{R} \setminus \{1\}$ and all $n \geq 0$ we have $\displaystyle\prod_{j=0}^{n}\left(1 + x^{2^j}\right) = \frac{1 - x^{2^{n+1}}}{1 - x}$.

(e) $\displaystyle\prod_{j=1}^{n-1}\left(1 + \frac{1}{j}\right) = n$

(f) $\displaystyle\prod_{j=1}^{n-1}\left(1 + \frac{1}{j}\right)^j = \frac{n^n}{n!}$

(g) $\displaystyle\sum_{j=1}^{2^n} \frac{1}{j} \geq 1 + \frac{n}{2}$

(h) **Bernoulli's Inequality.** For all real numbers $x > -1$, $x \neq 0$ and $n \geq 2$ we have that $(1+x)^n > 1 + nx$.

(i) If $p \geq 2$ then $p^n > n$ for all $n \geq 1$.

(j) For $-1 \leq x \leq 1$ and all $k \in \mathbb{N}$, we have $(1+x)^k \leq 1 + kx + 3^k x^2$.

3-56. **Binomial Formulas.** Let $(S, +)$ be a commutative semigroup and let $\cdot$ be an associative, commutative operation that is distributive over $+$. Prove each of the following. Assume that subtraction is defined as in Definition 3.41 and that it has the properties from Proposition 3.42.

(a) $(a+b)^2 = a^2 + 2ab + b^2$

(b) $(a-b)^2 = a^2 - 2ab + b^2$

(c) $(a+b)(a-b) = a^2 - b^2$

3-57. Let $(S, +)$ be a commutative semigroup, let $\cdot : S \times S \to S$ be an associative, commutative binary operation that is distributive over $+ : S \times S \to S$, and that has a neutral element, and let $n \in \mathbb{N}$. With subtraction defined as in Definition 3.41 and with the properties from Proposition 3.42, prove that $x^n - y^n = (x - y) \displaystyle\sum_{j=0}^{n-1} x^j y^{n-1-j}$.

3.6 Divisibility

After the repeated excursions into abstract realms of operations and semigroups, let us spend this section with the very concrete concept of divisibility. Formal quotients have already been introduced in Definition 3.43, because we needed them for binomial coefficients. The following definition is more customary when talking about dividing natural numbers.

Definition 3.78 *Let $n \in \mathbb{N}_0$ and let $d \in \mathbb{N}$. Then we say that d **divides** n, or, n is **divisible** by d iff there is a $k \in \mathbb{N}_0$ so that $n = kd$. If this is the case we also write $d \mid n$ and if $d \neq 1$ we call d a **factor** of n.*

Numbers that can only be divided by 1 and by themselves have a special place in the discussion of divisibility.

Definition 3.79 *Let $p \in \mathbb{N} \setminus \{1\}$. Then p is called **prime** or a **prime number** iff the only numbers that divide p are 1 and p. Numbers that are not prime numbers are called **composite numbers**.*

Prime numbers play an important role in public key encryption (see Section 3.9). But they have been interesting to people at least since the times of the ancient Greeks. In fact, the simplest algorithm for computing prime numbers is named after a Greek philosopher.

Example 3.80 The **sieve of Eratosthenes** is an elementary algorithm that computes all prime numbers up to a given threshold t. Starting with 2, for every $k \in \{2, \ldots, t\}$, we try to divide k by every prime number found so far. If k can be divided by one of these prime numbers, k is not prime. Otherwise, add k to the list of prime numbers.

k	divisions	result
2	none available	2 is prime
3	$2 \nmid 3$	3 is prime
4	$2 \mid 4$	4 is not prime
5	$2 \nmid 5, 3 \nmid 5$	5 is prime
6	$2 \mid 6$	6 is not prime
7	$2 \nmid 7, 3 \nmid 7, 5 \nmid 7$	7 is prime
8	$2 \mid 8$	8 is not prime
9	$2 \nmid 9, 3 \mid 9$	9 is not prime

Of course the algorithm can be improved. For example, we call a number **even** iff it is divisible by 2 and we call it **odd** otherwise. It is trivial that even numbers greater than 2 are not prime. Moreover, when we check for divisors of k, we do not need to check all prime numbers smaller than k. It suffices to check all prime numbers p with $p^2 \le k$. Finally, for hand computation, we can shorten the notation in the sieve of Eratosthenes to 2, 3, 5, 7, ~~9~~, 11, 13, ~~15~~, 17, 19, ~~21~~, 23, ~~25~~, ~~27~~, 29, 31, ~~33~~, ...[11]

The above improves the manual execution of the algorithm, as Exercise 3-58 shows. But Exercise 3-58 also shows that the sieve of Eratosthenes requires considerable effort. So to compute large prime numbers, say, with fifty digits, much more efficient algorithms are needed. For more on primality testing and the very important topic of integer factorization, the reader could consider, for example, [33]. □

With prime numbers, at least in principle, accessible, we can focus on why all natural numbers are composites of prime numbers. The product notation from Definition 3.69 is very helpful here.

Theorem 3.81 Fundamental Theorem of Arithmetic *or* **Unique Prime Factorization Theorem** *(existence and uniqueness of a* **prime factorization***). Let $n \in \mathbb{N}$. Then either $n = 1$ or there are unique distinct prime numbers $p_1, \ldots, p_k$ and unique (not necessarily distinct) exponents $q_1, \ldots, q_k \in \mathbb{N}$ so that*

$$n = \prod_{j=1}^{k} p_j^{q_j}.$$

Proof. The proof of the existence of a prime factorization is a (strong) induction on n.

Base step. Trivial. If $n = 1$, then $n = 1$.

Induction step. Let $n \in \mathbb{N}$ and assume that the existence of a prime factorization has been proved for all $m \in \{2, \ldots, n - 1\}$. If n is a prime number, choose $k = 1$, $p_1 = n$, $q_1 = 1$. If n is a composite number, then there is a prime number p so that $n = mp$ with $m \in \{2, \ldots, n - 1\}$. But then by induction hypothesis, there are prime numbers $p_1, \ldots, p_k$ and exponents $q_1, \ldots, q_k \in \mathbb{N}$ so that $m = \prod_{j=1}^{k} p_j^{q_j}$. Hence $n = p \prod_{j=1}^{k} p_j^{q_j}$ and the extra factor p is either equal to some p_j, in which case we add 1 to the exponent q_j, or it is not, in which case we set $p_{k+1} := p$ and $q_{k+1} := 1$.

---

[11] But remember that even though the notation makes things more efficient for humans, a computer program's efficiency would not be improved by our change in notation.

With existence established, the proof of the uniqueness of a prime factorization is another (strong) induction on n.

Base step. Trivial. If $n = 1$, then $n = 1$.

Induction step. Let $n \in \mathbb{N}$ and assume that uniqueness of the prime factorization has been proved for all $m \in \{2, \ldots, n-1\}$. In case n is a prime number, the uniqueness of the factorization is trivial. So we can assume that n is a composite number. Suppose for a contradiction that there are a number $k \in \mathbb{N}$ and prime numbers $p_1, \ldots, p_k$ and exponents $q_1, \ldots, q_k \in \mathbb{N}$ so that $n = \prod_{j=1}^{k} p_j^{q_j}$ as well as a number $\hat{k} \in \mathbb{N}$ and prime numbers $\hat{p}_1, \ldots, \hat{p}_{\hat{k}}$ and exponents $\hat{q}_1, \ldots, \hat{q}_{\hat{k}} \in \mathbb{N}$ so that $n = \prod_{j=1}^{\hat{k}} \hat{p}_j^{\hat{q}_j}$. Because p_1 divides $n = \prod_{j=1}^{\hat{k}} \hat{p}_j^{\hat{q}_j}$, p_1 must divide one of the $\hat{p}_j$ (see Exercise 3-59c). Hence we can assume without loss of generality that $p_1 = \hat{p}_1$. But then

$$\hat{p}_1 p_1^{q_1-1} \prod_{j=2}^{k} p_j^{q_j} = p_1 p_1^{q_1-1} \prod_{j=2}^{k} p_j^{q_j} = \prod_{j=1}^{k} p_j^{q_j} = n$$

$$= \prod_{j=1}^{\hat{k}} \hat{p}_j^{\hat{q}_j} = \hat{p}_1 \hat{p}_1^{\hat{q}_1-1} \prod_{j=2}^{\hat{k}} \hat{p}_j^{\hat{q}_j},$$

which implies that

$$p_1^{q_1-1} \prod_{j=2}^{k} p_j^{q_j} = \hat{p}_1^{\hat{q}_1-1} \prod_{j=2}^{\hat{k}} \hat{p}_j^{\hat{q}_j}.$$

But then by induction hypothesis, because the products are in $\{2, \ldots, n-1\}$ we conclude that the two factorizations must be the same. That is, $p_1 = \hat{p}_1$ (which we already knew), $q_1 = \hat{q}_1$, $k = \hat{k}$ and there is a bijective function $\sigma : \{2, \ldots, k\} \to \{2, \ldots, k\}$ so that for all $j \in \{2, \ldots, k\}$ we have $p_j = \hat{p}_{\sigma(j)}$ and $q_j = \hat{q}_{\sigma(j)}$. ∎

The proof of Theorem 3.81 shows that the uniqueness of the prime factorization is "uniqueness up to the order of the factors." This makes sense, because multiplication is commutative. Also note that, just as in Theorem 3.65, a bijective function is used to model the "scrambling" of the factors.

Standard Proof Technique 3.82 Just as in the proof of Theorem 3.65, existence and uniqueness are typically proved separately. □

Aside from the decomposition of numbers into prime factors, we are also interested in the largest factor that divides two numbers. This factor will be needed in Section 3.9. Moreover, it is related to the smallest number that is divisible by two numbers (see part 3 of Theorem 3.85 below), which is the "least common denominator" from the addition of fractions.

Definition 3.83 *Let $m, n \in \mathbb{N}$. The **greatest common divisor** (m, n) of m and n is the largest $d \in \mathbb{N}$ so that $d \mid m$ and $d \mid n$. The **least common multiple** $[m, n]$ of m and n is the smallest $c \in \mathbb{N}$ so that $m \mid c$ and $n \mid c$.*

For sufficiently simple numbers, the greatest common divisor and the least common multiple are easy to compute.

Example 3.84 Consider the numbers 2064 and 204. Because $2064 = 2^4 \cdot 3^1 \cdot 43^1$ and $204 = 2^2 \cdot 3^1 \cdot 17^1$, the greatest common divisor is $(2064, 204) = 2^2 \cdot 3^1 = 12$, and the least common multiple is $[2064, 204] = 2^4 \cdot 3^1 \cdot 17^1 \cdot 43^1 = 35088$. $\square$

In general, both the greatest common divisor as well as the least common multiple must exist for any two numbers: There are finitely many divisors of two numbers and picking the largest one of the set proves existence of the greatest common divisor. Moreover, by Theorem 3.51, every nonempty set of natural numbers has a smallest element. Thus the set of common multiples of two numbers has a smallest element, too. **Note that it is common practice to weave short arguments into the narrative of a mathematical presentation. This practice is preferable over explicitly stating simple observations as theorems or propositions.**

The next result gives a more concrete representation of the greatest common divisor and the least common multiple. Moreover, it shows that if we can compute one, then we can compute the other, too.

Theorem 3.85 Let $m, n \in \mathbb{N}$ and let $p_1, \ldots, p_k$ be the prime numbers that occur in the prime factorizations of m and n. Let $m = \prod_{j=1}^{k} p_j^{a_j}$ and $n = \prod_{j=1}^{k} p_j^{b_j}$, where a_j and b_j could be zero, too. Then the following hold.

1. $(m, n) = \prod_{j=1}^{k} p_j^{\min\{a_j, b_j\}}$, where $\min\{a_j, b_j\}$ is the smaller of the two numbers.

2. $[m, n] = \prod_{j=1}^{k} p_j^{\max\{a_j, b_j\}}$, where $\max\{a_j, b_j\}$ is the larger of the two numbers.

3. $mn = (m, n)[m, n]$

Proof. To prove part 1, first note that $\prod_{j=1}^{k} p_j^{\min\{a_j, b_j\}}$ clearly divides both m and n. Now let d be a common divisor of m and n. Then the prime factors of d must be among $p_1, \ldots, p_k$, so that $d = \prod_{j=1}^{k} p_j^{c_j}$. But if any c_j was greater than $\min\{a_j, b_j\}$, then d would be divisible by a higher power of p_j than m or n, which is not possible (see Exercise 3-61c). Hence for all $j \in \{1, \ldots, k\}$ we have $c_j \leq \min\{a_j, b_j\}$, which implies that $d = \prod_{j=1}^{k} p_j^{c_j} \leq \prod_{j=1}^{k} p_j^{\min\{a_j, b_j\}}$ and hence $(m, n) = \prod_{j=1}^{k} p_j^{\min\{a_j, b_j\}}$.
Parts 2 and 3 are left to Exercise 3-62.   ∎

Although Theorem 3.85 provides concrete representations of the greatest common divisor and the least common multiple, its computational use is limited (see also Exercise 3-63). This is because Theorem 3.85 assumes that we already have the prime factorization of the two numbers, and because it turns out that computing prime factorizations is very hard indeed. In fact, computing prime factorizations is so hard that the difficulty of this computation is the very reason why public key encryption is considered to be secure (see Section 3.9).

The strength of Theorem 3.85 is actually the very unimpressive looking part 3. Part 3 shows that if we have the greatest common divisor or the least common multiple, then we can compute the other one with a simple division. Euclid's Algorithm (see Theorem 4.51) will ultimately provide a fast way to compute the greatest common divisor.

Now that we have analyzed divisions that "work out," let us consider division with remainder. The next result is called the division algorithm, because the proof indeed provides an algorithm to compute quotient and remainder.

Theorem 3.86 *The **division algorithm**. Let $n \in \mathbb{N}_0$ and let $d \in \mathbb{N}$. Then there are unique numbers $q \in \mathbb{N}_0$ and $r \in \{0, \ldots, d-1\}$ so that $n = qd + r$. The number q is also called the **quotient** and the number r is also called the **remainder**.*

Proof. The proof that the numbers exist is a (strong) induction on n with an "extended base step" as presented in Exercise 3-38.

Base step. For $n \in \{0, \ldots, d-1\}$, the numbers $q := 0$ and $r := n$ are as desired.

Induction step. For $n \geq d$ note that $n - d \in \mathbb{N}_0$. By induction hypothesis, there are numbers $\tilde{q} \in \mathbb{N}_0$ and $\tilde{r} \in \{0, \ldots, d-1\}$ so that $n - d = \tilde{q}d + \tilde{r}$. Now $q := \tilde{q} + 1$ and $r := \tilde{r}$ are as desired.

For uniqueness, let $n = qd + r = \hat{q}d + \hat{r}$. Without loss of generality suppose that $r \leq \hat{r}$. Then $qd = \hat{q}d + \hat{r} - r$. Now suppose for a contradiction that $r \neq \hat{r}$. Then $r < \hat{r}$ and because $qd = \hat{q}d + \hat{r} - r$ we conclude (see Exercise 3-64b) that $\hat{r} - r \in \{1, \ldots, d-1\}$ is divisible by d. This is a contradiction, so $r = \hat{r}$. Therefore, by Proposition 3.19 we have $qd = \hat{q}d$ and then by Exercise 3-10b we obtain $q = \hat{q}$. Thus the numbers q and r are unique. ∎

Theorem 3.86 is called the division algorithm, because the existence proof provides a procedure to obtain q and r.[12] Repeatedly subtracting just the denominator to ultimately obtain the quotient is less efficient than the long division algorithm we typically use to divide numbers. But the idea is the same in both approaches. In the usual long division algorithm, we repeatedly subtract *multiples* of the denominator d from the numerator n until we have a remainder r left that is in $\{0, \ldots, d-1\}$. The usual long division algorithm works so well, because we take advantage of place value notation for natural numbers.

Example 3.87 *Compute the quotient and remainder obtained when 4736 is divided by 17.*

The left column below shows the customary division algorithm for natural numbers. The right column indicates the numbers that are actually subtracted. Clearly the idea is the same as for the proof of Theorem 3.86: We subtract something to reduce the problem to a division of a smaller number by the same divisor. But when we actually divide numbers in practice, we try to subtract as large a number as possible in each step, because anything less would drag out the process.

---

[12]To import and scale graphics, such as the graphs in Figure 2.13 on page 82, or in Figure 6.1 on page 253, the author needed a "quick and dirty" division routine. The one he wrote uses the algorithm from the proof of Theorem 3.86.

"Regular" long division.   Actual subtractions.

$$
\begin{array}{r|rrrr}
 & & 2 & 7 & 8 \\
17 & 4 & 7 & 3 & 6 \\
(-) & & 3 & 4 \\
\hline
 & & 1 & 3 & 3 \\
(-) & & 1 & 1 & 9 \\
\hline
 & & & 1 & 4 & 6 \\
(-) & & & 1 & 3 & 6 \\
\hline
 & & & & 1 & 0
\end{array}
$$

$$
\begin{array}{r|rrrrl}
 & & 2 & 7 & 8 \\
17 & 4 & 7 & 3 & 6 \\
(-) & & 3 & 4 & 0 & 0 & = 200 \cdot 17 \\
\hline
 & & 1 & 3 & 3 & 6 \\
(-) & & 1 & 1 & 9 & 0 & = 70 \cdot 17 \\
\hline
 & & & 1 & 4 & 6 \\
(-) & & & 1 & 3 & 6 & = 8 \cdot 17 \\
\hline
 & & & & 1 & 0
\end{array}
$$

So the quotient is 278 and the remainder is 10, or, in the form of Theorem 3.86, $4736 = 278 \cdot 17 + 10$. □

We all know how Example 3.87 works, but we have not formally defined the place value of digits yet. Therefore, we conclude this section with an exploration of place value notation. In base 10 notation, the rightmost digit has a place value of $1 = 10^0$, the digit to its left has a place value of $10 = 10^1$, and the digit that is k places to the left has a place value of 10^{k-1}. A number is actually decoded by multiplying each digit with its place value. This description may sound horribly abstract. But it only feels that way because we are very much accustomed to place 10 arithmetic. We automatically know that the "tens digit" carries more value than the "ones digit," etc.

However, the abstract description helps if we are told that a number, say 1011, is given in another base, say 2. Just like we know that in base 10 notation we have $1011 = 1 \times 10^3 + 0 \times 10^2 + 1 \times 10^1 + 1 \times 10^0$, we can make a mental switch to realize that in base 2 notation we have $1011 = 1 \times 2^3 + 0 \times 2^2 + 1 \times 2^1 + 1 \times 2^0$, which would be equal to 11 in base 10 notation.

Converting a number n from base 10 to another base is harder. First note that if $n = qb + r$ as in Theorem 3.86, then the remainder r is the rightmost digit d_0 of the base b representation of the number. Moreover, qb is "the rest" of the number. That is, the base b representation of q gives us the digits of n that are to the left of d_0. So, repeating this process gives the base b representation of n.

Example 3.88 Convert the number $d = 181$ from its base 10 representation as given to its base 2 representation.

1. $181 = 2 \cdot 90 + 1$, so the rightmost digit is $d_0 = 1$.

2. $90 = 2 \cdot 45 + 0$, so the next digit to the left is $d_1 = 0$.

3. $45 = 2 \cdot 22 + 1$, so the next digit to the left is $d_2 = 1$.

4. $22 = 2 \cdot 11 + 0$, so the next digit to the left is $d_3 = 0$.

5. $11 = 2 \cdot 5 + 1$, so the next digit to the left is $d_4 = 1$.

6. $5 = 2 \cdot 2 + 1$, so the next digit to the left is $d_5 = 1$.

7. $2 = 2 \cdot 1 + 0$, so the next digit to the left is $d_6 = 0$.

8. $1 = 2 \cdot 0 + 1$, so the next digit to the left is $d_7 = 1$.

So 181 (base 10) = 1011 0101 (base 2). This claim is easily verified, because we have $2^7 + 2^5 + 2^4 + 2^2 + 2^0 = 128 + 32 + 16 + 4 + 1 = 181$. $\qquad\square$

The above approach can easily be turned into an algorithm that, for any number and any base $b > 1$, produces the base b representation. But we can see once more that, when an intuitive idea is made precise, the actual result can be quite complex.

Theorem 3.89 Place value in base b. *Let* $b \in \mathbb{N} \setminus \{1\}$ *be fixed and let* $d \in \mathbb{N}$. *Recursively define* d_n *and* q_n *as follows. For* $n = 0$ *let* q_0 *and* d_0 *be the quotient and the remainder of the division of* d *by* b. *Once* q_n *is defined, let* q_{n+1} *and* d_{n+1} *be the quotient and remainder of the division of* q_n *by* d. *Then for all* $n \in \mathbb{N}$ *with* $b^n > d$ *we have that* $d_n = 0$ *and if* $N \in \mathbb{N}$ *is so that for all* $n \geq N$ *we have* $d_n = 0$, *then*

$$d = \sum_{n=0}^{N} d_n b^n.$$

Proof. Exercise 3-65. $\qquad\blacksquare$

Theorem 3.89 allows us to convert the familiar base 10 representation of a number into any other base representation. Moreover, $d = \sum_{n=0}^{N} d_n b^n$ allows us to convert back from any base to base 10. That means we can convert any base representation into any other base representation by, if necessary, using base 10 as an intermediate step (see Exercise 3-66). The bases 2, 10 and 16 are most commonly used for place value representations. The extra "digits" in base 16 notation are A (representing 10), B (representing 11), C (representing 12), D (representing 13), E (representing 14), and F (representing 15). Base 16 notation is used as an abbreviation for base 2 notation. Direct conversion between bases 2 and 16 is actually more efficient than going through base 10 as an intermediate step: Note that $16 = 2^4$, which means that blocks of four base 2 digits can be represented as one base 16 digit and vice versa. (This idea simplifies Exercises 3-66f and 3-66g.)

Exercises

3-58. The sieve of Eratosthenes.

   (a) Use the sieve of Eratosthenes to find the first 30 prime numbers.

   (b) Use the sieve of Eratosthenes to find all prime numbers that are less than or equal to 200.

3-59. Factors of products.

   (a) Let $a, b, m \in \mathbb{N}$ be so that a and m have no common factors. That is, the only number that divides both a and m is 1. Prove that if $m \mid ab$, then $m \mid b$.

   Hint. You may use the existence of a prime factorization.

   (b) Let $p \in \mathbb{N}$ be a prime number and let $a, b \in \mathbb{N}$. Prove that if $p \mid ab$, then $p \mid a$ or $p \mid b$.

(c) Let $p \in \mathbb{N}$ be a prime number and let $a_1, \ldots, a_k \in \mathbb{N}$. Prove that if $p \left| \prod_{j=1}^{k} a_j \right.$, then there is a $j \in \{1, \ldots, k\}$ so that $p \mid a_j$.

3-60. For each of the following numbers, find the prime factorization.

    (a) 2250

    (b) 91

    (c) 10403

3-61. Consider divisibility as a relation $\mid \, \subseteq \mathbb{N} \times \mathbb{N}$. In this exercise, we will prove that $\mid$ defines an order relation on $\mathbb{N}$.

    (a) Prove that $\mid$ is reflexive.

    (b) Prove that $\mid$ is antisymmetric.

    (c) Prove that $\mid$ is transitive.

3-62. Finishing the proof of Theorem 3.85.

    (a) Prove part 2 of Theorem 3.85.

    (b) Prove part 3 of Theorem 3.85.

    Hint. First prove that for all natural numbers a, b we have that $\max\{a, b\} + \min\{a, b\} = a + b$. Then use Exercise 3-46d.

3-63. For each of the following pairs of numbers m and n, find the greatest common divisor (m, n) and the least common multiple $[m, n]$.

    (a) 3 and 7

    (b) 45 and 105

    (c) 1309 and 1729

3-64. Let $a, b, d \in \mathbb{N}$

    (a) Prove that if $d \mid a$ and $d \mid b$, then $d \mid (a + b)$.

    (b) Prove that if $d \mid a$ and $d \mid (a + b)$, then $d \mid b$.

3-65. Prove Theorem 3.89 as follows. Let $b \in \mathbb{N}$ be a base and let $d \in \mathbb{N}$ be a number.

    (a) Let q and r being the quotient and remainder from the division algorithm dividing d by b so that $d = qb + r$.

        i. Prove that r is the rightmost digit d_0 in the base b representation of d.

        ii. Prove that the rightmost digit in the base b representation of qb is zero.

    (b) Prove that if $x = yb$, then the base b representation of x is obtained from the base b representation of y by appending a zero on the right.

    (c) Use parts 3-65a and 3-65b to prove Theorem 3.89 by induction.

3-66. Convert each of the following numbers from its given representation to the given new base.

    (a) 3521 (base 6) to base 10

    (b) 8497 (base 10) to base 4

    (c) 1121 (base 7) to base 5

    (d) 222 (base 10) to base 2

    (e) $A14F$ (base 16) to base 10

    (f) $3A5F$ (base 16) to base 2

(g) 1010111101100111 (base 2) to base 16

3-67. Let $a_1, \ldots, a_n \in \mathbb{N}$ each be divisible by $d \in \mathbb{N}$. Prove that $\sum_{j=1}^{n} a_j$ is divisible by d.

3-68. Divisibility of quotients.

(a) Let $a, b, d \in \mathbb{N}$ be so that $(a, d) = 1$ and $\dfrac{ab}{d}$ is defined. Prove that if $c \mid a$, then $c \mid \dfrac{ab}{d}$.

(b) Let $p \in \mathbb{N}$ be a prime number and let $k \in \{1, \ldots, p-1\}$. Prove that $\dbinom{p}{k}$ is divisible by p.

3-69. Prove each of the following by induction.

(a) For every $n \in \mathbb{N}$, the product $n(n+1)(n+2)$ is divisible by 6.

(b) For every $n \in \mathbb{N}$, the sum $2^{2n-1} + 3^{2n-1}$ is divisible by 5.

(c) For every $n \in \mathbb{N}$, the sum $n^5 - n$ is divisible by 30.

(d) For all $n \in \mathbb{N}$, the product $n\left(n^2 + 5\right)$ is divisible by 6.

(e) For all $n \in \mathbb{N}$, the difference $n^7 - n$ is divisible by 42.

(f) For all $n \in \mathbb{N}$, the product $n(n+1)(n+2)(n+3)$ is divisible by 24.

3-70. Let $n \in \mathbb{N}$ be divisible by $d \in \mathbb{N}$ and let $f \in \mathbb{N}$. Prove that the remainder of the division of n by df must be divisible by d.

3-71. Prove that if p is a prime factor of a square k^2, then p must be a prime factor of k, too.

3-72. Prove that every natural number that does not have any odd prime factors must be a power of 2.

3-73. Prove that a natural number is **divisible by 2** iff the last digit in its base 10 representation is divisible by 2.

3-74. Testing **divisibility by 3**.

(a) Prove that for all $j \in \mathbb{N}_0$ we have that $10^j - 1$ is divisible by 3.

(b) Prove that for any natural number $a = \sum_{j=0}^{n} a_j 10^j$, where the a_j are the digits in the base 10 representation of a, we have that $a - \sum_{j=0}^{n} a_j$ is divisible by 3.

(c) Prove that a natural number is **divisible by 3** iff the sum of the digits in its base 10 representation is divisible by 3.

3-75. Prove that a number is **divisible by 5** iff the last digit in its base 10 representation is 5 or 0.

3-76. Let $x \in \mathbb{N}$ be a number whose base 10 representation has at least 3 digits. Let a be two times the last digit of x. Let b be the number obtained by erasing the last digit of x (this leads to a number with one less digit than x had). Let $c := b - a$.

(a) Prove that x is **divisible by 7** iff c is divisible by 7.

(b) State an algorithm that tests for divisibility by 7.

3-77. Formulate and prove a divisibility test for divisibility by 6.

3-78. Formulate and prove a divisibility test for divisibility by 4.
    Hint. 100 is divisible by 4.

3-79. Revisiting Theorem 3.89.

(a) Devise an algorithm that produces the base b digits of the number d by constructing the digits from *left to right* rather than, as in Theorem 3.89, from right to left.

(b) Prove that your algorithm produces the correct digits.

3-80. **Multiplication in binary arithmetic.** Let $d \in \mathbb{N}$ and let $d_n d_{n-1} \ldots d_1 d_0$ be the base 2 representation of d.

(a) Prove that the the base 2 representation of $2d$ is $d_n d_{n-1} \ldots d_1 d_0 0$.

(b) Prove that the the base 2 representation of $2^k d$ is $d_n d_{n-1} \ldots d_1 d_0 \underbrace{0 \ldots 0}_{k \text{ times}}$.

(c) Let $c \in \mathbb{N}$ and let $c_m c_{m-1} \ldots c_1 c_0$ be the base 2 representation of c. Moreover, if $x \in \mathbb{N}$ and $x_m x_{m-1} \ldots x_1 x_0$ is its base 2 representation, let $[x_m x_{m-1} \ldots x_1 x_0]_2 := x$ (so the brackets "decode" the representation). Prove that $cd = \sum_{k=0}^{m} c_k \left[d_n d_{n-1} \ldots d_1 d_0 \underbrace{0 \ldots 0}_{k \text{ times}} \right]_2$.

Note. The formula in part 3-80c is literally how binary numbers are multiplied in computer operating systems. The operations are encoded in assembly language. Putting in zeros at the end is done by left shifting the bit strings, and the sum adds only the terms for which $c_k \neq 0$.

3-81. The last digit of a power.

(a) Prove that the last digit in the base 10 representation of 2^k is 2 if $k = 4m + 1$, 4 if $k = 4m + 2$, 8 if $k = 4m + 3$, and 6 if $k = 4m + 4$, respectively, for some $m \in \mathbb{N}_0$.

(b) State and prove a similar result for 3^k.

3-82. Consider the following algorithm. Let $d_n d_{n-1} \ldots d_1 d_0$ be the base b representation of the number d. Define $c_n := d_n$ and for $j \leq n$ define $c_{j-1} := c_j b + d_{j-1}$. The final output of the algorithm is the number c_1.

(a) Execute this algorithm with the number 4362, that is given in base its 10 representation.

(b) Execute this algorithm with the number 2122, that is given in base its base 3 representation.

(c) State a conjecture what the number c_1 is.

(d) Prove your conjecture from part 3-82c.

3.7   Equivalence Relations

In certain situations, it is sensible to classify objects as "equal" for the purpose at hand, even though the objects are not *actually* equal. For example, all children living in a certain area may be assigned to a particular school. Of course, no two of these children are *equal*, but they are all *equivalent* in the sense that any two of them are assigned to the same school. In mathematics, we typically say that two fractions, say $\frac{1}{2}$ and $\frac{2}{4}$ are equal because they represent the same number. This is usually not a problem, as we have already seen for fractions that represent natural numbers (see part 4 of Proposition 3.44). But formally we could say that the symbol combinations $\frac{1}{2}$ and $\frac{2}{4}$, which are not defined yet in this text[13], are not *equal*, because they do involve different symbols. We will ultimately say that they are *equivalent*, and this equivalence will be the key to constructing the rational numbers (see Section 5.1).

---

[13] And as long as they are not defined, we cannot be sure of their meaning.

The classification of objects as equivalent always depends on the purpose of the classification; that is, it depends on what properties we are interested in. The classification of children as equivalent when they live in the same school district helps determine how many children will be expected to be in a particular school in a particular year. It will not help determine how many will be in what grade. To determine grade level sizes, we need another notion of equivalence ("same school and same grade level"), but not equality. Because classifications as above typically relate objects to each other ("*A* lives in the same school district as *B*," "*A* lives in the same school district and is in the same grade as *B*"), a relation should be the right tool to describe the above. Equivalence relations are the abstract heart of such considerations.

Definition 3.90 *Let X be a set. A relation $\sim \subseteq X \times X$ is called an **equivalence relation** iff*

1. *$\sim$ is **reflexive**. That is, for all $x \in X$ we have $x \sim x$.*

2. *$\sim$ is **symmetric**. That is, for all $x, y \in X$ we have $x \sim y$ iff $y \sim x$.*

3. *$\sim$ is **transitive**. That is, for all $x, y, z \in X$ we have that $x \sim y$ and $y \sim z$ implies $x \sim z$.*

More descriptively, an equivalence relation retains some key properties of equality without actually demanding equality. Every object under consideration should be equivalent to itself (reflexivity), it should not matter in which order we state the relation (symmetry) and if two objects are related to a common third one, then the two objects should be related, too (transitivity). In terms of children being assigned to schools, if John is a unique child, then John goes to the same school as John. If John goes to the same school as Mary, then Mary goes to the same school as John. Finally, if John goes to the same school as Mary and Mary goes to the same school as Kelly, then John also goes to the same school as Kelly.

Example 3.91 For $m, n \in \mathbb{N}$ define $m \sim_1 n$ iff every prime number that divides m also divides n and every prime number that divides n also divides m. We claim that $\sim_1$ is an equivalence relation. So we must prove that the relation is reflexive, symmetric and transitive.

For reflexivity, note that for every number $n \in \mathbb{N}$, every prime number that divides n also divides n, so $n \sim_1 n$.

For symmetry, note that if $m \sim_1 n$, then every prime number that divides m also divides n and every prime number that divides n also divides m, which implies (upon switching m and n in the preceding statement) that $n \sim_1 m$.

Finally, for transitivity, if $m \sim_1 n$ and $n \sim_1 k$, then every prime number that divides m also divides n, every prime number that divides n also divides m, every prime number that divides n also divides k and every prime number that divides k also divides n. But then every prime number that divides m also divides k (see Exercise 3-61c) and every prime number that divides k also divides m, which means that $m \sim_1 k$.

To illustrate the relation, note that $6 \sim_1 24$, because both numbers have the same prime factors, but $6 \not\sim_1 9$, because $2 \mid 6$, but $2 \nmid 9$. $\qquad\square$

Example 3.92 For $m, n \in \mathbb{N}$ define $m \sim_2 n$ iff m and n have the same number of *distinct* prime factors. That is, m and n are related via $\sim_2$ iff the prime factorizations of m and n are $m = \prod_{j=1}^{k_m} p_j^{a_j}$ and $n = \prod_{j=1}^{k_n} q_j^{b_j}$ with all $a_j, b_j \neq 0$ and $k_n = k_m$. Then $\sim_2$ is an equivalence relation (see Exercise 3-83a).

For this relation, numbers such as $24 = 2^3 \cdot 3$ and $91 = 7 \cdot 13$ satisfy $24 \sim_2 91$, because both numbers have two distinct prime factors. On the other hand, $91 \not\sim_2 90$, because $90 = 2 \cdot 3^2 \cdot 5$ has three distinct prime factors, not two. $\qquad\square$

Example 3.93 For $m, n \in \mathbb{N}$ define $m \sim_3 n$ iff m and n have the same *total* number of prime factors, counted with multiplicity. That is, m and n are related via $\sim_3$ iff the prime factorizations of m and n are $m = \prod_{j=1}^{k_m} p_j^{a_j}$ and $n = \prod_{j=1}^{k_n} q_j^{b_j}$ with all $a_j, b_j \neq 0$, and $\sum_{j=1}^{k_m} a_j = \sum_{j=1}^{k_n} b_j$. Then $\sim_3$ is an equivalence relation (see Exercise 3-83b).

For this relation, numbers such as $24 = 2^3 \cdot 3$ and $90 = 2 \cdot 3^2 \cdot 5$ satisfy $24 \sim_3 90$, because both numbers have a total of four prime factors (counted with multiplicity). On the other hand, $91 \not\sim_3 90$, because $91 = 7 \cdot 13$ has a total of two prime factors (even when counted with multiplicity), not four. $\qquad\square$

Examples 3.92 and 3.93 show that even when equivalence relations look similar, they need not be the same. We started this section by talking about classifying objects into subsets. For an equivalence relation, the subsets in question are the equivalence classes.

Definition 3.94 *Let* $\sim \,\subseteq X \times X$ *be an equivalence relation on the set* X. *For each* $x \in X$, *the set* $[x] := \{y \in X : y \sim x\}$ *is called the* **equivalence class** *of* x.

Equivalence classes are subsets of X. But we call them *classes* to indicate that they are rather special subsets of X. Moreover, we will routinely work with sets of equivalence classes. It will be easier to keep track of what is what when we talk about sets of classes rather than about sets of sets.

Example 3.95 For the equivalence relation $\sim_1$ from Example 3.91, the singleton set $\{1\}$ is an equivalence class. The other equivalence classes are of the following form: Let $p_1, \ldots, p_k$ be fixed distinct prime numbers. Then $\left\{ p_1^{q_1} \cdots p_k^{q_k} : q_1, \ldots, q_k \in \mathbb{N} \right\}$ is an equivalence class of $\sim_1$. There is an equivalence class for every set $\{p_1, \ldots, p_k\}$ of k fixed distinct prime numbers. $\qquad\blacksquare$

Note that no two equivalence classes of $\sim_1$ overlap.

Example 3.96 For the equivalence relation $\sim_2$ from Example 3.92, the singleton set $\{1\}$ is an equivalence class. The other equivalence classes are of the following form: For each $k \in \mathbb{N}$, the set $\left\{ p_1^{q_1} \cdots p_k^{q_k} : q_1, \ldots, q_k \in \mathbb{N}, \, p_1, \ldots, p_k \text{ prime numbers} \right\}$ is an equivalence class of $\sim_2$ and there is an equivalence class for every $k \in \mathbb{N}$. $\qquad\square$

Note that no two equivalence classes of $\sim_2$ overlap. Moreover, each equivalence class for $\sim_1$ is contained in an equivalence class for $\sim_2$. So equivalence relations can give coarser or finer classifications, like in the example about school districts and grade levels. But this second observation is coincidence. The next example shows that such refinements only occur in special circumstances. There is no general containment relation between the equivalence classes of $\sim_3$ and those of $\sim_1$ or $\sim_2$.

Example 3.97 For the equivalence relation $\sim_3$ from Example 3.93, the singleton set $\{1\}$ is an equivalence class. The other equivalence classes are of the following form: Let $m \in \mathbb{N}$. Then

$$
\left\{ p_1^{q_1} \cdots p_k^{q_k} : k \in \mathbb{N}, q_1, \ldots, q_k \in \mathbb{N}, p_1, \ldots, p_k \text{ prime numbers}, \sum_{j=1}^{k} q_j = m \right\}
$$

is an equivalence class of $\sim_3$ and there is an equivalence class for every $m \in \mathbb{N}$. $\qquad \square$

Note that no two equivalence classes of $\sim_3$ overlap. This pairwise disjointness of equivalence classes, which we have observed repeatedly, is not an accident.

Definition 3.98 *Let X be a set. Then a family $\{X_i\}_{i \in I}$ of subsets of X is called a* **partition** *of X iff*

1. All X_i are nonempty.

2. The X_i are **pairwise disjoint***, that is, for $i \neq j$ we have that $X_i \cap X_j = \emptyset$.*

3. $\bigcup_{i \in I} X_i = X$.

Because non-indexed families of sets can be turned into indexed families of sets, a non-indexed family of sets is a partition iff all its sets are nonempty, pairwise disjoint and the union of the sets is the underlying set X. For a visualization of a partition, see Figure 3.4.

Of course, the family $\{X_i\}_{i \in I}$ is just an indexed set of subsets of X. But remember that the word "family" allows us to establish a mental hierarchy so that sets do not get confused with sets of sets. The next result shows that equivalence classes and partitions are two sides of the same coin.

Proposition 3.99 *Let X be a set. If $\sim$ is an equivalence relation on X, then the equivalence classes of $\sim$ form a partition of X. Conversely, if X is a set and $\{X_i\}_{i \in I}$ is a partition of X, then $a \sim b$ iff there is an $i \in I$ so that $a, b \in X_i$ defines an equivalence relation on X.*

Proof. First, let $\sim$ be an equivalence relation on X and let $\big\{[x] : x \in X\big\}$ be the family of equivalence classes of $\sim$. To prove that this family is a partition, we must prove that the equivalence classes are nonempty, pairwise disjoint and that their union is the whole set X.

All equivalence classes of $\sim$ are nonempty, because for every equivalence class $[x]$ we have $x \in [x]$.

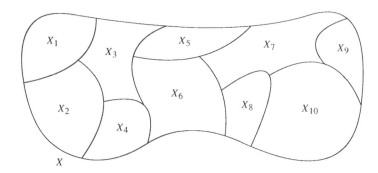

Figure 3.4 A partition $\{X_i\}_{i \in I}$ of the set X "cuts up" the set into pairwise disjoint, nonempty subsets that cover the whole set X. This is similar to how the assignment of areas to countries partitions the surface of the Earth, for example.

For pairwise disjointness, we prove the contrapositive, which is that $[x] \cap [y] \neq \emptyset$ implies $[x] = [y]$. So let $[x]$, $[y]$ be equivalence classes of $\sim$ so that there is an element $z \in [x] \cap [y]$. Then $x \sim z$ and $z \sim y$, which implies $x \sim y$. Now let $u \in [x]$. Then $u \sim x$ and $x \sim y$, so $u \sim y$ and $u \in [y]$. Hence $[x] \subseteq [y]$, and we prove $[y] \subseteq [x]$ similarly. Thus the equivalence classes of $\sim$ are pairwise disjoint.

Finally, for $\bigcup \{[x] : x \in X\} = X$ first note that $\bigcup \{[x] : x \in X\} \subseteq X$ is clear. For the reverse containment, let $y \in X$. Then $y \in [y]$ and $[y] \subseteq \bigcup \{[x] : x \in X\}$. Hence $y \in \bigcup \{[x] : x \in X\}$ and we conclude that $X \subseteq \bigcup \{[x] : x \in X\}$.

Conversely, let $\{X_i\}_{i \in I}$ be a partition of the set X, and define $a \sim b$ iff there is an $i \in I$ so that $a, b \in X_i$. **(We must prove that $\sim$ is an equivalence relation.)** Then for every $x \in X$ there is an X_i with $x \in X_i$ and hence $x \sim x$, so $\sim$ is reflexive. Moreover, if $x \sim y$, then there is an X_i with $x, y \in X_i$ and hence $y \sim x$, so $\sim$ is symmetric. Finally, if $x \sim y$ and $y \sim z$, then there are X_i and X_j so that $x, y \in X_i$ and $y, z \in X_j$. Therefore $X_i \cap X_j \ni y$, which implies $X_i = X_j$. But then $x, z \in X_i = X_j$ and hence $x \sim z$, so $\sim$ is transitive. ∎

Exercises

3-83. Prove that each of the following is an equivalence relation.

   (a) The relation in Example 3.92

   (b) The relation in Example 3.93

3-84. Prove that each of the following is an equivalence relation and compute the corresponding partition.

   (a) The relation $\sim \subseteq \mathbb{N} \times \mathbb{N}$ defined by $m \sim n$ iff $m - n$ or $n - m$ is divisible by 2.

   (b) The relation $\sim \subseteq \mathbb{R} \times \mathbb{R}$ so that $a \sim b$ iff the decimal expansions of a and b have the same number of digits behind the decimal point. (You may use what you know about real numbers for this exercise.)

   (c) For $m \in \mathbb{N}$ fixed, the relation $\sim \subseteq \mathbb{Z} \times \mathbb{Z}$ defined by $a \sim b$ iff $b - a$ is divisible by m. (You may use what you know about integers for this exercise.)

(d) The relation $\sim$ on $\mathbb{N} \times \mathbb{N}$ (that is $\sim \subseteq (\mathbb{N} \times \mathbb{N}) \times (\mathbb{N} \times \mathbb{N})$) defined by $(a, b) \sim (c, d)$ iff $a + d = b + c$.

(e) The relation $\sim$ on $\mathbb{N} \times \mathbb{N}$ (that is $\sim \subseteq (\mathbb{N} \times \mathbb{N}) \times (\mathbb{N} \times \mathbb{N})$) defined by $(a, b) \sim (c, d)$ iff $ad = bc$.

3-85. For each of the following relations, determine which of the three properties of an equivalence relation are violated.

(a) The relation $\sim \subseteq \mathbb{N} \times \mathbb{N}$ defined by $m \sim n$ iff $n = 1$.

(b) The relation $\sim \subseteq \mathbb{N} \times \mathbb{N}$ defined by $m \sim n$ iff $m + n \leq 3$.

(c) The relation $\sim \subseteq \mathbb{N} \times \mathbb{N}$ defined by $m \sim n$ iff $m - n = 2$.

(d) The relation $\sim \subseteq \mathbb{N} \times \mathbb{N}$ defined by $m \sim n$ iff 3 is a prime factor of m and of n.

3-86. Determine if each of the following is an equivalence relation. Justify your answer using the definition.

(a) The relation $\sim \subseteq \mathbb{N} \times \mathbb{N}$ defined by $m \sim n$ iff $m + n$ is even.

(b) The relation $\sim \subseteq \mathbb{N} \times \mathbb{N}$ defined by $m \sim n$ iff mn is even.

(c) The relation $\sim \subseteq \mathbb{N} \times \mathbb{N}$ defined by $m \sim n$ iff $m = n + 3$.

3-87. For each of the following relations determine if it is an equivalence relation. Justify your answer using the definition. If the relation is an equivalence relation, compute the corresponding partition. If it is not, determine which of the three properties of an equivalence relation are not satisfied.

(a) $\big\{ (1, 1), (2, 2), (3, 3), (4, 4), (2, 3), (3, 2) \big\}$ on the set $\{1, 2, 3, 4\}$

(b) $\big\{ (1, 1), (2, 2), (3, 3), (4, 4), (2, 3), (3, 2) \big\}$ on the set $\{1, 2, 3, 4, 5\}$

(c) $\big\{ (1, 1), (2, 2), (3, 3), (2, 3) \big\}$ on the set $\{1, 2, 3\}$

(d) $\big\{ (1, 1), (2, 2), (3, 3), (1, 2), (2, 1), (2, 3), (3, 2) \big\}$ on the set $\{1, 2, 3\}$

3-88. Let X be a set and let $V_n(X)$ be the sets from Definition 2.62 of superstructures.

(a) Let $\sim$ be an equivalence relation on X. Find the smallest n so that $\sim \in V_n(X)$.

(b) Let $\{X_i\}_{i \in I}$ be a partition of X, indexed by a subset I of X. Find the smallest n so that the function from the definition of the indexed family $\{X_i\}_{i \in I}$ (see Definition 2.61) is in $V_n(X)$.

3.8 Arithmetic Modulo m

The equivalence relations in Examples 3.91, 3.92 and 3.93 were merely presented to familiarize ourselves with equivalence relations. We will now investigate an equivalence relation with important practical applications: The equivalence of numbers in $\mathbb{N}_0$ modulo a given modulus m. We first prove that it is an equivalence relation and then we proceed with practical considerations.

Definition 3.100 *Let $m \in \mathbb{N}$ and let $x, y \in \mathbb{N}_0$. Then x and y are called **equivalent modulo** m iff $x = y$ or $x > y$ and $m \mid (x - y)$ or $x < y$ and $m \mid (y - x)$. Numbers that are equivalent modulo m are also called **congruent modulo** m. When x and y are equivalent modulo m, we write $x \equiv y$ (mod m) The number m is called the **modulus**.*

The definition becomes a little simpler when integers are available (see Example 4.15). But we usually think of integers modulo m as nonnegative, so we might as well consider them here. Although the following proofs will be a little more complicated than they could be, they provide good practice for working with case distinctions.

Proposition 3.101 *Let* $m \in \mathbb{N}$. *Equivalence modulo* m *is an equivalence relation on* $\mathbb{N}_0$.

Proof. To prove that equivalence modulo m is an equivalence relation, we must prove that the relation is reflexive, symmetric and transitive.

Reflexivity is trivial.

For *symmetry*, let $x, y \in \mathbb{N}_0$ be so that $x \equiv y \pmod{m}$. Note that there is nothing to prove if $x = y$. If $x > y$, then, by the second "or" in Definition 3.100, $m \mid (x - y)$. Hence, in this case we have $y < x$ and $m \mid (x - y)$, which means, by the third "or" in Definition 3.100, that $y \equiv x \pmod{m}$. If $x < y$, then $m \mid (y - x)$. Hence, in this case we have $y > x$ and $m \mid (y - x)$, which means that $y \equiv x \pmod{m}$. So in every possible case $x \equiv y \pmod{m}$ implies $y \equiv x \pmod{m}$, and the relation $\equiv$ is symmetric.

For *transitivity*, let $x, y, z \in \mathbb{N}_0$ be so that $x \equiv y \pmod{m}$ and $y \equiv z \pmod{m}$. Note that there is nothing to prove if $x = y$, $y = z$ or $x = z$. Hence we can assume that x, y and z are distinct. Without loss of generality assume that $x < z$ (the case $x > z$ is similar).

Case 1: $y < x < z$. In this case $m \mid (x - y)$ and $m \mid (z - y)$, that is, there are $j, k \in \mathbb{N}$ so that $x - y = jm$ and $z - y = km$. Because $z - y > x - y$, we infer (similar to the proof of part 1 of Proposition 3.42, see Exercise 3-89a) $z - x = (z - y) - (x - y) \in \mathbb{N}$ and $k > j$. Hence $z - x = (z - y) - (x - y) = km - jm = (k - j)m$. Therefore $m \mid (z - x)$ and hence $x \equiv z \pmod{m}$.

Case 2: $x < z < y$. In this case $m \mid (y - x)$ and $m \mid (y - z)$, that is, there are $j, k \in \mathbb{N}$ so that $y - x = jm$ and $y - z = km$. Because $y - x > y - z$, we infer (see Exercise 3-89b) $z - x = (y - x) - (y - z) \in \mathbb{N}$ and $k < j$. Hence the difference between z and x is $z - x = (y - x) - (y - z) = jm - km = (j - k)m$, that is, $m \mid (z - x)$. We conclude that $x \equiv z \pmod{m}$.

Case 3: $x < y < z$. In this case $m \mid (y - x)$ and $m \mid (z - y)$, that is, there are $j, k \in \mathbb{N}$ so that $y - x = jm$ and $z - y = km$. Adding the two equations yields (see Exercise 3-89c) $z - x = (z - y) + (y - x) = km + jm = (k + j)m$. Therefore $m \mid (z - x)$ and hence $x \equiv z \pmod{m}$.

In summary, in every possible case $x \equiv y \pmod{m}$ and $y \equiv z \pmod{m}$ implies that $x \equiv z \pmod{m}$. Hence the relation $\equiv$ is transitive.   ∎

Equivalence modulo m arises naturally in digital computation, where numbers are represented by binary digits. At the operating system level, the number of digits per number is fixed. That is, if an operating system works with 32-bit integers, then the computer can encode natural numbers[14] from 0 to $2^{32} - 1 = 4,294,967,295$. If an operating system works with 64-bit integers, then the computer can encode natural numbers from 0 to $2^{64} - 1 = 18,446,744,073,709,551,615$.

With these integers, a computer performs additions and multiplications. As long as the result does not exceed $2^{\text{number of bits}} - 1$, arithmetic works as usual. It was indicated in Exercise 1-13 that if the result of an addition is too large, then the carry bit of the last half-adder sets the "overflow flag." Figure 3.5 gives a visualization of the overflow: The numbers "wrap around" through $0 = m := 2^{\text{number of bits}}$. (In a computer, 0 and

---

[14]Natural numbers are typically called **unsigned integers** in programming languages and operating systems.

$$m = 2^8 = 256$$

1 (lost)   1 1   1 1 1   (carry overs)

$a =$ 0110 1010 $= 106$
$b =$ 1010 1110 $= 174$
"$a + b$"=   0001 1000 $= 24 \equiv 280 \pmod{256}$

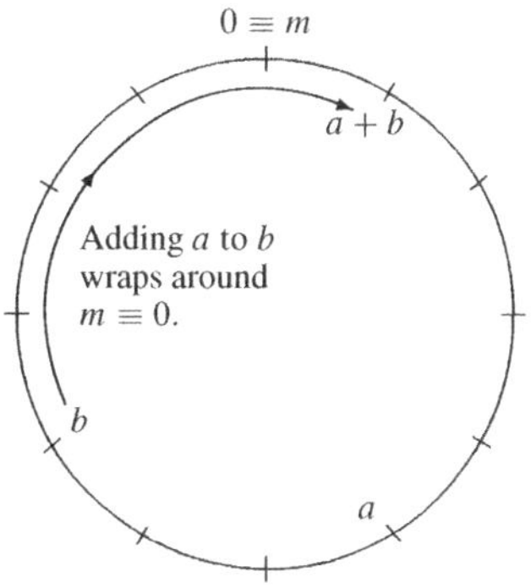

$0 \equiv m$

$a + b$

Adding a to b wraps around $m \equiv 0$.

b

a

Figure 3.5 An addition of 8 digit binary numbers (left) and a visualization of the wraparound of addition modulo $m = 2^8$ (right).

+	0	1	2	3	4	5	6	7
0	0	1	2	3	4	5	6	7
1	1	2	3	4	5	6	7	0
2	2	3	4	5	6	7	0	1
3	3	4	5	6	7	0	1	2
4	4	5	6	7	0	1	2	3
5	5	6	7	0	1	2	3	4
6	6	7	0	1	2	3	4	5
7	7	0	1	2	3	4	5	6

·	0	1	2	3	4	5	6	7
0	0	0	0	0	0	0	0	0
1	0	1	2	3	4	5	6	7
2	0	2	4	6	0	2	4	6
3	0	3	6	1	4	7	2	5
4	0	4	0	4	0	4	0	4
5	0	5	2	7	4	1	6	3
6	0	6	4	2	0	6	4	2
7	0	7	6	5	4	3	2	1

Figure 3.6 The addition and multiplication tables for arithmetic modulo 8.

m are literally equal, because they have the same bit representation.) But, because addition is carried out starting at the rightmost place and going right to left, the low bits of the sum are correct. Overall, the sum is just off by $m = 2^{\text{number of bits}}$, that is, the bits that we have computed are equivalent modulo m to the actual result! Therefore, even though the available bits of the sum are not correct in the sense of arithmetic with natural numbers, they are correct modulo m. The left table in Figure 3.6 illustrates these facts for $m = 2^3 = 8$, which would be an arithmetic with 3 bits.[15] Whenever the result of an addition exceeds $m - 1 = 7$, the recorded result is the remainder of the division of the actual result by $m = 8$. This is, in fact, a general phenomenon.

Proposition 3.102 *Let* $m \in \mathbb{N}$ *and let* $x \in \mathbb{N}_0$. *Then* $x \equiv d \pmod{m}$, *where* d *is the remainder of the division of* x *by* m.

Proof. Exercise 3-90. ∎

Because multiplication in a computer is achieved through repeated left shifts and additions (see Exercise 3-80), the result of a multiplication will be correct modulo m, too. In this case however, the result can exceed the actual product by more than m (also see the right table in Figure 3.6). Overall, because in a computer equality is replaced

---

[15]The addition and multiplication tables for 64 bit numbers were a little too large to include here.

with equivalence modulo $2^{\text{number of bits}}$, it is reasonable to investigate arithmetic for which equality is replaced by equivalence modulo m. If such an arithmetic is to be meaningful, then the result of an operation $a + b$ or $a \cdot b$ must be the same modulo m, no matter which of the infinitely many numbers that are equivalent to a and b modulo m are chosen as the input to the operation $+$ or $\cdot$. For example, if $m = 8$, then $3 \equiv 11 \pmod{8}$ and $9 \equiv 25 \pmod{8}$. These numbers could be intermediate results in a computation. If they are added, then it should not matter which representative modulo 8 is chosen. That is, our arithmetic will only be useful if $3 + 9 \equiv 11 + 25 \pmod{8}$. Of course this is true in our example, because $12 \equiv 36 \pmod{8}$. In fact, arithmetic modulo m "works out" in general.

Theorem 3.103 *Let $m \in \mathbb{N}$ and let $a, b, c, d \in \mathbb{N}_0$ be so that $a \equiv c \pmod{m}$ and $b \equiv d \pmod{m}$. Then $a + b \equiv c + d \pmod{m}$.*

Proof. Let $a, b, c, d \in \mathbb{N}_0$ be so that $a \equiv c \pmod{m}$ and $b \equiv d \pmod{m}$. We must prove that $a + b \equiv c + d \pmod{m}$.

Without loss of generality we can assume that $a + b \geq c + d$ and that $a \geq c$. (This is not a loss of generality, because we can always reach this situation by renaming the variables.) Then there is a $k_1 \in \mathbb{N}_0$ so that $a - c = k_1 m$. Moreover, there is a $k_2 \in \mathbb{N}_0$ so that, depending on whether $b \geq d$ or $b < d$, we have $b - d = k_2 m$ or $d - b = k_2 m$. In case $b \geq d$ we have that $b - d = k_2 m$ and then (by Proposition 3.42 and Exercise 3-91a)

$$
\begin{aligned}
(a + b) - (c + d) &= (a - c) + (b - d) \\
&= k_1 m + k_2 m \\
&= (k_1 + k_2) m.
\end{aligned}
$$

Similarly, in case $b < d$ we have that $d - b = k_2 m$, and because $a + b \geq c + d$ we infer $k_1 \geq k_2$. Therefore (see Exercise 3-91b)

$$
\begin{aligned}
(a + b) - (c + d) &= (a - c) - (d - b) \\
&= k_1 m - k_2 m \\
&= (k_1 - k_2) m.
\end{aligned}
$$

Hence, in either case, the difference $(a + b) - (c + d)$ is divisible by m, which means that $a + b \equiv c + d \pmod{m}$. ∎

Theorem 3.104 *Let $m \in \mathbb{N}$ and let $a, b, c, d \in \mathbb{N}_0$ be so that $a \equiv c \pmod{m}$ and $b \equiv d \pmod{m}$. Then $a \cdot b \equiv c \cdot d \pmod{m}$.*

Proof. Let $a, b, c, d \in \mathbb{N}_0$ be so that $a \equiv c \pmod{m}$ and $b \equiv d \pmod{m}$. We must prove that $a \cdot b \equiv c \cdot d \pmod{m}$.

Without loss of generality we can assume that $ab \geq cd$ and that $a \geq c$. Then there is a $k_1 \in \mathbb{N}_0$ so that $a - c = k_1 m$. Moreover, there is a $k_2 \in \mathbb{N}_0$ so that, depending on whether $b \geq d$ or $b < d$, we have $b - d = k_2 m$ or $d - b = k_2 m$. In case $b \geq d$ we have that $b - d = k_2 m$ and then (see Exercise 3-89c)

$$
\begin{aligned}
ab - cd &= (ab - cb) + (cb - cd) \\
&= (a - c)b + c(b - d) \\
&= k_1 m b + c k_2 m \\
&= (k_1 b + c k_2) m.
\end{aligned}
$$

The proof that in case $b < d$ the difference $ab - cd$ is divisible by m is similar and it is left to the reader as Exercise 3-92.

Hence, in either case, the difference $ab - cd$ is divisible by m, which implies that $a \cdot b \equiv c \cdot d \pmod{m}$. ∎

Theorem 3.103 says that as long as the summands in a sum are equivalent modulo m, the results will be equivalent modulo m, too. Theorem 3.104 shows the same result for products. So, independent of how many times we "wrap around" in a computation modulo m, or, in computing terms, how many times the overflow flag must be set in a computation on a processor, the result obtained in $\{0, \dots, m - 1\}$ is equivalent modulo m to the actual result. Aside from the fact that these results "make limited place binary arithmetic work," programmers take advantage of this fact when they construct arithmetics that can represent integers of arbitrary size. One way to encode integers of arbitrary size involves the Chinese Remainder Theorem (see Exercise 3-108).

From a mathematical point of view, arithmetic modulo m is the first time we naturally encounter operations that need not be well-defined. Recall that a binary operation on a set S is nothing but a function from $S \times S$ to S. When we add and multiply integers modulo m, we actually add and multiply their equivalence classes modulo m, because Theorems 3.103 and 3.104 show that it does not matter which representatives we choose as inputs. But, as we switch our point of view to addition and multiplication of equivalence classes, we note that the definition of these operations is based on choosing representatives from each equivalence class. Theorems 3.103 and 3.104 show that the results of these operations with classes do not depend on which representatives we choose for the operation. So, in the language of equivalence classes, Theorems 3.103 and 3.104 say the following.

Corollary 3.105 *Let $m \in \mathbb{N}$ and let $[a]_m, [b]_m$ be equivalence classes modulo m. If $[a]_m = [c]_m$ and $[b]_m = [d]_m$, then $[a + b]_m = [c + d]_m$ and $[a \cdot b]_m = [c \cdot d]_m$.* ∎

Corollary 3.105 says that the definitions $[a]_m + [b]_m := [a + b]_m$ for addition and $[a]_m \cdot [b]_m := [a \cdot b]_m$ for multiplication provide well-defined binary operations on the equivalence classes modulo m (recall that binary operations are functions). Hence we can define the following.

Definition 3.106 *Let $m \in \mathbb{N}$. We define $\mathbb{Z}_m$ to be the set of all equivalence classes modulo m.[16] For $[a]_m, [b]_m \in \mathbb{Z}_m$, we define addition by $[a]_m + [b]_m := [a + b]_m$ and multiplication by $[a]_m \cdot [b]_m := [a \cdot b]_m$.*

---

[16] At this stage the notation $\mathbb{N}_m$ may feel more natural, because we do not really have the integers yet. But $\mathbb{Z}_m$ is the commonly used name for the sets of these equivalence classes, and when we ultimately throw in the integers (see Example 4.15), the properties will not change.

Proving that an operation on equivalence classes is well-defined will also be very important as we construct the integers and the rational numbers within set theory. Arithmetic modulo m was included here to show that constructions using equivalence classes are not artificial and that they can have practical applications. We devote the next section to a very important application of arithmetic modulo m: public key encryption. Arithmetic modulo m will be revisited throughout the text to fill in further properties of these important, but "slightly different," number systems.

Exercises

3-89. Some details needed for the proof of Proposition 3.101. Let $k, m, n \in \mathbb{N}$ be so that $k < m < n$.

   (a) Prove that $n - m = (n - k) - (m - k)$.

   Hint. Start with $k + (m - k) + (n - m)$ on the left side and use the proof of part 1 of Proposition 3.42 as guidance.

   (b) Prove that $m - k = (n - k) - (n - m)$.

   (c) Prove that $n - k = (n - m) + (m - k)$.

3-90. Prove Proposition 3.102.

3-91. Some details that are needed for the proof of Theorem 3.103.

   (a) Let $k, m, n \in \mathbb{N}$ be so that $m < n$. Prove that $n - m = (n + k) - (m + k)$.

   (b) Let $k, l, m, n \in \mathbb{N}$ be natural numbers so that $l + k \leq n + m$ and $l \leq n$ and $m < k$. Prove that $(n + m) - (l + k) = (n - l) - (k - m)$.

3-92. Finish the proof of Theorem 3.104 by proving that for $ab \geq cd$, $a \geq c$ and $b < d$ the difference $ab - cd$ is divisible by m.

3-93. For each of the following moduli, state the addition and multiplication tables for arithmetic modulo m.

   (a) $m = 2$

   (b) $m = 3$

   (c) $m = 4$

   (d) $m = 5$

   (e) $m = 6$

   (f) $m = 7$

3-94. Let $m \in \mathbb{N}$ and let $a, b \in \mathbb{N}_0$ be so that $a < b$. Prove that $a \equiv b \pmod{m}$ iff $b - a \equiv 0 \pmod{m}$.

3-95. Some operations that are *not* well-defined.

   (a) On the set $\mathbb{N} \times \mathbb{N}$, consider the relation $\sim$ defined by $(a, b) \sim (c, d)$ iff $a + d = b + c$.

      i. Prove that $\sim$ is an equivalence relation.

      ii. Prove that the operation $*$ defined by $\big[(a, b) \big] * \big[(c, d) \big] := \big[(ac, bd) \big]$ is *not* well-defined.

      iii. Prove that the operation $*$ defined by $\big[(a, b) \big] * \big[(c, d) \big] := \big[(a + d, bc) \big]$ is *not* well-defined.

   (b) On the set $\mathbb{N} \times \mathbb{N}$, consider the relation $\sim$ defined by $(a, b) \sim (c, d)$ iff $ad = bc$.

      i. Prove that $\sim$ is an equivalence relation.

      ii. Prove that the operation $*$ defined by $\big[(a, b) \big] * \big[(c, d) \big] := \big[(a + c, b + d) \big]$ is *not* well-defined.

iii. Prove that the operation $*$ defined by $\big[(a,b)\big] * \big[(c,d)\big] := \big[(ad,bc)\big]$ is *not* well-defined.

3-96. Surprising injective and surjective functions.

    (a) Prove that the function $f : \mathbb{N} \to \mathbb{N}$ defined by $f(n) := 4n$ is injective.

    (b) Prove that the function $f : \mathbb{Z}_6 \to \mathbb{Z}_6$ defined by $f\big([n]_6\big) := [4]_6[n]_6$ is *not* injective.

    (c) Prove that the function $f : \mathbb{N} \to \mathbb{N}$ defined by $f(n) := 3n$ is not surjective.

    (d) Prove that the function $f : \mathbb{Z}_4 \to \mathbb{Z}_4$ defined by $f\big([n]_4\big) := [3]_4[n]_4$ *is* surjective.

3-97. Surprising unsolvable and solvable equations.

    (a) Prove that the equation $x + 6 = 2$ is not solvable in $\mathbb{N}$.

    (b) Prove that the equation $[x]_8 + [6]_8 = [2]_8$ is solvable in $\mathbb{Z}_8$.

    (c) Let $m \in \mathbb{N} \setminus \{1\}$. Prove that the equation $[x]_m + [6]_m = [2]_m$ is solvable in $\mathbb{Z}_m$.

3-98. **Addition of base b representations.** Let $b \in \mathbb{N}$ satisfy $b > 1$, let $d, e \in \mathbb{N}$ and let $d_n d_{n-1} \ldots d_1 d_0$ and $e_n e_{n-1} \ldots e_1 e_0$ be the base b representations of d and e, respectively, with leading zeros included as necessary so that both have the same number of digits. Define $s_0 :\equiv d_0 + e_0 \pmod{b}$, where it is understood that we take the representative that is in $\{0, \ldots, b-1\}$ as the result, and let $c_0 := 0$. For each $k \in \{1, \ldots, n+1\}$ let $c_k := \begin{cases} 1; & \text{if } d_{k-1} + e_{k-1} + c_{k-1} \geq b, \\ 0; & \text{if } d_{k-1} + e_{k-1} + c_{k-1} < b. \end{cases}$ For each $k \in \{1, \ldots, n\}$ let $s_k :\equiv c_k + d_k + e_k \pmod{b}$, where it is understood that we take the representative that is in $\{0, \ldots, b-1\}$ as the result. Finally, let $s_{n+1} := c_{n+1}$.

Prove that $s_{n+1} s_n s_{n-1} \ldots s_1 s_0$ is the base b representation of the sum $d + e$.

3-99. Let p be a prime number that is not equal to 2 or 3.

    (a) Prove that $p \equiv 1 \pmod 6$ or $p \equiv -1 \pmod 6$.
       Hint. Check the factors of the numbers that are equal to 0 through 5 modulo 6.

    (b) Prove that if $p + 2$ is a prime number, too, then $p + 1$ is divisible by 6.

3.9   Public Key Encryption

The idea of encryption and decryption is very simple. Two individuals want to send messages to each other and no one who intercepts the transmission is supposed to be able to read the message. Some of the earliest applications were in the transmission of written military orders. Julius Caesar communicated in writing with his commanders. Although most of the world (Romans as well as barbarians) was illiterate at the time, written orders were not secure. Some barbarians[17] could read Latin. If one of them intercepted the orders, then Roman armies would be at a disadvantage, because the enemy would know their next moves. So Caesar used a very simple encryption scheme, called a **Caesarian cipher**: Take the letters of the alphabet and scramble them. Formally, let $\mathcal{C}$ be a bijective function from the alphabet to itself.[18] Write the message in clear text. Then replace the letter "a" with the letter "$\mathcal{C}(a)$," the letter "b" with the letter "$\mathcal{C}(b)$," etc. The recipient must know the function $\mathcal{C}$, so this function must be transmitted in secure fashion. Ideally, in Caesar's times, this could have been done

---

[17]The term "barbarian" is not meant to be derogatory. The author himself is a proud descendant of the Teutonic barbarians who would not yield to the Romans.

[18]It is idle speculation to ask if Caesar would have approved of or even understood such language. It works for us, though.

in a face-to-face conversation that was not overheard and during which the key was committed to memory. Once both sender and recipient know the function C, encoded messages can be sent back and forth. If a message is intercepted, it looks quite meaningless.

Caesarian ciphers are only secure as long as the other side does not know how to encode or decode messages. So, barring treason, the Caesarian cipher was quite safe in the ancient world.[19] However, a more detailed analysis reveals that the Caesarian cipher is quite unsafe. Certain letters occur with certain frequencies in the dictionary. In the English language, the most frequent letter is "e." So by counting which letters occur most frequently, it should be possible to determine which letter stands for "e" and for other frequently occurring letters. Certain words, like "the," "a," "an," "and," "I" etc. are also more common than others. Some trial and error with the frequency of certain letters and with some short words ultimately reveals part of the cipher. Pretty soon, not yet identified letters can be filled into longer words, which in turn gives us more of the cipher. In many newspapers, readers can try their hand at breaking Caesarian ciphers. They're in the puzzle section and they are called "**cryptoquotes**."[20]

Of course, more sophisticated ciphers cannot be broken based on the frequency of letters. For example, if letters are grouped and the groups are assigned a code, then we would need to know the code to decrypt the message. But, for the next 2,000 years after Caesar, one pattern persisted in encryption and decryption: Anyone who knew how a message was encrypted, could decrypt it, too. Therefore encryption and decryption mechanisms were kept secret. For example, in the Second World War, the German military used a machine called the **ENIGMA** to send messages back and forth. This machine is well known for being used to transmit orders to and from submarines, but it was used in other military communications, too. The ENIGMA looked like a typewriter with letters only, plus a few extra dials. As a message was typed, for every typed letter, a letter at the back of the machine would light up. This letter would be recorded for the coded message. This sounds like a Caesarian cipher, but when the same letter was typed a second time, a *different* letter would light up. Which letter lit up depended on the setting of the dials, which was included in the message. The encoded message would be decoded on another ENIGMA machine by setting the dials to the appropriate settings, and then typing the coded message. This time the letters that lit up would be the letters of the original clear text message. So, possession of an ENIGMA machine would allow an individual to encode and decode messages. But without the machine, the code was believed to be unbreakable by the German military. Few individuals were given access to ENIGMA machines, and the whereabouts of the machines was tracked meticulously to guard against theft. Moreover, operators were under strict orders to destroy the machines if there was a significant chance of them falling in the other side's hands.

The British Intelligence Service had a special subdivision at Bletchley Park, specifically dedicated to breaking the ENIGMA code. Based on earlier work of Polish math-

---

[19]The author's barbarian ancestors were not terribly advanced in mathematics. Figuring out if enough men were available to vanquish the enemy and if there was enough food and mead to get through the winter just about did it. The Romans were not much further, by the way.

[20]The author has no information on his ancestors' attitude towards cryptoquotes, what with newspapers not available for another 1500+ years.

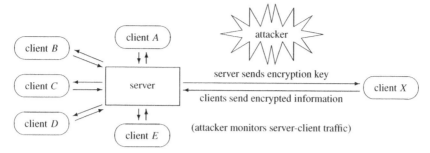

Figure 3.7 Schematic view of secure internet transactions. A server sends instructions how to encrypt information to its clients. Even if this information is intercepted by an attacker, the encrypted information sent by the clients to the server is considered safe, because public key codes cannot be decrypted easily.

ematicians, and assisted by some of these mathematicians, by late May 1940 a team of mathematicians that included **Alan Turing** designed a cryptanalytic machine that made it feasible to break the ENIGMA code. The decoding effort still was significant, on the order of hours, but it was manageable. Among the first decoded messages were German army instructions that related to the battles of the Wehrmacht with the British Expeditionary Force that supported the French army in 1940. These instructions confirmed other intercepted messages. Ultimately, the knowledge of the Wehrmacht's next moves helped plan the successful evacuation of the British Expeditionary Force from Dunkirk (see [17]). In a quite literal sense, breaking the ENIGMA code helped change the course of history for the better.

During the Cold War, the basic premise of encryption and decryption did not change. Anyone who knew how messages were encrypted could decrypt them, too. The effort may still have been tremendous, but at least breaking the code was possible. The ideal code that the intelligence community was looking for was one in which the ability to encode messages did not "automatically" provide the ability to decode messages, too. Such a code, called a **public key code** would be safe: Even if a coded message and instructions how it was encoded fell into the wrong hands,[21] the message could not be decoded.

Aside from applications in intelligence, such public key codes are now common in internet commerce (see Figure 3.7). Encoded transactions (purchases, credit card payoffs, tax information) are sent over the internet every second of every day. The algorithm that encodes the messages is sent by the server (which will ultimately receive the information) to the client (that can be the reader making a purchase or a payment). Internet transmissions are considered non-secure, that is, transmissions can be monitored without sender or recipient knowing about it. Consequently, the coded message and the encoding algorithm could be known to individuals other than server and client. But, because the ability to encode messages (which is what is needed to make a purchase or a payment) does not carry with it the ability to decode messages (which is what is needed to steal parts of the client's identity, such as personal infor-

---

[21]The US and the USSR had different definitions of "the wrong hands."

mation and credit card numbers), internet transactions have become a standard way to operate many businesses.

In the following, we describe the first public key code. We will use integer notation and equivalence modulo m for the description of the algorithm and equivalence classes modulo m and equality for proofs. Both notations are quite common, so this will train the reader to go back and forth between the two notations. We start with some mathematical results that were known for a few hundred years before public key encryption, mainly because mathematicians were interested in arithmetic modulo m.

Proposition 3.107 shows that under certain circumstances, numbers will have what we will call a *multiplicative inverse* in arithmetic modulo m. That is, under certain circumstances, for $x \in \mathbb{N}$ there will be a $y \in \mathbb{N}$ so that $xy \equiv 1 \pmod{m}$. This is a very nice property that we know from the nonzero rational, real and complex numbers, but it is equally clear that this property is out of the question for natural numbers (Exercise 3-21).

Proposition 3.107 *Let $x, m \in \mathbb{N}$ have no common factors. Then for all $c \in \mathbb{N}$ that are not divisible by m, we have that $cx \not\equiv 0 \pmod{m}$. Moreover, there is a $y \in \mathbb{N}$ so that $xy \equiv 1 \pmod{m}$.*

Proof. First consider a $b \in \mathbb{N}$ so that $bx \equiv 0 \pmod{m}$. Then bx is divisible by m. Because x and m do not have any common factors, Exercise 3-59a implies that b must be divisible by m.

Via contraposition, this means that if c is not divisible by m, then $cx \not\equiv 0 \pmod{m}$, which establishes the first part of the result.

We are now transitioning towards proofs as they are typically presented in mathematical texts. In particular, this means that assumptions need not always be re-stated and that the start of the proof may look surprising. For example, the first paragraph in this proof only becomes motivated once we read the second paragraph. When a proof has a surprising start, it is good advice to simply keep reading and verify that everything is correct. The idea should become clear later in the proof.

Now consider the set $\{xy : y = 1, \ldots, m - 1\}$. For any two distinct elements $y_1, y_2 \in \{1, \ldots, m - 1\}$ with $y_1 < y_2$ we have that m does not divide $y_2 - y_1$. Hence, by what we proved in the first paragraph, $x(y_2 - y_1) \not\equiv 0 \pmod{m}$, which means that $xy_1 \not\equiv xy_2 \pmod{m}$. Therefore $A := \{[xy]_m : y = 1, \ldots, m - 1\}$ contains $m - 1$ distinct elements and, moreover, it does not contain $[0]_m$. But there are exactly $m - 1$ equivalence classes modulo m that are not $[0]_m$, so $A = \{[z]_m : z = 1, \ldots, m - 1\}$ (see Theorem 3.54) and $[1]_m \in A$. Hence there is a $y \in \{1, \ldots, m - 1\}$ so that $[x]_m \cdot [y]_m = [xy]_m = [1]_m$, that is, so that $xy \equiv 1 \pmod{m}$.   ∎

The reader will prove a converse for Proposition 3.107 in Exercise 3-100. For arithmetic modulo prime numbers, Fermat's Little Theorem below says that the number y so that xy is equivalent to 1 has an especially simple form.

Theorem 3.108 Fermat's Little Theorem. *Let p be a prime number. Then for every $a \in \mathbb{N}$ we have that $a^p \equiv a \pmod{p}$. Moreover, for every $a \in \mathbb{N}$ that is not divisible by p we have that $a^{p-1} \equiv 1 \pmod{p}$.*

Proof. Let $a \in \mathbb{N}$. We prove $a^p \equiv a \pmod{p}$ by induction on a. The base step $a = 1$ is obvious, so we proceed to the induction step.

So, assuming that $a^p \equiv a \pmod{p}$, we must prove that $(a + 1)^p \equiv a + 1 \pmod{p}$. We start by using the Binomial Theorem. Note that all products in the sum are products of natural numbers.

$$(a + 1)^p = \sum_{k=0}^{p} \binom{p}{k} a^k 1^{p-k}$$

$$= 1 + a^p + \sum_{k=1}^{p-1} \binom{p}{k} a^k$$

> By Exercise 3-68b, because p is prime, $p \left| \binom{p}{k} \right.$ for all $k \in \{1, \ldots, p - 1\}$. By Exercise 3-67, this means that $p \left| \sum_{k=1}^{p-1} \binom{p}{k} a^k \right.$, which implies that the sum is 0 modulo p.

$$\equiv 1 + a^p \pmod{p}$$

> Now use the induction hypothesis.

$$\equiv 1 + a \pmod{p},$$

which proves that $(a + 1)^p \equiv a + 1 \pmod{p}$, thus completing the induction.

Now let $a \in \mathbb{N}$ be so that $p \nmid a$. By Proposition 3.107 there is a $b \in \mathbb{N}$ with $ab \equiv 1 \pmod{p}$. Hence $a^p \equiv a \pmod{p}$ implies $a^p b \equiv ab \pmod{p}$ and then $a^{p-1} \equiv 1 \pmod{p}$. ∎

We are now ready to talk about the RSA algorithm for public key encryption. **RSA encryption**, named after R. Rivest, A. Shamir, and L. Adleman, who first published the algorithm (see [28]), works as follows. Throughout, p and q are two fixed, distinct prime numbers and we set $n := pq$. The security of the algorithm depends on how hard it is to factor n. Therefore, the bigger p and q are, the better, and the large prime numbers used in encryption algorithms are typically kept secret. Nonetheless, the algorithm itself works for all choices of prime numbers p and q. Moreover, we set $\varphi(n) := (p - 1)(q - 1)$ and we let $e \in \{2, \ldots, \varphi(n) - 1\}$ be so that e and $\varphi(n)$ have no common factors. (The notation $\varphi(n)$ is explained in Exercise 4-48.) Finally, let d be the solution of $de \equiv 1 \pmod{\varphi(n)}$, which exists by Proposition 3.107, because e and $\varphi(n)$ do not have any common factors.

The ordered pair (n, e) is called the **public key**. These numbers are openly disseminated and they are used to encrypt data. The number d is called the **private key**. This number is kept secret and should only be known to the receiver of the data. It is used to decrypt the data encrypted with (n, e).

To encrypt data, turn the message into a natural number m (or a string of natural numbers) smaller than n. Of course, the encoding should not be done letter by letter, because otherwise this encryption scheme devolves into nothing but a glorified

Caesarian cipher. Instead, blocks of letters are typically grouped together using an agreed upon scheme that can be decoded to get the original letters back. Because the safety of the encoding depends on arithmetic modulo n "wrapping around," so called **padding functions** are also used to make the "message number" m large enough so that wraparounds will occur. The encrypted message is the number $c := m^e$ (mod n). So, the sender computes m^e, then computes the representative c of $\left[m^e\right]_n$ that is in $\{0, \ldots, n-1\}$ and sends c. If several blocks of data need to be sent, the procedure is repeated for all blocks.

The decryption is equally simple. To decrypt the message, raise c to d and find the representative of $\left[c^d\right]_n$ that is in $\{0, \ldots, n-1\}$. This representative is m because the following theorem says that $c^d \equiv m$ (mod n) and because m is the unique representative of $\left[c^d\right]_n = [m]_n$ in $\{0, \ldots, n-1\}$. Readers will execute the encryption and decryption process, as well as attempt to break the code in Exercises 3-103 to 3-105.

Theorem 3.109 RSA encryption *(see [28]). With notation as introduced in the preceding paragraphs, if $c := m^e$ (mod n), then $c^d \equiv m$ (mod n).*

Proof. Using modulo n arithmetic, we obtain $c^d \equiv \left(m^e\right)^d \equiv m^{ed}$ (mod n).

Now, $de \equiv 1 \left(\text{mod } (p-1)(q-1)\right)$, so $de \equiv 1 \left(\text{mod } p-1\right)$ and $de \equiv 1 \left(\text{mod } q-1\right)$ (see Exercise 3-107a), that is, there are x and y so that $ed = 1+x(p-1) = 1+y(q-1)$. We will now show that $c^d \equiv m$ modulo p and modulo q.

Case 1: $p \nmid m$. If m is not a multiple of p, then by Fermat's Little Theorem, used in the last step, we obtain $c^d \equiv m^{ed} = m^{1+x(p-1)} = \left(m^{p-1}\right)^x m \equiv m$ (mod p).

Case 2: $p \mid m$. If m is a multiple of p, then $c^d \equiv m^{ed} \equiv 0^{ed} \equiv 0 \equiv m$ (mod p).

Similarly, we prove that $c^d \equiv m$ (mod q), which implies $c^d \equiv m$ (mod n) (see Exercise 3-107b). ∎

Theorem 3.109 proves that RSA encryption and decryption works. Moreover, Euclid's Algorithm (see Theorem 4.51) is a fast algorithm to compute the greatest common divisor for two numbers, and, if applied to $\varphi(n)$ and e, it produces the number d along the way. So, even if we cannot directly generate e (remember that the factors $p-1$ and $q-1$ are large and factoring them completely may not be easy), at least there is a fast way to check if a given e works, and it generates d along the way.

With the keys reasonably easily generated from p and q, we must ask ourselves why the RSA encryption system is secure. Assuming that there is no way to break the code using a "cryptoquote style" approach, we need the number d to decode the message. As an attacker, we have e and n, we know that n is the product of two prime numbers and we know that $ed \equiv 1 \left(\text{mod } \varphi(n)\right)$. So, if we knew $\varphi(n)$, breaking the code amounts to arithmetic (use Euclid's Algorithm to generate d). We know that $\varphi(n) = (p-1)(q-1)$, so to break the code we need to be able to factor n into p and q. There are programming libraries, such as [10], and calculators, such as [34], which perform precise arithmetic with arbitrarily many places. Programs like these can be used to generate public key codes and to work with them. But at the time this text is written, there is no fast algorithm that factors large numbers into their prime factors. Readers should try to solve Exercise 3-60c without using a computer to get an impression how hard it is to factor the product of two prime numbers. There are much more sophisticated factorization schemes than trying prime numbers that were found with the sieve of Erathosthenes, plus computers are typically used for factorizations. But it should be clear that no matter how sophisticated the scheme is and no matter how fast the computer is, it takes inordinate amounts of time to factor the product of two prime numbers with fifty digits each when no further information is available.

It has not been shown that no fast factorization algorithm exists. But it is known in complexity theory that a fast factorization algorithm would also make a lot of other problems that are considered "hard" much faster to solve than is presently possible. For an introduction to complexity theory, consider, for example, [11]. It has been shown that a quantum computer could potentially factor large numbers (for an introduction, consider, for example, [23]), but, at the time this text is written, quantum computing is not feasible on a large scale. For the time being, as long as enough large prime numbers are available and people who write encryption software can use large prime numbers that are only known to them for their encryption algorithms, the RSA scheme is quite safe.

So, the answer to question 8 on page x is that publicly encrypted internet transactions are safe because the answer to question 13 on page x is negative: To date, there is no fast factorization algorithm for integers.

Remark 3.110 This section shows another instance of the power of abstract mathematics. Recall how in Section 1.3 we started by analyzing the behavior of abstract logical connectives, and in Example 1.22 we found that their digital counterparts are the foundation of all digital computing. In Section 3.8 we started analyzing the abstract properties of a certain equivalence relation, and one of our examples led us to the first public key encryption system. This system is still at the heart of making internet transactions secure. That is, just as the existence of the internet is founded on the logic of $\wedge$, $\vee$ and $\neg$, its security relies on the properties of arithmetic modulo m. $\qquad\square$

After this excursion into applications, we are ready to tackle the construction of the integers in the next chapter.

Exercises

3-100. A converse for Proposition 3.107. Prove that if $x, m \in \mathbb{N}$ have a common factor $k > 1$, then there is a $c \in \mathbb{N}$ that is not divisible by m so that $cx \equiv 0 \pmod{m}$.

3-101. Let $a, p \in \mathbb{N}$ be so that $(a, p) = 1$ and let $n, m \in \mathbb{N}$ be so that $a^{n+m} \equiv a^n \pmod{p}$. Prove that then $a^m \equiv 1 \pmod{p}$.

3-102. Use $p = 5$ to show that there is an $a \in \mathbb{N}$ so that the smallest number m so that $a^m \equiv 1 \pmod{p}$ is not equal to $p - 1$.

3-103. For each of the following messages, find the encrypted message and the decrypted message. Be careful with rounding errors for the larger powers.

    (a) Message: $m = 2$, public key: $(n, e) = (91, 31)$, private key: 7

    (b) Message: $m = 7$, public key: $(n, e) = (85, 13)$, private key: 5

3-104. The encrypted message $c = 54$ was sent with the public key $(n, e) = (119, 17)$. Find the decrypted message m.

Hint. First find d. Be careful with overflows on a computer arithmetic.

3-105. Consider the following public key encryption algorithm: Each message is a string of characters (letters and spaces). This string is broken up into blocks of two characters each and each character is then assigned a number: Space= 00, $A = 01$, $B = 02$, and so on. In this fashion each block of two characters becomes a four digit number (counting leading zeros), which is encoded using a public key encryption system.

(a) Use the public key $(n, e) = (30607, 323)$ to encode the clear text message "I WANT TO LEARN."

(b) Use the private key $d = 26027$ (and $n = 30607$) to decode the coded message 20560, 4539, 27643, 24309, 7922, 3605, 3232, 27788. (If a decoded number has fewer than four digits, add leading zeros.)

Hint. For the computations, you will need a calculator with modulo arithmetic and an arbitrary number of precise digits, such as [34].

3-106. For each of the following pairs of numbers, determine if it is a valid pair for a public key (n, e). If not, explain why.

(a) $(77, 13)$

(b) $(490, 133)$

(c) $(187, 91)$

(d) $(24341, 17)$

(e) $(1717, 15)$

3-107. **Equivalence modulo m and factorizations of the modulus.** Let $k, n \in \mathbb{N}$ and let $a, b \in \mathbb{N}_0$

(a) Prove that if $a \equiv b \pmod{kn}$, then $a \equiv b \pmod{k}$ and $a \equiv b \pmod{n}$.

(b) Prove that if k and n have no common factors, then $a \equiv b \pmod{k}$ and $a \equiv b \pmod{n}$ implies $a \equiv b \pmod{kn}$.

3-108. Let $n_1, \dots, n_k$ be natural numbers so that no two of them have a common factor.

(a) **Chinese Remainder Theorem**. Prove that for any $a_1, \dots, a_k \in \mathbb{N}$ there is an $x \in \mathbb{N}$ so that $x \equiv a_j \pmod{n_j}$ holds for $j = 1, \dots, k$.
Hints. First prove the result for $k = 2$. With y_1, y_2 being so that $n_1 y_1 \equiv 1 \pmod{n_2}$ and $n_2 y_2 \equiv 1 \pmod{n_1}$ (justify why such numbers exist), let $x = a_1 n_2 y_2 + a_2 n_1 y_1$.

(b) Prove that any two solutions $x \in \mathbb{N}$ of the system $x \equiv a_j \pmod{n_j}$ $(j = 1, \dots, k)$ are equivalent modulo the product $\displaystyle\prod_{j=1}^{k} n_j$.
Hint. Use Exercise 3-107b.

(c) Explain how the above can be used to encode numbers $a \in \left\{0, \dots, \prod_{j=1}^{k} n_j - 1\right\}$ and their arithmetic using k-tuples $(a_1, \dots, a_k)$ and componentwise arithmetic modulo n_j, that is, with addition and multiplication in the jth component being performed modulo n_j.

3-109. Fermat's **non-primality test**. Let $p \in \mathbb{N}$.

(a) Prove that if there is an $a \in \{2, 3, \dots, p - 1\}$ so that $a^{p-1} \not\equiv 1 \pmod{p}$, then p is not prime.

(b) Now suppose that there is a finite set $M \subseteq \mathbb{N}$ of possible messages that are supposed to be sent with a public-key encryption system. Prove that we can use the same steps as for RSA encryption with numbers p and q that need not necessarily be prime, as long as for all $m \in M$ we have that $m^{p-1} \equiv 1 \pmod{p}$ and $m^{q-1} \equiv 1 \pmod{q}$.

Chapter 4

Number Systems II: Integers

In this chapter, we construct the integers from the natural numbers. The idea that integers need to be constructed may seem strange at first, but it is part of the modern world, too: Computers work with natural numbers. Further routines in operating systems make them look and act like integers or floating point numbers. That is, in a computer, the representations of the other numbers are constructed from $\mathbb{N}$, too, albeit in ways quite different from the ones we are about to explore.

Integers will finally allow us to subtract without any need for the contortions exhibited in Proposition 3.42 or Exercises 3-89 and 3-91. The desire to subtract certain numbers from each other is inspired, for example, by attempts to solve certain equations. For example, the equation $x + 6 = 2$ was not solvable in $\mathbb{N}$ (see Exercise 3-97a), but it will be solvable in the integers.

Just as for the natural numbers, we will keep the construction separate from the properties. So in Section 4.1 the integers are constructed as equivalence classes of formal differences of natural numbers, and in Section 4.3 we prove that we can assume that the natural numbers are a subset of the integers. Sections 4.2, 4.4 and 4.5 provide properties of the integers in the context of rings. This more abstract investigation will assure that we won't need to re-prove the corresponding properties for the rational, real and complex numbers. We conclude in Section 4.6 with some size considerations for sets, and, in particular, with the somewhat surprising insight that there are as many integers as there are natural numbers.

4.1 Arithmetic With Integers

We know that the integers differ from the natural numbers in that they contain zero and negative numbers. But how do we cleanly introduce these new objects in the context of set theory? For that matter, these objects can be challenging to grasp in real life,

Fundamentals of Mathematics: An Introduction to Proofs, Logic, Sets, and Numbers.
By Bernd S. W. Schröder.

too. There are middle school students who have difficulties with negative numbers. The mathematical philosophers of ancient Greece stayed entirely away from negative numbers, despite the fact that they worked well with nonnegative rational numbers. Even in the 1500s, mathematicians looked upon negative numbers with suspicion.

Although adding a single element "zero" to obtain $\mathbb{N}_0$ worked well in Definition 3.72, the situation becomes more complicated when we want to add zero and the negative numbers. Adding another set of natural numbers to $\mathbb{N}$ and giving each of the new numbers a negative sign, is hard, because we would need to define what a "negative sign" is. At the very least, this approach would require many case distinctions,[1] and the proof of Proposition 3.101 has already shown that case distinctions can be cumbersome.

Instead of looking at what we want the integers to *be*, let us consider what the integers should allow us to *do*.[2] Subtraction is only a partial binary operation on the natural numbers. But we know that subtraction is a (totally defined) binary operation on the integers. This means that the differences that cannot be formed in $\mathbb{N}$ can be formed in the integers. So for any $n, m \in \mathbb{N}$ with $m \geq n$ the difference $n - m$ exists in the integers. Conversely, any integer is the difference $n - m$ of two natural numbers (which can satisfy $n < m, n = m$ or $n > m$). So integers should be differences, which we could represent as pairs of natural numbers. But differences are not defined for all pairs of natural numbers, and for every integer there are many pairs of natural numbers (n, m) so that the difference $n - m$ is the integer. Therefore, we must avoid using subtraction in the definition of the integers and we must work around the ambiguity in the differences.

We achieve this purpose by considering pairs (n, m) of natural numbers that "formally represent the difference $n - m$." Pairs of numbers are defined in set theory, and we do not use subtraction in the definition of pairs. The fact that several differences can lead to the same number is taken care of by realizing that if $a - b = c - d$, then both (a, b) and (c, d) represent the same integer. Here we can circumvent using differences by noting that $a - b = c - d$ iff $a + d = b + c$. So integers should be represented by pairs $(a, b) \in \mathbb{N} \times \mathbb{N}$ so that $(a, b), (c, d) \in \mathbb{N} \times \mathbb{N}$ represent the same integer denoted $(a, b) \sim (c, d)$ iff $a + d = b + c$. This relation $\sim$ on $\mathbb{N} \times \mathbb{N}$ turns out to be an equivalence relation.

Proposition 4.1 *The relation $\sim$ on $\mathbb{N} \times \mathbb{N}$ defined by $(a, b) \sim (c, d)$ iff $a + d = b + c$ is an equivalence relation.*

Proof. We must prove that $\sim$ is reflexive, symmetric and transitive. We have proved relations to be equivalence relations before. But the present relation is slightly more complicated because it relates ordered pairs to ordered pairs. As is often the case, rigorous use of the definition shows that the proof is manageable after all.

---

[1] Exercise 4-22 takes this approach, and especially Exercise 4-22c shows that the approach is cumbersome indeed.

[2] So, just like our abstract investigation of properties in semigroups allows us to apply results in many contexts, focusing on properties, rather than objects, will simplify our construction of the very objects themselves.

For reflexivity, note that for all $(a, b) \in \mathbb{N} \times \mathbb{N}$ we have $a + b = b + a$, which means that $(a, b) \sim (a, b)$.

For symmetry, let $(a, b), (c, d) \in \mathbb{N} \times \mathbb{N}$. Then $(a, b) \sim (c, d)$ is equivalent to $a + d = b + c$, which is equivalent to $c + b = d + a$, which is equivalent to $(c, d) \sim (a, b)$.

For transitivity, let $(a, b), (c, d), (e, f) \in \mathbb{N} \times \mathbb{N}$ be so that $(a, b) \sim (c, d)$ and $(c, d) \sim (e, f)$. Then $a + d = b + c$ and $c + f = d + e$. Adding these equations yields $a + d + c + f = b + c + d + e$. By Proposition 3.19 (and by commutativity of addition), we can cancel $c + d$ to obtain $a + f = b + e$, which means that $(a, b) \sim (e, f)$. ∎

The equivalence relation $\sim$ identifies which pairs of natural numbers "represent the same integer." Moreover, every integer can be represented by its corresponding equivalence class. This is how we ultimately will define $\mathbb{Z}$. But the integers are not just a set. The integers are equipped with binary operations $+$ (addition) and $\cdot$ (multiplication). We will first focus on addition. If $[(a, b)]$ and $[(c, d)]$ are equivalence classes representing integers $a - b$ and $c - d$, respectively, then their sum should be $(a-b)+(c-d) = (a+c)-(b+d)$, which would be represented by $[(a+c, b+d)]$. But this definition of addition for the equivalence classes of $\sim$ uses individual representatives. Therefore, just as for arithmetic with equivalence classes modulo m in Theorems 3.103 and 3.104, we must prove that the operation actually is well-defined. After all, it would be catastrophic if the results of arithmetic with differences depended on how we write the differences. Unlike in Section 3.8, this time we immediately work with equivalence classes and we do not include auxiliary results such as Theorems 3.103 and 3.104.

Proposition 4.2 *Let $\sim$ be the equivalence relation on $\mathbb{N} \times \mathbb{N}$ from Proposition 4.1 and for all $(x, y) \in \mathbb{N} \times \mathbb{N}$, let $[(x, y)]$ denote the equivalence class of (x, y) under $\sim$. Then the operation $[(a, b)] + [(c, d)] := [(a + c, b + d)]$ is well-defined.*

Proof. We will prove the corresponding result for multiplication, which is harder, in Proposition 4.3. The proof of Proposition 4.2 is left to the reader as Exercise 4-1. ∎

Of course, addition is not the only algebraic operation on the integers. We can multiply them, too. But integers are classes of pairs of natural numbers $[(a, b)]$ that represent differences $a - b$. So the product of two such classes $[(a, b)]$ and $[(c, d)]$ should represent $(a - b)(c - d) = ac - ad - bc + bd = (ac + bd) - (ad + bc)$, which would be represented by $[(ac+bd, ad+bc)]$. Once we understand this motivation, the definition of multiplication of integers is quite unsurprising, and, thankfully, multiplication is well-defined, too.

Proposition 4.3 *Let $\sim$ be the equivalence relation on $\mathbb{N} \times \mathbb{N}$ from Proposition 4.1 and for all $(x, y) \in \mathbb{N} \times \mathbb{N}$, let $[(x, y)]$ denote the equivalence class of (x, y) under $\sim$. Then the operation $[(a, b)] \cdot [(c, d)] := [(ac + bd, ad + bc)]$ is well-defined.*

Proof. Let $[(a, b)] = [(a', b')]$ and let $[(c, d)] = [(c', d')]$ be two equivalence classes of $\sim$. To prove that $[(ac + bd, ad + bc)] = [(a'c' + b'd', a'd' + b'c')]$, we must prove that $ac + bd + a'd' + b'c' = a'c' + b'd' + ad + bc$. From the definition of the equivalence relation we know that $a + b' = a' + b$ and $c + d' = c' + d$. We need to use these equations and "the right factorizations" to obtain the equation above. But no two terms on either side of the desired equation have a common factor. Thus we add a term $b'c$ to allow some factorizations. Note that we do not add and subtract the term, because subtraction may or may not be defined. Instead, we will use Proposition 3.19 to cancel the term $b'c$ at the end of the computation.

$$
\begin{aligned}
ac + bd + a'd' + b'c' + b'c &= (a + b')c + bd + a'd' + b'c' \\
&= (a' + b)c + bd + a'd' + b'c' \\
&= a'c + bc + bd + b'c' + a'd' \\
&= a'(c + d') + bc + bd + b'c' \\
&= a'(c' + d) + bc + bd + b'c' \\
&= a'c' + a'd + bc + bd + b'c' \\
&= a'c' + (a' + b)d + bc + b'c' \\
&= a'c' + (a + b')d + bc + b'c' \\
&= a'c' + ad + b'd + bc + b'c' \\
&= a'c' + ad + b'(d + c') + bc \\
&= a'c' + ad + b'(d' + c) + bc \\
&= a'c' + ad + b'd' + b'c + bc \\
&= a'c' + b'd' + ad + bc + b'c,
\end{aligned}
$$

which implies that $ac + bd + a'd' + b'c' = a'c' + b'd' + ad + bc$, that is, the equivalence classes satisfy $[(ac + bd, ad + bc)] = [(a'c' + b'd', a'd' + b'c')]$. Hence the multiplication of the equivalence classes is well-defined. ∎

With well-defined addition and multiplication operations, we can formally introduce the integers and their algebraic operations.

Definition 4.4 *The* **integers** $\mathbb{Z}$ *are defined to be the set of equivalence classes* $[(a, b)]$ *of elements of* $\mathbb{N} \times \mathbb{N}$ *under the equivalence relation* $\sim$ *of Proposition 4.1.* **Addition** *of integers is defined by* $[(a, b)] + [(c, d)] = [(a + c, b + d)]$ *and* **multiplication** *is defined by* $[(a, b)] \cdot [(c, d)] = [(ac + bd, ad + bc)]$.

Theorem 4.5 below shows that addition has exactly the properties that we expect and Theorem 4.6 does the same for multiplication. Properties are listed in the order in which they typically occur in the more abstract definitions of groups and rings (see Section 4.2).

In the remainder of this section, we use the symbols O and I for the zero and one of the natural numbers and for the zero and one of the integers, respectively. This should not cause confusion: Elements of the natural numbers only occur in the pairs of which we form equivalence classes, whereas the elements of the integers are equivalence classes of pairs of natural numbers. That means we can tell which symbol denotes which entity from the context in which the symbol is used. This approach is common practice in mathematics. It is usually preferable over distinguishing symbols that denote similar entities, say, with subscripts.

Theorem 4.5 *Addition $+$ of integers is associative, $0 := [(1, 1)]$ is a neutral element with respect to $+$, for every $x = [(a, b)] \in \mathbb{Z}$ there is an element $-x := [(b, a)]$ so that $x + (-x) = (-x) + x = 0$, and $+$ is commutative.*

Proof. Associativity of the addition operation follows from the associativity of addition of natural numbers as follows. Let $x, y, z \in \mathbb{Z}$ with $x = [(a, b)]$, $y = [(c, d)]$, and $z = [(e, f)]$. Then

$$
\begin{aligned}
(x + y) + z &= \left([(a, b)] + [(c, d)]\right) + [(e, f)] \\
&= [(a + c, b + d)] + [(e, f)] \\
&= [((a + c) + e, (b + d) + f)] \\
&= [(a + (c + e), b + (d + f))] \\
&= [(a, b)] + [(c + e, d + f)] \\
&= [(a, b)] + \left([(c, d)] + [(e, f)]\right) \\
&= x + (y + z).
\end{aligned}
$$

The equivalence class $0 := [(1, 1)]$ is a neutral element for addition, because for any $x = [(a, b)] \in \mathbb{Z}$ we have that

$$
\begin{aligned}
x + 0 &= [(a, b)] + [(1, 1)] \\
&= [(a + 1, b + 1)] \\
&= [(a, b)] \\
&= x \\
&= [(a, b)] \\
&= [(1 + a, 1 + b)] \\
&= [(1, 1)] + [(a, b)] \\
&= 0 + x.
\end{aligned}
$$

For each $x = [(a, b)] \in \mathbb{Z}$, the element $-x := [(b, a)] \in \mathbb{Z}$ is an "inverse element" under addition, because

$$
\begin{aligned}
x + (-x) &= [(a, b)] + [(b, a)] \\
&= [(a + b, b + a)] \\
&= [(1, 1)] \\
&= 0 \\
&= [(b + a, a + b)] \\
&= [(b, a)] + [(a, b)] \\
&= (-x) + x.
\end{aligned}
$$

Finally, commutativity follows from the commutativity of addition of natural numbers as follows. Let $x, y \in \mathbb{Z}$ with $x = \big[(a, b)\big]$ and $y = \big[(c, d)\big]$. Then

$$
\begin{aligned}
x + y &= \big[(a, b)\big] + \big[(c, d)\big] \\
&= \big[(a + c, b + d)\big] \\
&= \big[(c + a, d + b)\big] \\
&= \big[(c, d)\big] + \big[(a, b)\big] \\
&= y + x.
\end{aligned}
$$

■

Note that in the proof of Theorem 4.5 we could have saved ourselves some work if we had established the commutativity right after associativity.

Theorem 4.6 *Multiplication of integers is associative, distributive over addition, it has a neutral element* $1 := \big[(2, 1)\big]$, *and it is commutative.*

Proof. Exercise 4-2.   ■

Exercises

4-1. Prove Proposition 4.2.

Hint. To prove that the operation $\big[(a, b)\big] + \big[(c, d)\big] := \big[(a + c, b + d)\big]$ is well-defined on the equivalence classes of the equivalence relation $\sim$ in $\mathbb{N} \times \mathbb{N}$ from Proposition 4.1, you need to prove that if $(a, b) \sim (a', b')$ and $(c, d) \sim (c', d')$, then $(a + c, b + d) \sim (a' + c', b' + d')$.

4-2. Prove Theorem 4.6.

   (a) Prove that multiplication of integers is associative.

   (b) Prove that multiplication of integers is distributive over addition.

   (c) Prove that $1 := \big[(2, 1)\big]$ is a neutral element with respect to multiplication of integers.

   (d) Prove that multiplication of integers is commutative.

4-3. Let $x, y \in \mathbb{Z} \setminus \{0\}$. Prove that $xy \neq 0$.

4-4. Let $V(\mathbb{N}) = \bigcup\limits_{n=0}^{\infty} V_n(\mathbb{N})$ be the superstructure over $\mathbb{N}$. Find the smallest $n \in \mathbb{N}$ so that

   (a) $V_n(\mathbb{N})$ contains the set $\mathbb{Z}$.

   (b) $V_n(\mathbb{N})$ contains the addition operation on $\mathbb{Z}$.

   (c) $V_n(\mathbb{N})$ contains the multiplication operation on $\mathbb{Z}$.

   (d) $V_n(\mathbb{N})$ contains the triple $(\mathbb{Z}, +, \cdot)$.

   In each case, justify your answer.

4-5. Let $(a, b) \sim (c, d)$ with $\sim$ as in Proposition 4.1.

   (a) Prove that $a < b$ iff $c < d$.

   (b) Prove that $a = b$ iff $c = d$.

   (c) Prove that $a > b$ iff $c > d$.

4.2 Groups and Rings

Just as we did for the natural numbers, we want to investigate the properties of the integers. From the natural numbers, we remember that there are a good many properties that feel obvious to us, but which must be proved nonetheless. Therefore, it should be more efficient to focus once more on the properties of the operations, this time on the integers, and to derive consequences from them. In this fashion, the results we establish here will hold for $\mathbb{Z}$ *and* they will not need to be proved over and over for the other structures that will be introduced in this text. First, we focus on the properties of addition of integers (see Theorem 4.5), which are those of a commutative group.

Definition 4.7 *Let G be a set and let* $\circ : G \times G \to G$ *be a binary operation on G. Then* $(G, \circ)$ *is called a* **group** *iff the following hold.*

1. *The operation* $\circ$ *is* **associative**. *That is, for all* $a, b, c \in G$ *we have that* $(a \circ b) \circ c = a \circ (b \circ c)$.

2. *There is a* **neutral element** $e \in G$ *with respect to* $\circ$. *That is, there is an* $e \in G$ *so that for all* $a \in G$ *we have that* $e \circ a = a \circ e = a$.

3. *For every* $a \in G$, *there is an* **inverse element** $\tilde{a}$. *That is, for every* $a \in G$, *there is an* $\tilde{a} \in G$ *so that* $a \circ \tilde{a} = \tilde{a} \circ a = e$.

Moreover, a group is called **commutative** *iff* $\circ$ *is commutative. That is, iff for all* $a, b \in G$ *we have* $a \circ b = b \circ a$.

The notation $\circ$ can be a bit confounding when considering *addition* in $\mathbb{Z}$, because the group operation $\circ$ is more reminiscent of multiplication than of addition. This is unavoidable, because we want to work with the customary notation for groups. So for groups, we must pay close attention to which properties of an operation are given in its definition. Certain notations other than $\circ$ have become standard for groups.

Notation 4.8 In a (general) group G, the binary operation is usually denoted $\circ$, the neutral element is denoted e and the inverse element of $a \in G$ is denoted a^{-1}. That is, in a general group, the operation is pretty much viewed as something like multiplication of numbers or composition of functions.

In a *commutative* group G (and usually *only* in commutative groups), the binary operation is often denoted $+$, the neutral element is denoted 0 and the inverse element of $a \in G$ is denoted $-a$.

If the group operation specifically is a multiplication, the binary operation can also be denoted $\cdot$. In this case, the neutral element would be denoted 1 and the inverse element of $a \in G$ would be denoted a^{-1}. $\qquad\square$

In Chapters 4 and 5 we will almost exclusively work with commutative groups. Thus, once we have defined rings in Definition 4.11, we will not need to worry much about these notations, because the underlying group will use additive notation.

The following example shows how versatile the idea of a group is.

Example 4.9 Examples of groups.

1. Theorem 4.5 shows that the pair $(\mathbb{Z}, +)$, where $\mathbb{Z}$ denotes the integers and $+$ is addition of integers, is a commutative group. The neutral element is $0 := \big[(1, 1)\big]$ and for $\big[(a, b)\big] \in \mathbb{Z}$, the inverse element is $\big[(b, a)\big]$.

2. Let $m \in \mathbb{N}$. The pair $(\mathbb{Z}_m, +)$, with $\mathbb{Z}_m$ being the integers modulo m and $+$ being addition as defined in Definition 3.106, is a commutative group (see Exercise 4-6a). The neutral element is $0 := [0]_m$ and for $[a]_m \in \mathbb{Z}_m$ with $a \in \{0, \ldots, m-1\}$, the inverse element is $[m - a]_m$.

3. Let A be a set, let $\text{Bij}(A)$ be the set of all bijective functions $f : A \to A$ and let $\circ$ be composition of functions. Then $\big(\text{Bij}(A), \circ\big)$ is a group, but $\circ$ is not commutative. The neutral element is $\text{id}_A : A \to A$, defined by $\text{id}_A(a) = a$ for all $a \in A$, and for $f \in \text{Bij}(A)$, the inverse element is f^{-1}, the inverse function of f (see Exercise 4-6b).

   Composition of functions is one of the reasons why commutativity is treated as an additional property for groups. In this fashion, theorems about groups in general will be applicable to functions and composition, too.

4. Although every group is a semigroup, the pair $(\mathbb{N}, +)$ shows that not every semigroup is a group. $\qquad\qquad\square$

From here on, every time we introduce a more general structure, we will be exposed to an "avalanche of knowledge," because the structures are defined in such a way that the properties of the structures that were introduced earlier are retained. In this fashion, we will already have a good bit of information available for each new structure.

What do we already know about groups? Every group is easily seen to be a semigroup. Therefore, by Proposition 3.35 the neutral element e with respect to the group operation is unique. That is, there is no other element in the group with the same properties as e. Moreover, for commutative groups, we can define summations as in Definition 3.61 and the properties of sums given in Theorem 3.59 (no need for parentheses that group sums so that every addition has two summands), Theorem 3.64 (sums can be re-indexed and combined) and in Theorem 3.65 (sums can be reordered and added termwise) hold in commutative groups. (We demanded commutativity throughout, because summation notation is uncommon for non-commutative operations.)

As was recommended for all the abstract concepts in this text, we should also remind ourselves of what this knowledge provides in more concrete situations.

What does this say about $\mathbb{Z}$? For the integers, this means that 0 is the only neutral element with respect to addition in $\mathbb{Z}$ and that summations behave as they should: There is no need for extra parentheses in long sums and sums can be reordered and combined just like we are used to.

This was the first time we use abstract results to save us the proofs of the corresponding results for a new structure. This step of verifying that abstract results hold for the new structure under investigation will be repeated after we introduce rings in Definition 4.11, fields in Definition

5.8 and totally ordered fields in Definition 5.20. In this fashion, we will rapidly and efficiently learn a lot about these new structures.

In addition to the above, we also know that every integer has exactly one additive inverse. But we have not yet proved this fact. As it turns out, this uniqueness of inverses is actually a property of semigroups with neutral elements. Once more we prove the result in the abstract context, so that it will be available for the structures that will be introduced in the following.

Proposition 4.10 *Let* $(S, \circ)$ *be a semigroup with neutral element e and let* $a \in S$ *have an inverse element with respect to* $\circ$. *Then a has exactly one inverse element. That is, if* $\tilde{a}$ *and* $\overline{a}$ *both have the properties of an inverse element of a, then* $\tilde{a} = \overline{a}$.

Proof. Let $\tilde{a}, \overline{a} \in S$ be so that $\tilde{a} \circ a = a \circ \tilde{a} = e$ and $\overline{a} \circ a = a \circ \overline{a} = e$. Then

$$\tilde{a} = \tilde{a} \circ e = \tilde{a} \circ (a \circ \overline{a}) = (\tilde{a} \circ a) \circ \overline{a} = e \circ \overline{a} = \overline{a}.$$

∎

In particular, Proposition 4.10 proves that the additive inverse of every element of $\mathbb{Z}$ is unique. That is, every integer's "corresponding negative number" is unique. But note that, because Proposition 4.10 only relies on the structure of a semigroup with a neutral element (it does not even need commutativity), it also proves that for every bijective function $f : A \to A$, the inverse function is unique. We will use Proposition 4.10 later to prove that additive as well as multiplicative inverses are unique in $\mathbb{Q}$, $\mathbb{R}$ and $\mathbb{C}$. That is, Proposition 4.10 proves one result that we are immediately interested in (uniqueness of additive inverses in $\mathbb{Z}$); and it proves at least 7 (seven) more results that we would reasonably be interested in, and six of which we will encounter as we build the familiar number systems.

Just like addition of integers, multiplication of integers is associative and commutative and it has a neutral element, denoted 1 (see Theorem 4.6). Multiplication of integers does not admit multiplicative inverses for integers other than 1 (by definition) and -1 (see Proposition 4.18 below). Moreover, just like for the natural numbers, multiplication is distributive over addition. An abstract structure with these properties is called a commutative ring with unity. For rings, the notation reverts back to what we are familiar with.

Definition 4.11 *Let* R *be a set and let* $+ : R \times R \to R$ *and* $\cdot : R \times R \to R$ *be binary operations on* R. *Then the triple* $(R, +, \cdot)$ *is called a* **ring** *iff the following hold.*

1. *Addition is* **associative**, *that is, for all* $x, y, z \in R$ *we have*

   $(x + y) + z = x + (y + z).$

2. *Addition is* **commutative**, *that is, for all* $x, y \in R$ *we have*

   $x + y = y + x.$

3. *There is a* **neutral element** 0 *for addition, that is, there is an element* $0 \in R$ *so that for all* $x \in R$ *we have* $x + 0 = x.$

4. *For every element* $x \in R$, *there is an* **additive inverse** *element* $(-x)$ *so that*
   $x + (-x) = 0$.

5. *Multiplication is* **associative**, *that is, for all* $x, y, z \in R$ *we have*
   $(x \cdot y) \cdot z = x \cdot (y \cdot z)$.

6. *Multiplication is* **left distributive** *and* **right distributive** *over addition, that is, for all* $\alpha, x, y \in R$ *we have* $\alpha \cdot (x + y) = \alpha \cdot x + \alpha \cdot y$ *and* $(x + y) \cdot \alpha = x \cdot \alpha + y \cdot \alpha$.

As is customary for multiplication, the dot between factors is usually omitted. Moreover, we introduce the following special properties for rings.

7. *A ring is called* **commutative** *iff multiplication is* **commutative**, *that is, iff for all* $x, y \in R$ *we have* $x \cdot y = y \cdot x$.

8. *A ring is called a* **ring with unity** *iff there is a* **neutral element** $1 \neq 0$ *for multiplication, that is, iff there is an element* $1 \in R \setminus \{0\}$ *so that for all* $x \in R$ *we have* $1 \cdot x = x$.

9. *In a ring with unity, an element* b *is called a* **multiplicative inverse** *of the element* a *iff* $ab = ba = 1$.

To simplify notation, we often do not explicitly mention the addition and multiplication operation and speak of a ring R *instead of a ring* $(R, +, \cdot)$.

As with groups, which need not be commutative, commutativity of multiplication, the existence of a neutral element for multiplication and the existence of multiplicative inverses in rings with unity are additional properties that a general ring may or may not possess. Note, however, that the addition operation in a ring will always be commutative.

Notation 4.12 The notation for rings is much more standardized than that for groups. The operation with respect to which the ring is a commutative group is usually denoted by $+$, its neutral element is usually denoted by 0, the other operation is usually denoted by $\cdot$, and if there is a neutral element with respect to $\cdot$, it is denoted 1.

But note that, at this stage, the symbol $-x$ for the additive inverse of x is a *compound symbol*. These types of "negative signs" will have the properties we expect them to have, but we must prove these properties before we can use them.   □

What do we already know about rings? It follows directly from the definition that if $(R, +, \cdot)$ is a ring, then $(R, +)$ is a commutative group. Therefore, by Proposition 3.35 the neutral element 0 with respect to the addition operation is unique and by Proposition 4.10, additive inverses are unique. We can define summations as in Definition 3.61 and the properties of sums given in Theorem 3.59 (no need for parentheses that group sums so that every addition has two summands), Theorem 3.64 (sums can be re-indexed and combined) and in Theorem 3.65 (sums can be reordered and added termwise) hold in rings.

Regarding multiplication, we note that Theorem 3.59 also holds for the ring multiplication (so there is no need for parentheses in products to make sure every multiplication has two factors), that parentheses are multiplied out "as usual" because of Proposition 3.38 and Theorem 3.66, and that we can define products as in Definition 3.69 and powers as in Definition 3.70. Moreover, for commutative rings, power laws as in Theorem 3.71 hold, and if there is a neutral element 1 for multiplication, then the Binomial Theorem (Theorem 3.77) holds, too.

If the ring contains a neutral element 1 with respect to multiplication (but is not necessarily commutative), then by Proposition 3.35, the neutral element with respect to multiplication is unique. Finally, by Proposition 4.10 any inverse element with respect to multiplication is unique.

Note that in particular, all the above results hold for the integers $\mathbb{Z}$.

Once more, the above list shows the strength of establishing abstract results first. Read it carefully. We have barely introduced rings and we already know many of their properties, including, quite impressively, the Binomial Theorem for commutative rings with unity.

Example 4.13 Examples of rings.

1. Theorems 4.5 and 4.6 show that the triple $(\mathbb{Z}, +, \cdot)$, where $\mathbb{Z}$ denotes the integers, $+$ is addition of integers and $\cdot$ is multiplication of integers is a commutative ring with unity.

2. Let $m \in \mathbb{N}$. The triple $(\mathbb{Z}_m, +, \cdot)$, with $\mathbb{Z}_m$ being the integers modulo m and $+$ and $\cdot$ being addition and multiplication as defined in Definition 3.106, is a commutative ring with unity (see Exercise 4-7a).

   The structure of arithmetic modulo m (recall Figure 3.5 on page 149) is probably one of the reasons why rings are called "rings."

3. Let $(R, +, \cdot)$ be a commutative ring. For $n \in \mathbb{N}$ all $a_j \in R$ for all $j \in \mathbb{N}$, a function $f : R \to R$ of the form $f(x) = \sum_{j=0}^{n} a_j x^j$ is called a **polynomial**. Let $R[x]$ be the set of all functions $f : R \to R$ of the form $f(x) = \sum_{j=0}^{n} a_j x^j$, where $n \in \mathbb{N}$ and all $a_j \in R$. For any two functions $f, g \in R[x]$, define addition pointwise as $(f + g)(x) := f(x) + g(x)$ and define multiplication pointwise as $(f \cdot g)(x) := \big(f(x)\big) \cdot \big(g(x)\big)$. Then, with the representations $f(x) = \sum_{j=0}^{n} a_j x^j$, $g(x) = \sum_{j=0}^{m} b_j x^j$, with M being the larger of m and n, with $a_j := 0$ for $j \in \{n+1, \ldots, M\}$ and with $b_j := 0$ for $j \in \{m+1, \ldots, M\}$ we have that

$$(f+g)(x) = \sum_{j=0}^{M}(a_j + b_j)x^j \text{ and } (f \cdot g)(x) = \sum_{j=0}^{2M}\left(\sum_{k=\max\{0, j-M\}}^{\min\{j, M\}} a_k b_{j-k}\right) x^j.$$

   (The formula for sums is clear. The formula for products must be proved by induction, see Exercise 4-7(b)i.) Hence sums and products of polynomials are polynomials, too, and $\big(R[x], +, \cdot\big)$ is a commutative ring, called the **ring of polynomials** over R (see Exercise 4-7b). Moreover, if R is a ring with unity, then so is $R[x]$.

We will work extensively with rings of polynomials when we investigate the unsolvability of the quintic in radicals in Chapter 6. In algebra, rings of poly-nomials are defined over general (not necessarily commutative) rings as *formal* polynomials, not actual polynomial functions. In this fashion, cancelations of the variable x cannot occur in $R[x]$. For the rings of polynomials that we will investigate in Chapter 6, the two definitions lead to the same entities. There-fore we will stay with the more familiar definition. We pay a small price for this convenience: When we consider polynomials as functions, the usual algebra for polynomials works only if the underlying ring is commutative. So we will demand commutativity whenever we work with rings of polynomials.

4. The group $\text{Bij}(A)$ of bijective functions from a set A to itself shows that not every group is a ring. (Also see Figure 3.1 on page 107.)   □

The fact that every element of a ring has an additive inverse allows us to introduce "negative signs." With "negative signs" available, we can finally define subtraction as a (totally defined) binary operation.

Definition 4.14 *Let* $(R, +, \cdot)$ *be a ring. We define the binary operation of* **subtraction** *for all* $(x, y) \in R \times R$ *as* $x - y := x + (-y)$. *The element* $x - y$ *is also called the* **difference** *of* x *and* y.

With subtraction available, we can provide the more customary definition of the integers modulo m.

Example 4.15 Let $m \in \mathbb{N}$. For $x, y \in \mathbb{Z}$ we define $x \equiv y \pmod{m}$ iff there is a $k \in \mathbb{Z}$ so that $x - y = km$. Then the same proof as for Proposition 3.101 shows that $\equiv \pmod{m}$ is an equivalence relation on $\mathbb{Z}$. In fact, the proof is simpler, because we do not need to worry about which number is larger in the subtraction.

We define $\mathbb{Z}_m$ to be the set of all equivalence classes $[x]_m$ of $\mathbb{Z}$ with respect to $\equiv \pmod{m}$. Proofs similar to those of Theorems 3.103 and 3.104 show that the addition $[x]_m + [y]_m := [x + y]_m$ and the multiplication $[x]_m \cdot [y]_m := [x \cdot y]_m$ are well-defined.

Now $(\mathbb{Z}_m, +, \cdot)$ is a commutative ring with unity, because all the properties that work for integers also work for the representatives in the classes. For example, addition is commutative, because $[x]_m + [y]_m = [x + y]_m = [y + x]_m = [y]_m + [x]_m$, and the other properties are established similarly.

Moreover, we can formally establish that Definition 3.106 leads to "the same" mathematical entity as the one that was just given here: Map each equivalence class $[x]_m$ from Definition 3.106 to the equivalence class $[x]_m$ defined here. The resulting function is bijective and it does not matter if algebraic operations are performed be-fore or after applying the function. That means that the two entities are, for algebraic purposes, the same. We will investigate this idea of sameness (called isomorphism) further in Definition 4.19. Readers who want to make absolutely sure that Definition 3.106 and the above lead to isomorphic structures should consider Exercise 4-17 in the next section.

The commutative ring with unity $(\mathbb{Z}_m, +, \cdot)$ that we defined here is also called the ring of **integers modulo** m.   □

Now we are ready to prove some new results. We first note that the existence of additive inverses makes cancelations a lot easier.

Just as in our first encounters with abstract results in Chapter 3, you may first want to primarily think about integers as we prove the results below. But note that in every proof, we only use the structure of a ring.

Proposition 4.16 *Let $(R, +, \cdot)$ be a ring and let $a, b, c \in R$ be so that $a + c = b + c$. Then $a = b$.*

Proof. If $a + c = b + c$, then
$$a = a + 0 = a + (c - c) = (a + c) - c = (b + c) - c = b + (c - c) = b + 0 = b. \blacksquare$$

Note that, in general rings, we cannot establish the corresponding cancelation law for multiplication: $\mathbb{Z}_8$ is a ring and $[2]_8[4]_8 = [0]_8 = [6]_8[4]_8$, but $[2]_8 \neq [6]_8$.

Next we establish that "zero times anything is zero."

Proposition 4.17 *Let $(R, +, \cdot)$ be a ring and let $x \in R$. Then $0 \cdot x = x \cdot 0 = 0$.*

Proof. First note that $0x + 0x = (0 + 0)x = 0x$. But then

$$0 = 0x + (-0x) = (0x + 0x) + (-0x) = 0x + \big(0x + (-0x)\big) = 0x + 0 = 0x.$$

The proof that $x0 = 0$ is similar (see Exercise 4-9). $\blacksquare$

By Exercise 4-10, Proposition 4.17 implies that zero can never have a multiplicative inverse. This means, independent of how much we refine the mathematics of rings, we have just established that we will never be allowed to "divide by zero." So the answer to question 3 on page x is that we cannot divide by zero because "0/0" should be 1, but 0 times anything (including its fictitious multiplicative inverse) is 0, and of course "0/0" cannot be both.

The key property about working with negative signs is that "a negative times a negative gives a positive." Below we establish this insight for rings in general.

Proposition 4.18 *Let $(R, +, \cdot)$ be a ring and let $x, y \in R$. Then $(-x)(-y) = xy$.*

Proof. First note that $(-x)$ and $(-y)$ really are just compound symbols for the additive inverses of x and y. By the uniqueness of inverses in groups (see Proposition 4.10) the result is proved if we can prove that $(-x)(-y)$ is the additive inverse of $-(xy)$. We first note that because $(-x)y + xy = \big((-x) + x\big)y = 0y = 0$ we have that $-(xy) = (-x)y$. But then

$$(-x)(-y) + \big(-(xy)\big) = (-x)(-y) + (-x)y = (-x)\big((-y) + y\big) = (-x)0 = 0$$

implies that $(-x)(-y) = -\big(-(xy)\big) = xy.$ $\blacksquare$

Exercises

4-6. Examples of groups.

    (a) Let $m \in \mathbb{N}$. Prove that $(\mathbb{Z}_m, +)$, with $\mathbb{Z}_m$ being the integers modulo m and $+$ being addition as defined in Definition 3.106, is a commutative group.

    (b) Let A be a set. Prove that $\big(\mathrm{Bij}(A), \circ\,\big)$, with $\mathrm{Bij}(A)$ being the set of bijective functions $f : A \to A$ and $\circ$ being composition of functions, is a group that is not commutative.

    (c) Let $M(n \times n, \mathbb{R})$ be the set of all real invertible $n \times n$ matrices and let $\cdot$ denote matrix multiplication. Prove that $\big(M(n \times n, \mathbb{R}), \cdot\,\big)$ is a group that is not commutative for $n > 1$.

    (You may use what you know about real numbers and matrix multiplication for this part.)

4-7. Examples of rings.

    (a) Let $m \in \mathbb{N}$. Prove that the triple $(\mathbb{Z}_m, +, \cdot)$, with $\mathbb{Z}_m$ being the integers modulo m and $+$ and $\cdot$ being addition and multiplication as defined in Definition 3.106, is a commutative ring with unity.

    (b) Let $(R, +, \cdot)$ be a commutative ring. Let $R[x]$ be the set of all functions $f : R \to R$ of the form $f(x) = \sum_{j=0}^{n} a_j x^j$.

       i. Let $f(x) = \sum_{j=0}^{M} a_j x^j$, $g(x) = \sum_{j=0}^{M} b_j x^j$ be elements of $R[x]$. Prove that the product $f \cdot g$ satisfies $(f \cdot g)(x) = \sum_{j=0}^{2M} \left(\sum_{k=\max\{0, j-M\}}^{\min\{j, M\}} a_k b_{j-k} \right) x^j$, where $\min\{j, M\}$ denotes the smaller of j and M and $\max\{0, j - M\}$ denotes the larger of 0 and $j - M$.
       Hint. Induction on $M \in \mathbb{N}_0$.

       ii. Prove that $\big(R[x], +, \cdot\,\big)$ is a commutative ring.

       iii. Prove that if R is a ring with unity, then so is $R[x]$.

    (c) Prove that the set $\mathcal{F}(\mathbb{Z}, \mathbb{Z})$ of functions $f : \mathbb{Z} \to \mathbb{Z}$ with pointwise addition and multiplication, that is $(f + g)(x) := f(x) + g(x)$ and $(f \cdot g)(x) := f(x) \cdot g(x)$ for all $x \in \mathbb{Z}$, is a commutative ring with unity.

4-8. Pathological examples of rings. Aside from structures that we consider natural, such as the ones in Exercise 4-7, there are also some "stranger" structures that satisfy the axioms of a ring. Although these examples can (and possibly should) be considered pathological, examples of this type show that there are limitations to what can be concluded from a given definition. Hence, despite being pathological, examples like the ones below have a place in mathematics.

    (a) Let $\mathbb{Z}$ be the integers and let $+ : \mathbb{Z} \times \mathbb{Z} \to \mathbb{Z}$ be the usual addition of integers. For $x, y \in \mathbb{Z}$ define $x \odot y := 0$. Prove that $(\mathbb{Z}, +, \odot)$ is a commutative ring.

    (b) For $(a, b), (c, d) \in \mathbb{Z} \times \mathbb{Z}$ define $(a, b) + (c, d) := (a+c, b+d)$ and $(a, b) \odot (c, d) := (ac, 0)$. Prove that $(\mathbb{Z} \times \mathbb{Z}, +, \odot)$ is a commutative ring.

4-9. Let R be a ring and let $x \in R$. Prove that $x0 = 0$.

4-10. Let $(R, +, \cdot)$ be ring with unity 1. Prove that 0 cannot have a multiplicative inverse.
    Hint. Use Proposition 4.17 and the requirement that $1 \neq 0$ in a proof by contradiction.

4-11. A common mistake, and situations in which is not a mistake.

    (a) Let $(R, +, \cdot)$ be a ring so that for all $x \in R$ we have that $x + x \neq 0$ and let $a, b \in R$ be so that $ab \neq 0$. Prove that $(a + b)^2 \neq a^2 + b^2$.

    (b) Prove that in the ring $\mathbb{Z}_2$ the equality $a^2 + b^2 = (a + b)^2$ holds. Then explain *why* the equality holds.

(c) Prove that in the ring $\left(\mathcal{F}(\mathbb{Z}, \mathbb{Z}), +, \cdot\right)$ there are functions f and g with $(f + g)^2 = f^2 + g^2$. Then explain *why* the equality holds for your example and show that it does *not* hold for all functions f and g.

4-12. Let X be a set and $R \subseteq \mathcal{P}(X)$ be so that for any two $A, B \in R$ we have that $A \triangle B \in R$ (recall that $\triangle$ denotes the symmetric difference of two sets) and $A \cap B \in R$. Let $+ := \triangle$, let $\cdot := \cap$, and let $0 := \emptyset$.

(a) Prove that $(R, +, \cdot) = (R, \triangle, \cap)$ is a commutative ring.

(b) Prove that if $X \in R$, then $(R, +, \cdot) = (R, \triangle, \cap)$ is a commutative ring with unity.

(c) Prove that for any two $A, B \in R$ we have that $A \cup B \in R$.

   Note. Even though $(R, +, \cdot) = (R, \triangle, \cap)$ contains unions of any two elements of R, Exercises 4-13 and 4-14 below show that union and intersection do not lead to a ring structure in the algebraic sense.

(d) Prove that for any two $A, B \in R$ we have that $A \setminus B \in R$.

4-13. Let X be a set and let $R := \mathcal{P}(X)$, $+ := \cap$, $\cdot := \cup$, and $0 := X$.

(a) Prove that $\left(\mathcal{P}(X), \cap, \cup\right)$ does *not* satisfy part 4 (existence of additive inverses) of Definition 4.11 of rings.

(b) Prove that $\left(\mathcal{P}(X), \cap, \cup\right)$ satisfies all other parts of Definition 4.11 of rings.

(c) Find the neutral element for $\cup$.

(d) Does every element of $\mathcal{P}(X)$ have an inverse element with respect to $\cup$?

4-14. Let X be a set and let $R := \mathcal{P}(X)$, $+ := \cup$, $\cdot := \cap$, and $0 := \emptyset$.

(a) Prove that $\left(\mathcal{P}(X), \cup, \cap\right)$ does *not* satisfy part 4 (existence of additive inverses) of Definition 4.11 of rings.

(b) Prove that $\left(\mathcal{P}(X), \cup, \cap\right)$ satisfies all other parts of Definition 4.11 of rings.

(c) Find the neutral element for $\cap$.

(d) Does every element of $\mathcal{P}(X)$ have an inverse element with respect to $\cap$?

Note. Exercises 4-13 and 4-14 show that the most natural operations on sets don't quite allow us to get a ring in the algebraic sense. Exercise 4-12 shows that that's another (small) nice point about symmetric differences.

4-15. For this exercise, you may use what you know about $\mathbb{R}^3$. A function $f : \mathbb{R}^3 \to \mathbb{R}^3$ is called linear iff for all $x, y \in \mathbb{R}^3$ and all $\alpha, \beta \in \mathbb{R}$ we have that $f(\alpha x + \beta y) = \alpha f(x) + \beta f(y)$. Prove that the set of linear functions $f : \mathbb{R}^3 \to \mathbb{R}^3$ with pointwise addition and composition as multiplication is a noncommutative ring with unity.

4.3   Finding the Natural Numbers in the Integers

Strictly speaking, the natural numbers $\mathbb{N}$, as constructed in Sections 2.6 and 3.1, are not contained in the integers $\mathbb{Z}$, as constructed in Section 4.1. But we have seen that properties say a lot about a structure. Given this insight, structures with the *same* properties should be considered "the same." For our current purpose, which is to build the familiar number systems with their familiar arithmetic, all relevant properties relate to the operations $+$ and $\cdot$. If we can find a subset of $\mathbb{Z}$ that "acts like $\mathbb{N}$" with respect to $+$ and $\cdot$, then this subset may well *be* the set of natural numbers.

An isomorphism between two sets A and B, that both are equipped with binary operations, is a bijective function $f : A \to B$ that preserves the operations in the

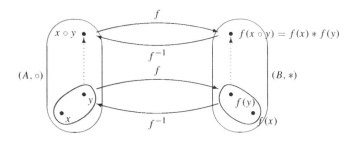

Figure 4.1 A and B are two sets with algebraic operations $\circ$ and $*$, respectively. An isomorphism is a bijective correspondence between A and B so that the algebraic operations can be carried out before or after mapping into the other set without affecting the result. That is, if $x, y \in A$, then the result is the same, independent of whether we first apply the operation $\circ$, which is native to A, to get $x \circ y$ and then map to B to get $f(x \circ y)$, or whether we first map x and y to $f(x)$ and $f(y)$ and then apply the operation $*$, which is native to B, to get $f(x) * f(y)$. Bijectivity also allows us to go from B to A if we choose to do so.

following way. For any two elements in A and any operation on A, we get the same result independent of whether we carry out the operation in A and map the result to B or whether we first map the elements to B and then carry out the corresponding operation in B (see Figure 4.1). Definition 4.19 below encodes this idea in mathematically precise terms. We will see in Theorem 4.20 that the integers contain a subset that, when equipped with addition and multiplication of integers, is isomorphic to $\mathbb{N}$ equipped with addition and multiplication of natural numbers.

Definition 4.19 *Let A and B be sets, let $n \in \mathbb{N}$, let $\circ_1, \ldots, \circ_n$ be binary operations on A and let $*_1, \ldots, *_n$ be binary operations on B. Then $(A, \circ_1, \ldots, \circ_n)$ is called* **isomorphic** *to $(B, *_1, \ldots, *_n)$ iff there is a bijective function $f : A \to B$ so that for all $k \in \{1, \ldots, n\}$ and all $x, y \in A$ we have that $f(x \circ_k y) = f(x) *_k f(y)$. The function f is called an* **isomorphism**.[3] *For a visualization, consider Figure 4.1.*

A (not necessarily bijective) function $f : A \to B$ so that for all $k \in \{1, \ldots, n\}$ and all $x, y \in A$ we have that $f(x \circ_k y) = f(x) *_k f(y)$ is called a **homomorphism**.

So an isomorphism assures that, even if two sets, both equipped with binary operations, are formally not the same, if the algebraic operations carry over via an isomorphism, we can treat the sets the same way. As an everyday example, consider the tasks of counting 10 apples and counting 10 bananas. These tasks should not be considered different from each other, because either way, we count. On the other hand, they are different tasks, because apples are not bananas. Isomorphisms provide the resolution: If we map apples bijectively to bananas, then the bijective function shows that counting is the same process for either set. As an abstract example, suppose $(A, \circ)$ and $(B, *)$

---

[3] Isomorphisms are often defined for sets that have just one operation. But because we typically will work with two operations here, the more general definition was given.

are isomorphic via the isomorphism $f : A \to B$. If $\circ$ is, say, commutative, then so is $*$, because for all $b_1, b_2 \in B$ there are $a_1, a_2 \in A$ so that

$$b_1 * b_2 = f(a_1) * f(a_2) = f(a_1 \circ a_2) = f(a_2 \circ a_1) = f(a_2) * f(a_1) = b_2 * b_1.$$

In similar fashion, isomorphisms preserve any of the algebraic properties we have discussed so far, thus showing that isomorphic structures are indeed "the same" for many intents and purposes. Returning to our construction of the integers, although we cannot find the *exact set* $\mathbb{N}$ in $\mathbb{Z}$, we can find an *isomorphic copy* of the natural numbers in the integers.

Theorem 4.20 *The set of natural numbers* $\mathbb{N}$, *equipped with addition and multiplication, is isomorphic to the subset* $\{[(n+1, 1)] : n \in \mathbb{N}\}$ *of the integers* $\mathbb{Z}$, *equipped with addition and multiplication. The set* $\{[(n + 1, 1)] : n \in \mathbb{N}\} \subseteq \mathbb{Z}$ *will be called* $\mathbb{N}$, *too, and we will use the customary digit and place value notation for these numbers.*

Proof. Define $f : \mathbb{N} \to \mathbb{Z}$ by $f(n) := [(n + 1, 1)]$. Then f is injective, because equality of the images $f(n) = f(m)$ implies $[(n+1, 1)] = [(m+1, 1)]$, which implies $(n + 1) + 1 = (m + 1) + 1$, which implies $n = m$. Moreover, f is clearly surjective onto the desired set. Finally, for all $n, m \in \mathbb{N}$ we have the following.

$$\begin{aligned}
f(n + m) &= [(n + m + 1, 1)] \\
&= [(n + m + 1 + 1, 1 + 1)] \\
&= [(n + 1) + (m + 1), 1 + 1)] \\
&= [(n + 1, 1)] + [(m + 1, 1)] \\
&= f(n) + f(m), \qquad \text{and} \\
f(nm) &= [(nm + 1, 1)] \\
&= [(nm + n + m + 1 + 1, n + m + 1 + 1)] \\
&= [(n + 1)(m + 1) + 1, (n + 1) + (m + 1)] \\
&= [(n + 1, 1)] \cdot [(m + 1, 1)] \\
&= f(n) f(m).
\end{aligned}$$

Therefore $\mathbb{N}$ is isomorphic to the subset $\{[(n + 1, 1)] : n \in \mathbb{N}\} \subseteq \mathbb{Z}$. ∎

As a consequence of Theorem 4.20, Theorem 4.21 shows that the integers as constructed are exactly what we were looking for at the start of this chapter: Natural numbers plus zero and negative numbers.

Theorem 4.21 *The subset* $\mathbb{N} \subseteq \mathbb{Z}$ *has the following properties.*

1. *For all* $x, y \in \mathbb{N}$, *we have* $x + y \in \mathbb{N}$ *and* $xy \in \mathbb{N}$,

2. *For all* $x \in \mathbb{Z}$, *exactly one of the following three properties holds:*

   Either $x \in \mathbb{N}$ *or* $-x \in \mathbb{N}$ *or* $x = 0$.

Proof. To prove part 1, let $\mathbb{N}_{\text{orig}}$ denote the "original" natural numbers from Sections 2.6 and 3.1 and let $f : \mathbb{N}_{\text{orig}} \to \mathbb{N}$ be the isomorphism from the proof of Theorem 4.20. Let $x, y \in \mathbb{N}$. Then there are $n, m \in \mathbb{N}_{\text{orig}}$ so that $f(n) = x$ and $f(m) = y$. Hence $x + y = f(n) + f(m) = f(n + m) \in \mathbb{N}$ and $x \cdot y = f(n) \cdot f(m) = f(n \cdot m) \in \mathbb{N}$.

To prove part 2 let $x = \big[(a, b)\big] \in \mathbb{Z}$. We first show that we have $x = 0$, or $x \in \mathbb{N}$ or $-x \in \mathbb{N}$. If $a = b$, then $x = \big[(a, a)\big] = \big[(1, 1)\big] = 0$. If $a > b$, first note that if $b = 1$, then $x = \big[(a, 1)\big] \in \mathbb{N}$. In case $b > 1$, with subtraction being the subtraction of natural numbers as in Definition 3.41, we obtain

$$
\begin{aligned}
\mathbb{N} \ni \big[(a - b + 1, 1)\big] &= \big[(a - b + 1, 1)\big] + 0 \\
&= \big[(a - b + 1, 1)\big] + \big[(b - 1, b - 1)\big] = \big[(a, b)\big] = x.
\end{aligned}
$$

Finally, if $a < b$, then an argument similar to the above shows that $-x = \big[(b, a)\big] \in \mathbb{N}$. (Exercise 4-16.)

To show that no element can satisfy two of the conditions at once, first note that 0 is not an element of $\mathbb{N}$. Hence -0 is not an element of $\mathbb{N}$ either. Finally, suppose for a contradiction that there is an integer $x \in \mathbb{N}$ so that $-x \in \mathbb{N}$. Then $x = \big[(a + 1, 1)\big]$ for some $a \in \mathbb{N}_{\text{orig}}$ and $x = \big[(1, b + 1)\big]$ for some $b \in \mathbb{N}_{\text{orig}}$. But then the equation $\big[(a + 1, 1)\big] = x = \big[(1, b + 1)\big]$ implies that $a + b + 1 + 1 = 1 + 1$. By Proposition 3.19 we infer $a + b + 1 = 1$, which cannot be, because 1 is not the successor of any natural number. ∎

Remark 4.22 Theorem 4.21 shows that every integer is either zero, a natural number, or the negative of a natural number. That means that from here on we can use the usual "digits and place value with negative signs" notation for integers. □

Exercises

4-16. Let $a, b \in \mathbb{N}$ be so that $a < b$ and let $x = \big[(a, b)\big] \in \mathbb{Z}$. Prove that $-x = \big[(b, a)\big] \in \mathbb{N}$.

4-17. Let $m \in \mathbb{N}$, let $\mathbb{Z}_m^{\text{old}}$, $[x]_m^{\text{old}}$, $+^{\text{old}}$ and $\cdot^{\text{old}}$ denote the equivalence classes modulo m and their addition and multiplication as in Definition 3.106 and let $\mathbb{Z}_m^{\text{new}}$, $[x]_m^{\text{new}}$, $+^{\text{new}}$ and $\cdot^{\text{new}}$ denote the equivalence classes modulo m and their addition and multiplication as in Example 4.15. Prove that the function $\varphi : \mathbb{Z}_m^{\text{old}} \to \mathbb{Z}_m^{\text{new}}$ defined by $\varphi\big([x]_m^{\text{old}}\big) := [x]_m^{\text{new}}$ is an isomorphism for addition and multiplication.

4-18. Let A and B be sets, let $n \in \mathbb{N}$, let $\circ_1, \ldots, \circ_n$ be binary operations on A, let $*_1, \ldots, *_n$ be binary operations on B and let $f : A \to B$ be an isomorphism from $(A, \circ_1, \ldots, \circ_n)$ to $(B, *_1, \ldots, *_n)$. Prove that $f^{-1} : B \to A$ is an isomorphism from $(B, *_1, \ldots, *_n)$ to $(A, \circ_1, \ldots, \circ_n)$.

4-19. Revisiting the **Fundamental Theorem of Arithmetic/Unique Prime Factorization Theorem**. Let $n \in \mathbb{Z} \setminus \{-1, 0, 1\}$. Prove that there are unique distinct prime numbers $p_1, \ldots, p_k \in \mathbb{N}$ and unique (not necessarily distinct) exponents $q_0, q_1, \ldots, q_k \in \mathbb{N}$ so that $n = (-1)^{q_0} \prod_{j=1}^{k} p_j^{q_j}$.

4-20. Prove that the set $\big\{ \big[(n + 1, 1)\big] : n \in \mathbb{N} \big\} \subseteq \mathbb{Z}$ satisfies the Peano Axioms.

4-21. "Bi-directional **induction**." Let $S \subseteq \mathbb{Z}$.

    (a) Prove that if $0 \in S$, and $k \in S$ implies $k + 1 \in S$ and $k - 1 \in S$, then $S = \mathbb{Z}$.

    (b) Prove that if $0 \in S$, and $k \in S$ implies $k + 1 \in S$ and $-k \in S$, then $S = \mathbb{Z}$.

    (c) Prove that if $S \neq \emptyset$, and $k \in S$ is equivalent to $k + 1 \in S$, then $S = \mathbb{Z}$.

4-22. Let $\tilde{\mathbb{N}}$ be an isomorphic copy of $\mathbb{N}$ which is disjoint from $\mathbb{N}_0$. Denote the element that corresponds to $n \in \mathbb{N}$ by $\tilde{n} \in \tilde{\mathbb{N}}$. For $n, m \in \mathbb{N}_0 \cup \tilde{\mathbb{N}}$ define $n \tilde{+} m :=$
$$\begin{cases} n + m; & \text{if } n, m \in \mathbb{N}_0 \text{ or } n, m \in \tilde{\mathbb{N}}, \\ \underline{n - k}; & \text{if } n \in \mathbb{N}, m = \tilde{k} \in \tilde{\mathbb{N}} \text{ and } n \geq k, \\ \widetilde{k - n}; & \text{if } n \in \mathbb{N}_0, m = \tilde{k} \in \tilde{\mathbb{N}} \text{ and } k > n. \end{cases}$$

(a) Prove that $\left(\mathbb{N}_0 \cup \tilde{\mathbb{N}}, \tilde{+} \right)$ is isomorphic to $(\mathbb{Z}, +)$.

(b) Use the fact that $\left(\mathbb{N}_0 \cup \tilde{\mathbb{N}}, \tilde{+} \right)$ is isomorphic to $(\mathbb{Z}, +)$ to prove that $\tilde{+}$ is associative and commutative, that 0 is the neutral element for $\tilde{+}$ and that every element of $\mathbb{N}_0 \cup \tilde{\mathbb{N}}$ has an additive inverse.

(c) Prove that $\tilde{+}$ is associative *without* using the isomorphism to $(\mathbb{Z}, +)$.

Note. This part shows that the idea of "just throwing in negative numbers" to construct the integers is cumbersome indeed, because in a construction from scratch, we could not use parts 4-22a and 4-22b.

(d) Define a multiplication operation $\tilde{\cdot}$ on $\mathbb{N}_0 \cup \tilde{\mathbb{N}}$ so that $\left(\mathbb{N}_0 \cup \tilde{\mathbb{N}}, \tilde{+}, \tilde{\cdot} \right)$ is isomorphic to $(\mathbb{Z}, +, \cdot)$.

4-23. Solving equations in the integers (Exercise 3-8 revisited).

(a) Solve the equation $x + 6 = 2$ in $\mathbb{Z}$. Explain any difference between your result and the result of Exercise 3-8b.

(b) Solve the equation $3x + 2 = 4$ in $\mathbb{Z}$. Explain any difference between your result and the result of Exercise 3-8d.

(c) Solve the equation $x^2 - 1 = 0$ in $\mathbb{Z}$. Explain any difference between your result and the result of Exercise 3-8e.

(d) Solve the equation $x^2 - 2 = 0$ in $\mathbb{Z}$. Explain any difference between your result and the result of Exercise 3-8f.

(e) Solve the equation $x^2 + 1 = 0$ in $\mathbb{Z}$. Explain any difference between your result and the result of Exercises 3-8g.

4-24. An elementary construction of the **Grothendieck group** $K(S)$ of a semigroup S. Let $(S, \circ)$ be a commutative semigroup so that if $a, b, c \in S$ satisfy $c \circ a = c \circ b$, then $a = b$.

(a) Prove that $(a, b) \sim (c, d)$ iff $a \circ d = c \circ b$ defines an equivalence relation on S.

(b) Let $K(S)$ be the set of equivalence classes for the equivalence relation $\sim$. Prove that the operation $\left[(a, b) \right] \bullet \left[(c, d) \right] := \left[(a \circ c, b \circ d) \right]$ is well-defined.

(c) Prove that $\left(K(S), \bullet \right)$ is a commutative group.

Note. $K(S)$ is called the Grothendieck group of S.

(d) Prove that there is an isomorphism from $(S, \circ)$ onto a subset of $\left(K(S), \bullet \right)$.

(e) Prove that $\mathbb{Z}$ is the Grothendieck group of $\mathbb{N}$.

(f) Prove that the Grothendieck group $K(G)$ of a commutative group G is isomorphic to the group G.

(g) Explain why the construction in this exercise cannot be used to construct a Grothendieck group for $(\mathbb{N}_0, \cdot)$. Explicitly state which parts of the construction will not work.

Note. There is a construction of Grothendieck groups that works for $(\mathbb{N}_0, \cdot)$, but it involves notions from algebra that we will not consider in this text.

4.4 Ordered Rings

The integers are more than just a ring. We know that the integers can be ordered. But it is not as easy to define the order relation as it was for the natural numbers in Definition 3.20. After all, $5 + (-3) = 2$ and we know that 5 is not smaller than 2. Thankfully, the adjustment is not too hard: We want to say that $x \leq y$ iff there is a number $n \in \mathbb{N}_0$ with $x + n = y$.

As we did with the algebraic properties of $\mathbb{N}$ and $\mathbb{Z}$, we want to view the idea in more generality. In this fashion, we can use the same idea to order the rational numbers and the real numbers.[4] To put an order relation on a ring, we need a subset so that sums and products of elements in the subset stay in the subset and so that, except for zero, no element and its negative are both in the subset.

Once more, the reader should feel free to think of integers throughout this section. But at the same time, it should be verified that the proofs do not use any special properties of $\mathbb{Z}$.

Definition 4.23 *A ring* $(R, +, \cdot)$ *is called an* **ordered ring** *iff there is a subset* $R^+ \subseteq R$ *so that*

1. For all $x, y \in R^+$ *we have* $x + y \in R^+$ *and* $xy \in R^+$, *and*

2. $0 \in R^+$, *and*

3. For all $x \in R \setminus \{0\}$, *at most one of* x *and* $-x$ *is in* R^+.

The subset R^+ *is also called the* **positive cone**[5] *of the ordered ring. The elements of* R^+ *are called* **nonnegative**, *the nonzero elements of* R^+ *are called* **positive** *and the elements* $x \in R \setminus \{0\}$ *so that* $-x \in R^+$ *are called* **negative**.

Example 4.24 Examples of ordered rings.

1. Choosing the positive cone $\mathbb{Z}^+ := \mathbb{N}_0$ makes the integers $\mathbb{Z}$ an ordered ring: Indeed, by Theorem 4.21 (and taking care of zero as a separate case), $\mathbb{N}_0$ is so that if $x, y \in \mathbb{N}_0$, then $x + y \in \mathbb{N}_0$ and $xy \in \mathbb{N}_0$, $0 \in \mathbb{N}_0$ and for all $x \in \mathbb{Z} \setminus \{0\}$ at most one of x and $-x$ is in $\mathbb{N}_0$.

2. The ring $\mathcal{F}(\mathbb{Z}, \mathbb{Z})$ of functions from $\mathbb{Z}$ to $\mathbb{Z}$ with the positive cone defined to be $\mathcal{F}^+(\mathbb{Z}, \mathbb{Z}) := \{f : [\forall x \in \mathbb{Z} : f(x) \geq 0]\}$ is an ordered ring. (Exercise 4-25.) $\square$

Once we have nonnegative elements, we can define comparability. Hence the term *ordered* ring in Definition 4.23 is an appropriate choice of language.

Definition 4.25 *Let* $(R, +, \cdot)$ *be an ordered ring with positive cone* R^+. *For* $x, y \in R$ *we define* $x \leq y$ *iff* $y - x \in R^+$. *As usual, we write* $x < y$ *for* $x \leq y$ *and* $x \neq y$.

Theorem 4.26 *Let* $(R, +, \cdot)$ *be an ordered ring with positive cone* R^+. *Then the relation* $\leq$ *from Definition 4.25 is an order relation.*

---

[4] In fact, this idea is also used to impose an order relation on spaces of real valued functions. So, this section presents a fairly standard construction. Readers may have seen the order relation for functions in action when talking about the "squeeze theorem" in calculus.

[5] It would be more accurate to call it the *nonnegative* cone, but "positive cone" has become customary.

Proof. We must prove that $\leq$ is reflexive, antisymmetric and transitive.
For reflexivity, let $x \in R$. Then $x - x = 0 \in R^+$, and hence $x \leq x$.

For antisymmetry, let $x, y \in R^+$ be so that $x \leq y$ and $y \leq x$ and suppose for a contradiction that $x \neq y$. Then $y - x \neq 0$, $y - x \in R^+$ and $-(y - x) = x - y \in R^+$, contradicting part 3 of Definition 4.23. Thus $x = y$.

For transitivity let $x, y, z \in R$ be so that $x \leq y$ and $y \leq z$. Then $y - x \in R^+$ and $z - y \in R^+$. Hence $z - x = (z - y) + (y - x) \in R^+$, which means that $x \leq z$. ∎

We can also establish that nonnegativity and positivity have their usual meanings.

Lemma 4.27 *Let* $(R, +, \cdot)$ *be an ordered ring with positive cone* R^+*. Then* $x \in R^+$ *iff* $x \geq 0$.

Proof. $x \in R^+$ iff $x - 0 \in R^+$ iff $x \geq 0$. ∎

Note that, unlike for the natural numbers, in ordered rings no trichotomy condition is needed to prove the antisymmetry of the order relation (see the proof of Theorem 3.23). On the other hand, the order relation for a ring need not be so that any two elements are comparable to each other. For example, in the ordered ring $\mathcal{F}(\mathbb{Z}, \mathbb{Z})$, the functions $f(x) = x^2$ and $g(x) = x^3$ are not comparable to each other, because $f(2) = 4 < 8 = g(2)$ and $g(-1) = -1 < 1 = f(-1)$. For the integers, we certainly would like to have a total order relation. It turns out that condition 2 of Theorem 4.21 plays a crucial role here.

Theorem 4.28 *Let* $(R, +, \cdot)$ *be an ordered ring with positive cone* R^+*. The order relation* $\leq$ *from Definition 4.25 is a total order relation iff for all* $x \in R \setminus \{0\}$ *we have either* $x \in R^+$ *or* $-x \in R^+$.

Proof. We first prove the "$\Rightarrow$" direction. So let $\leq$ be a total order relation and let $x \in R \setminus \{0\}$. Then x either satisfies $x > 0$ or $x < 0$. By definition, $x > 0$ iff $x = x - 0 \in R^+ \setminus \{0\}$ and $x < 0$ iff $-x = 0 - x \in R^+ \setminus \{0\}$. Hence we either have $x \in R^+$ or $-x \in R^+$.

For the "$\Leftarrow$" direction let either $z \in R^+$ or $-z \in R^+$ hold for all $z \in R \setminus \{0\}$ and let $x, y \in R$. In case $x = y$, there is nothing to prove. In case $x \neq y$, we have either $y - x \in R^+$, in which case $x \leq y$, or $x - y \in R^+$, in which case $y \leq x$. Thus in any case we conclude $x \leq y$ or $y \leq x$, and $\leq$ is a total order relation. ∎

To distinguish ordered rings like $\mathbb{Z}$ from ordered rings like $\mathcal{F}(\mathbb{Z}, \mathbb{Z})$, we introduce totally ordered rings.[6]

Definition 4.29 *An ordered ring is called a* **totally ordered ring** *iff the order relation* $\leq$ *from Definition 4.25 is a total order relation.*

Corollary 4.30 *With* $\mathbb{Z}^+ := \mathbb{N}_0$*, the integers are a totally ordered ring.*

Proof. Follows directly from Theorems 4.21 and 4.28. ∎

---

[6]In the literature, totally ordered rings are sometimes simply called "ordered rings," too.

Strangely enough, in an ordered ring with unity, the element 1 need not be greater than zero: For any ring R we can choose $R^+ := \{0\}$ and obtain an ordered ring with a trivial order ($x \leq y$ iff $x = y$).[7] However, in a *totally* ordered ring with unity, the elements 1 and -1 assume their familiar positions.

Proposition 4.31 *Let $(R, +, \cdot)$ be a totally ordered ring with unity and with positive cone R^+. Then $-1 < 0$ and $1 > 0$.*

Proof. To prove $-1 < 0$, suppose for a contradiction that $-1 \geq 0$, that is, $-1 > 0$. Then $-1 \in R^+$, which implies that $1 = (-1)(-1) \in R^+$, too. But this contradicts part 3 of Definition 4.23. Hence $-1 < 0$.

From $-1 < 0$ we conclude via Theorem 4.28 that $1 > 0$.   ∎

By Corollary 4.30, we know that $\mathbb{Z}$ can be totally ordered, and the discussion before Theorem 4.28 showed that the pointwise order on $\mathcal{F}(\mathbb{Z}, \mathbb{Z})$ is not a total order. Now let us consider $\mathbb{Z}_m$ once more. Proposition 4.31 implies that *if $\mathbb{Z}_m$ could be totally ordered, then the positive cone would would contain $[1]_m$. Consequently, by definition of positive cones, it would contain all of $\mathbb{Z}_m$, which is impossible. This rules out "the usual way" to possibly order $\mathbb{Z}_m$. Moreover, we also cannot circumvent this problem by defining an order as we did for $\mathbb{N}$: The next result shows that in $\mathbb{Z}_m$ there is no trichotomy condition as in Proposition 3.16.

Proposition 4.32 *Let $m \in \mathbb{N}$ and let $[x]_m, [y]_m \in \mathbb{Z}_m$. Then there are elements $[d]_m, [f]_m \in \mathbb{Z}_m$ so that $[x]_m + [d]_m = [y]_m$ and $[y]_m + [f]_m = [x]_m$.*

Proof. Let $[x]_m, [y]_m \in \mathbb{Z}_m$ and let $[x']_m, [y']_m \in \mathbb{Z}_m$ satisfy $[x]_m + [x']_m = [0]_m$ and $[y]_m + [y']_m = [0]_m$. Then $[d]_m := [x']_m + [y]_m$ and $[f]_m := [y']_m + [x]_m$ are as desired.   ∎

Now that we have a feel for which rings can be (totally) ordered, we can explore the structure of ordered rings. Ordered rings are not simply rings with an order on them. The crucial idea for ordered rings is that the order relation is compatible with the algebraic operations. The following result shows this compatibility. Essentially we want the order to connect to the algebraic operations just like the order for the natural numbers connected to addition and multiplication in Theorem 3.24.

Theorem 4.33 *Let $(R, +, \cdot)$ be an ordered ring with positive cone R^+ and order relation $\leq$. Then the following hold.*

1. *If $x \leq y$ and $z \in R$, then $x + z \leq y + z$.*

2. *If $x \leq y$ and $z \in R^+$, then $xz \leq yz$.*

Proof. For part 1 note that if $x \leq y$ and $z \in R$, then $(y+z)-(x+z) = y-x \in R^+$, so $x + z \leq y + z$.

For part 2 note that if $x \leq y$ and $z \in R^+$, then $yz - xz = (y - x)z \in R^+$, so $xz \leq yz$.   ∎

---

[7]This type of pathology is another indication why we must be careful and only work with what the definitions give us, no matter how tempting the "usual examples" are.

Theorem 4.33 clarifies the term "compatible order" from the preceding discussion. The order is compatible with the ring structure in the sense that neither addition of the same element on both sides, nor multiplication with elements of R^+, affects existing comparabilities. Once we have these properties, we can derive further familiar properties.

Proposition 4.34 *Let $(R, +, \cdot)$ be an ordered ring with positive cone R^+. If $x \leq y$ and $u \leq v$, then $x + u \leq y + v$.*

Proof. Exercise 4-26. ∎

Theorem 4.35 *Let $(R, +, \cdot)$ be an ordered ring and let $x, y, z \in R$. If $x \leq y$ and $z < 0$, then $xz \geq yz$.*

Proof. Because $x \leq y$ we have $y - x \in R^+$, and because $z < 0$ we have $-z \in R^+$. Therefore $xz - yz = (-z)(y - x) \in R^+$, so $xz \geq yz$. ∎

In a totally ordered ring in which products of nonzero elements do not zero out,[8] cancelation of positive or negative terms in inequalities is possible, too.

Theorem 4.36 *Let $(R, +, \cdot)$ be a totally ordered ring so that $ab = 0$ implies $a = 0$ or $b = 0$ and let $x, y, z \in R$.*

1. *If $xz = yz$ and $z \neq 0$, then $x = y$.*

2. *If $xz \leq yz$ and $z > 0$, then $x \leq y$.*

3. *If $xz \leq yz$ and $z < 0$, then $x \geq y$.*

Proof. The proof of part 1 is left to Exercise 4-29a.

To prove part 2, let $xz \leq yz$ and $z > 0$ and suppose for a contradiction that $x > y$. Then by part 2 of Theorem 4.33 we infer that $xz \geq yz$, and because $x \neq y$ we have by part 1 that $xz > yz$, contradicting $xz \leq yz$. Hence we must have $x \leq y$, which proves part 2.

The proof of part 3 is left to the reader (see Exercise 4-29c). ∎

Unlike in the natural numbers, in a totally ordered ring, not all elements are nonnegative. Thus it is sensible to investigate the idea of an absolute value. Note once more that we can think of $\mathbb{Z}$ if we want to, but that the construction will also provide absolute value functions for rational and real numbers.

Definition 4.37 *Let $(R, +, \cdot)$ be a totally ordered ring with positive cone R^+. For $x \in R$, we set*

$$|x| = \begin{cases} x; & \text{if } x \geq 0, \\ -x; & \text{if } x < 0, \end{cases}$$

*and we call it the **absolute value** of x.*

---

[8] It can indeed happen that the product of nonzero elements in a "natural" ring is zero, see Exercise 4-27. For the same thing to happen in a totally ordered ring, the examples need to be a bit more pathological, see Exercise 4-28.

Theorem 4.38 *Properties of the absolute value. Let $(R, +, \cdot)$ be a totally ordered ring with positive cone R^+.*

0. *For all $x \in R$, we have $|x| \geq 0$.*

1. *For all $x \in R$, we have $|x| = 0$ iff $x = 0$.*

2. *For all $x, y \in R$, we have $|xy| = |x||y|$.*

3. **Triangular inequality**. *For all $x, y \in R$, we have $|x + y| \leq |x| + |y|$.*

4. **Reverse triangular inequality**. *For all $x, y \in R$, we have $\big||x| - |y|\big| \leq |x - y|$.*

Proof. For part 0, let $x \in R$. In case $x \geq 0$, by Definition 4.37 we have $|x| = x \geq 0$. In case $x < 0$, we have $x \notin R^+$ and hence $-x \in R^+$. But then $|x| = -x > 0$, completing the proof of part 0.

For part 1, note that the direction "$\Leftarrow$" is trivial, because $|0| = 0$. For the direction "$\Rightarrow$," let $x \in R$ be so that $|x| = 0$ and suppose for a contradiction that $x \neq 0$. If $x > 0$, then $0 < x = |x| = 0$, which is not possible. Therefore $x < 0$. But then $0 < -x = |x| = 0$, a contradiction. Hence, x must be equal to 0.

For part 2, let $x, y \in R$. If $x \geq 0$ and $y \geq 0$, then by part 1 of Definition 4.23 we obtain $xy \geq 0$, and hence $|xy| = xy = |x||y|$. If $x \geq 0$ and $y < 0$, then by Theorem 4.35 we infer $xy \leq 0$. Hence, $|xy| = -xy = x(-y) = |x||y|$. The case $x < 0$ and $y \geq 0$ is similar and the reader will produce it in Exercise 4-30a. Finally, if $x < 0$ and $y < 0$, then by Proposition 4.18 and by part 1 of Definition 4.23 we obtain $xy = (-x)(-y) \geq 0$. Hence, $|xy| = xy = (-x)(-y) = |x||y|$.

To prove the triangular inequality, first note that for all $x \in R$ we have that $x \leq |x|$: This is clear for $x \geq 0$ and for $x < 0$ we simply note $x < 0 < -x = |x|$. Moreover, (see Exercise 4-30b) for all $x \in R$ we have $-x \leq |x|$. Now let $x, y \in R$. If the inequality $x + y \geq 0$ holds, then by part 1 of Theorem 4.33 at least one of x, y is greater than or equal to 0. (Otherwise $x < 0$ and $y < 0$ would imply $x + y < y < 0$.) Hence, by part 1 of Theorem 4.33 $|x + y| = x + y \leq |x| + y \leq |x| + |y|$. If $x + y < 0$, then at least one of x and y is less than 0. Hence, by part 1 of Theorem 4.33 we obtain $|x + y| = -(x + y) = -x + (-y) \leq |-x| + (-y) \leq |-x| + |-y| = |x| + |y|$.

Finally, for the reverse triangular inequality, let $x, y \in R$. Without loss of generality assume that $|x| \geq |y|$. (The proof for the case $|x| < |y|$ is left as Exercise 4-30c.) Then $|x| = |x - y + y| \leq |x - y| + |y|$, which implies $\big||x| - |y|\big| = |x| - |y| \leq |x - y|$. ∎

Exercises

4-25. Consider the ring $\mathcal{F}(\mathbb{Z}, \mathbb{Z})$ of functions from $\mathbb{Z}$ to $\mathbb{Z}$. Prove that the set of functions with nonnegative values $\mathcal{F}^+(\mathbb{Z}, \mathbb{Z}) := \{ f : [\forall x \in \mathbb{Z} : f(x) \geq 0] \}$ has the properties of a positive cone.

4-26. Prove Proposition 4.34.

4-27. Consider the ring $\mathcal{F}(\mathbb{Z}, \mathbb{Z})$ of functions from $\mathbb{Z}$ to $\mathbb{Z}$. Find two elements $f, g \in \mathcal{F}(\mathbb{Z}, \mathbb{Z})$ so that $fg = 0$.

4-28. Even in totally ordered rings, products of nonzero elements can be zero.

(a) Prove that in the ring $(\mathbb{Z}, +, \odot)$ from Exercise 4-8a the set $\mathbb{Z}^+ := \mathbb{N}_0$ is a positive cone that turns this ring into a totally ordered ring.

(b) Prove that the ring $(\mathbb{Z} \times \mathbb{Z}, +, \odot)$ from Exercise 4-8b is a totally ordered ring if we define the positive cone to be the set $(\mathbb{Z} \times \mathbb{Z})^+ := \big\{ (a, b) : a > 0 \vee (a = 0 \wedge b \geq 0) \big\}$.

4-29. Analyzing and finishing the proof of Theorem 4.36.

(a) Let $(R, +, \cdot)$ be a ring so that $ab = 0$ implies $a = 0$ or $b = 0$ and let $x, y, z \in R$. Prove that if $xz = yz$ and $z \neq 0$, then $x = y$.

(b) Where in the proof of part 2 of Theorem 4.36 was it used that $\leq$ is a total order?

(c) Prove part 3 of Theorem 4.36.

4-30. Finishing the proof of Theorem 4.38. Let $(R, +, \cdot)$ be a totally ordered ring with positive cone R^+.

(a) Let $x, y \in R$. Prove that if $x < 0$ and $y \geq 0$, then $|xy| = |x||y|$.

(b) Prove that for all $x \in R$ we have $-x \leq |x|$.

(c) Prove that if $|x| < |y|$, then $\big| |x| - |y| \big| \leq |x - y|$.

4-31. Let $x, y \in \mathbb{Z}$ be integers with $x < y$. Prove that the difference $y - x$ is at least 1.

4-32. Let $S \subseteq \mathbb{Z}$ be a nonempty subset of $\mathbb{Z}$.

(a) Prove that if there is an element $u \in \mathbb{Z}$ so that $u \geq s$ for all $s \in S$, then there is an element $m \in S$ so that $m \geq s$ for all $s \in S$.

(b) Prove that if there is an element $l \in \mathbb{Z}$ so that $l \leq s$ for all $s \in S$, then there is an element $i \in S$ so that $i \leq s$ for all $s \in S$.

4-33. Let R be a totally ordered ring and let $x \in R$. Prove that $x^2 \geq 0$.

4-34. Let $(R, +, \cdot)$ be a totally ordered ring with positive cone R^+.

(a) Let $x \in R$ and let $a \in R^+$. Prove that $|x| < a$ iff $-a < x < a$.

(b) Let $x \in R$ and let $a \in R^+$. Prove that $|x| > a$ iff $x > a$ or $x < -a$.

(c) Solve the inequality $|7x + 2| < 3$ in $\mathbb{Z}$.

(d) Solve the inequality $|6x + 5| > 2$ in $\mathbb{Z}$.

(e) Solve the inequality $3 \leq |2x + 4| \leq 11$ in $\mathbb{Z}$.

4.5 Division in Rings

The concept of divisibility is easily generalized to rings. Our two main examples for divisibility in rings will be the ring of integers and rings of polynomials. In the integers, we retain the familiar feel of dividing numbers. Moreover, Theorem 4.51 below will provide a fast way to compute greatest common divisors, for example, for RSA encryption. Rings of polynomials will provide some first results on polynomial equations. Polynomial equations will become more important in Chapter 5, and they will take center stage in Chapter 6.

Definition 4.39 *Let $(R, +, \cdot)$ be a ring and let $a, b \in R$. Then we say a **divides** b, or, b is **divisible** by a iff there is a $q \in R$ so that $b = qa$. In this case we write $a \mid b$. If a is not the unity element of R (in case R is a ring with unity) we say that a is a **factor** or a **divisor** of b.*

Example 4.40 Divisibility in rings extends the concept of divisibility for natural numbers.

1. For integers, Definition 4.39 for divisibility allows us to include negative numbers. For example, $-3 \mid -6$, because $-6 = 2 \cdot (-3)$.

2. For polynomials we do not have a simple analogy with numbers, but divisibility works the same way. For example, in $\mathbb{Z}[x]$, the ring of all polynomials with integer coefficients, we have that $(x - 2) \mid (x^2 - 4)$, because of the factorization $x^2 - 4 = (x - 2)(x + 2)$. $\qquad\qquad\square$

The idea of a greatest common divisor is harder to generalize, because in a ring there need not be an order relation. Ultimately we look at "size" in terms of divisibility. This should not be too surprising. After all, Exercise 3-61 showed that divisibility defines an order on $\mathbb{N}$. But note that, in rings, we lose uniqueness of the greatest common divisor. This loss of uniqueness is related to the fact that, on rings, divisibility only "almost" defines an order (see Exercise 4-35).

Definition 4.41 *Let* $(R, +, \cdot)$ *be a ring and let* $m, n \in R$. *Then* k *is called a* **greatest common divisor** *of* m *and* n *iff* $k \mid m$ *and* $k \mid n$ *and if* d *is another divisor of* m *and* n, *then* $d \mid k$.

Example 4.42 Greatest common divisors in rings need not be unique.

1. In $\mathbb{Z}$, 2 and -2 are both greatest common divisors of 4 and 6.

2. Similarly, in $\mathbb{Z}[x]$, $x + 2$ and $-x - 2$ are both greatest common divisors of the polynomials $x^2 - 4$ and $x^2 + 5x + 6$. $\qquad\qquad\square$

Exercise 5-14 will show that, in rings of polynomials, negative signs are not the only way in which the uniqueness of greatest common divisors can be lost. However, Proposition 4.43 below shows that, in the integers, negative signs are the only problem. Moreover, Proposition 4.54 will show that for polynomials, which are our other primary context for division, we can identify a canonical greatest common divisor, too.

Proposition 4.43 *Let* $m, n \in \mathbb{Z}$ *and let* $c, d \in \mathbb{Z}$ *be positive greatest common divisors of* m *and* n. *Then* $c = d$.

Proof. Exercise 4-36. $\qquad\qquad\blacksquare$

Definition 4.44 *Because the only possible ambiguity for greatest common divisors in* $\mathbb{Z}$ *is a negative sign, whenever we speak of* the **greatest common divisor** *in* $\mathbb{Z}$ *we mean the* positive *greatest common divisor. Similarly, the* **lowest common multiple** *of two integers is assumed to be the smallest* positive *integer that is divided by both numbers.*

Because divisibility for integers is so similar to divisibility for natural numbers, it is not surprising that for integers there is a division algorithm similar to that from Theorem 3.86. In essence, we do a division as in the natural numbers and then we adjust for negative signs.

Theorem 4.45 *The* **division algorithm** *for integers. Let* $n \in \mathbb{Z}$ *and let* $d \in \mathbb{Z} \setminus \{0\}$. *Then there are unique numbers* $q \in \mathbb{Z}$ *and* $r \in \{0, \ldots, |d| - 1\}$ *so that* $n = qd + r$. *The number* q *is also called the* **quotient** *and the number* r *is also called the* **remainder**.

Proof. We first establish that numbers as desired exist. By Theorem 3.86 we know that there are unique numbers $\tilde{q} \in \mathbb{N}_0$ and $\tilde{r} \in \{0, \ldots, |d| - 1\}$ so that $|n| = \tilde{q}|d| + \tilde{r}$. Now we will use this equation for each of the four possible ways in which n and d can take their signs.

Case 1: $n \geq 0$, $d > 0$. In this case, we choose $q := \tilde{q}$ and $r := \tilde{r}$.

Case 2: $n, d < 0$. In case $\tilde{r} = 0$, we have $n = -|n| = -(\tilde{q}|d| + \tilde{r}) = \tilde{q}d + 0$. Hence, for $\tilde{r} = 0$, we choose $q := \tilde{q}$ and $r := 0$. In case $\tilde{r} > 0$, we have $n = -|n| = -(\tilde{q}|d| + \tilde{r}) = \tilde{q}d - \tilde{r} = (\tilde{q} + 1)d - d - \tilde{r} = (\tilde{q} + 1)d + |d| - \tilde{r}$, and $|d| - \tilde{r} \in \{1, \ldots, |d| - 1\}$. Hence, for $\tilde{r} > 0$, we choose $q := \tilde{q} + 1$ and $r := |d| - \tilde{r}$.

The remaining cases are handled similarly. For $n \geq 0$ and $d < 0$, we choose $q := -\tilde{q}$ and $r := \tilde{r}$. Finally, for $n < 0$ and $d > 0$, in case $\tilde{r} = 0$, we choose $q := -\tilde{q}$ and $r := 0$, and in case $\tilde{r} > 0$, we choose $q := -(\tilde{q} + 1)$ and $r := d - \tilde{r}$. (See Exercise 4-37.)

For uniqueness, let $q, \hat{q} \in \mathbb{Z}$ and $r, \hat{r} \in \{0, \ldots, |d| - 1\}$ be so that the equation $qd + r = n = \hat{q}d + \hat{r}$ holds. Assume without loss of generality that $r \geq \hat{r}$. Then $(\hat{q} - q)d = r - \hat{r} \in \{0, \ldots, |d| - 1\}$. This can only happen when both sides of the equation are zero. Hence $r = \hat{r}$ and $(\hat{q} - q)d = 0$, which implies, by part 1 of Theorem 4.36 and Exercise 4-3, that $q = \hat{q}$. ∎

For the integers, it was obvious that we wanted the remainder of the division to be nonnegative and "closer to zero" than the divisor d was. For rings of polynomials, the algebraic notion of being "closer to zero" is a bit more complicated. In particular, it is *not* related to the pointwise order of the polynomials.

Definition 4.46 *Let* $(R, +, \cdot)$ *be a commutative ring and let* $p(x) = \sum_{j=0}^{n} a_j x^j$ *be a nonzero polynomial over* R. *Then the largest number* k *so that* $a_k \neq 0$ *is called the* **degree** $\deg(p)$ *of* p. *Moreover, if* $k = \deg(p)$, *then* a_k *is called the* **leading coefficient** *of* p. *Finally, we set the degree of the zero polynomial* $p(x) = 0$ *to be* $\deg(0) := -\infty$, *which is supposed to be smaller than any integer.*

Some properties of the degree function are proved in Exercise 4-38. With "closer to zero" being replaced with "smaller degree," we obtain a division algorithm for polynomials.

Theorem 4.47 *The* **division algorithm** *for polynomials. Let* $(R, +, \cdot)$ *be a commutative ring with unity so that there are no two* $a, b \in R \setminus \{0\}$ *with* $ab = 0$ *and let* $p, d \in R[x] \setminus \{0\}$ *be so that the leading coefficient of* d *has a multiplicative inverse. Then there are unique polynomials* $q, r \in R[x]$ *with* $\deg(r) < \deg(d)$ *so that* $p = qd + r$. *The polynomial* q *is also called the* **quotient** *and the polynomial* r *is also called the* **remainder**.

Proof. The proof of existence is a (strong) induction on the degree of p.

Base step. For $\deg(p) \in \{0, \ldots, \deg(d) - 1\}$ we set $q := 0$ and $r := d$.

Induction step. For $n := \deg(p) \geq \deg(d)$, let p_n be the leading coefficient of p, let m be the degree of d and let d_m be the leading coefficient of d. Then the polynomial $p_n d_m^{-1} x^{n-m} d(x)$ has degree $n = m + (n-m)$ and its leading coefficient is $p_n d_m^{-1} d_m = p_n$. Hence the degree of the polynomial $f(x) := p(x) - p_n d_m^{-1} x^{n-m} d(x)$ is less than $\deg(p)$. By induction hypothesis, there are polynomials $\tilde{q}, \tilde{r} \in R[x]$ so that $f = \tilde{q}d + \tilde{r}$ and $\deg(\tilde{r}) < \deg(d)$. Now set $r := \tilde{r}$ and set $q(x) := p_n d_m^{-1} x^{n-m} + \tilde{q}(x)$.

For uniqueness, let $q, \hat{q}, r, \hat{r} \in R[x]$ be so that $\deg(r) < \deg(d)$, $\deg(\hat{r}) < \deg(d)$ and $qd + r = p = \hat{q}d + \hat{r}$. If $r \neq \hat{r}$, then $\hat{q} \neq q$ and for the degrees we infer $\deg(d) > \deg(r - \hat{r}) = \deg\big((\hat{q} - q)d\big) \geq \deg(d)$, which is not possible. So $r = \hat{r}$, and then $q = \hat{q}$ (see Exercise 4-39), which completes the proof. ∎

Unlike the proof for the division algorithm for natural numbers in Theorem 3.86, the proof for the division algorithm for polynomials encodes *exactly* the long division for polynomials: In every step of the division algorithm, we eliminate the leading term.

Example 4.48 *Compute the quotient and the remainder of the division of the polynomial* $p(x) = 2x^4 - 7x^3 + 14x^2 - 12x - 2$ *by* $d(x) = x^3 - 4x^2 + 6x - 4$.

The division yields

$$
\begin{array}{r}
2x \quad + \quad 1 \\
x^3 \; - \; 4x^2 \; + \; 6x \; - \; 4 \;\big|\; 2x^4 \; - \; 7x^3 \; + \; 14x^2 \; - \; 12x \; - \; 2 \\
(-) \quad 2x^4 \; - \; 8x^3 \; + \; 12x^2 \; - \; 8x \\
\hline
x^3 \; + \; 2x^2 \; - \; 4x \; - \; 2 \\
(-) \; x^3 \; - \; 4x^2 \; + \; 6x \; - \; 4 \\
\hline
6x^2 \; - \; 10x \; + \; 2
\end{array}
$$

Therefore $q(x) = 2x + 1$, $r(x) = 6x^2 - 10x + 2$ and

$$2x^4 - 7x^3 + 14x^2 - 12x - 2 = (2x + 1)\left(x^3 - 4x^2 + 6x - 4\right) + \left(6x^2 - 10x + 2\right).$$

$\square$

The similarity between Theorems 4.45 and 4.47 motivates the definition of Euclidean rings.

Definition 4.49 *A ring* $(R, +, \cdot)$ *is called a* **Euclidean ring** *iff the ring* R *is commutative, there are no two elements* $a, b \in R \setminus \{0\}$ *so that* $ab = 0$ *and there is a function* $\delta : R \setminus \{0\} \to \mathbb{N}_0$ *so that for all elements* $a, b \in R$ *with* $b \neq 0$ *there are* $q, r \in R$ *so that* $a = qb + r$ *and* $\delta(r) < \delta(b)$ *or* $r = 0$. *That is, a Euclidean ring is a ring in which division with remainder works.*

Note that we did not explicitly demand that q and r in the definition of a Euclidean ring are unique. This is because uniqueness of q and r follow from the definition of a Euclidean ring. The proof (see Exercise 4-40) is very similar to the uniqueness of q and r for integers in Theorem 4.45 and for polynomials in Theorem 4.47.

Example 4.50 Examples of Euclidean rings

1. By Theorem 4.45 and Exercise 4-3 the integers $\mathbb{Z}$ are a Euclidean ring with $\delta(n) = |n|$.

2. By Theorem 4.47, if $\mathbb{F}$ is a commutative ring with unity in which every nonzero element has a multiplicative inverse (such rings are called fields, and we will investigate them in detail in Chapter 5), then the ring of polynomials $\mathbb{F}[x]$ is a Euclidean ring with $\delta(p) = \deg(p)$. $\qquad\square$

In Euclidean rings, there is a fast algorithm to compute a greatest common divisor of two elements. This is important for us, because we need a fast way to compute greatest common divisors for RSA encryption. Note that Theorem 4.51 is an "algorithm" in the same sense as the division algorithm is an algorithm: The proof provides a way to recursively construct the quantities in the theorem.

Theorem 4.51 Euclid's Algorithm. *Let $(R, +, \cdot)$ be a Euclidean ring and let $m, n \in R$. Then there are $s, t \in R$ so that $sm + tn$ is a greatest common divisor of m and n.*

Proof. Recursively construct a sequence of elements $a_j \in R$ as follows. Let $a_0 := m$, $a_1 := n$. Once the elements a_{j-1} and a_j are chosen, there are elements $q_j, a_{j+1} \in R$ so that $a_{j-1} = q_j a_j + a_{j+1}$ and $\delta(a_{j+1}) < \delta(a_j)$ or $a_{j+1} = 0$. Stop the construction at the value $j = k$ for which $a_{k+1} = 0$. Then $a_k \mid a_{k-1} = q_k a_k$, which implies $a_k \mid a_{k-2} = q_{k-1}a_{k-1} + a_k$, etc., showing that a_k divides all a_j. In particular, a_k is a common divisor of $m = a_0$ and $n = a_1$. Moreover, if $p \mid m$ and $p \mid n$, that is, $p \mid a_0$ and $p \mid a_1$, then $p \mid a_2 = a_0 - q_1 a_1$, which implies $p \mid a_3 = a_1 - q_2 a_2$, etc., showing that $p \mid a_k$. But this means that every common divisor of m and n divides a_k, thus showing that a_k is a greatest common divisor of m and n.

Finally, $a_k = a_{k-2} - q_{k-1}a_{k-1}$ and, for every $j \in \{2, \ldots, k-1\}$, we have the equality $a_j = a_{j-2} - q_{j-1}a_{j-1}$. Thus, after repeated backward substitution, a_k can be written in the form $a_k = sa_0 + ta_1$. $\qquad\blacksquare$

Because the computation of greatest common divisors is rather important, let us consider how Euclid's algorithm can be used to compute the greatest common divisor of two integers.

Example 4.52 *Compute the greatest common divisor of the numbers* 2064 *and* 204 *and find numbers* s *and* t *so that the greatest common divisor is* $s \cdot 2064 + t \cdot 204$.

1. Division of $a_0 = 2064$ by $a_1 = 204$ yields $2064 = 10 \cdot 204 + 24$.

2. Division of $a_1 = 204$ by $a_2 = 24$ yields $204 = 8 \cdot 24 + 12$.

3. Division of $a_2 = 24$ by $a_3 = 12$ yields $24 = 2 \cdot 12 + 0$.

So, by the proof of Theorem 4.51, the greatest common divisor of 2064 and 204 is 12 (compare with Example 3.84).

Moreover, as in the proof of Theorem 4.51, we obtain s and t as follows.

$$
\begin{aligned}
12 &= 1 \cdot 204 + (-8) \cdot 24 \\
&= 1 \cdot 204 + (-8) \cdot \big(2064 + (-10) \cdot 204\big) \\
&= 81 \cdot 204 + (-8) \cdot 2064
\end{aligned}
$$

So the numbers s and t are $s = -8$ and $t = 81$. $\qquad\qquad\square$

Aside from concrete computations as above, Euclid's algorithm also sheds some more light on the properties of the rings $\mathbb{Z}_m$. Proposition 3.107 has already shown that certain elements of $\mathbb{Z}_m$ have multiplicative inverses. Theorem 4.53 below gives an exact characterization which elements of $\mathbb{Z}_m$ have multiplicative inverses.

Theorem 4.53 *Let* $m \in \mathbb{N}$. *Then* $[x]_m \in \mathbb{Z}_m \setminus \big\{[0]_m\big\}$ *has a multiplicative inverse element iff x and m have no common factors.*

Proof. "$\Leftarrow$": Let $x \in \mathbb{Z}$ and $m \in \mathbb{N}$ have no common factors. By Euclid's Algorithm, there are $s, t \in \mathbb{Z}$ so that $1 = sx + tm$. But then $sx = 1 - tm$ and $[s]_m[x]_m = [sx]_m = [1 - tm]_m = [1]_m$. **(This part of the proof was already established in Proposition 3.107, but Euclid's Algorithm shows that we can explicitly and efficiently construct the multiplicative inverse.)**
"$\Rightarrow$": Let $[x]_m \in \mathbb{Z}_m$ be an element that has a multiplicative inverse $[y]_m$. Then $[x]_m[y]_m = [1]_m$. Hence there is a $k \in \mathbb{Z}$ so that $xy = 1 + km$. But then $xy - km = 1$ and if x and m had a common factor, then there would be a number in $\mathbb{N} \setminus \{1\}$ that divides 1, which cannot be. Hence x and m cannot have a common factor. $\qquad\blacksquare$

In particular, Theorem 4.53 shows that if $p \in \mathbb{N}$ is a prime number, then every nonzero element of $\mathbb{Z}_p$ has a multiplicative inverse. Number systems with this property, which also played a role in part 2 of Example 4.50, will be considered in more detail in Chapter 5. Proposition 4.54 shows another nice property of these number systems, namely that for them the greatest common divisors for their polynomials can be standardized. Note that the key step in the proof of Proposition 4.54 is the same as the key to Euclid's algorithm.

Proposition 4.54 *Let* $(R, +, \cdot)$ *be a commutative ring with unity so that every nonzero element has a multiplicative inverse. Let* $p, q \in R[x] \setminus \{0\}$ *and let* $c, d \in R[x]$ *be greatest common divisors of p and q so that the leading coefficients of c and d are both 1. Then* $c = d$.

Proof. The proof is an induction on $n := \deg(p) + \deg(q) \in \mathbb{N}_0$.
Base step, $n = 0$. In this case both p and q are constant polynomials. Hence c and d are constant polynomials with leading coefficient 1, which means that c and d are both equal to the constant function 1.
Induction step, $\{0, \dots, n-1\} \to n$. First consider the case that q divides p. In this case q is a greatest common divisor of p and q. Hence d divides q and q divides d, which means that q is a multiple of d. Therefore, if a is the leading coefficient of q, then $d = a^{-1}q$. Similarly we prove that $c = a^{-1}q$, which means that $c = d$.

The case in which p divides q is handled similarly. So for the remainder we can assume that $p \nmid q$, $q \nmid p$ and without loss of generality $\deg(p) \geq \deg(q)$.

In this case, by Theorem 4.47 there are polynomials a and r with $\deg(r) < \deg(q)$ and $r \neq 0$ so that $p = aq + r$. Because $r = p - aq$, every divisor of p and q is also a divisor of r (and trivially of q). Because $p = aq + r$, every divisor of q and r is also a divisor of p (and trivially of q). Hence every greatest common divisor of p and q is also a greatest common divisor of q and r and vice versa. In particular c and d are greatest common divisors of q and r. But $0 \leq \deg(r) < \deg(q) \leq \deg(p)$, so $\deg(q) + \deg(r) < n$ and by induction hypothesis we conclude that $c = d$. ∎

We conclude this section with some more results on rings of polynomials. These results will be useful when we start analyzing polynomial equations. Theorem 4.56 says that zeros of polynomials are connected to certain factorizations and Theorem 4.58 gives an upper bound on the number of zeros a polynomial can have. The interest in zeros of polynomials stems from the interest in solving polynomial equations, because zeros of polynomials and solutions of polynomial equations are two sides of the same coin.

Definition 4.55 *Let $(R, +, \cdot)$ be a commutative ring with unity so that there are no two $a, b \in R \setminus \{0\}$ with $ab = 0$, let $p : R \to R$ be a polynomial of degree $n > 0$ and let $z \in R$. Then z is called a* **solution** *of the polynomial equation $p(x) = 0$ iff $p(z) = 0$. We will also call z a* **zero** *of the polynomial p. Sometimes, a zero of a polynomial is also called a* **root**. *In this text, we will avoid calling zeros "roots" to avoid confusion with roots of numbers.*

Definition 4.55 allows us to give a first, partial, answer to question 4 on page x. The number 3 a solution of $x^3 - 5x^2 + 3x + 9 = 0$, because $3^3 - 5 \cdot 3^2 + 3 \cdot 3 + 9 = 0$.

Theorem 4.56 *Let $(R, +, \cdot)$ be a commutative ring with unity so that there are no two $a, b \in R \setminus \{0\}$ with $ab = 0$, let $p : R \to R$ be a polynomial of degree $n > 0$ and let $z \in R$ be a solution of the polynomial equation $p(x) = 0$. Then there is a polynomial $q : R \to R$ of degree $n - 1$ so that $p(x) = (x - z)q(x)$.*

Proof. By Theorem 4.47 there are polynomials $q, r : R \to R$ so that for all $x \in R$ we have that $p(x) = (x - z)q(x) + r(x)$ and so that $\deg(r) < \deg(x - z) = 1$. But then $\deg(r) \in \{-\infty, 0\}$ and hence r is a constant function. That is, there is a $c \in R$ so that $r(x) = c$ for all $x \in R$. But then $c = r(z) + (z - z)q(z) = p(z) = 0$, which means that $r = 0$. Hence $p(x) = (x - z)q(x)$ for all $x \in R$. Moreover, by Exercise 4-38c the degree of q must be $n - 1$. ∎

Theorem 4.56 shows that solutions of a polynomial equation $p(x) = 0$ split the polynomial into a linear factor and a factor q whose degree is one less than that of p. The remaining zeros of p must be zeros of q and every zero of q is a zero of p. In particular, Theorem 4.56 allows us answer question 4 on page x. Division of the polynomial $x^3 - 5x^2 + 3x + 9$ by $x - 3$ gives us that $x^3 - 5x^2 + 3x + 9 = (x - 3)(x^2 - 2x - 3)$. Once we note the factorization $x^2 - 2x - 3 = (x - 3)(x + 1)$, we obtain the factorization $x^3 - 5x^2 + 3x + 9 = (x - 3)^2(x - 1)$, and the solutions of $x^3 - 5x^2 + 3x + 9 = 0$

are 3 (apparently "twice") and -1. The next definition formalizes the idea that zeros can occur more than once.

Definition 4.57 *Let $(R, +, \cdot)$ be a commutative ring with unity so that there are no two $a, b \in R \setminus \{0\}$ with $ab = 0$. Let $p : R \to R$ be a polynomial and let $z \in R$ be a solution of the polynomial equation $p(x) = 0$. The largest number $m \in \mathbb{N}$ so that $(x - z)^m$ is a factor of p is called the* **multiplicity** *of the solution (or zero) z.* **Counting zeros/solutions with multiplicity** *means adding the multiplicities of the zeros of p.*

Looking once more at Theorem 4.56, we note that any time we factor out a term $(x - \langle \text{zero} \rangle)$ the degree of the quotient polynomial q is one less than that of the original polynomial. Therefore polynomials can have only finitely many zeros. In fact, even if we allow counting with multiplicity, polynomials of degree n have at most n zeros.

Theorem 4.58 *Let $(R, +, \cdot)$ be a commutative ring with unity so that there are no two $a, b \in R \setminus \{0\}$ with $ab = 0$ and let $p : R \to R$ be a polynomial of degree $n \in \mathbb{N}$. Then the polynomial equation $p(z) = 0$ has at most n solutions, counted with multiplicity.*

Proof. This proof is an induction on the degree n of p.

Base step. For $n = 1$ there are $a, b \in R$ with $a \neq 0$ so that $p(x) = ax + b$. To prove that the equation $p(z) = 0$ has at most one solution, let $x, \hat{x} \in R$ be solutions for the equation $ax + b = 0$. Then $a(x - \hat{x}) = ax - a\hat{x} = -b - (-b) = 0$. By assumption on R we conclude that $x = \hat{x}$. Hence $p(z) = 0$ has at most one solution. This completes the base step.

Induction step $n \to n + 1$. Let p be a polynomial of degree $n + 1$. If $p(z) = 0$ has no solutions, then there is nothing to prove. Otherwise, let $z \in R$ be so that $p(z) = 0$ and let $q \in R[x]$ be so that $p(x) = q(x)(x - z)$. For all $w \in R \setminus \{z\}$ and all $m \in \mathbb{N}$, the power $(x - w)^m$ is a factor of p iff it is a factor of q. Moreover, $(x - z)^m$ is a factor of p iff $(x - z)^{m-1}$ is a factor of q. Thus counting the zeros of p with multiplicity leads to a number that is at most one more than the number we obtain from counting zeros of q with multiplicity. Because q has at most $n = \deg(q)$ zeros counted with multiplicity, the result follows.   ■

Theorem 4.58 allows us to express equality of polynomials in terms of their coefficients.

Corollary 4.59 *Let $(R, +, \cdot)$ be a commutative ring with unity so that there are no two $a, b \in R \setminus \{0\}$ with $ab = 0$ and so that R has infinitely many elements. Let $p(x) = \sum_{j=0}^{n} p_j x^j$ be a polynomial of degree $n > 0$. Then $p \neq 0$. Consequently, if $f(x) = \sum_{j=0}^{n} f_j x^j$ and $g(x) = \sum_{j=0}^{n} g_j x^j$ are polynomials, then $f = g$ iff $f_j = g_j$ for all j.*

Proof. Suppose for a contradiction that there is a polynomial p with degree $n > 0$ so that $p(x) = 0$ for all $x \in R$. Let $r_1, \ldots, r_{n+1}$ be $n + 1$ distinct elements of R. Then $p(r_j) = 0$ for all $j \in \{1, \ldots, n + 1\}$, contradicting Theorem 4.58. Therefore no polynomial of positive degree is equal to zero.

Now let $f(x) = \sum_{j=0}^{n} f_j x^j$ and $g(x) = \sum_{j=0}^{n} g_j x^j$ be polynomials. Then $f = g$ iff $0 = (f - g)(x) = \sum_{j=0}^{n} (f_j - g_j) x^j$. By the above (and trivially if the degree of the difference is smaller than 1), this is the case iff $f_j = g_j$ for all j. ∎

By Exercise 4-41 the many hypotheses are needed for Corollary 4.59. But, although the interpretation of polynomials as functions feels very natural, abstract algebra often works with formal polynomials (see Exercise 5-16). For formal polynomials, equality is defined through the equality of the coefficients. Therefore, for formal polynomials the result of Corollary 4.59 is always true, independent of the number of elements of the underlying ring.

Exercises

4-35. Let $(R, +, \cdot)$ be a ring and consider divisibility as a relation $| \subseteq R \times R$. In this exercise, we will prove that $|$ almost defines an order relation on R.

    (a) Prove that $|$ is reflexive.

    (b) Prove that $|$ is transitive.

    Note. So the only thing that is missing for $|$ to be an order relation is antisymmetry. Such relations, that is, relations that are reflexive and antisymmetric, are called **preorders**.

4-36. Prove Proposition 4.43.

4-37. Finish the proof of Theorem 4.45.

    (a) Justify the choices of q and r in case $n \geq 0$ and $d < 0$.

    (b) Justify the choices of q and r in case $n < 0$ and $d > 0$.

4-38. The degree function. Let $(R, +, \cdot)$ be a commutative ring and let $p, q \in R[x]$.

    (a) Prove that $\deg(p + q) \leq \max \{ \deg(p), \deg(q) \}$.

    (b) Prove that $\deg(pq) \leq \deg(p) + \deg(q)$.

    (c) Prove that if the product of the leading coefficients of the polynomials p and q is not zero, then $\deg(pq) = \deg(p) + \deg(q)$.

    (d) Give an example of a commutative ring R and two polynomials $p, q \in R[x]$ so that we have the inequality $\deg(pq) < \deg(p) + \deg(q)$.

4-39. Let $(R, +, \cdot)$ be a commutative ring so that there are no two $a, b \in R \setminus \{0\}$ with $ab = 0$. Prove that for all $p, q \in R[x]$ we have $pq \neq 0$.

4-40. Prove that the elements q and r in Definition 4.49 are unique. That is, prove that if n, d are elements of a Euclidean ring R so that $d \neq 0$, and if $q, \hat{q}, r, \hat{r} \in R$ are so that $\delta(r) < \delta(d), \delta(\hat{r}) < \delta(d)$ and $qd + r = n = \hat{q}d + \hat{r}$, then $q = \hat{q}$ and $r = \hat{r}$.

4-41. Give an example of a commutative ring R with unity and two polynomials $f, g \in R[x]$ so that Corollary 4.59 does not hold.

    Hint. Fermat's Little Theorem.

4-42. For each of the following exercises, divide the polynomial p by the polynomial d. Work in $\mathbb{Z}[x]$.

    (a) $p(x) = 3x^3 - 13x^2 - 39x + 59, d(x) = x - 6$

    (b) $p(x) = 2x^5 + 3x^4 - 5x^3 + 2x^2 - 23x + 25, d(x) = x^2 + 2x - 3$

4-43. For each of the following pairs of elements m and n, compute a greatest common divisor g and elements s and t of the underlying ring (integers or polynomials) so that $g = sm + tn$.

    (a) $m = 352, n = 29$

(b) $m = 3526, n = 298$

(c) $m = 4978, n = 333$

(d) $m = 219, n = 352$

(e) $m = 2x^3 + 11x^2 + 23x + 9, n = 4x^3 - 4x^2 + 7x + 5$ (use rational numbers, if necessary)

(f) $m = 2x^5 + x^4 + 3x^3 + 5x^2 - 9x + 6, n = 4x^4 + 2x^3 + 7x^2 + 6x - 15$ (use rational numbers, if necessary)

4-44. For each of the following pairs of prime numbers p and q, compute a public key (n, e) and the corresponding private key d. (See Section 3.9 for public and private keys.)

(a) $p = 7, q = 17$

(b) $p = 19, q = 97$

(c) $p = 101, q = 103$

(d) $p = 1013, q = 2437$

4-45. For each of the following public keys (n, e), compute the private key d. (See Section 3.9 for public and private keys.)

(a) $(77, 7)$

(b) $(91, 5)$

(c) $(299, 85)$

(d) $(10403, 91)$

4-46. Prove each of the following divisibility results.

(a) For all $n \in \mathbb{Z}$, $n(n + 1)(n + 2)$ is divisible by 6.

(b) For all $n \in \mathbb{Z}$, $n\left(n^2 + 5\right)$ is divisible by 6.

(c) For all $n \in \mathbb{Z}$, $n^7 - n$ is divisible by 42.

(d) For all $n \in \mathbb{Z}$, $n(n + 1)(n + 2)(n + 3)$ is divisible by 24.

(e) For all $n \in \mathbb{Z}$, $n^5 - n$ is divisible by 30.

Hint. Use induction for $\mathbb{N}$, then generalize to $\mathbb{Z}$.

4-47. Let $m_1, \ldots, m_n \in \mathbb{Z}$. Define the greatest common divisor of $m_1, \ldots, m_n$ and prove that it exists and is unique.

4-48. For $n \in \mathbb{N}$ we define $\varphi(n)$ to be the number of elements x in $\{1, \ldots, n - 1\}$ so that $(x, n) = 1$. This function is called **Euler's φ function** or **Euler's totient function**.

(a) Prove that for all $m, n \in \mathbb{N}$ with $(m, n) = 1$ we have that $\varphi(mn) = \varphi(m)\varphi(n)$.

Hint. Consider $\mathbb{Z}_{mn}$ and $\mathbb{Z}_m \times \mathbb{Z}_n$ with componentwise addition and multiplication. Use the Chinese Remainder Theorem (Exercise 3-108a) to prove that $\varphi\left([x]_{mn}\right) = \left([x]_m, [x]_n\right)$ is an isomorphism from $\mathbb{Z}_{mn}$ to $\mathbb{Z}_m \times \mathbb{Z}_n$. Then use Theorem 4.53 and the fact that both rings must have the same number of elements with a multiplicative inverse.

(b) Prove that for all prime numbers $p \in \mathbb{N}$ we have that $\varphi\left(p^k\right) = (p - 1)p^{k-1}$.

(c) Prove that for all $m \in \mathbb{N}$ we have that $m = \sum_{d|m} \varphi(d)$.

Hint. Induction on the number of distinct prime factors of m. Use part 4-48a.

(d) Prove that for all $m \in \mathbb{N}$ we have that $\varphi(m) = m \prod_{\substack{p \,|\, m \\ p \text{ prime}}} \left(1 - \frac{1}{p}\right)$.

Hint. Induction on the number of distinct prime factors of m.

 (e) Let $a, m \in \mathbb{N}$ be so that $(a, m) = 1$. Prove that $a^{\varphi(m)} \equiv 1 \pmod{m}$

  Hint. For $m = p^k$ with p prime, use induction on k, and, in the induction step, that $a^{\varphi\left(p^{k+1}\right)} = \left(1 + p^k n\right)^p$ and apply an argument like in the proof of Fermat's Little Theorem to $\left(1 + p^k n\right)^p$. For numbers with more than one distinct prime factor, use part 4-48a and Exercise 3-107b.

4.6   Countable Sets

This section introduces some surprising results about how infinite sets can be equivalent to each other. The proofs are necessarily technical, because we must produce bijective functions between the sets. On first reading, it may suffice to understand the ideas behind the equivalences. Also consider Hilbert's Hotel (see Exercise 4-49) as an illustration how infinite sets can work.

Definition 2.59 introduced equivalence of sets via bijective functions between the sets and, at least for finite sets, Definition 3.53 introduced a numerical size. It is now time to revisit the idea of size, but this time for infinite sets. Theorem 3.54 has shown that if two finite sets A and B of the same size satisfy $A \subseteq B$, then $A = B$. Theorem 4.60 shows that the situation is more complicated for infinite sets.

Theorem 4.60 $\mathbb{N}$ *is equivalent to* $\mathbb{Z}$ *(also see Figure 4.2).*

 Proof. The function

$$f(n) := \begin{cases} \frac{n-1}{2}; & \text{if } n \text{ is odd,} \\ -\frac{n}{2}; & \text{if } n \text{ is even,} \end{cases}$$

is a bijective function from $\mathbb{N}$ to $\mathbb{Z}$ (see Exercise 4-50). ∎

Theorem 4.60 shows that an infinite set can be equivalent to another, seemingly "bigger" infinite set. Yet Theorem 2.60 says that no set is equivalent to its power set. The power set of an infinite set such as $\mathbb{N}$ is again infinite. Hence in mathematics "infinity comes in different sizes." Overall, this means we must be very careful in distinguishing "sizes" of infinite sets: There is more than one such size and just because one set contains another need not mean that it is "bigger."

Arguably, the most important size distinction for infinite sets is if they are countable or uncountable. Countable sets are, roughly speaking, sets for which it is possible to "count" the elements, where the counting process may stop or it may not. The natural numbers are the prototypical countable set. Because, aside from power sets, we do not have any examples of uncountable sets yet, we will focus on countable sets here, leaving uncountable sets to Section 5.6.

Definition 4.61 *A set C is called* **countably infinite** *iff there is a bijective function $f : \mathbb{N} \to C$. A set C is called* **countable** *iff C is finite or countably infinite.*

Subsets of a set cannot be larger than the set, so it is natural that subsets of countable sets are countable.

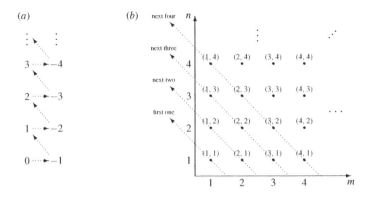

Figure 4.2 Visualization that $\mathbb{Z}$ is equivalent to $\mathbb{N}$ (a) and that $\mathbb{N} \times \mathbb{N}$ is countable (b).

Theorem 4.62 *If C is countable and $S \subseteq C$, then S is countable.*

Proof. If S is finite, there is nothing to prove. Hence, we can assume that S is infinite. Let $f : \mathbb{N} \to C$ be a bijection and let $n_1 := \min f^{-1}[S]$, where $\min(A)$ denotes the smallest element of a nonempty subset of $\mathbb{N}$, which is guaranteed to exist by Theorem 3.51. For $k \in \mathbb{N}$, we define n_{k+1} recursively. Once $n_1, \ldots, n_k$ are chosen, let $n_{k+1} := \min \left(f^{-1}[S] \backslash \{n_1, \ldots, n_k\} \right)$. Define $g : \mathbb{N} \to S$ by $g(k) := f(n_k)$. Because all n_k are in $f^{-1}[S]$, g maps $\mathbb{N}$ into S. Because no two n_k are equal and f is injective, g is injective.

Finally, suppose for a contradiction that g is not surjective. Let b be the smallest element of $f^{-1}\left[S \backslash g[\mathbb{N}] \right]$ and let the number of elements of $f^{-1}[S] \cap \{1, \ldots, b-1\}$ be k. Then by definition of n_1 we have $k \geq 1$, so $f^{-1}[S] \cap \{1, \ldots, b-1\} = \{n_1, \ldots, n_k\}$ and $b = \min \left(f^{-1}[S] \backslash \{n_1, \ldots, n_k\} \right)$, which means $b = n_{k+1}$, a contradiction.   ∎

Theorem 4.62 says that to prove that a set is countable, it is enough to embed a copy of it into a countable set. The reader will use this idea in Exercise 4-51 to produce a proof for Theorem 4.63 below that does not require the explicit construction of a bijective function between $\mathbb{N} \times \mathbb{N}$ and $\mathbb{N}$. But it is instructive to actually construct a bijective function between $\mathbb{N} \times \mathbb{N}$ and $\mathbb{N}$. The idea is given in part (b) of Figure 4.2. Rather than counting along rows or columns, which are all infinite, we can count $\mathbb{N} \times \mathbb{N}$ along its diagonals, which are all finite.

Theorem 4.63 *The set $\mathbb{N} \times \mathbb{N}$ is countable (also see Figure 4.2).*

Proof. First of all, because the inverse of a bijective function is bijective, too, it is admissible to construct a bijective function from $\mathbb{N} \times \mathbb{N}$ to $\mathbb{N}$, rather than the other way round. We claim that the function $f : \mathbb{N} \times \mathbb{N} \to \mathbb{N}$ defined by

$$f(m, n) := \frac{1}{2}(m + n - 1)(m + n - 2) + n$$

is a bijective function from $\mathbb{N} \times \mathbb{N}$ to $\mathbb{N}$.

Before we start the proof, it is instructive to verify that this function really encodes part (b) of Figure 4.2. Consider how counting along diagonals of slope -1 works. A line of slope -1 has the equation x+y=c for some c. That means, all points $(m,n) \in \mathbb{N} \times \mathbb{N}$ for which the sum of the entries is the same are on the same diagonal. Because every one of the diagonals along which we count must intersect the bottom row $\{(k,1):k \in \mathbb{N}\}$, we know that there will be one diagonal x+y=k+1 for each $k \in \mathbb{N}$. Moreover, on the diagonal x+y=k+1 there will be k elements of $\mathbb{N} \times \mathbb{N}$. That means the point $(m,n) \in \mathbb{N} \times \mathbb{N}$ will be on the $(m+n-1)^{\text{st}}$ diagonal and there will be at least $\sum_{j=1}^{m+n-2} j = \frac{1}{2}(m+n-1)(m+n-2)$ points that were counted before (m,n). That explains the first term of f(m,n): It assures that images of distinct diagonals do not overlap. The second term, n, simply gives the vertical position of n on the diagonal. Adding it assures that every point on each diagonal is counted exactly once.

Now back to the formal proof. The following equation for $m, n \in \mathbb{N}$ and $m > 1$ will be used in the proof of bijectivity:

$$
\begin{aligned}
f(m-1, n+1) &= \frac{1}{2}\big((m-1)+(n+1)-1\big)\big((m-1)+(n+1)-2\big)+n+1 \\
&= \frac{1}{2}(m+n-1)(m+n-2)+n+1 \\
&= f(m,n)+1.
\end{aligned}
$$

Graphically, this equation shows that going one step to the left and one step up on a diagonal leads us to the "next" element, as indicated in Figure 4.2.

To prove injectivity, we will prove by induction on m that $f(m,n) = f(a,b)$ implies $(m,n) = (a,b)$.

For the base step, let $m = 1$ and let $(1, n), (a, b) \in \mathbb{N} \times \mathbb{N}$ be so that we have $f(1, n) = f(a, b)$. We must prove that $a = 1$ and $b = n$. We first note the following, where each line implies the next.

$$
\begin{aligned}
f(1, n) &= f(a, b) \\
\frac{1}{2}(1+n-1)(1+n-2)+n &= \frac{1}{2}(a+b-1)(a+b-2)+b \\
n(n-1)+2n &= (a+b-1)(a+b-2)+2b \\
n^2+n &= (a+b-1)^2-(a+b-1)+2b \\
n^2+n &= (a+b-1)^2+(b-a+1)
\end{aligned}
$$

Now first suppose for a contradiction that $a+b-1 < n$. Then $b-a+1 \le b < n$ and the right side above is strictly smaller than n^2+n, contradicting the equality. Thus $a+b-1 \ge n$.

Now suppose for a contradiction that $a+b-1 > n$. Then $a+b-1 = n+k$ for some $k \in \mathbb{N}$. Therefore $(a+b-1)^2 = (n+k)^2 = n^2+2kn+k^2$, so that the above equality leads to $b-a+1 = -(2k-1)n-k^2$. Hence $n+k-b+1 = a = (2k-1)n+k^2+b+1$, which implies that $2b = (-2k+2)n+k-k^2 \le 0$, a contradiction.

Thus $a + b - 1 = n$. But then $b - a + 1 = n$, too. Hence $2b = 2n$, that is, $b = n$, and $a = 1$, as was to be proved. Therefore $f(1, n) = f(a, b)$ implies $(1, n) = (a, b)$.

For the induction step $(m - 1) \to m$, let $(m, n), (a, b) \in \mathbb{N} \times \mathbb{N}$ be so that we have $f(m, n) = f(a, b)$. Then we can assume that $m, a \neq 1$, because if either one was equal to 1, we would conclude $(m, n) = (a, b)$ from the base step. Because $m, a > 1$ we can infer

$$f(m - 1, n + 1) = f(m, n) + 1 = f(a, b) + 1 = f(a - 1, b + 1),$$

and by induction hypothesis we conclude that $m - 1 = a - 1$ and $n + 1 = b + 1$, that is, $(m, n) = (a, b)$. Hence f is injective.

For surjectivity, we prove by induction on k that for every $k \in \mathbb{N}$ there is a pair $(m, n) \in \mathbb{N} \times \mathbb{N}$ so that $f(m, n) = k$.

For the base step $k = 1$, note that $f(1, 1) = \frac{1}{2}(1 + 1 - 1)(1 + 1 - 2) + 1 = 1$.

For the induction step $k \to k + 1$, let $(m, n) \in \mathbb{N} \times \mathbb{N}$ be so that $f(m, n) = k$. If $m > 1$, then $f(m - 1, n + 1) = f(m, n) + 1 = k + 1$. So, as indicated in Figure 4.2, going one step to the left and one step up on a diagonal leads us to the "next" element.

On the other hand, if $m = 1$, then

$$
\begin{aligned}
f(n + 1, 1) &= \frac{1}{2}\big((n + 1) + (1) - 1\big)\big((n + 1) + (1) - 2\big) + 1 \\
&= \frac{1}{2}(n + 1)n + 1 \\
&= \frac{1}{2}(n - 1)n + n + 1 \\
&= \frac{1}{2}\big((1) + (n) - 1\big)\big((1) + (n) - 2\big) + (n) + 1 \\
&= f(1, n) + 1 \\
&= k + 1.
\end{aligned}
$$

This part says that, as indicated in Figure 4.2, at the end/top of a diagonal we continue counting at the bottom of the next diagonal.

So in either case $k + 1$ has a preimage under f. Hence f is surjective. ∎

Theorem 4.63 is the key to a result that shows that surprisingly "large looking" unions are still countable.

Definition 4.64 *Two sets A and B are called* **disjoint** *iff $A \cap B = \emptyset$. A family $\{C_i\}_{i \in I}$ is called* **pairwise disjoint** *iff for all $i \neq j$ we have $C_i \cap C_j = \emptyset$.*

Theorem 4.65 *Countable unions of countable sets are countable.*

Proof. Let $\{C_n\}_{n=1}^{\alpha}$ with $\alpha \in \mathbb{N} \cup \{\infty\}$ be a countable family of countable sets. (We use α in this proof to also cover finite unions with the same argument.) For each C_n, let $B_n := C_n \setminus \bigcup_{j=1}^{n-1} C_j$. We claim $\bigcup_{n=1}^{\alpha} B_n = \bigcup_{n=1}^{\alpha} C_n$. The containment "$\subseteq$" follows from $B_n \subseteq C_n$ for all $n \in \mathbb{N}$. For "$\supseteq$" let $x \in \bigcup_{n=1}^{\alpha} C_n$. Let $n \in \mathbb{N}$

be the smallest natural number so that $x \in C_n$. Then $x \in B_n$, which proves "$\supseteq$." Moreover, the B_n are pairwise disjoint, because if $m < n$, then $B_m = C_m \setminus \bigcup_{j=1}^{m-1} C_j \subseteq C_m$ and $B_n = C_n \setminus \bigcup_{j=1}^{n-1} C_j \subseteq C_n \setminus C_m$. Now for each n let $B_n = \{b_n^k : k \in I_n\}$, where I_n is $\mathbb{N}$ or a set of the form $\{1, \ldots, m_n\}$. Then $f(n, k) := b_n^k$ is a bijective function between $\{(n, k) \in \mathbb{N} \times \mathbb{N} : n \leq \alpha, k \in I_n\} \subseteq \mathbb{N} \times \mathbb{N}$ and $\bigcup_{n=1}^{\alpha} B_n = \bigcup_{n=1}^{\alpha} C_n$. Thus $\bigcup_{n=1}^{\alpha} C_n$ is countable. ∎

We will use Theorem 4.65 to prove, in Theorem 5.7, that the rational numbers are countable, a result which, at first, feels strange indeed.

Exercises

4-49. Hilbert's Hotel. Hilbert's Hotel (named after the mathematician David Hilbert, who introduced it to illustrate equivalences for countable sets) has countably many rooms, which are numbered by the natural numbers.

    (a) One evening, every room in Hilbert's Hotel is occupied when a new guest arrives. Explain how, *despite the fact that the hotel is fully booked*, this guest can be accommodated.
    Hint. Every guest could move from the guest's original room to the next room.

    (b) One evening, every room in Hilbert's Hotel is occupied when n new guests arrive. Explain how, *despite the fact that the hotel is fully booked*, all these guests can be accommodated.

    (c) One evening, every room in Hilbert's Hotel is occupied when a "Hilbert Bus" with countably infinitely many new guests arrives. Explain how, *despite the fact that the hotel is fully booked*, these countably infinitely many new guests can be accommodated.

    (d) Suppose for a mathematics convention, countably many buses with countably many guests in each one of them arrive at Hilbert's Hotel. Can all these guests be accommodated? If so, how? If not, why not?

4-50. Prove that the function $f : \mathbb{N} \to \mathbb{Z}$ defined by $f(n) := \begin{cases} \frac{n-1}{2}; & \text{if } n \text{ is odd,} \\ -\frac{n}{2}; & \text{if } n \text{ is even,} \end{cases}$ is bijective.

4-51. Another proof that $\mathbb{N} \times \mathbb{N}$ is countable.

    (a) Prove that the function $f : \mathbb{N} \times \mathbb{N} \to \mathbb{N}$ defined by $f(m, n) := 2^m 3^n$ is injective.

    (b) Use part 4-51a and Theorem 4.62 to prove that $\mathbb{N} \times \mathbb{N}$ countable.

4-52. Use Theorem 4.65 to prove Theorem 4.63.
    Note. This exercise establishes that Theorems 4.63 and 4.65 are equivalent. Nonetheless, for both of them to be true, one must be proved without using the other. In the text, we chose to establish Theorem 4.63 first.

4-53. Give a *direct* proof that the union of a countably infinite set and a finite set is countable. That is, for a countably infinite set C and a finite set F, define a bijective function $f : \mathbb{N} \to C \cup F$.

4-54. Give a *direct* proof that the union of *two* countably infinite sets is countable. That is, for two countably infinite sets C and D, define a bijective function $f : \mathbb{N} \to C \cup D$.

4-55. Prove that if $C_1, \ldots, C_n$ are countable, then $C_1 \times C_2 \times \cdots \times C_n$ is countable.
    Hint. First prove that the product of two countable sets is countable.

4-56. Let S be a set. Prove that S is countable iff there is an injective function $f : S \to \mathbb{N}$.

4-57. Let S be a finite set and let $f : S \to S$ be a function. Prove that f is injective iff f is surjective.

4-58. Let F be a finite set and let I be an infinite set. Prove that $I \setminus F$ is infinite.
    Hint. Assume for a contradiction that $I \setminus F$ is finite and use $I = F \cup (I \setminus F)$ to construct a bijective function between I and a finite set.

4-59. Prove that the set of **prime numbers** is countably infinite.
    Hint. Prove by contradiction that the set is infinite. Use one plus the "product of all prime numbers" to construct a contradiction.

Chapter 5

Number Systems III: Fields

The key property of the whole number systems in Chapters 3 and 4 is that whole numbers are discrete entities, each a distinct distance separated from its closest neighbor (see Exercise 4-31). These discrete, equidistant steps between consecutive whole numbers make the idea of counting sensible. Nonetheless, there are situations in which these discrete distances are a hindrance rather than an asset. The macroscopic world that surrounds us gives a distinctly continuous impression. The distance between two points can be halved, that half can be halved again, and so on, seemingly indefinitely. Although quantum mechanics tells us that ultimately certain entities do not split, continuous models are way too successful in the macroscopic world to be replaced with entirely discrete approaches — Not to mention that we do not have the computational capabilities to keep track of systems with on the order of 10^{23} particles.

For mathematics, the above means that there must be number systems in which equations such as $2x = 1$ have a solution. This chapter presents the major continuous number systems: The rational numbers, the real numbers and the complex numbers.

5.1 Arithmetic With Rational Numbers

The rational numbers are the smallest number system in which every integer has a multiplicative inverse. (Theorem 5.13 will make this statement more precise.) From our experience with number systems, we know that rational numbers are "fractions" $\frac{n}{d}$ with a numerator n and a denominator d. Of course, we face the same problem as when we introduced the integers: We know what the number system is supposed to do, but we must encode it with the mathematics that is available so far. Just as the definition of the integers relied on "formal differences" (see Proposition 4.1), the definition of the rational numbers relies on the introduction of "formal fractions" that are defined via the following equivalence relation.

Fundamentals of Mathematics: An Introduction to Proofs, Logic, Sets, and Numbers.
By Bernd S. W. Schröder.
Copyright © 2010 John Wiley & Sons, Inc.

Proposition 5.1 *The relation $(a, b) \sim (c, d)$ defined by $a \cdot d = b \cdot c$ is an equivalence relation on the set $\mathbb{Z} \times \left(\mathbb{Z} \setminus \{0\}\right)$.*

Proof. Exercise 5-1.   ∎

The pairs in Proposition 5.1 can rightly be viewed as formal fractions. After all, we know from arithmetic that $\frac{a}{b} = \frac{c}{d}$ iff $ad = bc$. So the equivalence relation of Proposition 5.1 encodes equality of fractions without explicitly talking about fractions and their properties. This is exactly what we need to define the rational numbers, because, similar to the integers, the properties of fractions/rational numbers will be derived from the definition. Moreover, this is exactly how we introduced the integers in Section 4.1, except that in Section 4.1 we used addition for the definition and subtraction as motivation, whereas here we use multiplication for the definition and division as motivation.

Before we continue, a word regarding the exclusion of 0 from the second component (the "formal denominator") is in order. We know from arithmetic that "you can't divide by zero," but this cannot be the reason why zero is excluded. After all, we are building the theory from the ground up, so any laws that forbid certain operations should follow from this process rather than being incorporated at the start.[1] It turns out that forbidding 0 as the second entry is necessary to build the theory. The relation $\sim$ in Proposition 5.1 would not be an equivalence relation if we allowed 0 in the second entry: With 0 allowed in the second entry we would need to consider $(0, 0) \in \mathbb{Z} \times \mathbb{Z}$. But $0 \cdot d = 0 \cdot c$ for all $(c, d) \in \mathbb{Z} \times \mathbb{Z}$, which would mean that $(0, 0) \sim (c, d)$ for all $(c, d) \in \mathbb{Z} \times \mathbb{Z}$. This leads to an unresolvable problem: We want $\sim$ to be transitive, because equivalence relations are our best (and possibly only) way to deal with the fact that the same fraction can have multiple representations. But if one object is related to all others, then either the relation is not transitive, which is not an option, or, transitivity would imply that all objects are equivalent to each other, leading to a rather useless relation. So $(0, 0)$ cannot be part of the set that we use to define the rational numbers.

What about elements $(a, 0)$ with $a \neq 0$? The relation $(a, 0) \sim (c, d)$ would mean that $a \cdot d = 0 \cdot c = 0$, that is, $a = 0$ (which had to be forbidden) or $d = 0$. Thus, if we exclude $(0, 0)$, elements with zero in their second component would only be equivalent to each other. That in itself would not exclude these elements from consideration. But using them would ultimately re-introduce the forbidden element $(0, 0)$: After all, we want to do algebra with these formal fractions, and we know what addition should look like. The sum $\frac{a}{b} + \frac{c}{d}$ should be $\frac{ad+bc}{bd}$. If both b and d were zero, the sum would be the forbidden element $(0, 0)$. Thus it is only sensible to exclude pairs whose second component is zero from consideration. So, a formal reason why "we cannot divide by zero" (see question 3 on page x) that is even more fundamental than the reason given after Proposition 4.17 is that $\frac{0}{0}$ destroys the equivalence relation that encodes equality of fractions, and that fractions of the form $\frac{a}{0}$ would ultimately introduce this forbidden element through addition.

Turning back to the development of the rational numbers, we see that algebraic operations must be introduced next. As for the integers, we must assure that the results of addition and multiplication do not depend on the representatives chosen for the

---

[1] Using Exercise 4-10 is out of the question, too, because we do not yet know the properties of the rational numbers.

summands or factors. Proposition 5.2 shows that this is indeed the case.

Proposition 5.2 *Let $\sim$ be the equivalence relation on $\mathbb{Z} \times (\mathbb{Z} \setminus \{0\})$ from Proposition 5.1 and let $\big[(a, b)\big]$ denote the equivalence class of (a, b) under $\sim$. Then the operations $\big[(a, b)\big] + \big[(c, d)\big] := \big[(ad + bc, bd)\big]$ and $\big[(a, b)\big] \cdot \big[(c, d)\big] := \big[(ac, bd)\big]$ are totally defined and well-defined.*

Proof. It is easy to see that both operations are totally defined. After all, for any pair of equivalence classes, we do assign at least one result. To show that we assign exactly one result, we will prove that the multiplication is well-defined and we will leave the proof for addition to Exercise 5-2.

The proof that addition is well-defined is harder than the proof that multiplication is well-defined. In Propositions 4.2 and 4.3, the harder proof was given and the easier proof was left as an exercise. Now that you have progressed further, it will be beneficial to fill in the harder proof.

Let $\big[(a, b)\big] = \big[(a', b')\big]$ and let $\big[(c, d)\big] = \big[(c', d')\big]$. Then we have $ab' = a'b$ and $cd' = c'd$. Therefore $acb'd' = ab'cd' = a'bc'd = a'c'bd$, which implies that $\big[(ac, bd)\big] = \big[(a'c', b'd')\big]$. Hence the multiplication of the equivalence classes is well-defined. ∎

Now that we have binary operations $+$ and $\cdot$ that do not depend on the representative we choose, we can formally define the rational numbers and their addition and multiplication.

Definition 5.3 *The **rational numbers** $\mathbb{Q}$ are defined to be the set of equivalence classes $\big[(a, b)\big]$ of elements of $\mathbb{Z} \times (\mathbb{Z} \setminus \{0\})$ under the equivalence relation $\sim$ of Proposition 5.1. **Addition** is defined by $\big[(a, b)\big] + \big[(c, d)\big] := \big[(ad + bc, bd)\big]$ and **multiplication** is defined by $\big[(a, b)\big] \cdot \big[(c, d)\big] := \big[(ac, bd)\big]$.*

As we would expect, the rational numbers have a host of nice properties. In particular, Theorem 5.4 below shows that the rational numbers are a commutative ring with unity in which every nonzero element has a multiplicative inverse. Such structures are called "fields" and we will investigate them in more detail in Section 5.2.

Theorem 5.4 *Addition of rational numbers is associative, commutative, has a neutral element $0 := \big[(0, 1)\big]$ and for every $\big[(a, b)\big] \in \mathbb{Q}$ the element $\big[(-a, b)\big]$ is its additive inverse. Multiplication of rational numbers is associative, commutative, has a neutral element $1 := \big[(1, 1)\big]$, and for every $\big[(a, b)\big] \in \mathbb{Q} \setminus \{0\}$ the element $\big[(b, a)\big]$ is its multiplicative inverse. Finally, multiplication is distributive over addition.*

Proof. Exercise 5-3. ∎

We know from experience that the rational numbers contain the integers. Indeed, because isomorphic structures are considered to be "the same," the integers are not hard to find in $\mathbb{Q}$.

Theorem 5.5 *The set $\big\{[(a, 1)] : a \in \mathbb{Z}\big\} \subseteq \mathbb{Q}$ is ring isomorphic to $\mathbb{Z}$. That is, there is an isomorphism from $\mathbb{Z}$ to $\big\{[(a, 1)] : a \in \mathbb{Z}\big\}$ that preserves addition and multiplication.*

Proof. The function $f : \mathbb{Z} \to \big\{[(a, 1)] : a \in \mathbb{Z}\big\} \subseteq \mathbb{Q}$ defined by $f(a) = \big[(a, 1)\big]$ is an isomorphism (see Exercise 5-4). ∎

Because isomorphic structures are, for all intents and purposes, "the same," we also say that the rational numbers contain $\mathbb{Z}$ and consequently, they contain $\mathbb{N}$, too.

Definition 5.6 *The set $\big\{[(a, 1)] : a \in \mathbb{Z}\big\} \subseteq \mathbb{Q}$ will also be called the set of* **integers** $\mathbb{Z}$. *Similarly, the set $\big\{[(a, 1)] : a \in \mathbb{N}\big\} \subseteq \mathbb{Q}$ will also be called the set of* **natural numbers** $\mathbb{N}$.

Theorem 5.5 shows that the rational numbers strictly contain the integers. But the integers and the natural numbers have already shown that, for infinite sets, strict containment need not mean that the sizes of the sets differ. It is slightly surprising that the rational numbers are countable. The reason is that the order in which the rational numbers are counted has nothing to do with their natural ordering.[2]

Theorem 5.7 *The rational numbers $\mathbb{Q}$ are countable.*

Proof. The set

$$\mathbb{Q} = \bigcup_{n \in \mathbb{Z}} \big\{[(n, d)] : d \in \mathbb{Z} \setminus \{0\}\big\}$$

is countable by Theorems 4.60 and 4.65. ∎

Exercises

5-1. Prove Proposition 5-1.

5-2. Prove that addition of rational numbers is well-defined.

5-3. Prove Theorem 5.4. That is, prove each of the following for $\mathbb{Q}$.

    (a) Addition is associative.

    (b) Addition is commutative.

    (c) $0 := \big[(0, 1)\big]$ is a neutral element for addition.

    (d) For every element $\big[(a, b)\big] \in \mathbb{Q}$ the element $\big[(-a, b)\big]$ is an additive inverse element.

    (e) Multiplication is associative.

    (f) Multiplication is commutative.

    (g) $1 := \big[(1, 1)\big]$ is a neutral element for multiplication.

    (h) For every element $\big[(a, b)\big] \in \mathbb{Q} \setminus \{0\}$ the element $\big[(b, a)\big]$ is a multiplicative inverse.

    (i) Multiplication is distributive over addition.

5-4. Proof of Theorem 5.5. Consider the function $f : \mathbb{Z} \to \big\{[(a, 1)] : a \in \mathbb{Z}\big\} \subseteq \mathbb{Q}$ defined by $f(a) = \big[(a, 1)\big]$.

    (a) Prove that f is injective.

    (b) Prove that f is surjective.

    (c) For all $x, y \in \mathbb{Z}$, prove that $f(x + y) = f(x) + f(y)$.

    (d) For all $x, y \in \mathbb{Z}$, prove that $f(xy) = f(x)f(y)$.

    (e) Conclude that f is an isomorphism.

---

[2] We have not yet introduced an order on $\mathbb{Q}$, but we will do so in Section 5.3.

5.2 Fields

As we did with the integers, now that we have introduced the formal structure of the rational numbers, we want to explore their properties. The abstract structure that best describes the rational numbers is that of a field. As with integers and rings, rational numbers are not the only example of a field, but they are an important field.

Definition 5.8 *Let $\mathbb{F}$ be a set and let $+ : \mathbb{F} \times \mathbb{F} \to \mathbb{F}$ and $\cdot : \mathbb{F} \times \mathbb{F} \to \mathbb{F}$ be binary operations on $\mathbb{F}$. Then the triple $(\mathbb{F}, +, \cdot)$ is called a* **field** *iff the following hold.*

1. *Addition is* **associative**, *that is, for all $x, y, z \in \mathbb{F}$ we have*

   $(x + y) + z = x + (y + z)$.

2. *Addition is* **commutative**, *that is, for all $x, y \in \mathbb{F}$ we have*

   $x + y = y + x$.

3. *There is a* **neutral element** 0 *for addition, that is, there is an element $0 \in \mathbb{F}$ so that for all $x \in \mathbb{F}$ we have $x + 0 = x$.*

4. *For every element $x \in \mathbb{F}$, there is an* **additive inverse** *element $(-x)$ so that*

   $x + (-x) = 0$.

5. *Multiplication is* **associative**, *that is, for all $x, y, z \in \mathbb{F}$ we have*

   $(x \cdot y) \cdot z = x \cdot (y \cdot z)$.

6. *Multiplication is* **commutative**, *that is, for all $x, y \in \mathbb{F}$ we have*

   $x \cdot y = y \cdot x$.

7. *There is a* **neutral element** $1 \neq 0$ *for multiplication, that is, there is an element $1 \in \mathbb{F} \setminus \{0\}$ so that for all $x \in \mathbb{F}$ we have $1 \cdot x = x$.*

8. *For every element $x \in \mathbb{F} \setminus \{0\}$, there is a* **multiplicative inverse** *element x^{-1} so that $x \cdot x^{-1} = 1$.*

9. *Multiplication is* **(left) distributive** *over addition, that is, for all $\alpha, x, y \in \mathbb{F}$ we have $\alpha \cdot (x + y) = \alpha \cdot x + \alpha \cdot y$.*

As is customary for multiplication, the dot between factors is usually omitted. Moreover, we will often shorten "$(\mathbb{F}, +, \cdot)$ is a field" to "$\mathbb{F}$ is a field."

What do we already know about fields? It follows directly from the definition that if $(\mathbb{F}, +, \cdot)$ is a field, then it is a commutative ring with unity in which every nonzero element has a multiplicative inverse. Therefore, by Proposition 3.35 the neutral element 0 with respect to the addition operation is unique and by Proposition 4.10, additive inverses are unique. We can define summations as in Definition 3.61 and the properties of sums given in Theorem 3.59 (no need for parentheses that group sums so that every addition has two summands), Theorem 3.64 (sums can be re-indexed and combined) and in Theorem 3.65 (sums can be reordered and added termwise) hold in fields.

Regarding multiplication, we note that Theorem 3.59 also holds for the ring multiplication (so there is no need for parentheses in products to make sure every multiplication has two factors), that parentheses are multiplied out "as usual" because of Proposition 3.38 and Theorem 3.66, and that we can define products as in Definition 3.69 and powers as in Definition 3.70. Moreover, power laws as in Theorem 3.71 and the Binomial Theorem (Theorem 3.77) hold.

By Proposition 3.35, the neutral element with respect to multiplication is unique and by Proposition 4.10 the multiplicative inverse of each element is unique.

Moreover, we can define subtraction as in Definition 4.14, we can cancel common summands as in Proposition 4.16, and zero and negative signs behave like they should by Propositions 4.17 and 4.18.

What do we already know about the rational numbers? Because of Theorem 5.4, which says that $\mathbb{Q}$ is a field, all the above results hold for the rational numbers $\mathbb{Q}$.

Once again we see how efficient the abstract approach is. Proving all the above results for fields or for $\mathbb{Q}$ would have taken quite a while.

Theorem 5.9 below finally gives us something like a converse to Proposition 4.17.

Theorem 5.9 *Let $(\mathbb{F}, +, \cdot)$ be a field and let $a, b \in \mathbb{F}$. Then $ab = 0$ implies $a = 0$ or $b = 0$.*

Proof. Let $a, b \in \mathbb{F}$ be so that $ab = 0$ and suppose that $a \neq 0$. Then a has a multiplicative inverse a^{-1} and we obtain $b = 1 \cdot b = a^{-1}ab = a^{-1}0 = 0$. ■

Note that Theorem 5.9 does not hold in mere rings: For example, in $\mathbb{Z}_4$ we have $[2]_4 \cdot [2]_4 = [0]_4$. The key is, of course, the presence or absence of multiplicative inverses.[3]

Example 5.10 Examples of fields.

1. By Theorem 5.4, $(\mathbb{Q}, +, \cdot)$ is a field.

2. Let $m \in \mathbb{N}$. Then $\mathbb{Z}_m$ is a field iff m is a prime number.

   Proof. For "$\Rightarrow$" suppose for a contradiction that $\mathbb{Z}_m$ is a field and m is not a prime number. Then there are $a, b \in \{2, \ldots, m - 1\}$ so that $ab = m$. But then $[a]_m[b]_m = [m]_m = [0]_m$, which, in a field, by Theorem 5.9 would imply that $[a]_m = [0]_m$ or $[b]_m = [0]_m$. But this is not true, because $a, b \in \{2, \ldots, m - 1\}$, a contradiction.

   For "$\Leftarrow$," let m be a prime number. We already know from part 2 of Example 4.13 that $\mathbb{Z}_m$ is a commutative ring with unity. Moreover, the lowest common divisor of any number in $\{2, \ldots, m - 1\}$ with m is 1. Hence by Proposition 3.107 every element of $\mathbb{Z}_m$ has a multiplicative inverse. Thus $\mathbb{Z}_m$ is a field.

3. The integers $\mathbb{Z}$ show that not every ring is a field. (Also see Figure 3.1 on page 107.) □

---

[3]Multiplicative inverses also provide a cancelation law for multiplication, see Exercise 5-6.

With multiplication by the multiplicative inverse being the same as division by the number, we can prove that, even in fields, division by zero is not allowed.

Proposition 5.11 *Let* $(\mathbb{F}, +, \cdot)$ *be a field. Then* 0 *does not have a multiplicative inverse.*

Proof. Suppose for a contradiction that $a \in \mathbb{F}$ was a multiplicative inverse of 0. By definition of multiplicative inverses and by Proposition 4.17 we would conclude that $1 = a0 = 0$, contradicting that $1 \neq 0$. ∎

Proposition 5.11 allows us to revisit question 3 on page x one final time: Ultimately, we want to work in a field, and therefore, by Proposition 5.11 we cannot divide by 0. That is, "division by zero" is not compatible with the properties of a field.

Part 2 of Example 5.10 shows that even though fields have many good properties, they need not be infinite. Therefore an abstract field need not contain a copy of the integers. The problem is that if, say, in $\mathbb{Z}_m$, we add 1 to itself sufficiently often, we get zero. This leads to the definition of the characteristic of a field.

Definition 5.12 *Let* $(\mathbb{F}, +, \cdot)$ *be a field. Then the* **characteristic** *of* $\mathbb{F}$ *is the smallest* $n \in \mathbb{N}$ *so that* $\sum_{j=1}^{n} 1 = 0$.[4] *If no such number exists, then we set the characteristic equal to* 0.

Fields of characteristic 0 behave like we may expect fields to behave. In particular, they contain copies of $\mathbb{N}$, $\mathbb{Z}$ and $\mathbb{Q}$.

Theorem 5.13 *Let* $(\mathbb{F}, +, \cdot)$ *be a field of characteristic* 0. *Then* $\mathbb{F}$ *contains an isomorphic copy of* $\mathbb{Q}$.

Proof. First note that

$$\varphi_{\mathbb{N}}(n) := \sum_{j=1}^{n} 1$$

defines an isomorphism from $\mathbb{N}$ to a subset of $\mathbb{F}$. Call this subset $\mathbb{N}_{\mathbb{F}}$.

Then note that $\mathbb{N}_{\mathbb{F}} \cup \{0\} \cup \{-n : n \in \mathbb{N}_{\mathbb{F}}\}$ is a subset of $\mathbb{F}$ that is isomorphic to $\mathbb{Z}$. Call this subset $\mathbb{Z}_{\mathbb{F}}$ and let $\varphi_{\mathbb{Z}} : \mathbb{Z} \to \mathbb{Z}_{\mathbb{F}}$ be an isomorphism.

Finally note that

$$\varphi_{\mathbb{Q}}\Big(\big[(n, d)\big]\Big) := \varphi_{\mathbb{Z}}(n)\big(\varphi_{\mathbb{Z}}(d)\big)^{-1}$$

is an isomorphism from $\mathbb{Q}$ to a subset $\mathbb{Q}_{\mathbb{F}}$ of $\mathbb{F}$.

Details are left to Exercise 5-7. ∎

It is tempting to hope that every infinite field has characteristic 0. However, Exercise 5-16 shows that this is not so.

Because the rational numbers have characteristic 0, the proof of Theorem 5.13 also shows once more (see also Theorem 5.5) that the rational numbers contain an

---

[4]With notation as in Definition 3.67, it is also correct to say that the characteristic is the smallest number n so that $n1 = 0$. Indeed, this is how the characteristic is often defined. But this definition could give the impression that the natural numbers are contained in the field, which need not be the case. (And which in fact *is* not the case for fields of nonzero characteristic.) Therefore, to avoid confusion, the author chose to use sums here.

isomorphic copy of the integers. As is customary (see Theorem 4.20 and Definition 5.6) we will say that the rational numbers contain the integers themselves. With integers thus available, it is natural to represent rational numbers as actual (not formal) fractions, that is, as quotients of integers. We first focus on the definition and properties of fractions in general and then we specifically show that rational numbers are quotients of integers.

Definition 5.14 *Let* $(\mathbb{F}, +, \cdot)$ *be a field and let* $n, d \in \mathbb{F}$ *with* $d \neq 0$. *Then*

$$\frac{n}{d} := nd^{-1}$$

is called a **fraction**. *The element* n *is called the* **numerator** *and the element* d *is called the* **denominator**. *The fraction is also called the* **quotient** *of* n *and* d.

Definition 5.14 defines fractions in general fields. Moreover, it does not restrict the numerator and denominator to be certain numbers, except that the denominator must not be zero, because zero does not have a multiplicative inverse. In this fashion, we will be able to use fractions in fields other than $\mathbb{Q}$ and also with more complicated numerators and denominators, such as π or sums of other fractions. The properties of fractions are now a consequence of the definition.

Proposition 5.15 Addition and multiplication of fractions. *Let* $(\mathbb{F}, +, \cdot)$ *be a field and let* $\frac{m}{c}$ *and* $\frac{n}{d}$ *be fractions in* $\mathbb{F}$. *Then*

$$\frac{m}{c} + \frac{n}{d} = \frac{md + nc}{cd} \quad \text{and} \quad \frac{m}{c} \cdot \frac{n}{d} = \frac{mn}{cd}.$$

Proof. Exercise 5-8.   ∎

With general properties of fractions in fields established, we turn once more to the rational numbers in particular.

Proposition 5.16 *Every rational number* q *is the quotient of an integer* $n_q \in \mathbb{Z}$ *and a natural number* $d_q \in \mathbb{N}$.

Proof. For this proof, to distinguish subsets of $\mathbb{Q}$ from the "original" natural numbers and integers, the copy of $\mathbb{N}$ in $\mathbb{Q}$ is denoted $\mathbb{N}_\mathbb{Q}$ and the copy of $\mathbb{Z}$ in $\mathbb{Q}$ is denoted $\mathbb{Z}_\mathbb{Q}$.

Let $q \in \mathbb{Q}$. Then $q = [(n, d)]$ for some $(n, d) \in \mathbb{Z} \times (\mathbb{Z} \setminus \{0\})$. On one hand, if $d \in \mathbb{N}$, we have $q = [(n, 1)][(1, d)] = [(n, 1)][(d, 1)]^{-1}$, where $[(n, 1)] \in \mathbb{Z}_\mathbb{Q}$ and $[(d, 1)] \in \mathbb{N}_\mathbb{Q}$. On the other hand, if $d \notin \mathbb{N}$, then $-d \in \mathbb{N}$ and we have $q = [(n, d)] = [(-n, -d)] = [(-n, 1)][(1, -d)] = [(-n, 1)][(-d, 1)]^{-1}$, where $[(-n, 1)] \in \mathbb{Z}_\mathbb{Q}$ and $[(-d, 1)] \in \mathbb{N}_\mathbb{Q}$.   ∎

In particular, Proposition 5.16 shows that rational numbers can be written in the familiar fraction notation. We conclude this section with a word on power laws. By Theorem 3.71 we know that, for positive exponents, "the usual power laws" hold in fields. But once we have multiplicative inverses, negative exponents can be introduced, too.

Definition 5.17 *Let* $(\mathbb{F}, +, \cdot)$ *be a field and let* $a \in \mathbb{F} \setminus \{0\}$. *Then we set* $a^0 := 1$ *and for* $n \in \mathbb{N}$ *we define*

$$a^{-n} := \left(a^{-1}\right)^n.$$

Note that $a^{-n} = \left(a^{-1}\right)^n$ is actually the *definition* of negative exponents. Specifically, the equality is *not* obtained by applying the law for raising a power to another power. With this definition, the usual power laws can be expanded to integer exponents.

Theorem 5.18 *Let* $(\mathbb{F}, +, \cdot)$ *be a field, let* $a, b \in \mathbb{F} \setminus \{0\}$ *and let* $m, n \in \mathbb{Z}$. *Then the following hold.*

1. $a^{m+n} = a^m \cdot a^n$

2. $(a \cdot b)^n = a^n \cdot b^n$

3. $\left(a^m\right)^n = a^{mn}$

Proof. For part 1, first note that there is nothing to prove if $n = 0$ or $m = 0$. Now we distinguish four cases.

Case 1: $m, n \in \mathbb{N}$. This is part 1 of Theorem 3.71. **(We will use this part repeatedly in the following.)**

Case 2: $m, -n \in \mathbb{N}$. If $m = -n$, then $1 = a^m \cdot a^{-m}$ can be proved with an easy induction (see Exercise 5-9a). Now $a^{m+n} = a^0 = 1 = a^m \cdot a^{-m} = a^m \cdot a^n$.

If $m > -n$, then $m + n \in \mathbb{N}$ and $a^{m+n} = a^{m+n}a^{-n}a^n = a^{m+n+(-n)}a^n = a^m a^n$.

If $m < -n$, then $-(m + n) \in \mathbb{N}$ and

$$
\begin{aligned}
a^{m+n} &= \left(a^{-1}\right)^{-(m+n)} = \left(a^{-1}\right)^{(-m)+(-n)} \left(a^{-1}\right)^m \left(a^{-1}\right)^{(-m)} \\
&= \left(a^{-1}\right)^{(-m)+(-n)+m} \left(a^{-1}\right)^{(-m)} \\
&= \left(a^{-1}\right)^{(-n)} \left(a^{-1}\right)^{(-m)} = a^n a^m = a^m a^n,
\end{aligned}
$$

where the equality $\left(a^{-1}\right)^{-m} = a^m$ requires a little thought (see Exercise 5-9b).

Case 3: $-m, n \in \mathbb{N}$. Simply interchange the roles of m and n in Case 2.

Case 4: $-m, -n \in \mathbb{N}$.

$$
\begin{aligned}
a^{m+n} &= a^{-\left((-m)+(-n)\right)} = \left(a^{-1}\right)^{(-m)+(-n)} \\
&= \left(a^{-1}\right)^{(-m)} \left(a^{-1}\right)^{(-n)} = a^m \cdot a^n.
\end{aligned}
$$

For parts 2 and 3, see Exercises 5-9c and 5-9d, respectively. ∎

Exercises

5-5. Let $(\mathbb{F}, +, \cdot)$ be a field. Prove that multiplication is **right distributive** over addition, that is, prove that for all $\alpha, x, y \in \mathbb{F}$ we have $(x + y) \cdot \alpha = x \cdot \alpha + y \cdot \alpha$.

5-6. Let $(\mathbb{F}, +, \cdot)$ be a field and let $a, b, c \in \mathbb{F}$ with $c \neq 0$. Prove that $ac = bc$ implies $a = b$.

5-7. Prove Theorem 5.13.

    (a) Prove that $\varphi_{\mathbb{N}}$ is an isomorphism.

    (b) State the definition of $\varphi_{\mathbb{Z}}$ and prove that $\varphi_{\mathbb{Z}}$ is an isomorphism.

    (c) Prove that $\varphi_{\mathbb{Q}}$ is an isomorphism.

5-8. Prove Proposition 5.15. Let $(\mathbb{F}, +, \cdot)$ be a field and let $\dfrac{m}{c}$ and $\dfrac{n}{d}$ be fractions in $\mathbb{F}$.

    (a) Prove that $\dfrac{m}{c} + \dfrac{n}{d} = \dfrac{md + nc}{cd}$.

    (b) Prove that $\dfrac{m}{c} \cdot \dfrac{n}{d} = \dfrac{mn}{cd}$.

5-9. Finish the proof of Theorem 5-9. That is, let $(\mathbb{F}, +, \cdot)$ be a field, let $a, b \in \mathbb{F}$, let $m, n \in \mathbb{Z}$ and prove each of the following.

    (a) $1 = a^m \cdot a^{-m}$

    (b) $\left(a^{-1}\right)^{-m} = a^m$

    (c) $(a \cdot b)^n = a^n \cdot b^n$

    (d) $\left(a^m\right)^n = a^{mn}$

5-10. Solving equations in the rational numbers (Exercises 3-8 and 4-23 revisited).

    (a) Solve the equation $3x + 2 = 4$ in $\mathbb{Q}$. Explain any difference between your result and the results of Exercises 3-8d and 4-23b.

    (b) Solve the equation $x^2 - 2 = 0$ in $\mathbb{Q}$. Explain any difference between your result and the results of Exercises 3-8f and 4-23d.

    (c) Solve the equation $x^2 + 1 = 0$ in $\mathbb{Q}$. Explain any difference between your result and the results of Exercises 3-8g and 4-23e.

5-11. Let $(\mathbb{F}, +, \cdot)$ be a field, $m, n \in \mathbb{F}$ and let $c, d \in \mathbb{F} \setminus \{0\}$.

    (a) Prove that $\dfrac{m + n}{d} = \dfrac{m}{d} + \dfrac{n}{d}$.

    (b) Prove that, in general, even if $c + d \neq 0$ we have $\dfrac{n}{c + d} \neq \dfrac{n}{c} + \dfrac{n}{d}$. That is, give an example in which the two sides are not equal.

5-12. Let $n \in \mathbb{N}$ be the characteristic of a field $(\mathbb{F}, +, \cdot)$. Prove that n must be a prime number.
Hint. Assume that n is not prime and obtain a contradiction to Theorem 5.9.

5-13. Let $(\mathbb{F}, +, \cdot)$ be a field of characteristic p. Prove that for all $a \in \mathbb{F}$ we have that $\displaystyle\sum_{j=1}^{p} a = 0$.

5-14. Find four polynomials p, q, d_1, d_2 in $\mathbb{Q}[x]$ so that d_1 and d_2 are both greatest common divisors of p and q and so that $d_1 \notin \{d_2, -d_2\}$.

5-15. Let $(\mathbb{F}, +, \cdot)$ be a field. Is $(\mathbb{F}, \cdot)$ a group? Justify your answer.

5-16. **Not every infinite field has characteristic** 0. However, to construct an example, we must abandon our view of polynomials as functions and work with formal polynomials. Let p be a prime number. A **formal polynomial** with coefficients in $\mathbb{Z}_p$ is an expression $f(x) = \displaystyle\sum_{j=0}^{n} a_j x^j$. Because this is a formal expression, there are no rules of cancelation. So for these expressions we have $x^p \neq x$, and in fact, the sum is not really a sum, just a formal reminder that things will work like for sums. Let $\mathbb{Z}_p[x]$ be the set of formal polynomials with coefficients in $\mathbb{Z}_p$ (so the notation remains the same). Define two formal polynomials to be equal iff their coefficients are equal (mimicking Corollary

4.59) and define addition and multiplication with the explicit formulas from part 3 of Example 4.13. But remember that the addition and multiplication of coefficients are done in $\mathbb{Z}_p$. Define a **formal rational function** with coefficients in $\mathbb{Z}_p$ to be a quotient $R(x) = \dfrac{N(x)}{D(x)}$, where $N(x)$ and $D(x)$ are formal polynomials in $\mathbb{Z}_p[x]$ and $D \neq 0$. Prove that the set of formal rational functions with addition and multiplication defined like for fractions is an infinite field of characteristic p.

5.3 Ordered Fields

Similar to what we found for integers in Theorem 4.21, the rational numbers contain a subset that should act like positive numbers.

Theorem 5.19 *The subset $P := \big\{[(n, d)] : n \in \mathbb{N}, d \in \mathbb{N}\big\} \subseteq \mathbb{Q}$ of the rational numbers is called the set of* **positive rational numbers** *and it has the following properties.*

1. *For all $x, y \in P$, we have $x + y \in P$ and $xy \in P$,*

2. *For all $x \in \mathbb{Q}$, exactly one of the following three properties holds.*

   Either $x \in P$ or $-x \in P$ or $x = 0$.

Proof. Exercise 5-17.   ∎

Similar to part 1 of Example 4.24, Theorem 5.19 can be used to prove that $P \cup \{0\}$ has the properties of a positive cone in a ring. Hence $\mathbb{Q}$ is an ordered ring. Moreover, Theorem 5.19 shows that for every $x \in \mathbb{Q} \setminus \{0\}$ either $x \in P \cup \{0\}$ or $-x \in P \cup \{0\}$. Hence, by Theorem 4.28, $\mathbb{Q}$ is a totally ordered ring with positive cone $\mathbb{Q}^+ := P \cup \{0\}$. Because $\mathbb{Q}$ is a field, we prefer to talk about it as a totally ordered field.

Definition 5.20 *A field $(\mathbb{F}, +, \cdot)$ is called a* **totally ordered field** *iff there is a subset $\mathbb{F}^+ \subseteq \mathbb{F}$ so that*

1. *For all $x, y \in \mathbb{F}^+$ we have $x + y \in \mathbb{F}^+$ and $xy \in \mathbb{F}^+$, and*

2. *$0 \in \mathbb{F}^+$, and*

3. *For all $x \in \mathbb{F} \setminus \{0\}$, either $x \in \mathbb{F}^+$ or $-x \in \mathbb{F}^+$ holds.*

The subset $\mathbb{F}^+$ is also called the **positive cone** *of the totally ordered field. The elements of $\mathbb{F}^+$ are called* **nonnegative**, *the elements of $\mathbb{F}^+ \setminus \{0\}$ are called* **positive**, *and the elements $x \in \mathbb{F} \setminus \{0\}$ so that $-x \in \mathbb{F}^+$ are called* **negative**.

We do not define ordered fields, because in fields, we are interested only in total order relations. In fact, it seems that fields that are not totally ordered are so rarely investigated that totally ordered fields are often referred to as **ordered fields**.

What do we already know about totally ordered fields? By definition, totally ordered fields are totally ordered rings. Therefore, all results from Section 4.4 are valid for totally ordered fields, too. In particular, there is an order relation $x \leq y$ iff $y - x \in \mathbb{F}^+$, that works as we expect it to work (multiplication with positive numbers preserves the inequality, multiplication with negative numbers reverses it), and there is an absolute value function that has the properties we expect it to have (it's nonnegative, it's zero iff the element is zero, it factors through products and it obeys the triangular inequality). **The usual "avalanche of knowledge" is a little bit smaller here, because we have already taken care of the algebraic properties of fields after Definition 5.8.**

Example 5.21 For completeness' sake, we record that by Theorem 5.19, the rational numbers are a totally ordered field with positive cone $\mathbb{Q}^+ := P \cup \{0\}$. $\qquad\qquad\square$

Because multiplicative inverses exist in totally ordered fields, we should investigate how multiplicative inversion interacts with the order relation. Theorem 5.22 is the abstract version of the rule that "when the reciprocal is taken on both sides, then the inequality is reversed."

Theorem 5.22 *Let $(\mathbb{F}, +, \cdot)$ be a totally ordered field and let $x, y \in \mathbb{F}$ be so that $0 < x \leq y$. Then $0 < y^{-1} \leq x^{-1}$.*

Proof. We first prove that $x > 0$ implies $x^{-1} > 0$. Suppose for a contradiction that $x^{-1} \leq 0$. Because 0 does not have a multiplicative inverse, we conclude that $x^{-1} < 0$. But then, by part 3 of Theorem 4.36, $x > 0$ implies that $1 = x \cdot x^{-1} \leq 0 \cdot x^{-1} = 0$, contradicting Proposition 4.31.

Now let $0 < x \leq y$. By the above, we know that $x^{-1}, y^{-1} > 0$. Now we can use part 2 of Theorem 4.36 to conclude $x^{-1} = x^{-1} \cdot 1 = x^{-1}yy^{-1} \geq x^{-1}xy^{-1} = y^{-1}$. $\qquad\blacksquare$

Finally, in totally ordered fields, squares are nonnegative.

Proposition 5.23 *Let $(\mathbb{F}, +, \cdot)$ be a totally ordered field let $x \in \mathbb{F}$. Then $x^2 \geq 0$.*

Proof. This was proved in Exercise 4-33. $\qquad\qquad\qquad\qquad\qquad\qquad\qquad\qquad\qquad\blacksquare$

Exercises

5-17. Prove Theorem 5.19. That is, let $P := \big\{ [(n, d)] : n \in \mathbb{N}, d \in \mathbb{N} \big\} \subseteq \mathbb{Q}$ and prove the following.

    (a) For all $x, y \in P$, we have $x + y \in P$ and $xy \in P$,

    (b) For all $x \in \mathbb{Q}$, exactly one of the following three properties holds.
        Either $x \in P$ or $-x \in P$ or $x = 0$.

5-18. Let $a, c \in \mathbb{Z} \subseteq \mathbb{Q}$ and let $b, d \in \mathbb{N}$. Prove that $\dfrac{a}{b} < \dfrac{c}{d}$ iff $abd^2 < b^2cd$.

5-19. Let $\mathbb{F}$ be a totally ordered field and let $x \leq y < 0$. Prove that $y^{-1} \leq x^{-1} < 0$.

5-20. Revisiting Exercise 4-34.

    (a) Solve the inequality $|7x + 2| < 3$ in $\mathbb{Q}$. Explain any difference between your result and that of Exercise 4-34c.

    (b) Solve the inequality $|6x + 5| > 2$ in $\mathbb{Q}$. Explain any difference between your result and that of Exercise 4-34d.

    (c) Solve the inequality $3 \leq |2x + 4| \leq 11$ in $\mathbb{Q}$. Explain any difference between your result and that of Exercise 4-34e.

5-21. Sums of fractions that are natural numbers.

    (a) Prove that for every $n \in \mathbb{Z}$ the number $\dfrac{n^7}{7} + \dfrac{n^3}{3} + \dfrac{11n}{21}$ is an integer.

    (b) Explain why part 5-21a is formally not appropriate as an exercise for Section 3.6.

    (c) Find $a \in \mathbb{Q}$ so that for every $n \in \mathbb{Z}$ the number $\dfrac{n^5}{5} + \dfrac{n^3}{3} + an$ is an integer.

    (d) Find $b \in \mathbb{Q}$ so that for every $n \in \mathbb{N}$ the number $\dfrac{n^5}{5} + \dfrac{n^4}{2} + bn$ is a natural number.

5-22. Let $(\mathbb{F}, +, \cdot)$ be a totally ordered field. Prove that the characteristic of $\mathbb{F}$ is 0.

    Hint. Suppose for a contradiction that the characteristic is not 0 and obtain a contradiction to Proposition 4.31.

5-23. Let C be a countable totally ordered set. Prove that C is isomorphic to a subset of $\mathbb{Q}$. That is, prove that there is a bijective function $f : C \to D$, where $D \subseteq \mathbb{Q}$, so that for all $x, y \in D$ we have that $x \leq y$ implies $f(x) \leq f(y)$.

    Hint. Construct the function inductively, map the nth point between the appropriate images of earlier points or to the right or left of all earlier points.

5.4   A Problem With the Rational Numbers

In the rational numbers, every number has an additive inverse and every nonzero number has a multiplicative inverse. Given that a multiplicative inverse for zero is impossible (see Proposition 5.11), this is as much as we can demand for the elementary algebraic operations. With all the algebraic properties that we can possibly want being firmly established, we can turn our full attention to solving equations. All linear equations $ax + b = 0$ with $a \neq 0$ can be solved in $\mathbb{Q}$, because we can subtract b and divide by a. It is natural to consider quadratic equations $ax^2 + bx + c = 0$ next. Unfortunately, we are immediately disappointed. The simplest nontrivial quadratic equation arguably is $x^2 - 2 = 0$, and this equation is not solvable in $\mathbb{Q}$.

Proposition 5.24 *There is no rational number r such that $r^2 = 2$.*

    Proof. First let $n \in \mathbb{N}$ be so that $n^2 = 2z$ for some $z \in \mathbb{N}$. Then 2 is a factor in the prime factorization of n^2. But every factor in the prime factorization of n^2 occurs with an even exponent. Thus $n^2 = 2^2 k^2$ for some k, and hence $n = 2k$. That is, if n^2 is divisible by 2, then so is n.

    Now suppose for a contradiction that there are $n \in \mathbb{Z}$ and $d \in \mathbb{N}$ so that $\left(\frac{n}{d}\right)^2 = 2$. Without loss of generality, we can assume that n and d have no common factors and that $n \in \mathbb{N}$. But by the above $n^2 = 2d^2$ implies $n = n_2 \cdot 2$. Consequently, $2d^2 = (n_2 \cdot 2)^2$, that is, $d^2 = n_2^2 \cdot 2$, which implies $d = d_2 \cdot 2$. But then $2 \mid n$ and $2 \mid d$, a contradiction to the assumption that n and d have no common factors. ∎

    Although this is a setback, we have been here before. After all, linear equations are fairly simple, and we needed just about everything in this text up to this point to construct a number system, the rational numbers $\mathbb{Q}$, or, in general, fields, in which linear equations are solvable. Thankfully, we will not need to spend as much time for quadratic equations. The existence of roots is firmly connected to the following order theoretical ideas.

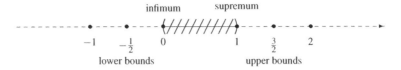

Figure 5.1 Visualization of various upper and lower bounds of a set versus the supremum and the infimum.

Definition 5.25 *Let X be a set and let $\leq$ be an order relation on X. Let $A \subseteq X$.*

1. *The element $u \in X$ is called an **upper bound** of A iff $u \geq a$ for all $a \in A$. If A has an upper bound, it is also called **bounded above**.*

2. *The element $l \in X$ is called a **lower bound** of A iff $l \leq a$ for all $a \in A$. If A has a lower bound, it is also called **bounded below**.*

*A subset $A \subseteq X$ that is bounded above and bounded below is also called **bounded**.*

It is important to note here that upper and lower bounds need not be elements of the set in question and that the bounds need not even be optimal in any sense. For example, consider the set $A := \{q \in \mathbb{Q} : 0 < q < 1\}$. (For a visualization, see Figure 5.1.) The number 2 is an upper bound of A and the number -1 is a lower bound. Neither one is optimal: $\frac{3}{2}$ is another upper bound and $-\frac{1}{2}$ is another lower bound. Both these new bounds are closer to A than 2 and -1, respectively. Apparently, the closest bounds we can get are 1 as an upper bound and 0 as a lower bound. Such "closest bounds" play a special role, when they exist.

Definition 5.26 *Let X be a set and let $\leq$ be an order relation on X. Let $A \subseteq X$.*

1. *The element $s \in X$ is called **lowest upper bound** of A or **supremum** of A, denoted $\sup(A)$, iff s is an upper bound of A and for all upper bounds u of A we have that $s \leq u$.*

2. *The element $i \in X$ is called **greatest lower bound** of A or **infimum** of A, denoted $\inf(A)$, iff i is a lower bound of A and for all lower bounds l of A we have that $l \leq i$.*

*A supremum of A that is an element of A is also called **maximum** of A, denoted $\max(A)$. An infimum of A that is an element of A is also called **minimum** of A, denoted $\min(A)$.*

We should note that, when they exist, suprema and infima need not be in the set that they bound. The set $A := \{q \in \mathbb{Q} : 0 < q < 1\}$ is a prime example. Its supremum is 1, its infimum is 0, and neither one is in A.

A priori, it is not even guaranteed that suprema and infima are unique. But the next result shows that this is indeed the case.

Proposition 5.27 Suprema are unique. *Let X be a set and let $\leq$ be an order relation on X. If the set $A \subseteq X$ is bounded above and $s, t \in X$ both are suprema of A, then $s = t$.*

Proof. Let $A \subseteq X$ and let $s, t \in X$ be as indicated. Then s is an upper bound of A and, because t is a supremum of A, we infer $s \geq t$. Similarly, t is an upper bound of A and, because s is a supremum of A, we infer $t \geq s$. This implies $s = t$. ∎

The proof that infima are unique is similar (see Exercise 5-24).

We started this section with the fact that the rational numbers do not contain the solution of the equation $x^2 - 2 = 0$. Then we sidestepped into some order properties that were claimed to be related to this problem. We can now show the connection. Although there is no rational number that solves $x^2 - 2 = 0$, there are rational numbers that come very close to solving this equation in the sense that $x^2 - 2$ will be really small. This leads to the idea that better and better approximations should ultimately give a solution. For example, we could try to find rational numbers whose squares are smaller than 2, but not by much. The smaller the difference, the better the approximation. Abstractly speaking, the largest of these numbers should be the solution of the equation. Of course there is no largest such number *in* $\mathbb{Q}$, as the following example shows.

Example 5.28 *The set* $\{x \in \mathbb{Q} : x^2 \leq 2\}$ *does not have a supremum in* $\mathbb{Q}$.

It seems that the supremum should solve the equation $s^2 = 2$. We will use exactly this idea in a proof by contradiction.

The proof is very much an "analytical" proof: We try to slip a number between two other numbers. To do so, we must make certain choices that seem unmotivated at the time, but which make sense once we see the whole proof. Typically, such proofs are conceived by working with the final estimate to determine the choices that need to be made. But, as all proofs, they are presented in linear fashion.

Suppose for a contradiction that the number $s \in \mathbb{Q}$ is the supremum of the set $S := \{x \in \mathbb{Q} : x^2 \leq 2\}$. First note that we must have $s^2 \leq 2$. Indeed, otherwise we would have $s^2 > 2$. In particular, $s \neq 0$ and because $(-x)^2 = x^2$ for all rational numbers x, we have $s > 0$. Moreover, there would be an $\varepsilon > 0$ so that $s^2 > 2 + \varepsilon$. Let $\delta := \frac{\varepsilon}{2s}$. Then for all $\nu > 0$ with $\nu < \delta$ we have $2s\nu - \nu^2 < 2s\frac{\varepsilon}{2s} = \varepsilon$. Consequently, for all ν with $0 < \nu < \delta$ we would have

$$(s - \nu)^2 = s^2 - 2s\nu + \nu^2 > 2 + \varepsilon - \left(2s\nu - \nu^2\right) > 2 + \varepsilon - \varepsilon = 2.$$

Hence no rational number between $s - \delta$ and s would be in S. This would mean that $s - \delta$ is an upper bound of S, contradicting the fact that s is the supremum of S.

The proof that s^2 cannot be strictly smaller than 2 is left to Exercise 5-25.

But this means that $s^2 = 2$ and $s \in \mathbb{Q}$, contradicting Proposition 5.24. □

It seems we have not made any progress. We still cannot solve $x^2 - 2 = 0$. But now we know the reason why: Some sets that could be used to approximate the answer do not have a supremum. This idea will be used in the construction of the real numbers. Recall that we defined integers as equivalence classes of "formal differences" because we wanted to introduce additive inverses. Then we introduced rational numbers as equivalence classes of "formal quotients" because we wanted to introduce multiplicative inverses for all nonzero elements. Now we need suprema for certain sets.

Figure 5.2 Visualization of real numbers. The real number D is defined as the indicated down set. This down set may or may not have a supremum in the rational numbers. But our standard visualization of the number line suggests (correctly) that the real number is the supremum of the down set that defines it.

In Section 5.5 we will define the real numbers in such a way that all sets for which a supremum should exist will have a supremum.

Exercises

5-24. **Infima are unique.** Let X be a set and let $\leq$ be an order relation on X. Prove that, if the set $A \subseteq X$ is bounded below and $s, t \in X$ both are infima of A, then $s = t$.

5-25. Prove that the number s in Example 5.28 cannot be so that $s^2 < 2$.
   Hint. Suppose there is an $\varepsilon > 0$ so that $s^2 + \varepsilon < 2$. Then find a $\delta > 0$ so that $(s + \delta)^2 < 2$.

5-26. Let $n \in \mathbb{N}$, let $p \in \mathbb{N}$ be prime and let $k \in \mathbb{N} \setminus \{0\}$ be so that $p \mid n^k$. Prove that $p \mid n$.

5.5   The Real Numbers

One of the problems with the natural numbers was that we could not form differences of arbitrary natural numbers. So, in the construction of the integers, we defined integers as equivalence classes of "formal differences," which allowed us to form differences of arbitrary integers. One of the problems with the integers was that we could not form quotients of arbitrary integers. So, in the construction of the rational numbers, we defined rational numbers as equivalence classes of "formal quotients," which allowed us to form quotients of rational numbers, except for zeros in the denominator. One of the problems with the rational numbers now is that not every subset that is bounded above has a supremum. So, it should not be surprising that real numbers will be defined as subsets of the rational numbers that are bounded above. This definition will ultimately imply that subsets of the real numbers that are bounded above have suprema.

Definition 5.29 *Let X be a set and let $\leq$ be an order relation on X. Then a subset $D \subseteq X$ is called a* **down set** *iff for all $x, y \in X$ we have that $y \in D$ and $x \leq y$ implies $x \in D$.*

Definition 5.30 *We define the set $\mathbb{R}$ to be the set of all nonempty down sets in $\mathbb{Q}$ that do not have a largest element and which are not equal to $\mathbb{Q}$. We call $\mathbb{R}$ the set of* **real numbers***. The construction of $\mathbb{R}$ presented in this section is also known as the construction via* **Dedekind cuts***.*

The real number that is represented by the down set D is best visualized as the supremum of that down set, see Figure 5.2. Of course, there are down sets for which there is no supremum in $\mathbb{Q}$. Our construction of $\mathbb{R}$ will close exactly these gaps.

For $\mathbb{Z}$ and $\mathbb{Q}$, we had to make sure that operations on the equivalence classes for the relations from Propositions 4.1 and 5.1 were well-defined. For $\mathbb{R}$, we will work

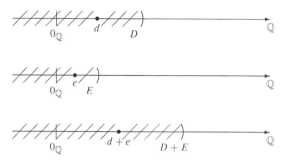

Figure 5.3 Visualization of the addition of real numbers. The numbers D and E depicted above are both semi-infinite intervals in $\mathbb{Q}$ which contain $0_{\mathbb{Q}}$. The sum is the set of all possible sums that can be formed with elements $d \in D$ and $e \in E$, as depicted with two sample elements in the figure. Consequently, because the sets in the figure both contain zero, in this example the right boundary of the sum is to the right of the right boundaries of the summands.

with the sets themselves. But we must make sure that the operations that we define on down-sets without a largest element will again lead to a down-set without a largest element.

Lemma 5.31 *For $D, E \in \mathbb{R}$ let $D + E := \{d + e \in \mathbb{Q} : d \in D, e \in E\}$. Then $D + E \in \mathbb{R}$.*

Proof. Let $D, E \in \mathbb{R}$. We first need to prove that $D + E$ is a down set in $\mathbb{Q}$. Let $x \in D + E$ and let $y \in \mathbb{Q}$ be so that $y \le x$. Then there are $d \in D$ and $e \in E$ so that $x = d + e$. Because $y - x \le 0$ we have that $e + y - x \in E$. But then $y = x + (y - x) = (d + e) + (y - x) = d + (e + y - x) \in D + E$.

To prove that $D + E$ does not have a largest element, suppose for a contradiction that $t \in D + E$ is the largest element of $D + E$. Then there are $d \in D$ and $e \in E$ so that $t = d + e$. Because D and E do not have largest elements, there are $d' \in D$, $e' \in E$ so that $d < d'$ and $e < e'$. But then $d' + e' \in D + E$ and $t = d + e < d' + e'$, a contradiction. ∎

Definition 5.32 *For any two real numbers D and E we define the **sum** of D and E to be the set $D + E$ from Lemma 5.31. The operation $+$ is called **addition**. For a visualization, consider Figure 5.3.*

As always, once we have an operation, we are interested in its properties.

Proposition 5.33 *Addition of real numbers is associative, commutative, there is a neutral element $0 := \{q \in \mathbb{Q} : q < 0_{\mathbb{Q}}\} \in \mathbb{R}$, where $0_{\mathbb{Q}}$ denotes the zero of the rational numbers, and for every $D \in \mathbb{R}$, the element*

$$E := \left\{ q \in \mathbb{Q} : -q \text{ is an upper bound of } D, \text{ but } -q \ne \sup(D) \right\}$$

is an additive inverse.

Proof. Associativity is straightforward. Let $D, E, F \in \mathbb{R}$. Then

$$
\begin{aligned}
(D + E) + F &= \{x + f : x \in D + E, f \in F\} \\
&= \{(d + e) + f : d \in D, e \in E, f \in F\} \\
&= \{d + (e + f) : d \in D, e \in E, f \in F\} \\
&= \{d + y : d \in D, y \in E + F\} \\
&= D + (E + F)
\end{aligned}
$$

Commutativity is left to Exercise 5-27a.

To prove that $0 = \{q \in \mathbb{Q} : q < 0_{\mathbb{Q}}\} \in \mathbb{R}$ is a neutral element with respect to addition, let $D \in \mathbb{R}$. We must prove that $D = D + 0$. Let $e \in D + 0$ be arbitrary. Then there are $d \in D$ and $q \in 0$ so that $e = d + q < d$. Hence, because D is a down-set, $e \in D$. Because $e \in D + 0$ was arbitrary, we infer $D + 0 \subseteq D$. Now let $d \in D$ be arbitrary. Then there is an $x \in \mathbb{Q}$ with $x > 0_{\mathbb{Q}}$ so that $d + x \in D$. But then $-x \in 0$ and hence $d = (d + x) + (-x) \in D + 0$. Because $d \in D$ was arbitrary, we infer $D \subseteq D + 0$. Hence $D = D + 0$.

The fact that E is an additive inverse for D is left to Exercise 5-27b. ∎

With addition established, we turn to multiplication. The definition of multiplication is complicated by the fact that multiplication with negative numbers reverses inequalities. Hence we will first focus on positive numbers and then treat negative numbers as "positive numbers with a negative sign."

Lemma 5.34 *Let $D, E \in \mathbb{R}$ be so that $0_{\mathbb{Q}} \in D$ and $0_{\mathbb{Q}} \in E$. Then the down-set $\{q \in \mathbb{Q} : (\exists d \in D, d > 0_{\mathbb{Q}}, e \in E, e > 0_{\mathbb{Q}} : q \leq de)\}$ does not have a largest element.*

Proof. Exercise 5-28. ∎

Definition 5.35 *Let $D, E \in \mathbb{R}$. If $D = 0$ or $E = 0$ we set $D \cdot E := 0$. For $D \neq 0$ and $E \neq 0$ we define $D \cdot E$ as follows.*

1. *If $0_{\mathbb{Q}} \in D$ and $0_{\mathbb{Q}} \in E$, we define*
   $D \cdot E := \{q \in \mathbb{Q} : (\exists d \in D, d > 0_{\mathbb{Q}}, e \in E, e > 0_{\mathbb{Q}} : q \leq de)\}$.

2. *If $0_{\mathbb{Q}} \in D$ and $0_{\mathbb{Q}} \notin E$, we define $D \cdot E := -(D \cdot (-E))$.*

3. *If $0_{\mathbb{Q}} \notin D$ and $0_{\mathbb{Q}} \in E$, we define $D \cdot E := -((-D) \cdot E)$.*

4. *If $0_{\mathbb{Q}} \notin D$ and $0_{\mathbb{Q}} \notin E$, we define $D \cdot E := (-D) \cdot (-E)$.*

*The operation $\cdot$ is called **multiplication**.*

Clearly, multiplication is totally defined and well-defined, because every pair of real numbers (D, E) is assigned a unique product in $\mathbb{R}$. Moreover, the multiplication from Definition 5.35 gives us the properties we want.

Theorem 5.36 *The real numbers $\mathbb{R}$ with addition and multiplication as defined above are a field.*

Proof. The properties for addition have been established in Proposition 5.33. We must establish the properties of multiplication, which is a bit more tedious than for addition, because of the case distinctions in Definition 5.35.

For associativity, let $D, E, F \in \mathbb{R}$. In case $0_\mathbb{Q}$ is an element of D, E and F, we argue as follows.

$$
\begin{aligned}
(D \cdot E) \cdot F \\
= \; & \left\{ q \in \mathbb{Q} : (\exists x \in D \cdot E, x > 0_\mathbb{Q}, f \in F, f > 0_\mathbb{Q} : q \le xf) \right\} \\
= \; & \left\{ q \in \mathbb{Q} : (\exists d \in D, d > 0_\mathbb{Q}, e \in E, e > 0_\mathbb{Q}, f \in F, f > 0_\mathbb{Q} : q \le (de)f) \right\} \\
= \; & \left\{ q \in \mathbb{Q} : (\exists d \in D, d > 0_\mathbb{Q}, e \in E, e > 0_\mathbb{Q}, f \in F, f > 0_\mathbb{Q} : q \le d(ef)) \right\} \\
= \; & \left\{ q \in \mathbb{Q} : (\exists d \in D, d > 0_\mathbb{Q}, y \in E \cdot F, y > 0_\mathbb{Q} : q \le dy) \right\} \\
= \; & D \cdot (E \cdot F)
\end{aligned}
$$

There are 7 cases left to consider and they are all handled similarly. For example, in case that $0_\mathbb{Q} \in D$, $0_\mathbb{Q} \in E$, and $0_\mathbb{Q} \notin F$, we argue as follows.

$$
(D \cdot E) \cdot F \;=\; -\big((D \cdot E) \cdot (-F)\big) = -\big(D \cdot (E \cdot (-F))\big)
$$

> Now use the definition of $X \cdot (-Y)$ twice.

$$
= \; -\big(D \cdot (-(E \cdot F))\big) = D \cdot (E \cdot F)
$$

The remaining cases are left to Exercise 5-29a.

The proof of commutativity is left to Exercise 5-29b.

For the neutral element with respect to multiplication, set $1 := \{ q \in \mathbb{Q} : q < 1_\mathbb{Q} \}$. Then for all $D \in \mathbb{R}$ with $0_\mathbb{Q} \in D$ we have that

$$
\begin{aligned}
D \cdot 1 \;=\; & \left\{ q \in \mathbb{Q} : (\exists d \in D, d > 0_\mathbb{Q}, e \in 1, e > 0_\mathbb{Q} : q \le de) \right\} \\
= \; & \left\{ q \in \mathbb{Q} : (\exists d \in D, d > 0_\mathbb{Q}, e \in \mathbb{Q}, e > 0_\mathbb{Q}, e < 1_\mathbb{Q} : q \le de) \right\}
\end{aligned}
$$

> For the next equality, the containment "$\subseteq$" is trivial: If $q \le de$ and $e < 1$, then $q \le d$. For the containment "$\supseteq$," let $q \le d$. Then there is an $\varepsilon > 0_\mathbb{Q}$ so that $\tilde{d} := d(1_\mathbb{Q} + \varepsilon) \in D$. But then $q \le \tilde{d} \frac{1_\mathbb{Q}}{1_\mathbb{Q} + \varepsilon}$ and so $q \le \tilde{d} e$ with $e := \frac{1_\mathbb{Q}}{1_\mathbb{Q} + \varepsilon} \in 1$.

$$
\begin{aligned}
= \; & \left\{ q \in \mathbb{Q} : (\exists d \in D, d > 0_\mathbb{Q} : q \le d) \right\} \\
= \; & D.
\end{aligned}
$$

For all $D \in \mathbb{R}$ with $0_\mathbb{Q} \notin D$ we have that $D \cdot 1 = -\big((-D) \cdot 1\big) = -(-D) = D$.

Regarding multiplicative inverses, for $D \in \mathbb{R} \setminus \{0\}$ with $0_\mathbb{Q} \in D$ we let D^{-1} be the down set of all $q \in \mathbb{Q}$ that are lower bounds of the set $\{ d^{-1} : d \in D, d > 0_\mathbb{Q} \}$ and which are not equal to the infimum of this set (if it exists). For $D \in \mathbb{R}$ with $0_\mathbb{Q} \notin D$ we set $D^{-1} := -(-D)^{-1}$. The proof that $D \cdot D^{-1} = 1$ is left to Exercise 5-29c.

The proof that multiplication is distributive over addition is similar to the proof for associativity of multiplication. It is left to Exercise 5-29d. ∎

Because $(\mathbb{R}, +, \cdot)$ is a field, everything we know about fields applies to $(\mathbb{R}, +, \cdot)$.

Think about all the information that gives us. Maybe make a list as
we did after the definitions of groups, rings and fields (see pp. 168, 170
and 205).

Establishing that $(\mathbb{R}, +, \cdot)$ is a field merely assures that we did not lose any algebraic properties. Next, we turn to the order properties, hoping that we have obtained
the missing suprema.

Proposition 5.37 *The set* $P := \{D \in \mathbb{R} : 0_{\mathbb{Q}} \in D\}$ *satisfies the following.*

1. For all $D, E \in P$ *we have that* $D + E \in P$ *and* $DE \in P$.

2. For every $D \in \mathbb{R}$ *exactly one of* $D \in P$, $D = 0$ *or* $-D \in P$ *holds.*

Proof. For part 1, let $D, E \in P$. Then $0_{\mathbb{Q}} = 0_{\mathbb{Q}} + 0_{\mathbb{Q}} \in D + E$, so $D + E \in P$.
Moreover, there are elements $d \in D$ with $d > 0_{\mathbb{Q}}$ and $e \in E$ with $e > 0_{\mathbb{Q}}$, which
means that $0_{\mathbb{Q}} < de \in DE$. Hence $0_{\mathbb{Q}} \in DE$ and $DE \in P$.

For part 2, if $D \in \mathbb{R}$, then either $0_{\mathbb{Q}} \in D$, that is, $D \in P$, or $0_{\mathbb{Q}} \notin D$. In the latter
case, either $D = 0$, or there is an $x \in \mathbb{Q}$ with $x > 0$ so that $-x$ is an upper bound of
D. But then (once more in the latter case) $x \in -D$, which implies $0_{\mathbb{Q}} \in -D$, that is,
$-D \in P$.   ∎

Given Proposition 5.37, similar to what we did for $\mathbb{Q}$, we note that $P \cup \{0\}$ is a
positive cone. Hence $\mathbb{R}$ is a totally ordered field, as the next result states. Maybe more
importantly, there is a set-theoretical way to encode the order, which will make the
proof that certain suprema exist much easier.

Theorem 5.38 *Let* $P := \{D \in \mathbb{R} : 0_{\mathbb{Q}} \in D\}$, *as in Proposition 5.37. The real numbers
are a totally ordered field with positive cone* $\mathbb{R}^+ := P \cup \{0\}$. *Moreover, for any two
real numbers* D *and* E *we have that* $D \leq E$ *iff* $D \subseteq E$.

Proof. The proof of the first sentence is left to Exercise 5-30.

For the second sentence, note that we are done if we can show $D < E$ iff $D \subset E$,
where $\subset$ denotes strict containment. Now $D < E$ iff $E - D \in P$ iff $0_{\mathbb{Q}} \in E - D$ iff
there are $e \in E$ and an upper bound x of D, that is not equal to the supremum of D, so
that $e + (-x) \geq 0_{\mathbb{Q}}$, which is the case iff $e \geq x$ for an $e \in E$ and an upper bound x of
D, which, because E is a down-set, can only happen iff $D \subset E$.   ∎

Because $\mathbb{R}$ is a totally ordered field, everything we know about totally ordered fields
applies to $\mathbb{R}$.

Think about all the information that gives us. Maybe make a list as
we did after the definition of totally ordered fields (see p. 211).

Now we have enough tools at hand to show that every set of real numbers that is
bounded above has a supremum in $\mathbb{R}$.

Theorem 5.39 *The totally ordered field* $\mathbb{R}$ *satisfies the following* **Completeness Axiom**[5]*: Any nonempty subset of* $\mathbb{R}$ *that is bounded above has a supremum.*

Proof. Let $A \subseteq \mathbb{R}$ be a nonempty subset that has an upper bound X. By construction of $\mathbb{R}$, every element of A is a nonempty down set in $\mathbb{Q}$ that is not equal to $\mathbb{Q}$ and that does not have a largest element. We can form the union $S := \bigcup A$ of the set of sets A. Then S is a subset of $\mathbb{Q}$ and because $\bigcup A \subseteq X \neq \mathbb{Q}$ we know that $S \neq \mathbb{Q}$. Moreover, S is a down set in $\mathbb{Q}$, because if $t \in S$ and $u \leq t$, then there is a $T \in A$ so that $t \in T$, which means that $u \in T$ and hence $u \in S$. Finally, S does not have a largest element, because if $l \in S$ was the largest element of S, then there would be an $L \in A$ with $l \in L$ and l would be the largest element of L, which cannot be. Therefore $S \in \mathbb{R}$.

Clearly S is an upper bound of A, because for all $T \in A$ we have $T \subseteq S$, that is, $T \leq S$. Finally, if $Y \in \mathbb{R}$ is any upper bound of A, then for all $T \in A$ we have that $T \subseteq Y$, which means that $Y \supseteq \bigcup A = S$, or, $Y \geq S$.   ∎

Now that we know that $\mathbb{R}$ has the desired properties, similar to our previous constructions, we don't want to work with the details of the construction any more. Proposition 5.40 establishes the last piece of the puzzle: The familiar number systems are subsets of $\mathbb{R}$. Therefore, similar to how we proceeded for $\mathbb{N}$, $\mathbb{Z}$ and $\mathbb{Q}$, once we have established that $\mathbb{R}$ is a totally ordered field that satisfies the Completeness Axiom and which contains a copy of $\mathbb{Q}$ (see Proposition 5.40), we will leave the details of the construction behind and we will only use the aforementioned facts.

Proposition 5.40 *The function* $f : \mathbb{Q} \to \mathbb{R}$ *defined by* $f(q) := \{x \in \mathbb{Q} : x < q\}$ *is an order-preserving field isomorphism from* $\mathbb{Q}$ *to a subset of* $\mathbb{R}$. *(That is, for all* $x, y \in \mathbb{Q}$ *we have* $f(x + y) = f(x) + f(y)$, $f(x \cdot y) = f(x) \cdot f(y)$ *and* $x \leq y$ *implies* $f(x) \leq f(y)$.) *We will refer to this subset of* $\mathbb{R}$ *as* $\mathbb{Q}$, *too. Moreover, via this embedding,* $\mathbb{R}$ *contains* $\mathbb{Z}$ *and* $\mathbb{N}$, *too.*

Proof. Exercise 5-31.   ∎

Although the Completeness Axiom formally only guarantees that nonempty subsets of $\mathbb{R}$ that are bounded above have suprema, existence of infima is a consequence.

Proposition 5.41 *Let* $S \subseteq \mathbb{R}$ *be nonempty and bounded below. Then* S *has a greatest lower bound.*

Proof. Let $L := \{x \in \mathbb{R} : x$ is a lower bound of $S\}$. Then $L \neq \emptyset$. Let $s \in S$. Then for all $l \in L$ we have that $l \leq s$. Because $S \neq \emptyset$ this means that L is bounded above. Because $L \neq \emptyset$, by the Completeness Axiom, L has a supremum $\sup(L)$. Every $s \in S$ is an upper bound of L, which means that $s \geq \sup(L)$. Hence $\sup(L)$ is a lower bound of S. By definition of suprema, $\sup(L)$ is greater than or equal to all elements of L, that is, it is greater than or equal to all lower bounds of S. By definition of infima, this means that $\sup(L) = \inf(S)$.   ∎

The next two theorems show how the familiar number systems $\mathbb{N}$ and $\mathbb{Q}$ are placed in the real numbers: Every real number is exceeded by a natural number (Theorem

---

[5]This property is also called Dedekind completeness

5.42), and the rational numbers are quite "dense"[6] in the real numbers (Theorem 5.43). For the proof of Theorem 5.42 note that we don't have qualms about using a fraction like $\frac{1}{2}$, because we know that $\frac{1}{2}$ is an element of $\mathbb{Q}$ and hence it is an element of $\mathbb{R}$, too. Similarly, there is no problem with using fractions in the proof of Theorem 5.43.

Theorem 5.42 *For every $x \in \mathbb{R}$, there is an $n \in \mathbb{N}$ so that $n \geq x$.*[7]

Proof. For a contradiction, suppose that $x \in \mathbb{R}$ is such that for all $n \in \mathbb{N}$ we have that $n < x$. Then $B := \left\{ y \in \mathbb{R} : \left(\forall n \in \mathbb{N} : n < y \right) \right\}$ is not empty. Moreover, B is bounded below by all $n \in \mathbb{N}$. By Proposition 5.41, B has an infimum, call it b. Then $b - \frac{1}{2} \notin B$, which means there is an $n \in \mathbb{N}$ with $n \geq b - \frac{1}{2}$. But then $n + 1 \geq b + \frac{1}{2}$ is a lower bound of B, a contradiction to $b = \inf(B)$. ∎

Theorem 5.43 *Let $a, b \in \mathbb{R}$ with $a < b$. Then there is a rational number $q \in \mathbb{Q}$ such that $a < q < b$.*

Proof. By Theorem 5.42, there is an $n \in \mathbb{N}$ so that $0 < \dfrac{1}{b - a} < n$. By Theorem 5.22, we obtain $\dfrac{1}{n} < b - a$. Now let $u := \min \left\{ m \in \mathbb{Z} : \dfrac{m}{n} \geq b \right\}$ and similarly let $l := \max \left\{ m \in \mathbb{Z} : \dfrac{m}{n} \leq a \right\}$ (see Exercise 4-32 for the existence of the maximum and the minimum). Then $\dfrac{u}{n} - \dfrac{l}{n} \geq b - a > \dfrac{1}{n}$, which means $\dfrac{l + 1}{n} < \dfrac{u}{n}$. Hence, by definition of l and u we infer $a < \dfrac{l + 1}{n} < b$. ∎

We are now ready to establish the existence of nth roots in $\mathbb{R}$, which was our main motivation to extend the rational numbers.

Theorem 5.44 *Let $n \in \mathbb{N}$. For every nonnegative real number a, there exists a unique nonnegative real number r such that $r^n = a$.*

Proof. We first prove the existence of r. Let $R := \left\{ x \in \mathbb{R} : x \geq 0 \text{ and } x^n \leq a \right\}$. Then $0 \in R$ and R is bounded above by $\max\{1, a\}$. Let $r := \sup(R)$. To show that $r^n = a$, we will show that $r^n \not< a$ and $r^n \not> a$. First, suppose for a contradiction that $r^n < a$. Then there is an $\varepsilon > 0$ so that $r^n + \varepsilon < a$. By Theorem 5.42 (or Exercise 5-32), for each $k \in \{1, \ldots, n\}$ we can find an $m_k \in \mathbb{N}$ so that $\binom{n}{k} r^{n-k} \frac{1}{m_k^k} < \frac{\varepsilon}{n}$. Let $m := \max\{m_1, \ldots, m_n\}$. Then by the Binomial Theorem we conclude

$$
\left(r + \frac{1}{m} \right)^n = \sum_{k=0}^{n} \binom{n}{k} r^{n-k} \left(\frac{1}{m} \right)^k \qquad \boxed{\text{Split off the zeroth term.}}
$$

$$
= r^n + \sum_{k=1}^{n} \binom{n}{k} r^{n-k} \frac{1}{m^k} \qquad \boxed{\text{Now use the definition of } m.}
$$

$$
< r^n + \sum_{k=1}^{n} \frac{\varepsilon}{n} = r^n + \varepsilon < a.
$$

---

[6]This term can be made precise in analysis.

[7]This property is sometimes called being an **Archimedean ordered field**.

The above shows that $r + \frac{1}{m} \in R$, contradicting the fact that $r = \sup(R)$. Hence, $r^n \not< a$. The proof that $r^n \not> a$ is similar and left to the reader as Exercise 5-33.

For uniqueness, suppose for a contradiction that there is another $b \geq 0$ with $b^n = a$. Then $b < r$ or $b > r$. But if $b > r$, then with $\delta := b - r$ we obtain

$$a = b^n = (r + \delta)^n = r^n + \sum_{k=1}^{n} \binom{n}{k} r^{n-k} \delta^k > a,$$

a contradiction. Hence, $b < r$. But then with $\delta := r - b$ we have

$$a = r^n = (b + \delta)^n = b^n + \sum_{k=1}^{n} \binom{n}{k} b^{n-k} \delta^k > a,$$

a contradiction. Therefore r is unique. ■

Note that the proof of Theorem 5.44 uses just about everything we have proved so far.[8] We can now introduce some more familiar language.

Definition 5.45 *Let $a > 0$ be a positive real number and let $n \in \mathbb{N} \setminus \{1\}$. Then the unique positive real number r so that $r^n = a$ is called the nth **root** of a, denoted $r = \sqrt[n]{a}$ or $r = a^{\frac{1}{n}}$. The second root of a is also called the **square root** of a, denoted $\sqrt{a}$.*

With the existence of nth roots proved, we know now that equations of the form $x^n - a = 0$ are solvable for $a > 0$. Unfortunately, the equation $x^2 + 1 = 0$ is still not solvable in $\mathbb{R}$. We will remedy this problem in Section 5.7. At least, roots allow us to extend powers to rational exponents. Note that if $a \in \mathbb{R}$, $a \geq 0$, $k, d \in \mathbb{N}$ and $n \in \mathbb{Z}$, then $\sqrt[kd]{a^{kn}}$ is the unique number r whose kd^{th} power is $r^{kd} = a^{kn}$. But then $\left(r^d\right)^k = \left(a^n\right)^k$, which implies (because k^{th} roots are unique) that $r^d = a^n$ and hence $r = \sqrt[d]{a^n}$. Therefore, the definition of fractional powers in part 2 of Definition 5.46 below is indeed well-defined.

Definition 5.46 *For all real numbers $a \geq 0$, all $n \in \mathbb{Z}$ and $d \in \mathbb{N}$ we define*

*1. $a^{\frac{1}{d}} := \sqrt[d]{a}$. That is, the $\left(\frac{1}{d}\right)^{\text{th}}$ **power** of a is the dth root of a.*

2. $a^{\frac{n}{d}} := \left(a^n\right)^{\frac{1}{d}}$.

Fractional powers satisfy the usual power laws.

---

[8] It would be an interesting scavenger hunt to find the results that were not used, directly or indirectly, in the proof.

Theorem 5.47 *Let $a, b \in \mathbb{R}$, $a, b \geq 0$ and let $p, q \in \mathbb{Q}$. Then the following hold.*

1. $(a \cdot b)^p = a^p \cdot b^p$

2. $\left(a^p\right)^q = a^{pq}$

3. $a^{p+q} = a^p \cdot a^q$

Proof. Let $m, n \in \mathbb{Z}$ and $d, e \in \mathbb{N}$ be so that $p = \frac{n}{d}$ and $q = \frac{m}{e}$. Then part 1 is proved as follows.

$$(a \cdot b)^{\frac{n}{d}} \quad = \quad \sqrt[d]{(a \cdot b)^n} = \sqrt[d]{a^n \cdot b^n}$$

By $\left(\sqrt[d]{a^n} \cdot \sqrt[d]{b^n}\right)^d = \left(\sqrt[d]{a^n}\right)^d \left(\sqrt[d]{b^n}\right)^d = a^n b^n$, the next equality follows from the uniqueness of dth roots.

$$= \quad \sqrt[d]{a^n} \cdot \sqrt[d]{b^n} = a^{\frac{n}{d}} b^{\frac{n}{d}}.$$

Part 2 is left to Exercise 5-34.

Part 3 now follows from parts 1 and 2. It is a good exercise to justify each step below.

$$\left(a^{\frac{n}{d}} a^{\frac{m}{e}}\right)^{de} \quad = \quad \left(a^{\frac{n}{d}}\right)^{de} \left(a^{\frac{m}{e}}\right)^{de} = a^{\frac{n}{d} de} a^{\frac{m}{e} de} = a^{ne} a^{md}$$

$$= \quad a^{ne+md} = \left(a^{\frac{ne+md}{de}}\right)^{de} = \left(a^{\frac{n}{d}+\frac{m}{e}}\right)^{de}$$

∎

Now that we have constructed the real numbers and established their properties, we might wonder if there is a way to continue these constructions and to obtain fields with ever more properties. Theorem 5.48 below says that this is not possible: Up to isomorphism, the real numbers are the *only* totally ordered field that satisfies the Completeness Axiom. Any further extensions of our number systems will not have all the properties of the real numbers. Philosophically, Theorem 5.48 shows that we only have one choice if we want all these properties to be true. That is, the real numbers are in some ways a "constant of nature." No matter from what axioms we build mathematics, there will always be only one set of real numbers (up to isomorphism).

Theorem 5.48 *Let $(\mathbb{F}, +, \cdot)$ be a totally ordered field that satisfies the Completeness Axiom. Then $\mathbb{F}$ is isomorphic to $\mathbb{R}$ (via an isomorphism that preserves order, addition and multiplication).*

Proof. By Exercise 5-22, the characteristic of $\mathbb{F}$ is 0. Hence by Theorem 5.13, $\mathbb{F}$ contains an isomorphic copy of $\mathbb{Q}$. Denote this copy of $\mathbb{Q}$ in $\mathbb{F}$ by $\mathbb{Q}_\mathbb{F}$. Because $\mathbb{F}$ is totally ordered, every $f \in \mathbb{F}$ is comparable to some element $q \in \mathbb{Q}_\mathbb{F}$.

Next we prove that for every $f \in \mathbb{F}$ there is a $q \in \mathbb{Q}_\mathbb{F}$ so that $q < f$. For a contradiction, suppose this is not the case. Then there is an $f \in \mathbb{F}$ so that $q \geq f$ for all $q \in \mathbb{Q}_\mathbb{F}$. Because $\mathbb{F}$ satisfies the Completeness Axiom and because $\mathbb{Q}_\mathbb{F}$ is bounded below by f, the set $\mathbb{Q}_{\mathbb{F}}$ has an infimum, call it a. Then $f \leq a \leq q$ for all $q \in \mathbb{Q}_{\mathbb{F}}$, which implies that for all $q \in \mathbb{Q}_{\mathbb{F}}$ we have that $q - 1 \geq a$ (because $q - 1 \in \mathbb{Q}_{\mathbb{F}}$) and hence $q \geq a + 1$. This means that $a + 1$ is a lower bound of $\mathbb{Q}_{\mathbb{F}}$ and $a + 1 > a$, contradicting that a is the infimum of $\mathbb{Q}_{\mathbb{F}}$. We conclude that for every $f \in \mathbb{F}$ we have that $\{q \in \mathbb{Q}_{\mathbb{F}} : q < f\} \neq \emptyset$.

Now let $\psi : \mathbb{Q}_{\mathbb{F}} \to \mathbb{Q} \subseteq \mathbb{R}$ be an isomorphism. The reader will show in Exercise 5-35 that the map $\varphi(f) := \{\psi(q) \in \mathbb{Q} \subseteq \mathbb{R} : q \in \mathbb{Q}_{\mathbb{F}}, q < f\}$ is an isomorphism from $\mathbb{F}$ to $\mathbb{R}$. ∎

Exercises

5-27. Finishing the proof of Theorem 5.33.

  (a) Prove that addition is commutative.

  (b) Prove that for every $D \in \mathbb{R}$, the element

$$E := \{q \in \mathbb{Q} : -q \text{ is an upper bound of } D, \text{ but } -q \neq \sup(D)\}$$

   is an additive inverse.

5-28. Prove Lemma 5.34.

5-29. Finish the proof of Theorem 5.36.

  (a) Finish the proof that multiplication is associative.

  (b) Prove that multiplication is commutative.

  (c) Prove that D^{-1} as defined in the proof satisfies $D \cdot D^{-1} = 1$.

  (d) Prove that multiplication is distributive over addition.

5-30. Prove that the real numbers are a totally ordered field with positive cone $P \cup \{0\}$, where, as in Proposition 5.37, $P = \{D \in \mathbb{R} : 0_{\mathbb{Q}} \in D\}$.

5-31. Prove that the function $f : \mathbb{Q} \to \mathbb{R}$ defined by $f(q) := \{x \in \mathbb{Q} : x < q\}$ is an order preserving field isomorphism from $\mathbb{Q}$ to a subset of $\mathbb{R}$.

  (a) Prove that for all $q \in \mathbb{Q}$ we have that $f(q) \in \mathbb{R}$.

  (b) Prove that f is injective.

  (c) Prove that for all $p, q \in \mathbb{Q}$ we have $f(p + q) = f(p) + f(q)$.

  (d) Prove that for all $p, q \in \mathbb{Q}$ we have $f(p \cdot q) = f(p) \cdot f(q)$.

  (e) Prove that for all $p, q \in \mathbb{Q}$ we have that $p \leq q$ implies $f(p) \leq f(q)$.

5-32. Prove that for any real numbers $x, \varepsilon > 0$ there is an $n \in \mathbb{N}$ so that $\dfrac{x}{n} < \varepsilon$.

  Hint. Theorem 5.42.

5-33. Finish the proof of Theorem 5.44 by showing that $r^n \not> a$.

  Hint. Suppose $r^n > a$ and prove that then for some $\delta > 0$ and *all* v with $0 < v < \delta$ we have $r - v \notin R$.

5-34. Prove part 2 of Theorem 5.47.

5-35. Finish the proof of Theorem 5.48 as follows. Let the map $\varphi : \mathbb{F} \to \mathbb{R}$ be as defined at the end of the proof of Theorem 5.48.

  (a) Prove that between any two elements of $\mathbb{F}$ there is an element of $\mathbb{Q}_{\mathbb{F}}$.

  (b) Prove that if $x < y$ then $\varphi(x) < \varphi(y)$.

  (c) Prove that if $A \subseteq \mathbb{F}$ is bounded above, then $\varphi\left(\sup(A)\right) = \sup\left(\varphi[A]\right)$.

(d) Prove that φ is injective.

(e) Prove that φ is surjective.

   Hint. Use that $\mathbb{F}$ satisfies the Completeness Axiom.

(f) Prove that for all $x, y \in \mathbb{F}$ we have $\varphi(x + y) = \varphi(x) + \varphi(y)$.

   Hint. Use that for all $z \in \mathbb{F}$ we have $z = \sup\{q \in \mathbb{Q}_{\mathbb{F}} : q < z\}$.

(g) Prove that for all $x \in \mathbb{F}$ we have $\varphi(-x) = -\varphi(x)$.

(h) Prove that for all $x, y \in \mathbb{F}$ we have $\varphi(x \cdot y) = \varphi(x) \cdot \varphi(y)$.

   Hint. Prove the result for $x, y > 0$, then use part 5-35g.

5-36. Solving equations in the real numbers (Exercises 3-8, 4-23 and 5-10 revisited).

(a) Solve the equation $x^2 - 2 = 0$ in $\mathbb{R}$. Explain any difference between your result and the results of Exercises 3-8f, 4-23d and 5-10b.

(b) Solve the equation $x^2 + 1 = 0$ in $\mathbb{R}$. Explain any difference between your result and the results of Exercises 3-8g, 4-23e and 5-10c.

5-37. Revisiting Exercises 4-34 and 5-20.

(a) Solve the inequality $|7x + 2| < 3$ in $\mathbb{R}$. Explain any difference between your result and those of Exercises 4-34c and 5-20a.

(b) Solve the inequality $|6x + 5| > 2$ in $\mathbb{R}$. Explain any difference between your result and those of Exercises 4-34d and 5-20b.

(c) Solve the inequality $3 \le |2x + 4| \le 11$ in $\mathbb{R}$. Explain any difference between your result and those of Exercises 4-34e and 5-20c.

5-38. Sketch figures similar to Figure 5.3

(a) To illustrate the addition of a positive and a negative real number.

(b) To illustrate the addition of two negative real numbers.

5-39. Prove that the function $f : \mathbb{R}^+ \to \mathbb{R}^+$ defined by $f(x) := \sqrt[n]{x}$ is bijective.

5-40. Prove that the square root function is a bijective function from $\mathbb{R}^+$ to $\mathbb{R}^+$, but that it is not a bijective function from $\mathbb{R}$ to $\mathbb{R}$.

5-41. Prove that the cube root function $f(x) = \sqrt[3]{x}$ is a bijective function from $\mathbb{R}$ to $\mathbb{R}$.

5.6   Uncountable Sets

Aside from providing roots of positive numbers, the real numbers also are the first truly familiar example of an infinite set that is not countable.[9] We start by considering intervals.

Definition 5.49 *An* **interval** *is a set* $I \subseteq \mathbb{R}$ *so that for all* $c, d \in I$ *and* $x \in \mathbb{R}$ *the inequalities* $c < x < d$ *imply* $x \in I$. *In particular for* $a, b \in \mathbb{R}$ *with* $a < b$ *we define*

1. $[a, b] := \{x \in \mathbb{R} : a \le x \le b\}$,

2. $(a, b) := \{x \in \mathbb{R} : a < x < b\}$, $(a, \infty) := \{x \in \mathbb{R} : a < x\}$,
   $(-\infty, b) := \{x \in \mathbb{R} : x < b\}$, $(-\infty, \infty) := \mathbb{R}$,

---

[9]Of course, by Theorem 2.60 the power set $\mathcal{P}(\mathbb{N})$ is another such set. But we should be much more familiar with real numbers than with power sets.

3. $[a, b) := \{x \in \mathbb{R} : a \le x < b\}$, $[a, \infty) := \{x \in \mathbb{R} : a \le x\}$,

4. $(a, b] := \{x \in \mathbb{R} : a < x \le b\}$, $(-\infty, b] := \{x \in \mathbb{R} : x \le b\}$.

The points a and b are also called the **endpoints** *of the interval. An interval that does not contain either of its endpoints (where $\pm\infty$ are also considered to be "endpoints that are not in $\mathbb{R}$") is called* **open**. *An interval that contains exactly one of its endpoints is called* **half-open** *and an interval that contains both its endpoints is called* **closed**.

To prove that a set is not countable, by Theorem 4.62 it suffices to prove that a subset is not countable. Therefore, we will first prove that $(0, 1) \subseteq \mathbb{R}$ is not countable.

Theorem 5.50 *The interval* $(0, 1)$ *is not equivalent to* $\mathbb{N}$.

Proof. Let S be the set of all sequences $\{x_n\}_{n=1}^{\infty}$ of numbers[10] in the set $\{0, 1, \ldots, 9\}$ so that there is an $m \in \mathbb{N}$ with $x_m \ne 0$ and so that for every $n \in \mathbb{N}$ there is an $m \ge n$ so that $x_m \ne 9$. We will first prove that $(0, 1)$ contains a subset that is equivalent to S. To do so, define $\varphi : S \to \mathbb{R}$ by

$$\varphi\left(\{x_n\}_{n=1}^{\infty}\right) := \sup\left\{\sum_{j=1}^{k} x_j 10^{-j} : k \in \mathbb{N}\right\}.$$

For each element $\{x_n\}_{n=1}^{\infty} \in S$, the image $\varphi\left(\{x_n\}_{n=1}^{\infty}\right)$ is well-defined, because all sums in the set are smaller than 1 and because every nonempty bounded subset of $\mathbb{R}$ has a supremum. The function φ maps into $(0, 1)$, because each $\{x_n\}_{n=1}^{\infty} \in S$ has at least one $m \in \mathbb{N}$ with $x_m \ne 0$, which means that $\varphi\left(\{x_n\}_{n=1}^{\infty}\right) \ge 10^{-m} > 0$ and because each $\{x_n\}_{n=1}^{\infty} \in S$ has at least one $m \in \mathbb{N}$ with $x_m \ne 9$, which means that $\varphi\left(\{x_n\}_{n=1}^{\infty}\right) \le 1 - 10^{-m} < 1$ (also see Exercise 5-42).

To prove that φ is injective, let $\{x_n\}_{n=1}^{\infty}, \{y_n\}_{n=1}^{\infty} \in S$ be so that $\{x_n\}_{n=1}^{\infty} \ne \{y_n\}_{n=1}^{\infty}$. Let $m \in \mathbb{N}$ be the smallest natural number so that $x_m \ne y_m$ and without loss of generality assume that $x_m < y_m$. Then there is an $l > m$ so that $x_l \ne 9$. Hence

$$\begin{aligned}
\varphi\left(\{y_n\}_{n=1}^{\infty}\right) &= \sup\left\{\sum_{j=1}^{k} y_j 10^{-j} : k \in \mathbb{N}\right\} \ge \sum_{j=1}^{m} y_j 10^{-j} \\
&= \sum_{j=1}^{m-1} y_j 10^{-j} + (y_m - 1)10^{-m} + 10^{-m} \\
&\ge \sum_{j=1}^{m-1} x_j 10^{-j} + x_m 10^{-m} + 10^{-m}
\end{aligned}$$

Use Exercise 5-42.

---

[10]Formally, a sequence is a function $x : \mathbb{N} \to \mathbb{R}$. But it is customary to denote $x_n := x(n)$ and the sequence as $\{x_n\}_{n=1}^{\infty}$.

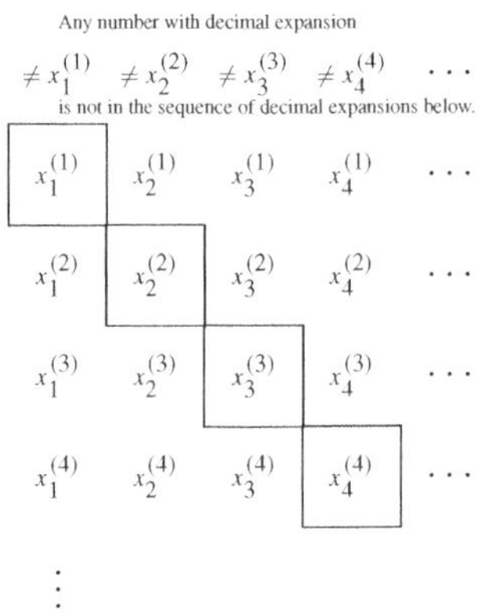

Any number with decimal expansion

$$\neq x_1^{(1)} \quad \neq x_2^{(2)} \quad \neq x_3^{(3)} \quad \neq x_4^{(4)} \quad \cdots$$

is not in the sequence of decimal expansions below.

$$
\begin{array}{cccc}
x_1^{(1)} & x_2^{(1)} & x_3^{(1)} & x_4^{(1)} & \cdots \\
x_1^{(2)} & x_2^{(2)} & x_3^{(2)} & x_4^{(2)} & \cdots \\
x_1^{(3)} & x_2^{(3)} & x_3^{(3)} & x_4^{(3)} & \cdots \\
x_1^{(4)} & x_2^{(4)} & x_3^{(4)} & x_4^{(4)} & \cdots
\end{array}
$$

$$\vdots$$

Figure 5.4 The idea behind the proof that $(0, 1)$ is not countable.

$$
> \quad \sum_{j=1}^{m-1} x_j 10^{-j} + x_m 10^{-m} + \sum_{j=m+1}^{l} 9 \cdot 10^{-j}
$$

$$
\geq \quad \sum_{j=1}^{l-1} x_j 10^{-j} + (x_l + 1)10^{-l}
$$

Use Exercise 5-42 again.

$$
\geq \quad \sup \left\{ \sum_{j=1}^{l} x_j 10^{-j} + \sum_{j=l+1}^{k} x_j 10^{-j} : k \in \mathbb{N}, k > l \right\}
$$

$$
\geq \quad \sup \left\{ \sum_{j=1}^{k} x_j 10^{-j} : k \in \mathbb{N} \right\} = \varphi \left(\{x_n\}_{n=1}^{\infty} \right).
$$

Therefore φ is injective, and the set $\varphi[S]$ is equivalent to S and contained in $(0, 1)$.

Finally, suppose for a contradiction that $(0, 1)$ was countable. Then S, being equivalent to a subset of $(0, 1)$ would be countable, too. Let $\psi : \mathbb{N} \to S$ be a bijective function and for every $k \in \mathbb{N}$, let $\{x_n^{(k)}\}_{n=1}^{\infty} := \psi(k)$. For each $n \in \mathbb{N}$, let y_n be a number in the set $\{1, 2, 3, 4, 5, 6, 7, 8\} \setminus \{x_n^{(n)}\}$. Then $\{y_n\}_{n=1}^{\infty} \in S$, but for all $n \in \mathbb{N}$ we have that $y_n \neq x_n^{(n)}$. Hence for all $k \in \mathbb{N}$ we have $\{y_n\}_{n=1}^{\infty} \neq \psi(k)$ (see Figure 5.4 for a visualization), contradiction. ∎

Note that the function φ in the proof of Theorem 5.50 is actually surjective (see Exercise 5-43) and that the sequence $\{x_n\}_{n=1}^{\infty}$ actually is the decimal expansion of the number $x = \varphi \left(\{x_n\}_{n=1}^{\infty} \right)$. To actually treat this sequence as the decimal expansion, we would need to learn about the convergence of infinite series, which would take more time than just this one section. The interested reader can find this approach in Proposition 6.6 in [30].

Rather than calling infinite sets that are not equivalent to $\mathbb{N}$ "not countable" we call them "uncountable." Theorem 5.50 easily implies that $\mathbb{R}$ and $\mathbb{R} \setminus \mathbb{Q}$ are uncountable.

Definition 5.51 *An infinite set that is not countable is called* **uncountable**.

Corollary 5.52 $\mathbb{R}$ *is uncountable.*

Proof. Suppose for a contradiction that $\mathbb{R}$ was countable. Then by Theorem 4.62, the interval $(0, 1)$ would be countable, too, contradicting Theorem 5.50. Hence $\mathbb{R}$ is uncountable. ∎

Definition 5.53 *The set* $\mathbb{R} \setminus \mathbb{Q}$ *is called the set of* **irrational numbers**.

Corollary 5.54 *The irrational numbers* $\mathbb{R} \setminus \mathbb{Q}$ *are uncountable.*

Proof. Suppose for a contradiction that the set $\mathbb{R} \setminus \mathbb{Q}$ was countable. Then the set $\mathbb{R} = \mathbb{Q} \cup (\mathbb{R} \setminus \mathbb{Q})$ would be the union of two countable sets. Hence by Theorem 4.65 $\mathbb{R}$ would be countable, contradicting Corollary 5.52. ∎

Corollary 5.54 is quite interesting. The numbers we work with in daily life are all rational. So are the numbers that can be represented in a computer. In fact, from the point-of-view of this text, we presently know only one irrational number: $\sqrt{2}$. In the remainder of this text, we will encounter countably many more irrational numbers that we might be able to visualize (see Exercises 6-2a and 6-9b). That means most numbers, namely the uncountable "rest," remain beyond visualization, even though they do exist.

Exercises

5-42. Let $n, m \in \mathbb{N}$. Prove that $\displaystyle\sum_{j=n+1}^{m} 9 \cdot 10^{-j} < 10^{-n}$.

5-43. Prove that the function $\varphi : S \to (0, 1)$ defined by $\varphi\left(\{x_n\}_{n=1}^{\infty}\right) := \sup\left\{\displaystyle\sum_{j=1}^{k} x_j 10^{-j} : k \in \mathbb{N}\right\}$ (see the proof of Theorem 5.50) is surjective.

Hint. For $x \in (0, 1)$ recursively construct $x_n := \max\left\{a \in \{0, \ldots, 9\} : \displaystyle\sum_{j=1}^{n-1} x_j 10^{-j} + a 10^{-n} < x\right\}$.

5-44. **Nested Interval Principle.** For each $n \in \mathbb{N}$, let $[a_n, b_n]$ be a nonempty closed interval in $\mathbb{R}$ so that $[a_{n+1}, b_{n+1}] \subseteq [a_n, b_n]$. Prove that $\displaystyle\bigcap_{n=1}^{\infty} [a_n, b_n] \neq \emptyset$.

Hint. Consider the supremum of the a_n.

5-45. Intersections of intervals.

    (a) Prove that $\displaystyle\bigcap_{n=1}^{\infty} \left(0, \frac{1}{n}\right) = \emptyset$.

    (b) Prove that $\displaystyle\bigcap_{n=1}^{\infty} \left(\left[\sqrt{2} - \frac{1}{n}, \sqrt{2} + \frac{1}{n}\right] \cap \mathbb{Q}\right) = \emptyset$.

5-46. Let $a \in \mathbb{R}$. Prove that if $a \geq 0$ and $a < p$ for all $p > 0$, then $a = 0$.

5-47. For each $n \in \mathbb{N}$ let $[a_n, b_n]$ be an interval with $[a_{n+1}, b_{n+1}] \subseteq [a_n, b_n]$ and $b_n - a_n < \dfrac{1}{n}$. Prove

that $\bigcap\limits_{n=1}^{\infty} [a_n, b_n]$ is a singleton.

Hint. Use Exercises 5-44 and 5-46.

5-48. Let $a, b, x, y \in \mathbb{R}$. Prove that if $x, y \in [a, b]$, then $|x - y| \le b - a$.

5-49. Prove that $\mathcal{P}(\mathbb{N})$ is equivalent to the set of all sequences $\{x_n\}_{n=1}^{\infty}$, where $x_n \in \{0, 1\}$.

5-50. Prove that if $I, J \subseteq \mathbb{R}$ are intervals, then $I \cap J$ is an interval, too.

5-51. Let X be a set and let $\le$ be an order relation on X. An up-set is a subset $U \subseteq X$ so that if $u \in U$ and $u \le x$, then $x \in U$. Prove that every interval on the real line (see Definition 5.49) is the intersection of an up-set and a down-set.

5.7   The Complex Numbers

As noted after Definition 5.45, real numbers allow us to solve equations of the form $x^2 - a = 0$ for $a > 0$. But to solve an equation such as $x^2 + 1 = 0$, we would need "square roots of negative numbers." Such entities exist in the complex numbers, which we will introduce in this section. But we must be a bit cautious. Because the solutions of $x^2 + 1 = 0$ cannot be real numbers (see Proposition 5.23), and because we do not want to abandon any real numbers, the complex numbers must properly contain the real numbers. Moreover, we certainly want the complex numbers to be a field. (Otherwise algebra would be rather tedious.) But Theorem 5.48 has shown that $\mathbb{R}$ is the only field with the properties of the real numbers. So, even though the complex numbers will have some properties that $\mathbb{R}$ does not have, the complex numbers cannot have *all* properties of $\mathbb{R}$. Exercise 5-52 shows the price we pay for "square roots of negative numbers:" The complex numbers cannot be ordered to become a totally ordered field. Thankfully the lack of a compatible order relation is usually not a problem.

Compared to the definitions of $\mathbb{N}$, $\mathbb{Z}$, $\mathbb{Q}$ and $\mathbb{R}$, the definition of the complex numbers is quite simple: We introduce a "second dimension," which contains the "square roots of negative numbers."

Definition 5.55 *The **complex numbers** $\mathbb{C}$ are the set $\mathbb{R} \times \mathbb{R}$ equipped with addition and multiplication defined as follows. For all complex numbers $(a, b), (c, d) \in \mathbb{C}$, we set $(a, b) + (c, d) := (a + c, b + d)$ and $(a, b) \cdot (c, d) := (ac - bd, ad + bc)$. We define $0 := (0_{\mathbb{R}}, 0_{\mathbb{R}})$, $1 := (1_{\mathbb{R}}, 0_{\mathbb{R}})$ and $i := (0_{\mathbb{R}}, 1_{\mathbb{R}})$, where the subscript denotes elements in $\mathbb{R}$, and we also write complex numbers in the form $(a, b) = a \cdot 1 + b \cdot i = a + ib$. For $z = a + ib \in \mathbb{C}$, the number a is also called the **real part** of z, denoted $\Re(z)$, and the number b is also called the **imaginary part** of z, denoted $\Im(z)$.*

The algebraic properties of $\mathbb{C}$ are summarized in Theorems 5.56 and 5.57.

Theorem 5.56 *The complex numbers $\mathbb{C}$ with addition, multiplication, 0 and 1 as defined above are a field.*

Proof. The field axioms are verified in Exercise 5-53. For $a + ib \in \mathbb{C}$, the additive inverse is $-(a + bi) = (-a) + (-b)i$ and for $a + ib \in \mathbb{C} \setminus \{0\}$, the multiplicative inverse is $(a + ib)^{-1} = \dfrac{a}{a^2 + b^2} - \dfrac{b}{a^2 + b^2} i$. ∎

Because $(\mathbb{C}, +, \cdot)$ is a field, everything we know about fields applies to $(\mathbb{C}, +, \cdot)$.

Think about all the information that gives us. Maybe make a list as we did after the definitions of groups, rings and fields.

The special element i serves as the closest we can get to a "square root of (-1)."

Theorem 5.57 $i^2 = -1$.

Proof. $i^2 = (0 + 1i) \cdot (0 + 1i) = (0 \cdot 0 - 1 \cdot 1) + (0 \cdot 1 + 1 \cdot 0)i = -1$. ∎

The **imaginary unit** i is usually not called the square root of -1: Because have $(-i)^2 = -1$ and because there is no canonical order, there is no natural way to distinguish one of the two solutions of $x^2 + 1 = 0$ as *the* square root. Thus we avoid arguments about language by not using the terminology that may be in doubt. Instead, we simply call i the imaginary unit, one of whose properties is that $i^2 = -1$.

Aside from the above algebraic properties, we need to know how to measure distances in the complex numbers. This is done via the absolute value function.

Definition 5.58 *For $z = a + ib \in \mathbb{C}$, the **absolute value** of z is $|z| := \sqrt{a^2 + b^2}$.*

Thankfully, the absolute value behaves like the absolute value in a totally ordered ring (see Theorem 4.38). But because the complex numbers are not a totally ordered ring, we cannot quote Theorem 4.38 to establish the properties.

Theorem 5.59 *Properties of the absolute value.*

0. *For all $z \in \mathbb{C}$, we have $|z| \geq 0$. (Note that because $|z| \in \mathbb{R}$, it is permissible to use inequalities here.)*

1. *For all $z \in \mathbb{C}$, we have $|z| = 0$ iff $z = 0$.*

2. *For all $z_1, z_2 \in \mathbb{C}$, we have $|z_1 z_2| = |z_1||z_2|$.*

3. *The **triangular inequality** holds.*
   That is, for all $z_1, z_2 \in \mathbb{C}$ we have $|z_1 + z_2| \leq |z_1| + |z_2|$.

Proof. Parts 0 to 2 are left to Exercise 5-54. For part 3, let $z_1 = a + ib$ and $z_2 = c + id$ with $a, b, c, d \in \mathbb{R}$. Then $0 \leq (ad - bc)^2 = a^2 d^2 - 2abcd + b^2 c^2$, which means that

$$
\begin{aligned}
2abcd &\leq a^2 d^2 + b^2 c^2 \\
a^2 c^2 + 2abcd + b^2 d^2 &\leq a^2 c^2 + a^2 d^2 + b^2 c^2 + b^2 d^2 \\
(ac + bd)^2 &\leq \left(a^2 + b^2\right)\left(c^2 + d^2\right) \\
2ac + 2bd &\leq 2\sqrt{a^2 + b^2}\sqrt{c^2 + d^2} \\
a^2 + 2ac + c^2 + b^2 + 2bd + d^2 &\leq a^2 + b^2 + 2\sqrt{a^2 + b^2}\sqrt{c^2 + d^2} + c^2 + d^2 \\
(a + c)^2 + (b + d)^2 &\leq \left(\sqrt{a^2 + b^2} + \sqrt{c^2 + d^2}\right)^2 \\
|z_1 + z_2|^2 &\leq \left(|z_1| + |z_2|\right)^2 \\
|z_1 + z_2| &\leq |z_1| + |z_2|,
\end{aligned}
$$

which proves part 3.   ∎

If we switch the sign of the imaginary part of a complex number, we obtain the complex conjugate.

Definition 5.60 *For $z = a + ib \in \mathbb{C}$, the* **complex conjugate** *of z is $\overline{z} := a - ib$.*

Absolute value and complex conjugate are related via a simple equation. This equation can be used to express multiplicative inverses.

Proposition 5.61 *For all $z \in \mathbb{C}$, the equalities $z + \overline{z} = 2\Re(z)$ and $|z|^2 = z\overline{z}$ hold. Moreover, for all $z \in \mathbb{C} \setminus \{0\}$ the multiplicative inverse is*

$$\frac{1}{z} = \frac{\overline{z}}{|z|^2}.$$

Proof. Exercise 5-55.   ∎

Exercises

5-52. Prove that $\mathbb{C}$ cannot be made into a totally ordered field.

   Hint. Suppose there was a subset P of $\mathbb{C}$ so that $\mathbb{C}$ is a totally ordered field with positive cone P. Prove that P must contain both 1 and -1, which is not possible. Use that P must contain i or $-i$.

5-53. Prove that $\mathbb{C}$ is a field. That is, prove each of the following.

   (a) Prove that for all $x, y, z \in \mathbb{C}$ we have $(x + y) + z = x + (y + z)$.

   (b) Prove that for all $x, y \in \mathbb{C}$ we have $x + y = y + x$.

   (c) Prove that, with $0 = 0 + 0i$, for all $x \in \mathbb{C}$ we have $x + 0 = x$.

   (d) Prove that for every element $x = a + ib \in \mathbb{C}$ the element $(-x) = (-a) + i(-b)$ is so that $x + (-x) = 0$.

   (e) Prove that for all $x, y, z \in \mathbb{C}$ we have $(x \cdot y) \cdot z = x \cdot (y \cdot z)$.

   (f) Prove that for all $x, y \in \mathbb{C}$ we have $x \cdot y = y \cdot x$.

   (g) Prove that the element $1 := 1 + i0$ is so that for all $x \in \mathbb{R}$ we have $1 \cdot x = x$.

   (h) Prove that for every element $x = a + ib \in \mathbb{C}$ the element $x^{-1} := \dfrac{a}{a^2 + b^2} - i \dfrac{b}{a^2 + b^2}$ is so that $x \cdot x^{-1} = 1$.

   (i) Prove that for all $\alpha, x, y \in \mathbb{C}$ we have $\alpha \cdot (x + y) = \alpha \cdot x + \alpha \cdot y$.

5-54. Prove the remaining parts of Theorem 5.59. That is, prove each of the following.

   (a) Prove that for all $z \in \mathbb{C}$ we have $|z| \geq 0$.

   (b) Prove that for all $z \in \mathbb{C}$ we have $|z| = 0$ iff $z = 0$.

   (c) Prove that for all $z_1, z_2 \in \mathbb{C}$ we have $|z_1 z_2| = |z_1||z_2|$.

5-55. Prove Proposition 5.61.

5-56. **Reverse triangular inequality.** Prove that for all $z_1, z_2 \in \mathbb{C}$, we have $\big| |z_1| - |z_2| \big| \leq |z_1 - z_2|$.

5-57. (This exercise requires experience with matrix operations.) Prove that $\mathbb{C}$ is ring isomorphic to the set of matrices of the form $\begin{pmatrix} a & b \\ -b & a \end{pmatrix}$ with matrix addition and matrix multiplication.

   Hint. The isomorphism φ assigns each complex number $z = x + iy$ the matrix $\begin{pmatrix} x & y \\ -y & x \end{pmatrix}$, that is,

$$\varphi(z) = \begin{pmatrix} \Re(z) & \Im(z) \\ -\Im(z) & \Re(z) \end{pmatrix}.$$

5.8 Solving Polynomial Equations

One reason to introduce complex numbers is the desire to solve quadratic equations of the form $x^2 + c = 0$, where c is a positive real number. Once such simple equations can be solved, it is natural to inquire about more general quadratic equations, as well as about polynomial equations of order higher than 2. Indeed, the solution of polynomial equations has occupied mathematicians throughout the centuries. Ultimately, there is good news and there is bad news. The good news is that, for polynomial equations up to order 4, the complex solutions can be explicitly determined. That is, in addition to what is known as the quadratic formula, there are procedures that would deserve the names "cubic formula" and "quartic formula." We will investigate these procedures in this section. The bad news is that, for polynomial equations of order 5 or greater, it can be proved that no such procedure can be expressed with finitely many root symbols. We will investigate this significant (and a bit scary) insight in Chapter 6.

The simplest polynomial equations are equations of the form $x^n + c = 0$. Solutions of these equations are called roots of c. But language conventions for the use of the word "root" differ, depending on which field we are working in. In the real numbers, roots are typically assumed to be positive (see Definition 5.45). In a general field, we typically do not have an order relation. Thus, unlike in Definition 5.45 for $\mathbb{R}$, in a general field, *any* solution of the equation $x^n = a$ is called a root of a.

Definition 5.62 *Let $(\mathbb{F}, +, \cdot)$ be a field, let $a \in \mathbb{F}$ and let $n \in \mathbb{N} \setminus \{1\}$. A number $r \in \mathbb{F}$ so that $r^n = a$ is called an nth* **root** *of a, often denoted $r = \sqrt[n]{a}$ or $r = a^{\frac{1}{n}}$. Second roots will be called* **square roots** *and they will be denoted $\sqrt{a}$.*

With the above notation in place, we can tackle quadratic equations.

Theorem 5.63 *The* **quadratic formula**. *Let $(\mathbb{F}, +, \cdot)$ be a field and let $a, b, c \in \mathbb{F}$ with $a \neq 0$. Then the number of solutions in $\mathbb{F}$ of the equation $ax^2 + bx + c = 0$ equals the number of square roots of $b^2 - 4ac$ in $\mathbb{F}$. Moreover, if $b^2 - 4ac$ has at least one square root in $\mathbb{F}$, then the solutions of $ax^2 + bx + c = 0$ are*

$$x_{1,2} = \frac{-b \pm \sqrt{b^2 - 4ac}}{2a}.$$

Proof. To prove the claim about the number of solutions, let $x \in \mathbb{F}$ be a solution of $ax^2 + bx + c = 0$. Then the following must hold.

$$
\begin{aligned}
0 &= ax^2 + bx + c \\
0 &= a\left(x^2 + \frac{b}{a}x\right) + c \\
0 &= a\left(x^2 + \frac{b}{a}x + \left(\frac{b}{2a}\right)^2 - \left(\frac{b}{2a}\right)^2\right) + c \\
0 &= a\left(x + \frac{b}{2a}\right)^2 - a\left(\frac{b}{2a}\right)^2 + c
\end{aligned}
$$

$$a\left(x + \frac{b}{2a}\right)^2 = a\left(\frac{b}{2a}\right)^2 - c$$

$$\left(x + \frac{b}{2a}\right)^2 = \left(\frac{b}{2a}\right)^2 - \frac{c}{a}$$

$$\left(x + \frac{b}{2a}\right)^2 = \frac{b^2 - 4ac}{(2a)^2}$$

> At this stage we see, because every solution of the equation $ax^2 + bx + c = 0$ in $\mathbb{F}$ must be a solution of the equation above and vice versa, that there are exactly as many solutions in $\mathbb{F}$ as $b^2 - 4ac$ has square roots. Moreover, when $b^2 - 4ac$ has a square root, then, because $(-1)^2 = 1$, there are two possibilities: x_1 (associated with the $+$) and x_2 (associated with the $-$). Of course, it could be that $x_1 = x_2$.

$$x_{1,2} + \frac{b}{2a} = \pm\sqrt{\frac{b^2 - 4ac}{(2a)^2}}$$

$$x_{1,2} = -\frac{b}{2a} \pm \frac{\sqrt{b^2 - 4ac}}{2a}$$

$$x_{1,2} = \frac{-b \pm \sqrt{b^2 - 4ac}}{2a}$$

The above establishes the claim about the number of solutions and also that the solutions must be among x_1 and x_2. To prove that, in case $b^2 - 4ac$ has a square root, the numbers x_1 and x_2 are solutions, we can either re-trace the above argument in the opposite direction, or we can substitute x_1 and x_2 into the equation. The substitution is not hard at all, and we demonstrate it for x_1.

$$ax_1^2 + bx_1 + c = a\left(\frac{-b + \sqrt{b^2 - 4ac}}{2a}\right)^2 + b\left(\frac{-b + \sqrt{b^2 - 4ac}}{2a}\right) + c$$

$$= a\frac{b^2 - 2b\sqrt{b^2 - 4ac} + b^2 - 4ac}{4a^2} + \frac{-b^2 + b\sqrt{b^2 - 4ac}}{2a} + c$$

$$= \frac{b^2 - 2b\sqrt{b^2 - 4ac} + b^2 - 4ac}{4a} + \frac{-2b^2 + 2b\sqrt{b^2 - 4ac}}{4a} + c$$

$$= \frac{-4ac}{4a} + c$$

$$= 0.$$

The proof that x_2 is a solution, too, is left to Exercise 5-58.   ∎

With quadratic equations solved, the next question in an infinite sequence of natural questions to pose is "Can we solve cubic equations?" After that we could consider equations of fourth order, fifth order, and so on. For any such sequence of natural questions, there is always the hope that there may be some underlying formula or principle that solves all of them.

As soon as we tackle higher order equations, we realize how strong a result the quadratic formula actually is. Not only does it give us the solutions, but the quadratic formula also tells us how many solutions there are under what circumstances. It turns out that such a characterization would be very tedious for higher order equations in general fields. Theorem 4.58 gives at least an upper bound on the number of solutions for a general polynomial equation in a general field. Because of Theorem 4.58, we can stop looking for solutions when we have found n solutions for an nth order equation. But we would still have problems if we are working in a field in which n^{th} order equations can have fewer than n solutions. This is why, customarily, polynomial equations are tackled in $\mathbb{C}$. The Fundamental Theorem of Algebra guarantees that every n^{th} order polynomial with complex coefficients has exactly n zeros, counted with multiplicity. Thus, as long as we work with complex polynomials we will know exactly when we are done. Unfortunately, any proof of the Fundamental Theorem of Algebra uses methods that will not be covered in this text.[11] In the following, we will not need the full strength of the Fundamental Theorem of Algebra. But we will need the following result.

Theorem 5.64 *Every nonzero complex number has exactly n distinct complex roots.*

Instead of a proof: This result is easy to prove if we can represent complex numbers in the form $z = re^{i\theta}$, where $\theta \in [0, 2\pi)$. With this representation, the n roots are the n distinct complex numbers

$$w_k = \sqrt[n]{r}\, e^{i\left(\frac{\theta}{n} + k\frac{2\pi}{n}\right)},$$

where $k \in \{0, \ldots, n-1\}$. But for the exponential representation of complex numbers, we would need the complex exponential function. A formal introduction of this function, once more, uses methods that will not be covered in this text.[12] □

Although there is no intent to re-create the scruples mathematicians had before the complex numbers were accepted as viable mathematical entities, Theorem 5.64 can help us appreciate what they went through. Theorem 5.64 tells us that every real number has n distinct complex roots.[13] But does it really? We do not have a complete proof here, which should make us at least a little leery. Now imagine how much harder it must have been at a time when there was not just no complete proof, but no clear notion of what these roots actually might *be*.

Focusing back on cubic equations, we realize that there are a lot of coefficients in the equation $ax^3 + bx^2 + cx + d = 0$. One of these coefficients can be chosen to be equal to 1, because we could divide by it without affecting the equation. Because a must be nonzero, it is customary to divide by a. In fact, it is not uncommon to do this even for quadratic equations before proceeding to solve them (see Exercise 5-59). Aside from this rather trivial insight, it is natural to want to get rid of as many other

---

[11] The reader is referred to Exercise 16-74 of [30] for a proof that uses analysis. This proof does not use any of the results that follow in this section, so we are not falling victim to circular reasoning here.

[12] The interested reader could consider Exercise 15-53 in [30] for a proof.

[13] And again, just for emphasis, this can be proved without using anything that is presented in the rest of this text, so there is no circular reasoning.

nonzero coefficients as possible. The general idea is that, the fewer coefficients there are, the easier the equation should be.

So the first step for cubic equations is to get rid of the quadratic term. The idea is quite simple. Replacing x with $x =: y + \alpha$, where α is constant, is a minor modification. But, after the parentheses are multiplied out, this new constant α will be part of all coefficients, except for the coefficient of y^3. Thus we should be able to find a value for α that makes one coefficient equal to zero. The easiest target is the coefficient of the term whose order is one less than that of the equation, because that term will have the simplest coefficient. The same idea works for the third order term in quartic equations (see Theorem 5.67), the fourth order term in quintic equations (see Exercise 5-65a) and the $(n-1)^{st}$ order term in equations of nth order (see Exercise 5-65c). For cubic equations, we show the full transformation from $ax^3 + bx^2 + cx + d = 0$ to a simpler equation. Note that this transformation still works in arbitrary fields.

Theorem 5.65 *The* **cubic formula**, *step I:* **Canonical form***. Let $(\mathbb{F}, +, \cdot)$ be a field and let $a, b, c, d \in \mathbb{F}$ with $a \neq 0$. Then the number $x \in \mathbb{F}$ solves the cubic equation $ax^3 + bx^2 + cx + d = 0$ iff*

$$y = x + \frac{b}{3a}$$

solves the equation

$$y^3 + py + q = 0,$$

where

$$p = -\frac{b^2}{3a^2} + \frac{c}{a} \quad \text{and} \quad q = \frac{2b^3}{27a^3} - \frac{bc}{3a^2} + \frac{d}{a}.$$

Proof. Let $x \in \mathbb{F}$ be a solution of the equation $ax^3 + bx^2 + cx + d = 0$. For any $y, \alpha \in \mathbb{F}$ so that $x = y + \alpha$ we have the following.

$$
\begin{aligned}
ax^3 &+ bx^2 + cx + d \\
&= a(y+\alpha)^3 + b(y+\alpha)^2 + c(y+\alpha) + d \\
&= ay^3 + 3a\alpha y^2 + 3a\alpha^2 y + a\alpha^3 + by^2 + 2b\alpha y + b\alpha^2 + cy + c\alpha + d \\
&= ay^3 + (3a\alpha + b)\, y^2 + \left(3a\alpha^2 + 2b\alpha + c\right) y + \left(a\alpha^3 + b\alpha^2 + c\alpha + d\right).
\end{aligned}
$$

The above says that x solves $ax^3 + bx^2 + cx + d = 0$ iff $y = x - \alpha$ solves $ay^3 + (3a\alpha + b) y^2 + \left(3a\alpha^2 + 2b\alpha + c\right) y + \left(a\alpha^3 + b\alpha^2 + c\alpha + d\right) = 0$. This statement is similar to the result we want to prove. Setting $3a\alpha + b := 0$ leads to $\alpha = -\dfrac{b}{3a}$, and now we know from the above that x solves $ax^3 + bx^2 + cx + d = 0$ iff $y = x + \dfrac{b}{3a}$ solves $ay^3 + \left(3a\alpha^2 + 2b\alpha + c\right) y + \left(a\alpha^3 + b\alpha^2 + c\alpha + d\right) = 0$, where the choice of α eliminated the quadratic term. Division by a and substitution of $\alpha = -\dfrac{b}{3a}$ into the remaining terms shows that x solves $ax^3 + bx^2 + cx + d = 0$ iff $y = x + \dfrac{b}{3a}$ solves

$$y^3 + py + q = 0, \text{ where } p = \frac{3a\alpha^2 + 2b\alpha + c}{a} = \frac{3b^2}{9a^2} - \frac{2b^2}{3a^2} + \frac{c}{a} = -\frac{b^2}{3a^2} + \frac{c}{a} \text{ and}$$

$$q = \frac{a\alpha^3 + b\alpha^2 + c\alpha + d}{a} = -\frac{b^3}{27a^3} + \frac{b^3}{9a^3} - \frac{bc}{3a^2} + \frac{d}{a} = \frac{2b^3}{27a^3} - \frac{bc}{3a^2} + \frac{d}{a}. \quad \blacksquare$$

Theorem 5.65 reduces the task of solving the cubic equation $ax^3 + bx^2 + cx + d = 0$ to the task of solving the cubic equation $x^3 + px + q = 0$. It is reasonable to hope that $x^3 + px + q = 0$ is easier to solve than $ax^3 + bx^2 + cx + d = 0$ because $x^3 + px + q = 0$ has one fewer term and it also has only two coefficients, p and q, as opposed to the previous four. To avoid trivialities, we note that if $q = 0$ the equation $x^3 + px + q = 0$ is reduced to $0 = x^3 + px = x(x^2 + p)$, which is easy to solve. Similarly, if $p = 0$, then solving the equation $x^3 + px + q = 0$ amounts to finding the cube roots of q. Thus, the only case we really need to consider is the case in which both p and q are not zero. For this case, we need to work in $\mathbb{C}$, because we need the existence of square roots and cube roots to obtain a reasonably readable statement.

Theorem 5.66 *The* **cubic formula**, *step 2:* **Cardano's formula**.[14] *Let $p, q \in \mathbb{C} \setminus \{0\}$*

and consider the equation $x^3 + px + q = 0$. Let $\alpha_{1,2,3} := \left(-\dfrac{q}{2} + \sqrt{\dfrac{q^2}{4} + \dfrac{p^3}{27}} \right)^{\frac{1}{3}}$ be

the three cube roots of $-\dfrac{q}{2} + \sqrt{\dfrac{q^2}{4} + \dfrac{p^3}{27}}$, where it does not matter which of the two

numbers whose square is $\dfrac{q^2}{4} + \dfrac{p^3}{27}$ is chosen as the square root. Then the solutions of

$x^3 + px + q = 0$ are $x_{1,2,3} = \alpha_{1,2,3} - \dfrac{p}{3\alpha_{1,2,3}}$.

Proof. To prove that all solutions must be of the indicated form (and also to see how anyone might come up with the formula), let x be a solution of $x^3 + px + q = 0$. The elimination of the quadratic term has shown that it helps to represent the solution as something other than just x: Because the extra constant α in the proof of Theorem 5.65 could be adjusted to suit our needs, we were able to reduce arbitrary cubic equations to cubic equations without a quadratic term.

In this proof, we use a substitution $x = \alpha + \beta$ to ultimately obtain an equation without a linear or a quadratic term. But this time, unlike in the proof of Theorem 5.65, we must work with both summands α and β. If we left one summand alone, as we did with y in the proof of Theorem 5.65, then we might get rid of the linear term, but we would reintroduce the quadratic term. After we substitute $x = \alpha + \beta$ into the equation, we will get an idea how to choose α and β to simplify the equation.

$$\begin{aligned}
0 &= x^3 + px + q \\
&= (\alpha + \beta)^3 + p(\alpha + \beta) + q \\
&= \alpha^3 + 3\alpha^2\beta + 3\alpha\beta^2 + \beta^3 + p\alpha + p\beta + q
\end{aligned}$$

---

[14]For the story behind this name, see Section A.2.

> Here is the crucial step. Eliminating terms clearly would sim-
> plify the equation. We will most likely not be able to eliminate
> α^3 and β^3, because the equation is a third order equation in the
> variable x. Also, because the constant term q has no obvious
> connection to the other terms, it seems to be here to stay. But
> both $3\alpha^2\beta$ and $3\alpha\beta^2$ have $3\alpha\beta$ in common. If we *could* choose
> α and β so that $3\alpha\beta = -p$, then all the other terms would can-
> cel. Exercise 5-60 shows that such a choice is possible in $\mathbb{C}$.
> (Which is yet another reason why we need to prove this re-
> sult with our field being $\mathbb{C}$.) So from now on we assume that
> $\alpha + \beta = x$ *and* $3\alpha\beta = -p$.

$$= \alpha^3 - p\alpha - p\beta + \beta^3 + p\alpha + p\beta + q$$
$$= \alpha^3 + \beta^3 + q$$

> It would be easier if the equation had only one variable rather
> than two. Because $p \neq 0$ and $3\alpha\beta = -p$, both α and β must
> be nonzero. Via $3\alpha\beta = -p$ we obtain that $\beta = -\dfrac{p}{3\alpha}$.

$$= \alpha^3 - \frac{p^3}{27\alpha^3} + q.$$

The above equation only looks complicated until we multiply by α^3 and realize that
it is in fact a quadratic equation in $u := \alpha^3$. Hence we can solve for u.

$$\alpha^6 + q\alpha^3 - \frac{p^3}{27} = 0$$

$$u^2 + qu - \frac{p^3}{27} = 0$$

$$u_{1,2} = \frac{-q \pm \sqrt{q^2 + 4\frac{p^3}{27}}}{2}$$

$$= -\frac{q}{2} \pm \sqrt{\frac{q^2}{4} + \frac{p^3}{27}}.$$

Therefore α must be one of the three third roots of $u_1 = -\frac{q}{2} + \sqrt{\frac{q^2}{4} + \frac{p^3}{27}}$ or of
$u_2 = -\frac{q}{2} - \sqrt{\frac{q^2}{4} + \frac{p^3}{27}}$.

Now recall that we are not primarily interested in representing α. Our overall goal
is to find $x = \alpha + \beta$. We know that α and β must satisfy $27\alpha^3\beta^3 = -p^3$. But it is
easy to check that $27u_1u_2 = -p^3$. Hence, if $\alpha^3 = u_1$, then β^3 must be equal to u_2 and
vice versa. Thus $x = \alpha + \beta$ is always the sum of a cube root of u_1 and a cube root of
u_2, which means there is no loss of generality if we assume that $\alpha^3 = u_1$. But then we
must have that

$$\alpha = \sqrt[3]{u_1} = \left(-\frac{q}{2} + \sqrt{\frac{q^2}{4} + \frac{p^3}{27}} \right)^{\frac{1}{3}}$$

and from $3\alpha\beta = -p$ we obtain that $\beta = -\frac{p}{3\alpha}$. (Our derivation also proves that for every possible choice for α, the expression for β is one of the three cube roots of u_2, a fact that seems daunting to prove directly.) Hence the solutions of the equation $x^3 + px + q = 0$ must be of the form $x_{1,2,3} = \alpha_{1,2,3} + \beta_{1,2,3} = \alpha_{1,2,3} - \frac{p}{3\alpha_{1,2,3}}$, where $\alpha_{1,2,3}$ denote the three cube roots of u_1 (or of u_2).

We leave the verification that these numbers are indeed solutions of the cubic equation $x^3 + px + q = 0$ to the reader (see Exercise 5-61). ∎

The process outlined in Theorems 5.65 and 5.66 provides another way to answer the questions listed under number 4 on page x: Apply the theorems to the polynomial equation and obtain all solutions from them. Of course, for a simple equation as given there, this approach may be overkill. But this process will work for all cubic equations, not just the simple ones.

We can briefly note that a version of Theorem 5.66 for the real solutions of polynomial equations with real coefficients can be proved as indicated, and without using Theorem 5.64: Indeed, the proof of Theorem 5.66 requires the solvability of certain quadratic equations, which means the existence of certain square roots, and it requires the existence of certain cube roots. If we restrict considerations to real coefficients and real solutions, then all roots that need to be taken are roots of real numbers. But the roots of any positive real number a are $\pm\sqrt{a}$, and the roots of any negative real number b are $\pm i\sqrt{|b|}$. Similarly, (see Exercise 5-62) we can construct the cube roots of nonzero real numbers without referring to Theorem 5.64. The necessary excursions into the complex numbers are allowed, because we have constructed the complex numbers as valid mathematical entities. So, readers interested in a self-contained presentation could rephrase Theorem 5.66 by demanding that the coefficients are real numbers and that the solutions are only to be considered in case they are real.[15]

For quartic equations $x^4 + ax^3 + bx^2 + cx + d = 0$, we follow the template from above. We first eliminate the cubic term and then we consider the simpler equation $x^4 + px^2 + qx + r = 0$. Also note that, because the division by the nonzero leading coefficient is so natural, we assume right away that the leading coefficient is 1. In this fashion, our formulas will be a bit more bearable. The solution method for solving quartic equations is also called the **Ferrari method** after its discoverer, L. Ferrari.[16]

Theorem 5.67 *The **quartic formula**, step I: Elimination of the cubic term. Let $(\mathbb{F}, +, \cdot)$ be a field and let $a, b, c, d \in \mathbb{F}$. Then the number $x \in \mathbb{F}$ solves the quartic equation $x^4 + ax^3 + bx^2 + cx + d = 0$ iff $y = x + \frac{a}{4}$ solves the equation $y^4 + py^2 + qy + r = 0$ where $p = -\frac{3a^2}{8} + b$, $q = \frac{a^3}{8} - \frac{ab}{2} + c$ and $r = -\frac{3a^4}{256} + \frac{a^2 b}{16} - \frac{ac}{4} + d$.*

Proof. Exercise 5-63. ∎

---

[15] Historically, complex solutions with nonzero imaginary part were not considered "actual" solutions for centuries. During that time, complex numbers were not well understood at all, except as a convenient formalism to "make certain equations work." So by thinking about how to obtain a proof that makes the text self-contained, we get an idea about the mental contortions mathematicians went through before complex numbers became accepted in mathematics.

[16] As far as the author knows, this gentleman did not make carriages, horseless or otherwise.

Unlike for the cubic equation, for which we obtained an actual formula, for the quartic equation, we will derive an equivalent solvable equation. Indeed, the equation

$$\left(x^2 + \frac{p}{2} + \alpha_0\right)^2 - 2\alpha_0(x - z_0)^2 = 0$$

in Theorem 5.68 below splits into two quadratic equations,

$$x^2 + \frac{p}{2} + \alpha_0 = \sqrt{2\alpha_0}(x - z_0) \quad \text{and} \quad x^2 + \frac{p}{2} + \alpha_0 = -\sqrt{2\alpha_0}(x - z_0),$$

which will provide the solutions of the quartic equation. Closer analysis of Theorem 5.68 shows that we could write down formulas for the solutions, but they would be a bit on the lengthy side. Once more, for the proof that we will obtain all solutions, we need the existence of solutions of certain quadratic and cubic equations. Therefore, once more we work in $\mathbb{C}$.

Theorem 5.68 *The* **quartic formula**, *step 2:* **Splitting the quartic.** *Let* $p, q, r \in \mathbb{C}$. *Then* $x \in \mathbb{C}$ *solves the equation* $x^4 + px^2 + qx + r = 0$ *iff* x *solves the equation*

$$\left(x^2 + \frac{p}{2} + \alpha_0\right)^2 - 2\alpha_0(x - z_0)^2 = 0,$$

where α_0 *is a solution of the cubic equation*

$$q^2 - 4 \cdot 2\alpha\left(\alpha^2 + p\alpha + \frac{p^2}{4} - r\right) = 0, \qquad \text{and} \qquad z_0 = \frac{q}{4\alpha_0}.$$

Proof. For "$\Leftarrow$" note that the proof that the numbers x in question solve the equation is a computation that uses the properties of x and the right expansions of the equations.

$x^4 + px^2 + qx + r$

> Expanding the leading term into the "squared square"
> $$x^4 = \left(\left(x^2 + \frac{p}{2} + \alpha_0\right) - \left(\frac{p}{2} + \alpha_0\right)\right)^2 \text{ leads to the following.}$$

$$= \left(x^2 + \frac{p}{2} + \alpha_0\right)^2 - 2x^2\left(\frac{p}{2} + \alpha_0\right) - 2\left(\frac{p}{2} + \alpha_0\right)^2 + \left(\frac{p}{2} + \alpha_0\right)^2 + px^2 + qx + r$$

$$= \left(x^2 + \frac{p}{2} + \alpha_0\right)^2 - \left[2\alpha_0 x^2 - qx + \left(\alpha_0^2 + p\alpha_0 + \frac{p^2}{4} - r\right)\right]$$

> Because $\quad q^2 - 4 \cdot 2\alpha_0\left(\alpha_0^2 + p\alpha_0 + \dfrac{p^2}{4} - r\right) = 0,\quad$ by
> the quadratic formula (in x), the term in square brackets is a perfect square, namely, it is equal to
> $$2\alpha_0\left(x - \frac{q}{4\alpha_0}\right)^2 = 2\alpha_0(x - z_0)^2.$$

$$= \left(x^2 + \frac{p}{2} + \alpha_0\right)^2 - 2\alpha_0(x - z_0)^2$$

$$= 0$$

The proof of the direction "$\Rightarrow$" is left to the reader as Exercise 5-64.   ∎

In both directions, the expansion $x^4 = \left(\left(x^2 + \frac{p}{2} + \alpha\right) - \left(\frac{p}{2} + \alpha\right)\right)^2$ is the key to the proof of Theorem 5.68. Although this expansion looks as if it materializes "out of the blue," (after someone smart figured it out) it can be considered to be quite natural. We have seen in earlier proofs that substitutions $x = y - \alpha$ work well. But for the quartic, such a substitution would reintroduce the cubic term that we removed using Theorem 5.67. A substitution $x^4 = \left(x^2 + \gamma - \gamma\right)^2$ is similar, and it will not introduce cubic terms. Making $\frac{p}{2}$ part of γ is natural, because it cancels the px^2. To have some further freedom, using $\gamma = \frac{p}{2} + \alpha$, where α still needs to be determined, now seems quite reasonable.

Similar to Theorem 5.66, a version of Theorem 5.68 for the real solutions of quartics with real coefficients can be proved without referring to Theorem 5.64: The proof of Theorem 5.68 requires the existence of one solution of a certain cubic equation. This cubic equation has real coefficients if we start with real coefficients, and hence the version of Theorem 5.66 that was discussed after the proof of Theorem 5.66 can be used to obtain a solution α_0. Any generalization of Theorem 5.68 to arbitrary fields is made cumbersome by the fact that we don't know if this cubic equation has a solution in an arbitrary field.

Going from quadratic equations to cubic equations to quartic equations showed a discouraging and an encouraging trend. On the down side, the solutions become more involved as the order of the equation increases. This is to be expected, because as the order of the equation increases, the equation itself becomes more complicated. It would be naive to hope that the equations could increase in their complexity, while the solutions stayed simple. On the up side, it remained possible to write down exact solutions of the equations.

Unfortunately, the encouraging trend stopped here. The above results were all known by the year 1545, when both Cardano's formula (which is actually due to Tartaglia, who confided it to Cardano) and the Ferrari method (Ferrari was a student of Cardano's) were published by Cardano. After that, mathematicians were looking for ways to solve quintic equations and, for centuries, they did not get very far. The reason was discovered by Abel in 1824: It is not possible to represent the solutions of an arbitrary quintic equation with formulas (or procedures) that involve only finitely many roots. Or, in simpler words, there is no quintic formula. The problem is not that we have not found it yet. It simply does not exist.

This result still has implications from a modern point of view. Solutions of polynomial equations are important in many applications. With computers we can find approximate solutions to any degree of accuracy. But as computations become large, rounding errors and numerical instabilities continue to be a challenge. So with symbolic processors in computer algebra systems becoming increasingly sophisticated, availability of solution formulas for higher order equations would be useful, even today. But, because they don't exist, this line of investigation is closed, at least when it comes to attempts to find general solutions.

Exercises

5-58. Let $(\mathbb{F}, +, \cdot)$ be a field and let $a, b, c \in \mathbb{F}$ with $a \neq 0$. Prove that $x_2 = \dfrac{-b - \sqrt{b^2 - 4ac}}{2a}$ solves the equation $ax^2 + bx + c = 0$.

5-59. Let $(\mathbb{F}, +, \cdot)$ be a field and consider the equation $x^2 + px + q = 0$.

    (a) Prove that the number of solutions of the equation in $\mathbb{F}$ is equal to the number of square roots of $\dfrac{p^2}{4} - q$ in $\mathbb{F}$, and that if $\dfrac{p^2}{4} - q$ has at least one square root in $\mathbb{F}$, then the solutions of the equation are given by $x_{1,2} = -\dfrac{p}{2} \pm \sqrt{\dfrac{p^2}{4} - q}$.

    (b) In some curricula the formula in part 5-59a is referred to as the **quadratic formula**. Explain how the formula can be used to solve arbitrary quadratic equations.

    (c) Use the formula in part 5-59a to solve each equation below.

        i. $x^2 + 4x - 12 = 0$

        ii. $2x^2 + x - 10 = 0$

        iii. $\dfrac{1}{2}x^2 - 4x + 8 = 0$

5-60. Let $x, p \in \mathbb{C}$. Prove that there are $\alpha, \beta \in \mathbb{C}$ so that $\alpha + \beta = x$ and $3\alpha\beta = -p$. *Hint.* Quadratic formula.

5-61. Finish the proof of Theorem 5.66: Let $p, q \in \mathbb{C} \setminus \{0\}$ and let $\alpha_{1,2,3} := \left(-\dfrac{q}{2} + \sqrt{\dfrac{q^2}{4} + \dfrac{p^3}{27}} \right)^{\frac{1}{3}}$ be

the three cube roots of $-\dfrac{q}{2} + \sqrt{\dfrac{q^2}{4} + \dfrac{p^3}{27}}$, where it does not matter which of the two numbers whose

square is $\dfrac{q^2}{4} + \dfrac{p^3}{27}$ is chosen as the square root. Prove that each of $x_{1,2,3} = \alpha_{1,2,3} - \dfrac{p}{3\alpha_{1,2,3}}$ solves

the equation $x^3 + px + q = 0$.

5-62. Let $c \in \mathbb{R}$.

    (a) Prove, without using exponential representations of complex numbers, that if $c > 0$, then $-\sqrt[3]{c}$, $\sqrt[3]{c}\dfrac{1}{2}\left(1 + i\sqrt{3}\right)$, and $\sqrt[3]{c}\dfrac{1}{2}\left(1 - i\sqrt{3}\right)$ are solutions of the equation $x^3 + c = 0$.

    (b) For $c < 0$, find the three solutions of the equation $x^3 + c = 0$. Do not use exponential representations of complex numbers when stating the result.

5-63. Prove Theorem 5.67.

5-64. Let $p, q, r \in \mathbb{C}$. Prove that if $x \in \mathbb{C}$ solves the equation $x^4 + px^2 + qx + r = 0$ then x solves the equation $\left(x^2 + \dfrac{p}{2} + \alpha_0\right)^2 - 2\alpha_0(x - z_0)^2 = 0$, where α_0 is a solution of the cubic equation $q^2 - 4 \cdot 2\alpha\left(\alpha^2 + p\alpha + \dfrac{p^2}{4} - r\right) = 0$ and $z_0 = \dfrac{q}{4\alpha_0}$.

Hints. First replace x^4 with $\left(\left(x^2 + \dfrac{p}{2} + \alpha\right) - \left(\dfrac{p}{2} + \alpha\right)\right)^2$. Expand and collect terms. Note that α can be chosen so that the sum of all terms other than $\left(x^2 + \dfrac{p}{2} + \alpha\right)^2$ is a perfect square.

5-65. Reducing the quintic.

    (a) Find an α so that x is a solution of the equation $x^5 + ax^4 + bx^3 + cx^2 + dx + e = 0$ iff $y = x + \alpha$ is a solution of the equation $x^5 + px^3 + qx^2 + rx + s = 0$.

    (b) Find p, q, r and s from part 5-65a.

(c) Explain how, in an nth order equation, the term of order $(n-1)$ can be eliminated.

5-66. Solve each of the following equations in $\mathbb{C}$.

(a) $x^3 + x^2 + x + 1 = 0$

(b) $x^4 + x^3 + x^2 + x + 1 = 0$

(c) $3x^3 + 2x^2 = 1$

(d) $x^3 - 5x^2 + 3x + 9 = 0$

5-67. Prove that the function $f : \mathbb{R} \to \mathbb{R}$ defined by $f(x) = x^3 - x$ is surjective, but not injective.

5-68. **Viète's Theorem**, quadratic case. Let $(\mathbb{F}, +, \cdot)$ be a field. Let x_1 and x_2 be the two solutions of the equation $ax^2 + bx + c = 0$, or, if the equation has only one solution, let $x_1 = x_2$ be that solution. Prove the following.

(a) $ax_1x_2 = c$,

(b) $ax_1 + ax_2 = -b$,

(c) Solve the equation $3x^2 - 4x - 12 = 0$ in $\mathbb{R}$. Then use 5-68a and 5-68b to check if your solution is correct.

(d) Prove that any two numbers that satisfy parts 5-68a and 5-68b also solve the quadratic equation $ax^2 + bx + c = 0$.

5-69. **Viète's Theorem/Girard's Theorem/Newton's Theorem.** Let $z_1, \ldots, z_n$ be all the roots of the equation $x^n + a_{n-1}x^{n-1} + \cdots + a_1x + a_0 = 0$, counted/listed with multiplicity.

(a) Prove that for every $k \in \{0, \ldots, n-1\}$ we have $a_k = (-1)^{n-k} \sum_{j_1 < j_2 < \cdots < j_{n-k}} z_{j_1} z_{j_2} \cdots z_{j_{n-k}}$.

(b) Prove that the solutions of the above system of equations must be the solutions of the polynomial equation.

Hint. The proof for each part starts by multiplying out the product $\prod_{j=1}^{n} (x - z_j)$.

5.9 Beyond Fields: Vector Spaces and Algebras

Abel's Theorem (Theorem 6.78) will show that there is no quintic formula. That is, unlike for $\mathbb{R}$ and quadratic equations, there is no way to expand $\mathbb{C}$ and obtain a field in which all quintic equations can be solved with field operations and finitely many radicals. The question beckons if there is a way to expand $\mathbb{C}$ at all? Comparing the situation once more to that for $\mathbb{R}$, we see that, by Theorem 5.48, the answer is negative for $\mathbb{R}$, as long as we want to keep the total order and the Completeness Axiom. For the complex numbers, the answer is negative, too, as long as we want to keep the analytical property of completeness.[17] We will not go deeper into analysis in this text[18], but we can explain two important mathematical structures that use the field structure and help answer the question whether $\mathbb{C}$ can be expanded. The ideas we introduce, specifically the idea of a base, will be useful in Chapter 6, too.

Vector spaces are used to explore higher dimensions. Note that the idea of the dimension of a vector space can only be introduced after we have laid a substantial amount of groundwork.

---

[17] This property states that every Cauchy sequence must converge.

[18] Readers may wish to consider the text [30], which the author likes for obvious and slightly selfish reasons.

Definition 5.69 *Let* $(\mathbb{F}, +, \cdot)$ *be a field. A* **vector space** *over the field* $\mathbb{F}$ *is a triple* $(X, +, \cdot)$ *of a set* X *and two operations,* **vector addition** $+ : X \times X \to X$ *and* **scalar multiplication** $\cdot : \mathbb{F} \times X \to X$, *so that the following hold.*

1. *Vector addition is* **associative.**
   That is, for all $x, y, z \in X$ *we have* $(x + y) + z = x + (y + z)$.

2. *Vector addition is* **commutative.**
   That is, for all $x, y \in X$ *we have* $x + y = y + x$.

3. *There is a* **neutral element** $0 \in X$ *for addition.*
   That is, there is an element $0 \in X$ *so that for all* $x \in X$ *we have* $x + 0 = x$.

4. *For every element* $x \in X$, *there is an* **additive inverse** *element.*
   That is, for every $x \in X$ *there is a* $(-x) \in X$ *so that* $x + (-x) = 0$.

5. *Scalar multiplication is* **(left) distributive** *over vector addition.*
   That is, for all $x, y \in X$ *and* $\alpha \in \mathbb{F}$ *we have* $\alpha \cdot (x + y) = \alpha \cdot x + \alpha \cdot y$.

6. *Scalar multiplication is* **(right) distributive** *over scalar addition.*
   That is, for all $x \in X$ *and* $\alpha, \beta \in \mathbb{F}$ *we have* $(\alpha + \beta) \cdot x = \alpha \cdot x + \beta \cdot x$.

7. *Scalar multiplication is* **"associative."**
   That is, for all $x \in X$ *and* $\alpha, \beta \in \mathbb{F}$ *we have* $\alpha \cdot (\beta \cdot x) = (\alpha\beta) \cdot x$.

8. *The number* $1 \in \mathbb{F}$ *is a* **neutral element** *for scalar multiplication.*
   That is, for all $x \in X$ *we have* $1 \cdot x = x$.

An element of a vector space is also called a **vector** *and an element of* $\mathbb{F}$ *is also called a* **scalar** *in this context. We will usually refer to the set* X *as the vector space, implicitly assuming addition and multiplication are denoted as usual. As is customary for multiplications, the dot is usually omitted.*

The standard example and visualization for a vector space is d-dimensional space.

Example 5.70 Let $d \in \mathbb{N}$ and let $\mathbb{F}$ be a field. The set $\mathbb{F}^d := \{(x_1, \ldots, x_d) : x_i \in \mathbb{F}\}$ with componentwise addition and scalar multiplication is a vector space. $\square$

We will not directly prove the vector space properties for d-dimensional space. Instead, we note that d-dimensional space can be interpreted as the space of functions $f : \{1, \ldots, d\} \to \mathbb{F}$. The fact that d-dimensional space is a vector space then follows from Example 5.71 below.

Example 5.71 *Let* D *be a set and let* $\mathbb{F}$ *be a field. The set* $\mathcal{F}(D, \mathbb{F})$ *of all functions* $f : D \to \mathbb{F}$ *with addition defined pointwise by* $(f + g)(x) := f(x) + g(x)$ *and scalar multiplication defined pointwise by* $(\alpha \cdot f)(x) := \alpha f(x)$ *is a vector space.*

All properties follow from the corresponding pointwise properties for $\mathbb{F}$. For example, addition of functions is commutative, because for all $x \in D$ we have that $(f + g)(x) = f(x) + g(x) = g(x) + f(x) = (g + f)(x)$. The neutral element with respect to addition is the function that is equal to $0 \in \mathbb{F}$ for all $x \in D$. $\square$

By identifying each vector $(x_1, \ldots, x_d) \in \mathbb{F}^d$ with the function that maps each index $i \in \{1, \ldots, d\}$ to x_i, we see that $\mathbb{F}^d$ is the vector space $\mathcal{F}(\{1, \ldots, d\}, \mathbb{F})$. Spaces of functions are very important for all branches of analysis. For example, many numerical solutions for partial differential equations are formulated and justified in function spaces.

It is often useful to represent vectors in terms of certain standard vectors. The most familiar example is the standard coordinate system in d-dimensional space.

Definition 5.72 *Let X be a vector space. A subset $S \subseteq X \setminus \{0\}$ of nonzero vectors is called* **linearly independent** *iff for all finite subsets $\{x_1, \ldots, x_n\} \subseteq S$ and all sets of scalars $\{\alpha_1, \ldots, \alpha_n\} \subseteq \mathbb{F}$,*

$$\sum_{i=1}^{n} \alpha_i x_i = 0 \qquad \text{implies} \qquad \alpha_1 = \alpha_2 = \cdots = \alpha_n = 0.$$

A sum $\sum_{i=1}^{n} \alpha_i x_i$ with $\alpha_i \in \mathbb{F}$ and $x_i \in X$ is also called a **linear combination** of $x_1, \ldots, x_n$.

Definition 5.73 *Let X be a vector space. A linearly independent set $B \subseteq X$ such that for every $x \in X$ there are a finite subset $\{b_1, \ldots, b_n\} \subseteq B$ and a set of scalars $\{\alpha_1, \ldots, \alpha_n\} \subseteq \mathbb{F}$ so that $x = \sum_{i=1}^{n} \alpha_i b_i$ is called a* **base** *of the vector space.*

Linear independence and bases are usually investigated in linear algebra.[19] For our purposes, they are important to define the dimension of a vector space. The following example shows that the space that we call "d-dimensional space" has a base that consists of d vectors. These d vectors are exactly the d "directions" that exist in this space.

Example 5.74 *In $\mathbb{F}^d$, let e_i denote the vector such that the ith component is 1 and all other components are zero. Then $\{e_1, \ldots, e_d\}$ is a base of $\mathbb{F}^d$.*

To prove linear independence, for each $i = 1, \ldots, d$ let $e_i^{(j)}$ denote the jth component of e_i. Then for any $\alpha_1, \ldots, \alpha_d$ the vector equation $\sum_{i=1}^{d} \alpha_i e_i = 0$ leads with $j = 1, \ldots, d$ to the scalar equations $\sum_{i=1}^{d} \alpha_i e_i^{(j)} = 0$, which, for each j, simply state that $\alpha_j = 0$, as was to be proved.

Regarding the representation of elements, for each $x = (x_1, \ldots, x_d) \in \mathbb{F}^d$ we have that $x = (x_1, \ldots, x_d) = \sum_{i=1}^{d} x_i e_i$. $\qquad\qquad \square$

Because $\mathbb{F}^d$ has a base with d vectors, it seems reasonable to define the dimension of a vector space with a finite base to be the number of elements in that base. But that only works if all bases have *the same* number of vectors. The next theorem proves that this is the case.

Theorem 5.75 *Let X be a vector space with a finite base F. Then every linearly independent subset L of X has at most as many elements as F. Moreover, all bases of X have as many elements as F.*

---

[19] And just, like for analysis, we can only barely touch this subject in this text.

Proof. Let $F = \{f_1, \ldots, f_n\}$ be a finite base of X. Suppose for a contradiction that there is a linearly independent set $L \subseteq X$ that has more elements than F. Without loss of generality we can assume that L is such that $|L \cap F|$ is maximal. That is, if $\tilde{L}$ is another linearly independent subset of X with more elements than F, then $\left|\tilde{L} \cap F\right| \leq |L \cap F|$. Now let $b \in L \setminus F$ and consider the linearly independent sets $L \setminus \{b\}$ and F. By Exercise 5-70a, there is a subset $H \subseteq F \setminus L$ so that $C := (L \setminus \{b\}) \cup H$ is a base of X. By Exercise 5-70b, $L \setminus \{b\}$ is not a base of X, so $H \neq \emptyset$. Hence, C has at least as many elements as L and in particular it has more elements than F. Because $b \notin F$, we obtain $\left|[(L \setminus \{b\}) \cup H] \cap F\right| = |(L \cap F) \cup H| > |L \cap F|$, a contradiction.

Thus no linearly independent subset of X has more elements than F. Now let B be another base of X. Then $|B| \leq |F|$. Therefore B is finite, and, with the same argument as above, if $L \subseteq X$ is linearly independent, then $|L| \leq |B|$. Because F is linearly independent, we obtain $|F| \leq |B|$, and hence $|F| = |B|$.   ∎

Theorem 5.75 shows that the **dimension** of a vector space with a finite base F can be defined to be $|F|$, because any two bases have the same number of elements. Moreover, Theorem 5.75 shows that a vector space is either finite dimensional (that is, it has a finite base) or it is not (that is, it does not have a finite base). Other than that, we introduced Theorem 5.75 here, because it will be needed to prove that the index of a field over another (see Definition 6.18) is well-defined.

With $\mathbb{F} := \mathbb{C}$, Examples 5.70 and 5.71 show that there are many algebraic entities that strictly contain $\mathbb{C}$: Indeed, except for the multiplication operation, $\mathbb{C}$ is isomorphic to the set $\{ze_1 : z \in \mathbb{C}\}$ in $\mathbb{C}^d$, and, more generally, for any $f \in \mathcal{F}(D, \mathbb{C})$, $\mathbb{C}$ is isomorphic to $\{zf : z \in \mathbb{C}\}$. Why should there not be a field among all these entities that properly contain $\mathbb{C}$? Vector spaces cannot be fields, because there is no multiplication available for the elements of a vector space. But the fact that complex valued functions can be multiplied pointwise shows that this is merely a limitation of the definition. There are useful examples of vector spaces whose elements can be multiplied. Hence it makes sense to give these entities a separate name.

Definition 5.76 *Let X be a vector space over the field $\mathbb{F}$ and let $\cdot : X \times X \to X$ be a binary operation. Then $(X, \cdot)$ is called an* **algebra** *iff*

1. *The multiplication operation on X is associative*

2. *Multiplication on X is left- and right distributive over addition on X.*

3. *For all $\alpha \in \mathbb{F}$ and all $x, y \in X$ we have that $\alpha(xy) = (\alpha x)y = x(\alpha y)$.*

Example 5.77 *Let D be a set and let $\mathbb{F}$ be a field. The vector space $\mathcal{F}(D, \mathbb{F})$ of all functions $f : D \to \mathbb{F}$ is an algebra, with the multiplication operation defined pointwise by $(f \cdot g)(x) := f(x) \cdot g(x)$.*

All properties follow from the corresponding pointwise properties for elements of fields.   □

It follows directly from the definitions that every field is an algebra (over itself) and that every algebra is a ring. Moreover, the functions $\mathcal{F}(\{0, 1\}, \mathbb{R})$ show that not every algebra is a field (because $(0, 1) \cdot (1, 0) = (0, 0)$, which is impossible in a field), and the integers modulo m with m not prime show that not every ring is an algebra. In relation to vector spaces, it follows directly from the definition that every algebra is a vector space. Some vector spaces can be turned into algebras. For example, $\mathbb{R}^3$ with the familiar cross product is an algebra over $\mathbb{R}$. But the space $\mathbb{R}^5$ shows that not every vector space is an algebra, because there is no five dimensional cross product.[20] (Also see Figure 3.1 on page 107 for the relation between these different structures.)

But if a vector space can be equipped with a multiplication to become an algebra, is there anything that could prevent it from becoming a field? After all, we "only" need commutativity of the multiplication, a unit element for multiplication and multiplicative inverses. It turns out that we cannot just turn any algebra with multiplicative inverses that strictly contains $\mathbb{C}$ into a field with an absolute value function in which Cauchy sequences converge[21]: The Gelfand-Mazur Theorem (see Theorem 12 in Chapter 18 of [16])[22] says that any complete (in the sense of analysis) normed (see Exercise 5-72e) algebra over $\mathbb{C}$ with a multiplicative unit element and so that every nonzero element has a multiplicative inverse must be isomorphic to $\mathbb{C}$. But if $\mathbb{F}$ was a complete normed field that contains $\mathbb{C}$, then $\mathbb{F}$ would be a complete normed algebra over $\mathbb{C}$ with a multiplicative unit element and multiplicative inverses, and hence it would be isomorphic to $\mathbb{C}$.

Although it seems that our journey to ever better number systems has thus ended on a negative note, results such as Theorem 5.48 and the one discussed in the preceding paragraph are actually very positive: They prove that entities such as the real and complex numbers are, in some ways, "constants of nature." No matter how we build mathematics, the real and complex numbers will always be uniquely defined by their properties. Hence we can stop looking for new number systems and we can start doing mathematics with the number systems we have. And that was the purpose of this text: To provide the foundation for the courses that will use the proof methods, logic, sets and numbers introduced here. Chapter 6, which is quite demanding, will show one such use of the machinery we have developed so far. Chapter 7 will conclude this text by completing our exposure to the axioms of set theory.

Exercises

5-70. The span of linearly independent sets. Let X be a vector space and let $W \subseteq X$. Define the **span** of W to be

$$\text{span}(W) := \left\{ v = \sum_{i=1}^{n} a_i w_i : a_i \in \mathbb{F}, w_i \in W \right\}.$$

(a) Let B and F be linearly independent sets and let F be finite. Prove that if $\text{span}(B \cup F) = X$, then there is a subset H of F so that $B \cup H$ is a base of X.

Hint. Induction on the size of F. In the induction step, let $f \in F$ and distinguish the cases $f \in \text{span}(B)$ and $f \notin \text{span}(B)$.

(b) Let B be a linearly independent set and let $b \in B$. Prove that $\text{span}(B) \neq \text{span}\left(B \setminus \{b\} \right)$.

---

[20] It is not customary to consider componentwise multiplication on spaces $\mathbb{F}^d$. So, technically, $\mathbb{R}^5$ is not *considered* to be an algebra, but with good reason: The componentwise multiplication has little use on this space.

[21] And these are properties that, especially in analysis, we would certainly want a field to have.

[22] The proof relies on major results from the preceding chapters in [16], which itself is best read after reading a fundamental text on analysis. This is why the proof was not included.

5-71. Let X be a vector space and let $B \subseteq X$ be a base. Prove that for each $x \in X$ the $b_1, \ldots, b_n \in B$ and $\alpha_1, \ldots, \alpha_n \in \mathbb{F}$ in Definition 5.73 are unique, except that any vector of $B \setminus \{b_1, \ldots, b_n\}$ can be added to the b_i with a coefficient zero and that any b_i with $\alpha_i = 0$ can be omitted.

5-72. Consider the set $\mathbb{H} := \{a1 + bi + cj + dk : a, b, c, d \in \mathbb{R}\}$, where $1, i, j, k$ are considered to be four linearly independent vectors (like e_1, e_2, e_3, e_4), with multiplication being defined by $1x = x1 = x$ for all $x \in \{i, j, k\}$, $i^2 = j^2 = k^2 = -1$, $ij = k = -ji$, $jk = i = -kj$, and $ki = j = -ik$, and parentheses being multiplied out in the customary fashion.

   (a) Prove that $\mathbb{H}$ is a four dimensional algebra over $\mathbb{R}$.

   (b) Prove that $\mathbb{H}$ has a multiplicative unit element.

   (c) Prove that every $a + bi + cj + dk \neq 0$ has a multiplicative inverse.

   Hint. Use the "conjugate" $a - bi - cj - dk$ similar to how the conjugate in complex numbers is used.

   (d) Explain why $\mathbb{H}$ is not a field.

   (e) A complex **normed vector space** is a vector space X over $\mathbb{C}$ with a **norm** $\| \cdot \| : X \to [0, \infty)$ so that

      i. For all $x \in X$, we have $\|x\| = 0$ iff $x = 0$.

      ii. For all $\alpha \in \mathbb{C}$ and $x \in X$, we have $\|\alpha x\| = |\alpha| \|x\|$.

      iii. The **triangular inequality** holds.
         That is, for all $x_1, x_2 \in X$ we have $\|x_1 + x_2\| \leq \|x_1\| + \|x_2\|$.

   Prove that, with $\|a + bi + cj + dk\| := \sqrt{a^2 + b^2 + c^2 + d^2}$, $\mathbb{H}$ is a two dimensional complex normed vector space.

   (f) Prove that for all $x, y \in \mathbb{H}$ we have $\|xy\| = \|x\| \|y\|$.

   (g) It can be proved that $\mathbb{H}$ is complete (in the sense of analysis). That means $\mathbb{H}$ is dangerously close to being a complete normed complex algebra with unit element and multiplicative inverses (which, by the Gelfand-Mazur Theorem, see Theorem 12 in Chapter 18 of [16], would mean it would be isomorphic to $\mathbb{C}$).

   Prove that part 3 of Definition 5.76 is violated if we try to prove that $\mathbb{H}$ is an algebra over $\mathbb{C}$.

   Note. This last part shows that the results of this exercise do not contradict the Gelfand-Mazur Theorem. Note however, by how little we miss satisfying the hypotheses of the Gelfand-Mazur Theorem.

Note. The set $\mathbb{H}$ with the indicated operations is called the set of **hypercomplex numbers** or the set of **quaternions**.

Chapter 6

Unsolvability of the Quintic by Radicals

I do not consider this chapter a standard part of a first course on proofs. Rather, this chapter gives you an outlook on abstract algebra, as well as an overview of how and why mathematicians recognized the power of the currently used levels of abstraction. If you have just completed your first proof class, do not be frustrated by the level of sophistication in this chapter. Fully understanding this chapter is something to aspire to, even long after taking a class that covers most of this text. My shelf holds several books with certain chapters that I have not yet made my way through. They remind me that there will always be something more to learn. That is what keeps mathematics (indeed, once we tear ourselves away from books, life itself) interesting. So maybe this chapter will be a reason to revisit this book every so often.

After the discovery of Cardano's formula and the Ferrari method, for hundreds of years, there were failed attempts to find a "quintic formula." Given this state of affairs, it was reasonable to investigate the possibility that there is no such formula. But how would we set up a proof for such a conjecture? How would we precisely *state* the conjecture, for that matter? A constructive approach is out of the question. There are infinitely many ways to process the coefficients of the quintic, and we cannot explicitly rule out every one of them. That leaves a proof by contradiction as the only feasible avenue. So, after formulating what solvability by radicals means, we start with the assumption that a solution by radicals is possible. Then we need to derive properties of quintic polynomials and of the ring $\mathbb{Z}[x]$ of polynomials with integer coefficients. Deriving such structural properties actually is a win-win approach that is typical for the mathematical investigation of hard problems: As we derive more

Fundamentals of Mathematics: An Introduction to Proofs, Logic, Sets, and Numbers.
By Bernd S. W. Schröder.

properties, our knowledge of the problem grows. This knowledge can be helpful in the ultimate resolution of the problem, independent of whether the answer is positive or negative. In the context of the quintic, the structural properties we derive ultimately facilitate the proof by contradiction. Aside from that, these structural properties have also contributed to a deeper understanding of the properties of quintics. For example, insights from the theory presented here have been used to solve special types of quintic equations. More generally, Galois' proof of the unsolvability of the quintic by radicals, which is quite different from Abel's original proof, spawned the idea of a group, which is fundamental in algebra, as well as the more abstract approach to mathematics that is used today. (Consider Appendix A for a little more on the history, or [25] for a lot more on the history.) So, although the ultimate result is negative, the proof that quintics cannot be solved by radicals has many constructive and positive consequences.

Because we already "know the answer," this chapter is geared towards answering question 9 on page x by proving the unsolvability of the quintic by radicals[1] in the shortest possible fashion.

6.1   Irreducible Polynomials

When working on a hard problem, it is always reasonable to identify which parts of the problem are harder than others. To prove that the quintic cannot be solved by radicals, a polynomial of the form $p(x) = x(x - 1)(x - 2)(x - 3)(x - 4)$ is not a good object to study. If we already know the roots of the polynomial, it is quite likely that, for this polynomial, we can find a way to obtain all roots using only field operations and root extractions (a "formula"). Indeed, even if we do not know the exact zeros of a polynomial, there may be ways to find them. For example, as long as we know that all roots are rational, the Rational Zeros Theorem (see Exercise 6-1) is a procedure that provides all rational roots using nothing but field operations.

Thus, we must first identify the "truly hard" polynomials, that is, polynomials that most likely will "resist" any attempts of extracting their zeros using field operations and radicals. Because rational zeros are easy to find, we should avoid polynomials with rational zeros. More generally, any way to factor a polynomial would simplify the task of finding the roots. In fact, for quintic polynomials, any nontrivial factorization reduces the task of finding the roots to an application of the quadratic formula and Cardano's formula or to an application of the Ferrari method. Thus, polynomials that do not factor in $\mathbb{Z}[x]$ or $\mathbb{Q}[x]$ are the natural targets for our investigation. The definition of irreducibility captures this idea by identifying polynomials without nontrivial factorizations.

Definition 6.1 *Let $(R, +, \cdot)$ be a commutative ring and let $p \in R[x]$ be a polynomial of positive degree. Then p is called* **irreducible** *iff there is no factorization $p = yz$ in $R[x]$ so that both y and z have positive degree. Otherwise, p is called* **reducible**.

---

[1]Note that non-existence of a quintic formula does not mean that the zeros of certain polynomials are forever hidden. There are approximation schemes that can produce the zeros to arbitrary precision. It's just that, for some polynomials, there is no *exact* formula.

We should note that irreducibility of a polynomial depends on the underlying ring or field. The polynomial $p(x) = x^2 - 2$ is irreducible in the ring $\mathbb{Q}[x]$, but it factors as $p(x) = (x + \sqrt{2})(x - \sqrt{2})$ in the ring $\mathbb{R}[x]$. Ultimately, the Fundamental Theorem of Algebra says that in $\mathbb{C}[x]$ the only irreducible polynomials are those of degree at most 1. However, in $\mathbb{Q}[x]$, there are many irreducible polynomials, and we will see that they need not just be of degree ≤ 2. First we note that for reducibility of polynomials with integer coefficients it does not matter if we work in $\mathbb{Z}[x]$ or $\mathbb{Q}[x]$. This is useful because irreducibility in $\mathbb{Z}[x]$ should be easier to establish than irreducibility in $\mathbb{Q}[x]$.

Lemma 6.2 Lemma of Gauss. *Let $p(x) = \sum_{j=0}^{n} a_j x^j \in \mathbb{Z}[x]$ be a polynomial. If p is reducible in $\mathbb{Q}[x]$, then p is reducible in $\mathbb{Z}[x]$.*

Proof. We will first prove the result for the case that the greatest common divisor of the coefficients of p is 1. Let $p(x) = y(x)z(x)$ be a factorization of p into polynomials of positive degree $y(x) = \sum_{j=0}^{s} y_j x^j$ and $z(x) = \sum_{j=0}^{t} z_j x^j$ in $\mathbb{Q}[x]$, where $s, t \geq 1$, $y_s \neq 0$ and $z_t \neq 0$. Without loss of generality, assume that the rational numbers y_j and z_j are written as fractions in lowest terms. Let d_y be the lowest common multiple of the denominators of the y_j and write y as $y = \frac{1}{d_y} \tilde{y}$. Then the coefficients of $\tilde{y}$ are integers. Now let n_y be the greatest common divisor of the numerators of the coefficients of $\tilde{y}$ and write y as $y(x) = \frac{n_y}{d_y} Y(x) = \frac{n_y}{d_y} \sum_{j=0}^{s} Y_j x^j$. Then the greatest common divisor of the integers Y_j is 1. Similarly, rewrite z as $z(x) = \frac{n_z}{d_z} Z(x) = \frac{n_z}{d_z} \sum_{j=0}^{t} Z_j x^j$, where the Z_j are integers whose greatest common divisor is 1. Thus $p(x) = \frac{n_y n_z}{d_y d_z} Y(x) Z(x)$ or $d_y d_z p(x) = n_y n_z Y(x) Z(x)$. Because the greatest common divisor of the coefficients of p is 1, we infer that $n_y n_z \mid d_y d_z$.

If we can prove that the greatest common divisor of the coefficients of YZ is 1, too, then $d_y d_z \mid n_y n_z$, and hence $d_y d_z = n_y n_z$, and $p = YZ$ is a factorization of p in $\mathbb{Z}[x]$. To prove that the greatest common divisor of the coefficients of YZ is 1, let q be a prime number and suppose, for a contradiction, that q divides all coefficients of YZ. Because the greatest common divisor of the coefficients of Y is 1, there is a $k \in \mathbb{N}_0$ so that $q \mid Y_0, \ldots, Y_{k-1}$, but $q \nmid Y_k$. Similarly, there is an $m \in \mathbb{N}_0$ so that $q \mid Z_0, \ldots, Z_{m-1}$, but $q \nmid Z_m$. But the coefficient of x^{k+m} in YZ is

$$c_{k+m} = \sum_{j=0}^{k-1} Y_j Z_{k+m-j} + Y_k Z_m + \sum_{j=k+1}^{k+m} Y_j Z_{k+m-j}$$

(where we assume that $Y_j = 0$ for $j > s$ and that $Z_j = 0$ for $j > t$). The first sum is divisible by q, because the Y_j are divisible by q. The second sum is divisible by q, because the Z_{k+m-j} are divisible by q. By assumption, c_{k+m} is divisible by q. But this would imply that $Y_k Z_m$ is divisible by q, a contradiction (see Exercise 3-59b). Hence the greatest common divisor of the coefficients of YZ is 1, which completes the proof in case the greatest common divisor of the coefficients of p is 1.

Now consider the case that the greatest common divisor d of the coefficients of p is greater than 1 and let $p(x) = y(x)z(x)$ be a factorization of p into polynomials $y, z \in \mathbb{Q}[x]$ of positive degree. Then the coefficients of $\tilde{p}(x) := \frac{1}{d} p(x)$ have greatest common divisor 1, and $\tilde{p}(x) = \frac{1}{d} y(x)z(x)$ is a factorization of $\tilde{p}$ into polynomials

$\frac{1}{d}y \in \mathbb{Q}[x]$ and $z \in \mathbb{Q}[x]$ of positive degree. By what we have proved above, $\tilde{p}$ can be factored as $\tilde{p} = \tilde{Y}\tilde{Z}$ with $\tilde{Y}, \tilde{Z} \in \mathbb{Z}[x]$ having positive degree. But then the representation $p = d\tilde{p} = d\tilde{Y}\tilde{Z}$ proves the result.   ∎

The next result now establishes a sufficient criterion for irreducibility in $\mathbb{Q}[x]$ for polynomials with integer coefficients.

Theorem 6.3 Eisenstein's Irreducibility Criterion. *Let $p(x) = \sum_{j=0}^{n} a_j x^j \in \mathbb{Z}[x]$ be a polynomial and let q be a prime number so that $a_0, \ldots, a_{n-1}$ are divisible by q, but q does not divide a_n and q^2 does not divide a_0. Then p is irreducible in $\mathbb{Q}[x]$.*

Proof. By Lemma 6.2 we only need to prove that p is irreducible in $\mathbb{Z}[x]$.

Suppose for a contradiction that there are two polynomials $y(x) = \sum_{j=0}^{s} y_j x^j$ and $z(x) = \sum_{j=0}^{t} z_j x^j$ in $\mathbb{Z}[x]$ so that $p = yz$, $s, t \geq 1$, $y_s \neq 0$ and $z_t \neq 0$. Because $a_0 = y_0 z_0$ is divisible by q, but not by q^2, exactly one of y_0 and z_0 is divisible by q. Without loss of generality, assume that $q \mid y_0$ and $q \nmid z_0$. Because $a_n = y_s z_t$ is not divisible by q, q does not divide y_s. Let $k \in \{1, \ldots, s\}$ be the smallest number so that q divides each of $y_0, \ldots, y_{k-1}$, but not y_k. Because $k \leq s < n$, we have that $q \mid a_k$, that is $q \mid \sum_{j=0}^{k} y_j z_{k-j}$. But then $y_k z_0 = a_k - \sum_{j=0}^{k-1} y_j z_{k-j}$, which is divisible by q. Because $q \nmid z_0$, this implies that $q \mid y_k$, a contradiction. Thus the polynomial p is irreducible in $\mathbb{Z}[x]$.   ∎

The next example shows how Eisenstein's criterion is applied. The polynomial p below will also be the one for which we will explicitly show that the equation $p(x) = 0$ cannot be solved by radicals.

Example 6.4 *The polynomial $p(x) = x^5 - 4x + 2 \in \mathbb{Q}[x]$ is irreducible.*

All coefficients, except the leading coefficient 1, are divisible by 2 (0 is by default divisible by anything), and the constant coefficient 2 is not divisible by $4 = 2^2$. Thus by Eisenstein's irreducibility criterion, p is irreducible in $\mathbb{Q}[x]$.   □

Remark 6.5 At this stage we must make an excursion into analysis: We need to use the **Intermediate Value Theorem**, which states that if $f : \mathbb{R} \to \mathbb{R}$ is a continuous function so that $f(a)$ and $f(b)$ have opposite signs, then f has a zero in (a, b). Based on this result, we conclude from $p(0) = 2$, $p(1) = -1$, $p(2) = 26$ and the fact that polynomials are continuous (another result we must quote) that the polynomial p from Example 6.4 has at least two positive zeros. Moreover, because $p(-2) = -22$, it has at least one negative zero.

Of course, by now we accept that whenever a result is used to prove something else, we must assure that the result that we use does not depend on what will be proved. In the present case, it feels obvious that this should be so. It does not seem likely that a general result such as the Intermediate Value Theorem should depend on the properties of a special polynomial like the one in Example 6.4. The interested reader can find the proof of the Intermediate Value Theorem, for example, as the proof of Theorem 3.34 in [30], and the polynomial from Example 6.4 is not used in the proof. Similarly, the continuity of polynomials is Example 3.28 in [30].

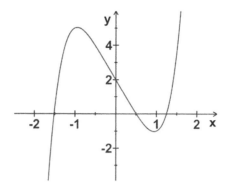

Figure 6.1 The polynomial $p(x) = x^5 - 4x + 2$ from Example 6.4. Note that all real roots are "clearly visible."

By Descartes' Rule of Signs (see Exercise 6-8) p has at most two positive zeros and at most one negative zero. Thus p has exactly two positive zeros, exactly one negative zero and exactly two complex zeros that are not real numbers (Conjugate Pairs Theorem, see Exercise 6-6). We will need this insight when we prove that, for this polynomial, the equation $p(x) = 0$ is not solvable by radicals.

The polynomial is depicted in Figure 6.1. ■

Exercises

6-1. **Rational Zeros Theorem.** Let p and q be integers with greatest common divisor one so that $\dfrac{p}{q}$ is a rational zero of the polynomial $f(x) = \sum_{j=0}^{n} a_j x^j \in \mathbb{Z}[x]$.

   (a) Prove that p is a factor of the constant term a_0.

   (b) Prove that q is a factor of the leading term a_n.

   Note. The Rational Zeros Theorem can be used to extract the rational zeros of polynomials with integer coefficients by setting up a list of possible zeros $z_1, \ldots, z_k$, factoring out a term $(x - z_m)$ for an actual zero z_m of f and then applying this process to the other factor, until all rational zeros have been found.

6-2. Use the Rational Zeros Theorem (see Exercise 6-1) to prove each of the following.

   (a) For every prime number p and every $n \in \mathbb{N} \setminus \{1\}$, the number $\sqrt[n]{p}$ is not rational.

   (b) If $k \in \mathbb{N}$ and $n \in \mathbb{N} \setminus \{1\}$ are so that $\sqrt[n]{k}$ is rational, then k is the nth power of another natural number.

6-3. Let $(R, +, \cdot)$ be a commutative ring, let $p \in R[x]$ and let $a \in R$. Prove that p is irreducible in $R[x]$ iff $g(x) := p(x - a)$ is irreducible in $R[x]$.

6-4. Let $p \in \mathbb{N}$ be a prime number. Prove that the polynomial $f(x) = x^{p-1} + x^{p-2} + \cdots + x^2 + x + 1$ is irreducible in $\mathbb{Q}[x]$.

   Hint. Use Exercise 3-32d, the Binomial Theorem and Exercise 3-68b to rewrite $f(x + 1)$ so that Eisenstein's Irreducibility Criterion can be applied. Then use Exercise 6-3.

6-5. Prove that the polynomial $f(x) = 8x^3 - 6x - 1$ is irreducible in $\mathbb{Q}[x]$.

   Hint. Use Exercise 6-3.

6-6. **Conjugate Pairs Theorem.** Let $p(x) = \sum_{j=0}^{n} a_j x^j \in \mathbb{R}[x]$ be a polynomial with real coefficients so that $a + bi$ with $b \neq 0$ is a zero of p. Prove that $a - bi$ is a zero of p.

6-7. Explain why, despite the factorization $p(x) = 2\left(x^2 - 2\right)$, the polynomial $p(x) = 2x^2 - 4$ is irreducible in $\mathbb{Q}[x]$.

6-8. **Descartes' Rule of Signs.** Let $f(x) = a_0 x^n + \cdots + a_{n-1} x + a_n$ be a polynomial with real coefficients.

   Note. The reason for the uncommon indexing of the coefficients will become clear in part 6-8d.

   (a) Prove that if $a_n a_0 > 0$, then f has an even number of positive zeros and if $a_n a_0 < 0$ then f has an odd number of positive zeros.

   Note. Above, as well as throughout this exercise, the zeros are counted with multiplicity.

   Hint. Use the Intermediate Value Theorem, the continuity of f and the fact that for large $x > 0$ all values of $f(x)$ will have the same sign. (We definitely need to draw on some calculus/analysis for this part.)

   (b) Throughout the rest of this exercise, for $n + 1$ real numbers $a_0, \ldots, a_n$, let $V(a_0, \ldots, a_n)$ be the number of sign changes in the sequence, reading the numbers from left to right, omitting zeros and not counting the transitions to the first nonzero number and from the last nonzero number as sign changes.

   Prove that if $a_0 a_n > 0$, then $V(a_0, \ldots, a_n)$ is even and if $a_0 a_n < 0$, then $V(a_0, \ldots, a_n)$ is odd.

   Hint. Induction on n.

   (c) Let $r_0, \ldots, r_{n-1}$ be positive real numbers and let $a_0, \ldots, a_n, b_0, \ldots, b_n$ satisfy $a_0 = b_0$, $a_0 \neq 0$, $a_n \neq 0$, and $b_j = a_j + r_{j-1} b_{j-1}$ for $j = 1, \ldots, n$.

   Note. These definitions will become more motivated when we prove part 6-8d.

   i. Prove that if $V(a_0, \ldots, a_n) = V(b_0, \ldots, b_n)$, then a_n and b_n have the same sign or b_n is zero.
   Hint. By part 6-8b, if $b_n \neq 0$, then $a_0 a_n$ and $b_0 b_n$ have the same sign.

   ii. Prove that if any $a_j = 0$, then a_j and b_j can be erased from their respective sequences and the resulting (reindexed) sequences $a_0', \ldots, a_{n-1}', b_0', \ldots, b_{n-1}'$ satisfy $V\left(a_0', \ldots, a_{n-1}'\right) = V(a_0, \ldots, a_n)$, $V\left(b_0', \ldots, b_{n-1}'\right) = V(b_0, \ldots, b_n)$, $a_0' = b_0'$, $a_0' \neq 0$, $a_{n-1}' \neq 0$, and $b_j' = a_j' + r_{j-1}' b_{j-1}'$ for $j = 1, \ldots, n - 2$ and positive r_j'.
   Note. So from now on we can assume without loss of generality that all a_j are nonzero.

   iii. Prove that $V(b_0, \ldots, b_n) \leq V(a_0, \ldots, a_n)$ and that in case $b_n = 0$, we even have $V(b_0, \ldots, b_n) < V(a_0, \ldots, a_n)$.
   Hint. Induction on n. For the induction step $n \to n + 1$, distinguish the cases $b_{n+1} \neq 0$ and $b_{n+1} = 0$.
   In case $b_{n+1} \neq 0$, suppose, for a contradiction, $V(b_0, \ldots, b_{n+1}) > V(a_0, \ldots, a_{n+1})$. Conclude that we must have $V(b_0, \ldots, b_n) = V(a_0, \ldots, a_n)$, that a_n, b_n and a_{n+1} have the same sign, that b_{n+1} has the opposite sign and obtain a contradiction by showing that b_{n+1} must also have the same sign as b_n.
   In case $b_{n+1} = 0$, either $V(b_0, \ldots, b_n)$ is smaller than $V(a_0, \ldots, a_n)$ or they are equal. Use $a_{n+1} = -r_n b_n$ and part 6-8(c)i in the latter case.

   (d) Prove that the number of positive real zeros is equal to the number of sign changes in the coefficients of $f(x)$ or it is less than that number by an even integer.

   Hint. Let $\rho > 0$ be a root of $f(x) = \sum_{j=0}^{n} a_j x^{n-j}$. Then $f(x) = (x - \rho) \sum_{j=0}^{n-1} b_j x^{n-1-j}$. Use part 6-8c to show $V(b_0, \ldots, b_{n-1}) \leq V(a_0, \ldots, a_n) - 1$. Recursively obtain the inequality $V(a_0, \ldots, a_n) - t \geq 0$, where t is the number of positive zeros. Then use parts 6-8a and 6-8b.

   (e) Prove that the number of negative real zeros is equal to the number of sign changes in the coefficients of $f(-x)$ or is less than that number by an even integer.

Note. In process control (see [18]), Descartes' Rule of Signs can be used to quickly conclude if certain systems are unstable. A system is stable iff its characteristic equation (which is a polynomial equation) only has zeros with nonpositive real part. So an odd number of sign changes in the coefficients of the characteristic equation guarantees a positive zero and hence instability. Only systems with an even number of sign changes in the coefficients *can* be stable. But even if there are no sign changes in the coefficients, the system is not *guaranteed* to be stable, because there could be complex zeros with positive real part.

6.2 Field Extensions and Splitting Fields

Eisenstein's Irreducibility Criterion shows that there are polynomials with integer coefficients and of arbitrarily high degree that are irreducible in $\mathbb{Q}[x]$. On the other hand, by the Fundamental Theorem of Algebra, every polynomial in $\mathbb{C}[x]$ is reducible.[2] But when discussing solvability by radicals, going from $\mathbb{Q}$ to $\mathbb{R}$ and then to $\mathbb{C}$ is too much. Exercise 6-9 shows that most of the numbers we newly introduced in $\mathbb{R}$ are not needed to investigate roots of polynomials with integer coefficients. For solvability by radicals, we will thus focus only on the numbers that actually arise in a solution by radicals.

Simply speaking, solutions by radicals (or, evaluations of a "formula") rely on computations within a field, which are now and then interrupted by a root extraction. Because we work with polynomials in $\mathbb{Z}[x]$, the field we start with is $\mathbb{Q}$, the smallest field that contains $\mathbb{Z}$ (see Theorem 5.13). Because $\mathbb{Q}$ does not contain roots for all of its elements, the result of a root extraction may or may not be in $\mathbb{Q}$. When a root is not in $\mathbb{Q}$, we must extend the focus of our arithmetic to include that root. After the root extraction, we still want to keep working in a field, and we want this new field to be as small as possible. Moreover, the results of subsequent root extractions may not be in this new field. So every root extraction may require us to work in a larger field. Every time we expand our focus in this fashion, we want to use the smallest field that allows all the operations that we have carried out so far.

In this section, we will focus on how much root extractions will add to the "current" field, and on the smallest possible "new field" that we need after the root extraction. Because it is hard to start with a small field and "somehow" obtain a larger one (see the constructions of $\mathbb{R}$ and $\mathbb{Q}$, for example), we start with the definition of a subfield. For our investigation of quintics, this is not a problem, because all fields that we are interested in are subfields of $\mathbb{C}$.

Definition 6.6 *Let $(\mathbb{E}, +, \cdot)$ be a field and let $\mathbb{F} \subseteq \mathbb{E}$ be a subset so that the restricted operations $+|_{\mathbb{F} \times \mathbb{F}}$ and $\cdot|_{\mathbb{F} \times \mathbb{F}}$ both map into $\mathbb{F}$ and so that $\left(\mathbb{F}, +|_{\mathbb{F} \times \mathbb{F}}, \cdot|_{\mathbb{F} \times \mathbb{F}}\right)$ is a field whose identity element for addition is $0 \in \mathbb{E}$ and whose identity element with respect to multiplication is $1 \in \mathbb{E}$. Then $\mathbb{F}$ is called a **subfield** of $\mathbb{E}$ and $\mathbb{E}$ is called an **extension** of $\mathbb{F}$. We denote this situation more briefly by $\mathbb{F} \subseteq \mathbb{E}$.*

With this notion, we can give a precise definition of solvability by radicals. The definition is inspired by the step-by-step evaluation of "formulas."

---

[2]That was the reason to introduce complex numbers in the first place.

Definition 6.7 *Let* $(\mathbb{F}, +, \cdot)$ *be a field of characteristic* 0 *and let* $p \in \mathbb{F}[x]$ *be a polynomial of positive degree. Then the equation* $p(x) = 0$ *is* **solvable by radicals** *(over* $\mathbb{F}$*) iff all its solutions can be calculated from its coefficients and elements of* $\mathbb{F}$ *in a finite number of steps using field operations (addition, multiplication, additive and multiplicative inversion) and root extractions. The root extractions are allowed to yield elements that are in an extension* $\mathbb{E}$ *of* $\mathbb{F}$, *and not necessarily in* $\mathbb{F}$.

Note that the definition of solvability by radicals refers to a specific polynomial. We will ultimately prove that a specific polynomial equation is not solvable by radicals. This means that not only is there no quintic formula that would work for all polynomials (which is conceivable, because a method that works for one class of quintics may not work for another), but there is a *specific polynomial* whose roots cannot be represented in a finite number of steps that involve only field operations and root extractions.[3] In some ways this situation is similar to the fact that $\sqrt{2}$ is not rational. We can see the roots of the polynomial from Example 6.4 in Figure 6.1, just as we can "see $\sqrt{2}$ in front of us" as the hypothenuse of an isosceles right triangle whose other sides have length 1. But we cannot write a simple, finite, expression for either. In Exercises 6-5, 6-16, 6-31, 6-32 and 6-33, an approach similar to proving unsolvability by radicals will show that there is no straightedge-and-compass construction that trisects arbitrary angles.

To start our journey, we investigate the properties of field extensions. The results are all stated in terms of abstract fields. This is necessary, because in our constructions we will work with fields that arise through the abstract adjoining of new elements to another field. To better visualize and accept some of the hypotheses in the results, it is helpful to recall that, for the context of solvability of polynomial equations with integer coefficients by radicals, all fields in question will be subfields of $\mathbb{C}$. That is, for our main focus, there will always be a surrounding universe with very nice properties. So, for example, for our next result, the last part of the proof depends on everything being contained in a surrounding universe. But this is not a problem, because, for what we have in mind, there will be such a surrounding universe.

Proposition 6.8 *Let* $(\mathbb{E}, +, \cdot)$ *be a field and let* $\{\mathbb{F}_j\}_{j \in J}$ *be a family of subfields of* $\mathbb{E}$. *Then* $\bigcap_{j \in J} \mathbb{F}_j$ *is a subfield of* $\mathbb{E}$.

Proof. By assumption, $\bigcap_{j \in J} \mathbb{F}_j$ contains the two elements 0 and 1. Because the $\mathbb{F}_j$ are subfields, sums and products of elements of $\mathbb{F}_j$ are in $\mathbb{F}_j$, too. Hence the restricted operations $+|_{\bigcap_{j \in J} \mathbb{F}_j \times \bigcap_{j \in J} \mathbb{F}_j}$ and $\cdot|_{\bigcap_{j \in J} \mathbb{F}_j \times \bigcap_{j \in J} \mathbb{F}_j}$ both map into $\bigcap_{j \in J} \mathbb{F}_j$. Because the binary operations on $\bigcap_{j \in J} \mathbb{F}_j$ are the restrictions of the binary operations in $\mathbb{E}$, associativity and commutativity will hold for addition and multiplication, and distributivity of multiplication over addition will hold, too. Moreover, because 0 and 1 are identity elements for addition and multiplication, respectively, in $\mathbb{E}$, they will also be identity elements for addition and multiplication, respectively, in $\bigcap_{j \in J} \mathbb{F}_j$. For the existence of inverses, let $x \in \bigcap_{j \in J} \mathbb{F}_j$. Then x has a unique additive inverse $-x$ in $\mathbb{E}$.

---

[3] On the brighter side, Exercise 6-78e will show which polynomial equations are solvable by radicals. But we must be careful. This result will not mean that there is a quintic formula for these polynomial equations. It simply means that, for each of these polynomial equations, there is *some* way to express the solutions in finitely many steps that involve only field operations and radicals.

Let $j \in J$ and let y be the additive inverse of x in $\mathbb{F}_j$. Then

$$y = y + 0 = y + \big(x + (-x)\big) = (y + x) + (-x) = 0 + (-x) = -x.$$

Because $j \in J$ was arbitrary, we conclude that $-x \in \bigcap_{j \in J} \mathbb{F}_j$. Thus $\bigcap_{j \in J} \mathbb{F}_j$ contains additive inverses. The proof that the intersection contains multiplicative inverses is similar. $\blacksquare$

Proposition 6.8 says that intersections of subfields are subfields, too. Thus we can now think about the smallest subfield that contains certain elements, which is exactly what we need for solvability by radicals: We only want to add as many elements as absolutely necessary to the scope of our computations. Recall once more that, for the polynomials we are interested in, $\mathbb{C}$ is a surrounding universe in which the polynomial factors into linear factors.

Definition 6.9 *Let $(\mathbb{F}, +, \cdot)$ be a field, let $p \in \mathbb{F}[x]$ be a polynomial over $\mathbb{F}$ and let $\mathbb{E}$ be an extension of $\mathbb{F}$. Then f* **splits** *in the extension field $\mathbb{E} \supseteq \mathbb{F}$ iff p can be factored into linear factors with coefficients in $\mathbb{E}[x]$.*

Now let $\mathbb{F}$ be a field, let $p \in \mathbb{F}[x]$ and let $\mathbb{E}$ be an extension field in which p splits. Then the field

$$\mathbb{S} := \bigcap \{\mathbb{D} : \mathbb{D} \text{ is an extension field of } \mathbb{F}, \mathbb{D} \subseteq \mathbb{E}, p \text{ splits in } \mathbb{D}\}$$

is called the **splitting field** *for p over $\mathbb{F}$.*

The name "splitting field" is justified, because (see Exercise 6-10) the polynomial p splits in its splitting field.

Because in Definition 6.9 there is a formal dependence on the extension field $\mathbb{E}$, formally we would need to call $\mathbb{S}$ *a* splitting field of p over $\mathbb{F}$ *and in* $\mathbb{E}$. But Theorem 6.24 will show that any two splitting fields of a polynomial over $\mathbb{F}$ are isomorphic. Therefore no matter in what surrounding universe we construct a splitting field, it always "works the same way" and we might as well talk about *the* splitting field.

Constructively speaking, the splitting field should contain the zeros of p that are not in $\mathbb{F}$, plus the bare minimum of elements needed to get a field once more. Of course, for our idea of consecutive root extractions, we have specific elements that we want to add, and which are not necessarily roots of the polynomial under investigation. Either of the two observations above leads to the idea of adjoining finitely many elements to a field. In the remainder of this section we will think about adjoining the roots of the polynomial one by one. The idea of a root tower (see Definition 6.26) will get us back to thinking about solvability by radicals.

Definition 6.10 *Let $(\mathbb{F}, +, \cdot)$ be a field, let $\mathbb{E}$ be an extension of the field $\mathbb{F}$ and let $\theta_1, \ldots, \theta_n \in \mathbb{E} \setminus \mathbb{F}$. We define $\mathbb{F}(\theta_1, \ldots, \theta_n)$ to be the intersection of all subfields of $\mathbb{E}$ that contain $\mathbb{F}$ and $\theta_1, \ldots, \theta_n$. The field $\mathbb{F}(\theta_1, \ldots, \theta_n)$ is called the field $\mathbb{F}$ with the elements $\theta_1, \ldots, \theta_n$* **adjoined**.

Definition 6.10 does not look very constructive. But the next example and our construction of $\mathbb{C}$ in Section 5.7 show that there should be a constructive representation of the elements of a field that is obtained by adjoining new objects to a smaller field.

Example 6.11 $\mathbb{C} = \mathbb{R}(i)$, because $\mathbb{C}$ is an extension of $\mathbb{R}$ and every subfield of $\mathbb{C}$ that contains $\mathbb{R}$ and the imaginary unit i must contain the real and the imaginary numbers and all sums of real and imaginary numbers, that is, it must contain $\mathbb{C}$.   ∎

Adjoining elements is not always as easy as it is for the imaginary unit i. But the next result proves that it is close to that simple.

Theorem 6.12 *Let $(\mathbb{F}, +, \cdot)$ be a field, let $\mathbb{E}$ be an extension of the field $\mathbb{F}$ and let $\theta_1, \ldots, \theta_n \in \mathbb{E} \setminus \mathbb{F}$. Then the elements of $\mathbb{F}(\theta_1, \ldots, \theta_n)$ are rational combinations of the θ_j, where a rational combination is formed from elements of $\mathbb{F}$ and the $\theta_1, \ldots, \theta_n$ using sums, products, additive inversions and multiplicative inversions (except divisions by zero).*

Proof. We first claim that every rational combination r is a quotient of two polynomial combinations, that is, $r = \frac{p}{q}$, where p and q are formed using sums, products and additive inversions, and $q \neq 0$. This claim is proved by induction on the total number k of operations (sums, products, additive and multiplicative inversions) needed to form r. For $k = 0$, the claim is trivial, because $r \in \mathbb{F} \cup \{\theta_1, \ldots, \theta_n\}$ and $r = \frac{r}{1}$. For the induction step with $k > 0$, let r be a rational combination. We distinguish cases by what operation was performed last. If $r = r_1 + r_2$, where r_1 and r_2 are rational combinations, then both r_1 and r_2 were formed using fewer than k operations. By induction hypothesis, for $j = 1, 2$ we have $r_j = \frac{p_j}{q_j}$, where p_j and q_j are polynomial combinations. Now $r = r_1 + r_2 = \frac{p_1}{q_1} + \frac{p_2}{q_2} = \frac{p_1 q_2 + p_2 q_1}{q_1 q_2}$ is the quotient of two polynomial combinations. The arguments for $r = r_1 \cdot r_2, r = -r_1$ and $r = (r_1)^{-1}$ (for $r_1 \neq 0$) are similar (see Exercise 6-11).

It is now easy to verify that the rational combinations are a field and that every subfield of $\mathbb{E}$ that contains $\mathbb{F} \cup \{\theta_1, \ldots, \theta_n\}$ contains all rational combinations of the θ_j. (Exercise 6-12.)   ∎

Although rational combinations are a bit more complicated than what we did in the construction of $\mathbb{C}$, the next result confirms our earlier idea what a splitting field should look like: We add the zeros of p and the bare minimum number of elements needed to get a field.

Proposition 6.13 *Let $(\mathbb{F}, +, \cdot)$ be a field, let $p \in \mathbb{F}[x]$ be a polynomial over $\mathbb{F}$, let $\mathbb{E}$ be an extension of $\mathbb{F}$ in which p splits and let $\theta_1, \ldots, \theta_n \in \mathbb{E} \setminus \mathbb{F}$ be the zeros of p that are not in $\mathbb{F}$. Then $\mathbb{F}(\theta_1, \ldots, \theta_n)$ is the splitting field for p over $\mathbb{F}$.*

Proof. Let $a_d \in \mathbb{F}$ be the leading coefficient of p, let $\theta_1, \ldots, \theta_n$ be the zeros of p in $\mathbb{E} \setminus \mathbb{F}$, let m_j be the multiplicity of θ_j, let $\nu_1, \ldots, \nu_l$ be the zeros of p in $\mathbb{F}$ and let M_k be the multiplicity of ν_k. Then

$$p(x) = a_d \prod_{j=1}^{n} (x - \theta_j)^{m_j} \prod_{k=1}^{l} (x - \nu_k)^{M_k}.$$

Because $a_d, \theta_1, \ldots, \theta_n, \nu_1, \ldots, \nu_l \in \mathbb{F}(\theta_1, \ldots, \theta_n)$, the polynomial p splits in the field $\mathbb{F}(\theta_1, \ldots, \theta_n)$. Moreover, every field $\mathbb{G}$ with $\mathbb{F} \subseteq \mathbb{G} \subseteq \mathbb{E}$ in which p splits must contain

$\theta_1, \ldots, \theta_n$ and hence it must contain $\mathbb{F}(\theta_1, \ldots, \theta_n)$. Thus $\mathbb{F}(\theta_1, \ldots, \theta_n)$ is the splitting field for p over $\mathbb{F}$. ∎

Example 6.14 By Proposition 6.13, $\mathbb{Q}(\sqrt{2})$ is the splitting field for $p(x) = x^2 - 2$ over $\mathbb{Q}$. It is instructive to consider the elements of $\mathbb{Q}(\sqrt{2})$. By the proof of Theorem 6.12, because $(\sqrt{2})^2 = 2$, the elements of $\mathbb{Q}(\sqrt{2})$ are of the form $\frac{a+b\sqrt{2}}{c+d\sqrt{2}}$, with $a, b, c, d \in \mathbb{Q}$. In particular, for each of these elements we have $c \in \mathbb{Q}$ and $d\sqrt{2} \notin \mathbb{Q}$. Therefore $c \notin \{\pm d\sqrt{2}\}$, and hence $c^2 - 2d^2 \neq 0$. Therefore

$$\frac{a+b\sqrt{2}}{c+d\sqrt{2}} = \frac{a+b\sqrt{2}}{c+d\sqrt{2}} \cdot \frac{c-d\sqrt{2}}{c-d\sqrt{2}} = \frac{ac - 2bd + (bc - ad)\sqrt{2}}{c^2 - 2d^2}$$
$$= \frac{ac - 2bd}{c^2 - 2d^2} + \frac{bc - ad}{c^2 - 2d^2}\sqrt{2},$$

where $x := \frac{ac-2bd}{c^2-2d^2} \in \mathbb{Q}$ and $y := \frac{bc-ad}{c^2-2d^2} \in \mathbb{Q}$. Hence the elements of $\mathbb{Q}(\sqrt{2})$ are of the form $x + y\sqrt{2}$ with $x, y \in \mathbb{Q}$. ☐

Example 6.14 shows that the denominators in the rational combinations of Theorem 6.12 can be quite trivial. (For another example, consider Exercise 6-17.) It would be nice if we could find such a simpler representation for all fields to which elements are adjoined. Theorems 6.20 and 6.25 will show that this is possible when roots of polynomials are adjoined.

Exercises

6-9. A real number is called **algebraic** iff it is a zero of a polynomial with integer coefficients. Otherwise it is called **transcendental**.

    (a) Prove that the ring of polynomials $\mathbb{Z}[x]$ is countable.

    (b) Prove that the set of algebraic numbers is countable.

    (c) Prove that the set of transcendental numbers is uncountable.

6-10. Let $(\mathbb{F}, +, \cdot)$ be a field, let $p \in \mathbb{F}[x]$ be a polynomial with coefficients in $\mathbb{F}$ and let $\mathbb{S}$ be the splitting field for p over $\mathbb{F}$. Prove that p splits in $\mathbb{S}$.

6-11. Let $(\mathbb{F}, +, \cdot)$ be a field, let $\theta_1, \ldots, \theta_n$ be elements of an extension $\mathbb{E}$ of $\mathbb{F}$ and let rational and polynomial combinations of the $\theta_1, \ldots, \theta_n$ be defined as in Theorem 6.12 and its proof. Let $r_1 = \dfrac{p_1}{q_1}$ and $r_2 = \dfrac{p_2}{q_2}$ be quotients of polynomial combinations of the $\theta_1, \ldots, \theta_n$.

    (a) Prove that $r_1 \cdot r_2$ is a quotient of polynomial combinations of the $\theta_1, \ldots, \theta_n$.

    (b) Prove that $-r_1$ is a quotient of polynomial combinations of the $\theta_1, \ldots, \theta_n$.

    (c) Prove that, for $r_1 \neq 0$, $(r_1)^{-1}$ is a quotient of polynomial combinations of the $\theta_1, \ldots, \theta_n$.

6-12. Finish the proof of Theorem 6.12.

    (a) Prove that the rational combinations of the θ_j form a field.

    (b) Prove that every subfield of $\mathbb{E}$ that contains $\mathbb{F} \cup \{\theta_1, \ldots, \theta_n\}$ must contain the rational combinations of the θ_j.

    Hint. First prove (by induction on the number of needed operations) that it must contain the polynomial combinations.

6-13. Let $(\mathbb{F}, +, \cdot)$ be a field and let $\mathbb{D} \subseteq \mathbb{F}$. Prove that $\left(\mathbb{D}, +|_{\mathbb{D} \times \mathbb{D}}, \cdot|_{\mathbb{D} \times \mathbb{D}} \right)$ is a subfield of $\mathbb{F}$ iff

- $\mathbb{D}$ is not empty.

- $\mathbb{D}$ is closed under $+$ and $\cdot$, that is, for all $x, y \in \mathbb{D}$ we have that $x + y \in \mathbb{D}$ and $x \cdot y \in \mathbb{D}$.

- For all $x \in \mathbb{D}$ we have $-x \in \mathbb{D}$ and for all $x \in \mathbb{D} \setminus \{0\}$ we have that $x^{-1} \in \mathbb{D}$.

6-14. Prove that for every complex number $a + bi$ there is a polynomial p with real coefficients so that $p(a + bi) = 0$.

6-15. Let $(\mathbb{F}, +, \cdot)$ be a field, let $\mathbb{E}$ be an extension of $\mathbb{F}$ and let $\theta \in \mathbb{E}$ be a zero of a polynomial $p \in \mathbb{F}[x]$ with coefficients in $\mathbb{F}$. Prove that θ is a zero of an *irreducible* polynomial $q \in \mathbb{F}[x]$ with coefficients in $\mathbb{F}$.

6-16. **Constructible lengths in geometry**. A length is called constructible iff there is a procedure that can be carried out with straightedge and compass and which produces a segment of the desired length. By designating a length to be equal to 1, we can always construct the length 1.

  (a) Lengths can be duplicated by putting the points of the compass on the endpoints of a line segment of the requisite length and then placing the compass at the starting point of the desired duplicate. Prove that for each $n \in \mathbb{N}$ the length n is constructible.

  (b) Because lengths can be duplicated in any direction and perpendicular lines can be constructed with straightedge and compass, we can construct an $\mathbb{N}_0 \times \mathbb{N}_0$ grid. Prove that within this grid any length $l \in \mathbb{Q}$ with $l > 0$ is constructible.

  Hint. A line through two points can be drawn with a straightedge. For the length $l = \dfrac{a}{b}$, consider the triangle comprised of the line L through the points $(0, 0)$ and $(1, 0)$, the line perpendicular to L that goes through $(1, 0)$ and the line through the vertices $(0, 0)$ and (b, a).

  (c) Let $\mathbb{F}_0 := \mathbb{Q}$ and let $\mathbb{F}_j$ be a field that contains all lengths that have been constructed so far. Let L_1 be a line that is determined by the points $(x_{11}, y_{11}), (x_{12}, y_{12}) \in \mathbb{F}_j \times \mathbb{F}_j$ and let L_2 be a line that is determined by the points $(x_{21}, y_{21}), (x_{22}, y_{22}) \in \mathbb{F}_j \times \mathbb{F}_j$. Similarly, let C_1 be a circle with center $(h_1, k_1) \in \mathbb{F}_j \times \mathbb{F}_j$ and radius $r_1 \in \mathbb{F}_j$ and let C_2 be a circle with center $(h_2, k_2) \in \mathbb{F}_j \times \mathbb{F}_j$ and radius $r_2 \in \mathbb{F}_j$.

   i. Prove that the intersection of L_1 and L_2 is empty or in $\mathbb{F}_j \times \mathbb{F}_j$.

   ii. Prove that the intersection of L_1 and C_1 is empty or in $\mathbb{F}_j(w_{j+1})$ for some number w_{j+1} with $w_{j+1}^2 \in \mathbb{F}_j$.

   iii. Prove that the intersection of C_1 and C_2 is empty or in $\mathbb{F}_j(w_{j+1})$ for some number w_{j+1} with $w_{j+1}^2 \in \mathbb{F}_j$.

  Hint. For every part, write equations for the geometric objects and compute the intersection like you would in calculus. If two objects are equal, we consider their intersection to be in $\mathbb{F}_j \times \mathbb{F}_j$.

  (d) The only entities that would allow the construction of new lengths are new points that are constructed via intersections of lines or circles. Prove that for every constructible length c there are sequences $\mathbb{F}_1, \ldots, \mathbb{F}_k$ and $w_1, \ldots, w_k$ so that $\mathbb{F}_j = \mathbb{F}_{j-1}(w_j)$ and $w_j^2 \in \mathbb{F}_{j-1}$ and so that $c \in \mathbb{F}_k$.

6.3   Uniqueness of the Splitting Field

The solution of equations by radicals will be investigated in the splitting fields from Section 6.2. Therefore it is only appropriate to show that splitting fields are unique up to isomorphism, which we will do in this section. Example 6.14 has shown that the elements of some splitting fields have an even simpler representation than the rational combinations from Theorem 6.12. These simpler representations are the key to numerous results about splitting fields. We start with the idea of linear independence.

Definition 6.15 *Let* $(\mathbb{F}, +, \cdot)$ *be a field, let* $\mathbb{E}$ *be an extension of the field* $\mathbb{F}$ *and let* $a_1, \ldots, a_n \in \mathbb{E}$. *Then a sum* $\sum_{j=1}^{n} \lambda_j a_j$ *with* $\lambda_1, \ldots, \lambda_n \in \mathbb{F}$ *is called an* $\mathbb{F}$-**linear combination** *of* $a_1, \ldots, a_n \in \mathbb{E}$ *over* $\mathbb{F}$.

The set $\{a_1, \ldots, a_n\} \in \mathbb{E}$ *is called* **linearly independent** *over* $\mathbb{F}$ *iff the only choice of elements* $\lambda_1, \ldots, \lambda_n \in \mathbb{F}$ *so that* $\sum_{j=1}^{n} \lambda_j a_j = 0$ *is* $\lambda_1 = \cdots = \lambda_n = 0$. *The set* $\{a_1, \ldots, a_n\} \subseteq \mathbb{E}$ *is called a* **basis** *for* $\mathbb{E}$ *over* $\mathbb{F}$ *iff* $\{a_1, \ldots, a_n\}$ *is linearly independent over* $\mathbb{F}$ *and for each* $e \in \mathbb{E}$ *there are* $\lambda_1, \ldots, \lambda_n$ *so that* $\lambda_1 a_1 + \cdots + \lambda_n a_n = e$.

If there is a basis for $\mathbb{E}$ *over* $\mathbb{F}$, *then* $\mathbb{E}$ *is called a* **finite extension** *of* $\mathbb{F}$.

Example 6.16 $\mathbb{C}$ is a finite extension of $\mathbb{R}$, because $\{1, i\}$ is a basis of $\mathbb{C}$ over $\mathbb{R}$. □

Example 6.17 $\mathbb{R}$ is *not* a finite extension of $\mathbb{Q}$: If $\{a_1, \ldots, a_n\}$ was a basis of $\mathbb{R}$ over $\mathbb{Q}$, then $\mathbb{R}$ would be the set of all $\mathbb{Q}$-linear combinations of the a_j. But this set is countable and $\mathbb{R}$ is not. □

The similarity between Definition 6.15 and Definitions 5.72 and 5.73 is not accidental. If $\mathbb{E}$ has a basis B over $\mathbb{F}$, then we can interpret $\mathbb{E}$ as a vector space over $\mathbb{F}$ with base B: For addition, we use the addition in $\mathbb{E}$. For scalar multiplication of elements of $\mathbb{F}$ with elements of $\mathbb{E}$, we use the multiplication in $\mathbb{E}$. Of course the multiplication in $\mathbb{E}$ has more properties, because elements of $\mathbb{E}$ can be multiplied with each other, too. But there is no problem with using fewer properties than are available. In particular, we can use the results proved about vector spaces. Hence, by Theorem 5.75, all bases of a finite extension $\mathbb{E}$ of a field $\mathbb{F}$ have the same number of elements, which means the following definition is sensible.

Definition 6.18 *Let* $(\mathbb{F}, +, \cdot)$ *be a field and let* $\mathbb{E}$ *be a finite extension of* $\mathbb{F}$ *with basis* B. *If the basis* B *has* n *elements, we set* $[\mathbb{E} : \mathbb{F}] := n$ *and call it the* **index** *of* $\mathbb{E}$ *over* $\mathbb{F}$.

Just as for vector spaces, the coefficients in a basis representation are unique (see Exercise 6-18). Throughout our work with splitting fields, we will work with extensions of extensions, that is, with extensions that are nested into each other. The next result shows that the indices of the extensions behave as they should.

Theorem 6.19 *Let* $(\mathbb{F}, +, \cdot)$ *be a field, let* $\mathbb{E}$ *be a finite extension of* $\mathbb{F}$ *and let* $\mathbb{D}$ *be another extension of* $\mathbb{F}$ *so that* $\mathbb{F} \subseteq \mathbb{D} \subseteq \mathbb{E}$. *Then* $\mathbb{D}$ *is a finite extension of* $\mathbb{F}$, $\mathbb{E}$ *is a finite extension of* $\mathbb{D}$ *and* $[\mathbb{E} : \mathbb{F}] = [\mathbb{E} : \mathbb{D}][\mathbb{D} : \mathbb{F}]$.

Proof. Let $\{v_1, \ldots, v_k\}$ be a basis of $\mathbb{E}$ over $\mathbb{F}$. Then by Theorem 5.75 no $\mathbb{F}$-linearly independent subset of $\mathbb{D} \subseteq \mathbb{E}$ has more than k elements. Hence $\mathbb{D}$ is a finite extension of $\mathbb{F}$. Moreover, any $\mathbb{D}$-linearly independent subset of $\mathbb{E}$ is $\mathbb{F}$-linearly independent, too. Thus no $\mathbb{D}$-linearly independent subset of $\mathbb{E}$ can have more than k elements. Hence $\mathbb{E}$ is a finite extension of $\mathbb{D}$.

Now let $m := [\mathbb{E} : \mathbb{D}]$, let $n := [\mathbb{D} : \mathbb{F}]$, let $\{e_1, \ldots, e_m\}$ be a basis for $\mathbb{E}$ over $\mathbb{D}$ and let $\{d_1, \ldots, d_n\}$ be a basis for $\mathbb{D}$ over $\mathbb{F}$. Then $\{e_i d_j : i = 1, \ldots, m, j = 1, \ldots, n\}$ is $\mathbb{F}$-linearly independent: Indeed, let $\lambda_{ij} \in \mathbb{F}$, $i = 1, \ldots, m$, $j = 1, \ldots, n$ be so that

$$0 = \sum_{i=1}^{m} \sum_{j=1}^{n} \lambda_{ij} e_i d_j = \sum_{i=1}^{m} \underbrace{\left(\sum_{j=1}^{n} \lambda_{ij} d_j \right)}_{\in \mathbb{D}} e_i.$$

Then, because the e_i are $\mathbb{D}$-linearly independent, for all $i = 1, \ldots, m$ we have that

$$0 = \sum_{j=1}^{n} \underbrace{\lambda_{ij}}_{\in \mathbb{F}} d_j.$$

Now, because the d_j are $\mathbb{F}$-linearly independent, this means that for all $i = 1, \ldots, m$ and all $j = 1, \ldots, n$ we have that $\lambda_{ij} = 0$. Hence $\{e_i d_j : i = 1, \ldots, m, j = 1, \ldots, n\}$ is $\mathbb{F}$-linearly independent.

To see that the set $\{e_i d_j : i = 1, \ldots, m, j = 1, \ldots, n\}$ is a basis for $\mathbb{E}$ over $\mathbb{F}$, let $x \in \mathbb{E}$. Then there are $\mu_1, \ldots, \mu_m \in \mathbb{D}$ so that $x = \sum_{i=1}^{m} \mu_i e_i$. Moreover, for each index $i \in \{1, \ldots, m\}$ there are $\lambda_{ij} \in \mathbb{F}$ so that $\mu_i = \sum_{j=1}^{n} \lambda_{ij} d_j$. Hence we can represent x as

$$x = \sum_{i=1}^{m} \mu_i e_i = \sum_{i=1}^{m} \sum_{j=1}^{n} \lambda_{ij} d_j e_i.$$

Therefore $\{e_i d_j : i = 1, \ldots, m, j = 1, \ldots, n\}$ is a basis for $\mathbb{E}$ over $\mathbb{F}$, and the indices satisfy $[\mathbb{E} : \mathbb{F}] = mn = [\mathbb{E} : \mathbb{D}][\mathbb{D} : \mathbb{F}]$. ∎

The next result shows that Example 6.14 is not an accident. When a zero of an irreducible polynomial is adjoined to $\mathbb{F}$, then $\mathbb{F}(\theta)$ consists of linear combinations of powers of θ.

Theorem 6.20 *Let* $(\mathbb{F}, +, \cdot)$ *be a field, let* $\mathbb{E}$ *be an extension of* $\mathbb{F}$, *let* $p \in \mathbb{F}[x]$ *be an irreducible polynomial of degree n and let* $\theta \in \mathbb{E}$ *be a zero of p. Then* $\{1, \theta, \ldots, \theta^{n-1}\}$ *is a basis for* $\mathbb{F}(\theta)$ *over* $\mathbb{F}$.

Proof. Let $z \in \mathbb{F}(\theta)$. By Theorem 6.12, z is a rational combination of θ with elements of $\mathbb{F}$. From the proof of Theorem 6.12 we obtain that z is a quotient $\frac{m(\theta)}{d(\theta)}$, where m and d are polynomials in $\mathbb{F}[x]$. Moreover, because $p(\theta) = 0$ and because of the division algorithm for polynomials, the degrees of m and d can be chosen to be less than n (divide numerator and denominator by p and continue with the remainders of these divisions). Finally, because $d(\theta)$ is the denominator, we have that $d(\theta) \neq 0$.

Because p is irreducible in $\mathbb{F}$, it has no nontrivial factors (in $\mathbb{F}[x]$) of degree less than n. Hence the only common divisors of d and p are the constant polynomials. By Euclid's Algorithm, there are polynomials $s, t \in \mathbb{F}[x]$ so that $s(x)d(x) + t(x)p(x) = 1$. But then $s(\theta)d(\theta) = s(\theta)d(\theta) + 0 = s(\theta)d(\theta) + t(\theta)p(\theta) = 1$, which means that $s(\theta)$ is the multiplicative inverse of $d(\theta)$. Hence

$$z = \frac{m(\theta)}{d(\theta)} = m(\theta)\big(d(\theta)\big)^{-1} = m(\theta)s(\theta),$$

so z is a linear combination of powers of θ. Once more by the division algorithm for polynomials, there are polynomials $a, r \in \mathbb{F}[x]$ so that $m(x)s(x) = a(x)p(x) + r(x)$ and so that the degree of r is less than n. But then $z = a(\theta)p(\theta) + r(\theta) = r(\theta)$ is a linear combination of $1, \theta, \ldots, \theta^{n-1}$.

Finally, suppose for a contradiction that $\{1, \theta, \ldots, \theta^{n-1}\}$ is not linearly independent. Then there is a polynomial $q \in \mathbb{F}[x] \setminus \{0\}$ of degree less than n so that $q(\theta) = 0$.

Choose $q \in \mathbb{F}[x] \setminus \{0\}$ so that $q(\theta) = 0$ and so that no polynomial in $\mathbb{F}[x] \setminus \{0\}$ with degree less than $\deg(q)$ has θ as a zero. By the division algorithm for polynomials, there are polynomials $b, c \in \mathbb{F}[x]$ so that $p(x) = b(x)q(x) + c(x)$, the degree of c is less than that of q and $c \neq 0$, because p is irreducible. But then $0 = p(\theta) = b(\theta)q(\theta) + c(\theta) = c(\theta)$, a contradiction to the choice of q. ∎

Now we are ready to prove that the splitting field of a polynomial over $\mathbb{F}$ is unique up to isomorphism. Because field isomorphisms will play an important role, we state the definition explicitly.

Definition 6.21 *Let $(\mathbb{F}, +, \cdot)$ and $(\mathbb{E}, +, \cdot)$ be fields and let $\sigma : \mathbb{F} \to \mathbb{E}$ be a function. Then σ is called a **field isomorphism** iff σ is bijective and for all $a, b \in \mathbb{F}$ we have $\sigma(a+b) = \sigma(a)+\sigma(b)$ and $\sigma(a \cdot b) = \sigma(a) \cdot \sigma(b)$. In case $\mathbb{F} = \mathbb{E}$, a field isomorphism $\sigma : \mathbb{F} \to \mathbb{E}$ will also be called a **field automorphism**.*

To prove the uniqueness of the splitting field, we need Lemmas 6.22 and 6.23 below. Although Theorem 6.24 is a quick consequence of Lemma 6.23, it is necessary to use two "starting fields" $\mathbb{F}_1$ and $\mathbb{F}_2$ in Lemmas 6.22 and 6.23 and an isomorphism between them. The reason is that the induction step in the proof of Lemma 6.23 can have two different "starting fields." Even if this was not the case, Lemmas 6.22 and 6.23 will be needed in their present forms in Section 6.5.

If Lemmas 6.22 and 6.23 feel a bit too technical, it may help to read them first assuming that $\mathbb{F}_1 = \mathbb{F}_2$. In general, it pays to not be unduly impressed with isomorphisms. All they say is that operations work out the same way on either side of the isomorphism.

Lemma 6.22 *Let $(\mathbb{F}_1, +, \cdot)$ and $(\mathbb{F}_2, +, \cdot)$ be fields, let $\varphi : \mathbb{F}_1 \to \mathbb{F}_2$ be a field isomorphism, let $\mathbb{E}_1, \mathbb{E}_2$ be extensions of $\mathbb{F}_1$ and $\mathbb{F}_2$, respectively, and let $p_1 \in \mathbb{F}_1[x]$ be an irreducible polynomial over $\mathbb{F}_1$. Let p_2 be the polynomial over $\mathbb{F}_2$ obtained by replacing the coefficients of p_1 with their images under φ. If θ_1 is a root of p_1 in $\mathbb{E}_1$ and θ_2 is a root of p_2 in $\mathbb{E}_2$, then there is a field isomorphism $\varphi^* : \mathbb{F}_1(\theta_1) \to \mathbb{F}_2(\theta_2)$ so that $\varphi^*(\theta_1) = \theta_2$ and so that for all $z \in \mathbb{F}_1$ we have that $\varphi^*(z) = \varphi(z)$.*

Proof. Let n be the degree of p_1. Then n is also the degree of p_2. Moreover, p_2 is irreducible over $\mathbb{F}_2$, because any nontrivial factorization of p_2 in $\mathbb{F}_2[x]$ could be mapped back to a nontrivial factorization of p_1 in $\mathbb{F}_1[x]$ (see Exercise 6-19). By Theorem 6.20, $\{1, \theta_1, \ldots, \theta_1^{n-1}\}$ is a basis for $\mathbb{F}_1(\theta_1)$ over $\mathbb{F}_1$ and $\{1, \theta_2, \ldots, \theta_2^{n-1}\}$ is a basis for $\mathbb{F}_2(\theta_2)$ over $\mathbb{F}_2$. The function

$$\varphi^* \left(\sum_{j=0}^{n-1} a_j \theta_1^j \right) := \sum_{j=0}^{n-1} \varphi(a_j) \theta_2^j$$

is the desired isomorphism (see Exercise 6-20). ∎

Lemma 6.23 *Let $(\mathbb{F}_1, +, \cdot)$ and $(\mathbb{F}_2, +, \cdot)$ be fields, let $\varphi : \mathbb{F}_1 \to \mathbb{F}_2$ be a field isomorphism, and let $p_1 \in \mathbb{F}_1[x]$ be a nonconstant polynomial over $\mathbb{F}_1$. Let p_2 be the polynomial over $\mathbb{F}_2$ obtained by replacing the coefficients in p_1 with their images under φ, and let $\mathbb{S}_1$ and $\mathbb{S}_2$ be splitting fields of p_1 over $\mathbb{F}_1$ and of p_2 over $\mathbb{F}_2$, respectively. Then there is a field isomorphism $\Phi : \mathbb{S}_1 \to \mathbb{S}_2$ so that for all $z \in \mathbb{F}_1$ we have $\Phi(z) = \varphi(z)$.*

Proof. The proof is an induction on the index $[\mathbb{S}_1 : \mathbb{F}_1]$. If $[\mathbb{S}_1 : \mathbb{F}_1] = 1$, then $\mathbb{S}_1 = \mathbb{F}_1$, that is, p_1 splits over $\mathbb{F}_1$ as $p_1(x) = \prod_{j=1}^{n}(x - a_j)$. Therefore p_2 splits over $\mathbb{F}_2$ as $p_2(x) = \prod_{j=1}^{n}\left(x - \varphi(a_j)\right)$, which means that $\mathbb{S}_1$ is isomorphic to $\mathbb{S}_2$.

For the induction step, let $[\mathbb{S}_1 : \mathbb{F}_1] = n > 1$ and assume that the result holds when $[\mathbb{S}_1 : \mathbb{F}_1] < n$. Let q_1 be an irreducible factor of p_1 over $\mathbb{F}_1$. (If p_1 is irreducible, we choose $q_1 = p_1$.) Let q_2 be the polynomial over $\mathbb{F}_2$ obtained by replacing the coefficients in q_1 with their images under φ. Then the field $\mathbb{S}_1$ contains a root θ_1 of q_1 and the field $\mathbb{S}_2$ contains a root θ_2 of q_2. By Lemma 6.22, there is a field isomorphism $\varphi^* : \mathbb{F}_1(\theta_1) \rightarrow \mathbb{F}_2(\theta_2)$ so that $\varphi^*(\theta_1) = \theta_2$ and so that for all $z \in \mathbb{F}_1$ we have that $\varphi^*(z) = \varphi(z)$. By the definition of splitting fields, $\mathbb{S}_1$ is also a splitting field of p_1 over $\mathbb{F}_1(\theta_1)$ and $\mathbb{S}_2$ is a splitting field of p_2 over $\mathbb{F}_2(\theta_2)$. By Theorem 6.19 we have that $\left[\mathbb{S}_1 : \mathbb{F}_1(\theta_1)\right] = \frac{[\mathbb{S}_1:\mathbb{F}_1]}{\left[\mathbb{F}_1(\theta_1):\mathbb{F}_1\right]} < n$. Hence by induction hypothesis there is a field isomorphism $\Phi : \mathbb{S}_1 \rightarrow \mathbb{S}_2$ with $\Phi(z) = \varphi^*(z)$ for all $z \in \mathbb{F}_1(\theta_1)$, which means in particular that $\Phi(z) = \varphi(z)$ for all $z \in \mathbb{F}_1$. ∎

Theorem 6.24 *Let $(\mathbb{F}, +, \cdot)$ be a field and let $p \in \mathbb{F}[x]$ be a polynomial over $\mathbb{F}$. Then the splitting field for p over $\mathbb{F}$ is unique up to isomorphism.*

Proof. Use $\mathbb{F}_1 := \mathbb{F}_2 := \mathbb{F}$ and $\varphi = \mathrm{id}_{\mathbb{F}}$ in Lemma 6.23. ∎

The proof of Theorem 6.24 shows that the representation of elements as $\mathbb{F}$-linear combinations of powers of *one* new element, as in Theorem 6.20, is very useful. It turns out that this representation is also available when more than one root of a polynomial is adjoined to the field.

Theorem 6.25 *Let $\mathbb{F}$ be a field of characteristic 0, let $\mathbb{E}$ be an extension of $\mathbb{F}$ and let $\theta_1, \ldots, \theta_k \in \mathbb{E}$ be zeros of polynomials in $\mathbb{F}[x]$. Then there is a $\theta \in \mathbb{E}$ so that $\mathbb{F}(\theta_1, \ldots, \theta_k) = \mathbb{F}(\theta)$. Moreover, every $z \in \mathbb{F}(\theta)$ is a zero of a polynomial in $\mathbb{F}[x]$.*

Proof. Because $\theta_1, \ldots, \theta_k$ can be adjoined one by one (see Exercise 6-21), it is enough to prove the result for $k = 2$. To simplify notation, let α and β be the adjoined elements and let $p_\alpha, p_\beta \in \mathbb{F}[x]$ be so that $p_\alpha(\alpha) = 0$ and $p_\beta(\beta) = 0$. By Exercise 6-22b we can assume that p_α and p_β both are irreducible and that both have leading coefficient 1. By Exercise 6-23, the zeros $\alpha = \alpha_1, \alpha_2, \ldots, \alpha_n$ of p_α (which are in some splitting field for p_α) are distinct, and the zeros $\beta = \beta_1, \beta_2, \ldots, \beta_{n'}$ of p_β (which are in some splitting field for p_β) are distinct. Because $\mathbb{F}$ has characteristic 0, there is an element $a \in \mathbb{F} \setminus \{0\}$ so that $a \neq \frac{\alpha_i - \alpha_j}{\beta_l - \beta_m}$ for all $i \neq j$ and $l \neq m$.

Now let $\theta := \alpha + a\beta$. Then $\theta \in \mathbb{F}(\alpha, \beta)$, so $\mathbb{F}(\theta) \subseteq \mathbb{F}(\alpha, \beta)$.

To prove the reverse inclusion, we first prove that $\beta \in \mathbb{F}(\theta)$. Define the polynomial $h(x) := p_\alpha(\theta - ax)$. Then $h \in \mathbb{F}(\theta)[x]$ and $h(\beta) = p_\alpha(\theta - a\beta) = p_\alpha(\alpha) = 0$. Hence h and p_β have β as a common zero. Let g be a greatest common divisor of h and p_β in $\mathbb{F}(\theta)[x]$ and assume that the leading coefficient of g is 1. By Euclid's algorithm, there are polynomials $a, b \in \mathbb{F}(\theta)[x]$ so that $g(x) = a(x)h(x) + b(x)p_\beta(x)$. Arguing in $\mathbb{F}(\alpha, \beta)$ for a moment, we note that $g(\beta) = a(\beta)h(\beta) + b(\beta)p_\beta(\beta) = 0$. This means that if g was constant, it would be equal to zero. Because g cannot be the constant polynomial 0, we infer that g has positive degree. Moreover, $x - \beta$ divides g in $\mathbb{F}(\alpha, \beta)[x]$.

Now suppose for a contradiction that some $x - \beta_j$ with $j > 1$ also divides $h(x)$. Then $0 = h(\beta_j) = p_\alpha(\theta - a\beta_j)$, so there is an $i > 1$ so that $\alpha_i = \theta - a\beta_j$. We conclude $\theta = \alpha_i + a\beta_j$ and, by definition, $\theta = \alpha + a\beta$, which leads to the contradiction $a = \frac{\alpha - \alpha_i}{\beta_j - \beta}$. Hence no other factor $x - \beta_j$ divides both h and p_β. Because all factors of p_β are products of factors $(x - \beta_j)$, we conclude that g is a power of $x - \beta$. By Exercise 6-23, all factors of p_β have multiplicity 1. Hence $g(x) = x - \beta$. But then, using $a, b \in \mathbb{F}(\theta)[x]$ from above, $\beta = -g(0) = -a(0)h(0) - b(0)p_\beta(0) \in \mathbb{F}(\theta)$.

Moreover, $\alpha = \theta - a\beta \in \mathbb{F}(\theta)$. We conclude $\mathbb{F}(\alpha, \beta) \subseteq \mathbb{F}(\theta)$, and therefore $\mathbb{F}(\alpha, \beta) = \mathbb{F}(\theta)$.

The fact that every $z \in \mathbb{F}(\theta)$ is a zero of a polynomial in $\mathbb{F}[x]$ is proved in Exercise 6-26. ∎

Together with Proposition 6.13, Theorem 6.25 assures that every splitting field over a field of characteristic 0 can be obtained by adjoining a single element. This insight will simplify many abstract arguments about splitting fields. In the proof of Theorem 6.25 we had to use, for the first time in this chapter, that the field $\mathbb{F}$ is of characteristic 0. For our investigation of polynomial equations for polynomials in $\mathbb{Z}[x]$, this is not a problem, because the rational numbers have characteristic 0. In the future, we will feel free to demand characteristic 0 when needed. **But, in a general investigation, fields of characteristic p $\neq$ 0 are interesting, too. Therefore, we will keep track of when we use characteristic 0 and how. Although this will make us work a little harder, it is a good exercise for at least two reasons. First, analyzing a proof to determine where and how certain hypotheses were used will make us even more familiar with the proof. The following results are quite demanding, so more familiarity with the proofs certainly is desirable. Second, you may ultimately revisit the results presented here in an abstract algebra course. In such a course, the results are typically stated with minimal hypotheses. So tracking how we used the hypothesis that our field has characteristic 0 will show which proofs are valid in a general presentation and which proofs would need to be replaced. It turns out that, in this presentation, characteristic 0 is only required for Corollary 4.59, for Theorem 6.25 and for guaranteeing that the zeros of irreducible polynomials are distinct (see Exercise 6-23). Two of these three cornerstones of our development do not truly require characteristic 0: When working with formal polynomials, the result of Corollary 4.59 is part of the definition, so it's a given. Moreover, Theorem 6.25 can also be proved without requiring $\mathbb{F}$ to have characteristic 0. (We will not do so here, because it would require a substantial excursion into group theory.) Thus, any proofs that only refer to Corollary 4.59 and Theorem 6.25 generalize directly to fields of arbitrary characteristic. Only the distinctness of the roots of irreducible polynomials (Exercise 6-23) requires characteristic 0 as a hypothesis (see Exercise 6-24). So the only proofs that would need to be replaced in a general presentation for arbitrary characteristic are those that use that roots of irreducible**

polynomials are distinct.

In conclusion, this section has shown that as zeros of polynomials are adjoined to a field, there is always a *unique* smallest possible field that contains the original field and the new zero. Moreover, in case of characteristic 0, this field has a pretty simple representation. Coming back to solvability by radicals, we note that this idea can be used to follow the computation of the solutions of an equation $p(\theta) = 0$. Steps in a solution by radicals that involve field operations do not require any changes to the underlying field, but root extractions might. Essentially, a root is a solution of the polynomial equation $\theta^n - f = 0$ for some element f in the field that currently "houses" the computation. If the equation $\theta^n - f = 0$ has no solutions in the current field, then a solution must be adjoined to the field to continue the computation. By adding only what is necessary at each step, we construct a nested sequence of fields that contain the original field and which (as we will see in the following sections) will allow further inferences about the splitting field of the polynomial. The properties of this nested sequence of fields will ultimately allow us to prove that there are polynomial equations that are not solvable by radicals. The needed formal structure is that of a root tower.

Definition 6.26 *Let $(\mathbb{F}, +, \cdot)$ be a field of characteristic 0 and let $p \in \mathbb{F}[x]$ be a polynomial of positive degree so that the equation $p(x) = 0$ is solvable by radicals. Then there is a sequence of field extensions $\mathbb{F} = \mathbb{F}_1 \subseteq \mathbb{F}_2 \subseteq \cdots \subseteq \mathbb{F}_q$ and $n_1, \ldots, n_{q-1} \in \mathbb{N} \setminus \{1\}$ so that for $j = 2, \ldots, q$ we have $\mathbb{F}_j = \mathbb{F}_{j-1}(w_j)$, where $w_j^{n_{j-1}} = f_{j-1} \in \mathbb{F}_{j-1}$ and $\mathbb{F}_q$ contains all solutions of $p(x) = 0$. The sequence of field extensions is called a **root tower** for $\mathbb{F}_q$ over $\mathbb{F}$.*

In a root tower, field extensions are nested into each other. The next section will investigate how simpler structures can be associated with nested field extensions. The properties of these simpler structures will ultimately allow us to decide if a polynomial equation is solvable by radicals.

Note that, although we are not quite halfway through the proof that certain polynomial equations are not solvable by radicals, we already have enough machinery to prove that there is no straightedge-and-compass construction to trisect angles (see Exercise 6-33f). That is quite impressive, considering that the question if there is such a construction was posed by the ancient Greeks and remained unsolved for close to 2000 years.

Exercises

6-17. Prove directly that $\left\{ a + b\sqrt[3]{2} + c\left(\sqrt[3]{2}\right)^2 \in \mathbb{C} : a, b, c \in \mathbb{Q} \right\}$ is a subfield of $\mathbb{C}$. You may use Exercise 6-13 but not Theorem 6.20.

   Hint. If you get stuck on multiplicative inverses, consider the proof of Theorem 6.20.

6-18. Let $\mathbb{F}$ be a field, let $\mathbb{E} \supseteq \mathbb{F}$ be an extension field of $\mathbb{F}$ and let $\{a_1, \ldots, a_n\} \subseteq \mathbb{E}$ be a basis of $\mathbb{E}$ over $\mathbb{F}$. Prove that for each $e \in \mathbb{E}$ the coefficients $\lambda_j \in \mathbb{F}$ so that $e = \sum_{j=1}^{n} \lambda_j a_j$ are unique.

6-19. Let $\mathbb{F}_1$ and $\mathbb{F}_2$ be fields and let the function $\varphi : \mathbb{F}_1 \to \mathbb{F}_2$ be a field isomorphism. Define the function $\tilde{\varphi} : \mathbb{F}_1[x] \to \mathbb{F}_2[x]$ by $\tilde{\varphi}\left(\sum_{k=0}^{n} c_k x^k \right) := \sum_{k=0}^{n} \varphi(c_k) x^k$

     (a) Prove that $\tilde{\varphi}$ is a ring isomorphism from $\mathbb{F}_1[x]$ to $\mathbb{F}_2[x]$.

        Hint. For $\tilde{\varphi}(pq) = \tilde{\varphi}(p)\tilde{\varphi}(q)$ consider Exercise 4-7(b)i.

     (b) Prove that $p \in \mathbb{F}_1[x]$ is reducible in $\mathbb{F}_1[x]$ iff $\tilde{\varphi}(p)$ is reducible in $\mathbb{F}_2[x]$.

6-20. Prove that the function $\varphi^* : \mathbb{F}_1(\theta_1) \to \mathbb{F}_2(\theta_2)$ in the proof of Lemma 6.22 is an isomorphism.

6-21. Let $(\mathbb{F}, +, \cdot)$ be a field, let $\mathbb{E}$ be an extension field of $\mathbb{F}$ and let $\theta_1, \ldots, \theta_n \in \mathbb{E}$. Prove that for all $k \in \{2, \ldots, n\}$ we have that $\mathbb{F}(\theta_1, \ldots, \theta_k) = \left(\mathbb{F}(\theta_1, \ldots, \theta_{k-1}) \right)(\theta_k)$.

6-22. **A "minimality property" of irreducible polynomials.** Let $(\mathbb{F}, +, \cdot)$ be a field, let $\mathbb{E}$ be an extension of $\mathbb{F}$ and let $\theta \in \mathbb{E}$ be a zero of a polynomial $f \in \mathbb{F}[x] \setminus \{0\}$.

     (a) Prove that there is an irreducible polynomial $q \in \mathbb{F}[x]$ so that $q(\theta) = 0$.

        Hint. Induction on the degree of f.

     (b) Prove that if a is the leading coefficient of q and we set $p(x) := \dfrac{1}{a}q(x)$, then p is the only irreducible polynomial with leading coefficient 1 so that $p(\theta) = 0$.

        Hint. For a contradiction, suppose there are two, suppose that one of them has smallest possible degree and construct one with smaller degree.

     (c) Prove that p divides every polynomial $f \in \mathbb{F}[x]$ that satisfies $f(\theta) = 0$.

        Hint. Proof by contradiction. Construct an irreducible polynomial with degree smaller than that of p, which has θ as a zero.

6-23. **Zeros of irreducible polynomials are distinct if we demand characteristic 0.** Let $\mathbb{F}$ be a field of characteristic 0 and let $p \in \mathbb{F}[x]$ be irreducible with leading coefficient 1. Prove that the zeros of p must be distinct.

     Hint. For a contradiction, suppose otherwise, that is, suppose $p(x) = (x - \theta)b(x)$ and θ is a zero of b. Prove that $b(x) = x^{n-1} + \sum_{j=0}^{n-2} b_j(\theta)x^j$, where the degree of b_j in θ is $n - 1 - j$ (start with $j = n - 2$ and work backwards to $j = 0$) and the coefficients of b_j are in $\mathbb{F}$. Conclude that $b(\theta) = 0$ implies that θ is a zero of a polynomial in $\mathbb{F}[x]$ of degree $n - 1$, contradicting Exercise 6-22c.

6-24. **In general, zeros of irreducible polynomials need not be distinct.** For any variable y, let $R_p[y]$ be the field of formal rational functions in the variable y with coefficients in $\mathbb{Z}_p$ for p a prime number (see Exercise 5-16). Let a be a variable and consider the field $R_p[a]$.

     (a) Prove that $R_p\left[a^p\right]$, the set of formal rational functions in which the powers of the variable are multiples of p, is a subfield of $R_p[a]$.

     (b) Prove that if $\mathbb{F}$ is a field of characteristic $p > 0$ and $a \in \mathbb{F}$ is an arbitrary element of $\mathbb{F}$, then the polynomial $f(x) = x^p - a^p$ in $\mathbb{F}[x]$ is equal to $g(x) = (x - a)^p$.

        Hint. Binomial Theorem and Exercises 3-68b and 5-13.

     (c) Prove that the polynomial $f(x) = x^p - a^p$ is irreducible in $R_p\left[a^p\right][x]$, the ring of polynomials with coefficients in $R_p\left[a^p\right]$.

        Hint. Use that all factors in $R_p\left[a^p\right][x]$ are products of factors in the surrounding ring $R_p[a][x]$.

     Note. This proves that $f(x) = x^p - a^p$ is an irreducible polynomial that has repeated zeros. So the hypothesis in Exercise 6-23 that the underlying field has characteristic 0 cannot be dropped.

6-25. **Zeros of irreducible polynomials revisited.** Let $\mathbb{F}$ be a field and let $f(x) = \sum_{j=0}^{n} a_j x^j$ be any polynomial in $\mathbb{F}[x]$. Define the (formal) **derivative** of f to be $f'(x) = \sum_{j=1}^{n} j a_j x^{j-1}$.

     Note. At this stage, this is a purely formal definition, because we have not introduced any analysis (and will not do so). Nonetheless, this formal derivative will (naturally) act like the derivative we know from calculus. We will establish the necessary facts and put them to good use.

(a) Prove that for any $f, g \in \mathbb{F}[x]$ and $a, b \in \mathbb{F}$ we have $(af + bg)' = af' + bg'$.

(b) Prove that for any $f, g \in \mathbb{F}[x]$ we have $(fg)' = f'g + g'f$.

(c) Prove that a polynomial $f \in \mathbb{F}[x]$ has repeated roots, which may lie in an extension field $\mathbb{E}$, iff f and f' have a nonconstant common divisor in $\mathbb{F}[x]$.

   Hint. For "$\Rightarrow$" suppose the polynomials had only constant common divisors in $\mathbb{F}[x]$. Then $s(x)f(x) + t(x)f'(x) = 1$ in $\mathbb{E}[x]$. Now use the representation $f(x) = (x - a)^2 g(x)$ in $\mathbb{E}[x]$ and the product rule from part 6-25b to obtain a contradiction. For "$\Leftarrow$," let f, f' have a nonconstant greatest common divisor $g \in \mathbb{F}[x]$. Use the product rule from part 6-25b to prove that every zero of g must be a repeated zero of f.

(d) Prove that an irreducible polynomial $f \in \mathbb{F}[x]$ has repeated zeros iff $f'(x) = 0$.

   Hint. Exercise 6-22c and part 6-25c.

(e) Prove that if the characteristic of $\mathbb{F}$ is zero and $f \in \mathbb{F}[x]$ is an irreducible polynomial, then f does not have any repeated zeros.

(f) Prove that if the characteristic of $\mathbb{F}$ is $p > 0$ and $f \in \mathbb{F}[x]$ is an irreducible polynomial, then all powers of x in f must be of the form x^{np} with $n \in \mathbb{N}_0$.

   Note. The above shows that irreducible polynomials with repeated roots are of a very special form indeed.

6-26. Prove that every $z \in \mathbb{F}(\theta)$ in Theorem 6.20 is a zero of a polynomial in $\mathbb{F}[x]$.

   Hint. Use that some set $\{1, z, \ldots, z^n\}$ is linearly dependent over $\mathbb{F}$.

6-27. Let B be a basis of $\mathbb{E}$ over $\mathbb{F}$ and let $L \subseteq \mathbb{E}$ be a linearly independent set with $|L| = |B|$. Prove that L is a basis of $\mathbb{E}$.

   Hint. Proof by contradiction. Use Exercise 5-70a to construct a basis for $\mathbb{E}$ with more than $|B|$ elements.

6-28. Find the simplest possible representation for the elements of the splitting field of $p(x) = x^2 - 3$ over $\mathbb{Q}$.

6-29. Let $\mathbb{F}$ be a field, let $\mathbb{E}$ be an extension of $\mathbb{F}$ and let $\theta \in \mathbb{E}$ be so that there is a polynomial $p \in \mathbb{F}[x]$ so that $p(\theta) = 0$. Let q be a polynomial so that $q(\theta) = 0$ and so that θ is not a root of any polynomial whose degree is smaller than that of q. Prove that if $\tilde{q}$ is another polynomial so that $\tilde{q}(\theta) = 0$, then $\tilde{q} = fq$ for some polynomial $f \in \mathbb{F}[x]$.

   Hint. Divide by q and analyze the remainder.

   Note. Let a be the leading coefficient of q. The polynomial $\dfrac{q}{a}$ is called the **minimal polynomial** of θ over $\mathbb{F}$.

6-30. Let $\mathbb{F}$ be a field of characteristic 0, let $\mathbb{E}$ be an extension of $\mathbb{F}$ so that $[\mathbb{E} : \mathbb{F}] = n$. Prove that there is a $\theta \in \mathbb{E}$ so that $\mathbb{E} = \mathbb{F}(\theta)$ and θ is a zero of a polynomial in $\mathbb{F}[x]$.

6-31. Let $\mathbb{F}_0$ be a field and let there be a sequence $w_1, \ldots, w_k$ so that $\mathbb{F}_j = \mathbb{F}_{j-1}(w_j)$ and $w_j^2 \in \mathbb{F}_{j-1}$. (Compare with Exercise 6-16d.) Prove that $[\mathbb{F}_k : \mathbb{F}_0]$ is a power of 2.

6-32. Let c be a zero of the irreducible polynomial $f(x) = 8x^3 - 6x - 1$ from Exercise 6-5. Prove that the index of $\mathbb{Q}(c)$ over $\mathbb{Q}$ is $\left[\mathbb{Q}(c) : \mathbb{Q} \right] = 3$.

   Hint. Suppose the index was 2 and use Exercise 6-22c.

6-33. **Trisection of angles with straightedge and compass is impossible.** (In this exercise we assume knowledge of the sine and cosine function and of their properties.)

(a) Prove that equilateral triangles can be constructed with straightedge and compass.

(b) Prove that $60°$ angles can be constructed with straightedge and compass.

(c) Prove that if there was a straightedge and compass construction to trisect angles, then the length $\cos(20°)$ could be constructed.

(d) Prove that $\cos(3\alpha) = 4\cos^3(\alpha) - 3\cos(\alpha)$ for all $\alpha \in \mathbb{R}$.

   Hint. Use the identities for $\cos(x + y)$ and $\sin(x + y)$, as well as the Pythagorean identity for sine and cosine.

(e) Prove that $\cos(20°)$ must be a zero of the polynomial $f(x) = 8x^3 - 6x - 1$ from Exercise 6-5.

Hint. Use the known value for $\cos(60°)$.

(f) Use Exercises 6-31 and 6-32 to prove that the length $\cos(20°)$ is not constructible. Conclude that angles cannot be trisected with straightedge and compass.

Note. This concludes the answer to question 10 on page x.

6.4 Field Automorphisms and Galois Groups

Field extensions, splitting fields and root towers provide the framework in which to track how the underlying field expands when constructing a solution by radicals for an equation $p(\theta) = 0$. Proposition 6.13 and Theorems 6.20, 6.24 and 6.25 show that the new information is contained in a finite number of new elements and, as noted after the proof of Theorem 6.25, that the elements can be obtained by adjoining a single new element. To reflect this "finitary" nature of the splitting fields, it would be nice to remove some of the extra detail, and to obtain a finite structure associated with finite field extensions. This structure is the group of field automorphisms of the field extension $\mathbb{E}$, which fix the field $\mathbb{F}$. By fixing $\mathbb{F}$, we fix so much of the essential structure of $\mathbb{E}$, that the resulting automorphism group is finite. Once we are more familiar with these groups, we can more closely follow the effects of adjoining roots during a solution by radicals.

We start with a characterization of subgroups. By Exercise 4-6b, the bijective self maps $\sigma : \mathbb{E} \to \mathbb{E}$ are a group. Theorem 6.28 will make it simpler to establish that a set of bijective self maps is a group.

Definition 6.27 *Let* $(G, \circ)$ *be a group and let* $S \subseteq G$. *Then* S *is called a* **subgroup** *of* G *iff* $(S, \circ|_{S \times S})$ *is a group.*

The following characterization of subgroups is similar to that of subfields in Exercise 6-13.

Theorem 6.28 *Let* $(G, \circ)$ *be a group and let* $S \subseteq G$. *Then* S *is a subgroup of* G *iff*

1. *S is nonempty.*

2. *S is closed under $\circ$, that is, for all $x, y \in S$ we have that $x \circ y \in S$.*

3. *For any $x \in S$ we have that x^{-1}, the inverse element of x in G, is in S, too.*

Proof. Exercise 6-34. ∎

We can now investigate how field automorphisms that fix subfields determine certain subgroups of the group of field automorphisms and, conversely, how subgroups of the group of field automorphisms determine certain subfields.

Definition 6.29 *Let* $(\mathbb{F}, +, \cdot)$ *be a field and let* $\mathbb{E}$ *be an extension of* $\mathbb{F}$. *We define* $G(\mathbb{E}/\mathbb{F})$ *to be the set of all automorphisms of* $\mathbb{E}$ *that fix every element of* $\mathbb{F}$. *For a visualization, consider Figure 6.2.*

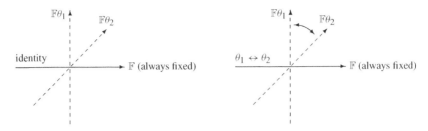

Figure 6.2 The groups $G(\mathbb{E}/\mathbb{F})$ consist of automorphisms of $\mathbb{E}$ that fix every element in the "$\mathbb{F}$-direction" and which, for $\mathbb{E}$ being a splitting field of an irreducible polynomial, may permute the other "directions" (see Theorem 6.34). The figure shows the ever-present identity function and one more possible automorphism.

Proposition 6.30 *Let $(\mathbb{F}, +, \cdot)$ be a field and let $\mathbb{E}$ be an extension of $\mathbb{F}$. Then the following hold.*

1. *$\big(G(\mathbb{E}/\mathbb{F}), \circ\big)$ is a group.*

2. *If $\mathbb{D}$ is a field so that $\mathbb{F} \subseteq \mathbb{D} \subseteq \mathbb{E}$, then $G(\mathbb{E}/\mathbb{D})$ is a subgroup of $G(\mathbb{E}/\mathbb{F})$.*

3. *Let H be a subgroup of $G(\mathbb{E}/\mathbb{F})$. Then the set $\mathbb{D}^H$ of all elements of $\mathbb{E}$ that are fixed by all elements of H is a subfield of $\mathbb{E}$ so that $\mathbb{F} \subseteq \mathbb{D}^H \subseteq \mathbb{E}$. This subfield is also called the **fixed field** of H.*

Proof. Part 1 is left to Exercise 6-35.

For part 2, first note that, by part 1, $\big(G(\mathbb{E}/\mathbb{D}), \circ\big)$ is a group. The group operation in $G(\mathbb{E}/\mathbb{D})$ and in $G(\mathbb{E}/\mathbb{F})$ is composition. Moreover, $G(\mathbb{E}/\mathbb{D}) \subseteq G(\mathbb{E}/\mathbb{F})$. Hence $G(\mathbb{E}/\mathbb{D})$ is a subset of $G(\mathbb{E}/\mathbb{F})$ that becomes a group when equipped with the restriction of the group operation on $G(\mathbb{E}/\mathbb{F})$. Therefore, $G(\mathbb{E}/\mathbb{D})$ is a subgroup of $G(\mathbb{E}/\mathbb{F})$.

For part 3, first note that the containment $\mathbb{F} \subseteq \mathbb{D}^H \subseteq \mathbb{E}$ is trivial. We will use Exercise 6-13 to show that $\mathbb{D}^H$ is a subfield of $\mathbb{E}$. Clearly, $0, 1 \in \mathbb{F} \subseteq \mathbb{D}^H$, so $\mathbb{D}^H$ is not empty. To prove that $\mathbb{D}^H$ is closed under addition, let $x, y \in \mathbb{D}^H$ and let $h \in H$. Then $h(x + y) = h(x) + h(y) = x + y$. Because $h \in H$ was arbitrary, $x + y$ is fixed by all elements of H and hence $x + y \in \mathbb{D}^H$. The proof that $\mathbb{D}^H$ is closed under multiplication is similar. To see that for each $x \in \mathbb{D}^H$ the additive inverse $-x$ is in $\mathbb{D}^H$, too, let $x \in \mathbb{D}^H$ and let $h \in H$. Then $h(-x) + h(x) = h(-x + x) = h(0) = 0$. By the uniqueness of additive inverses in fields, we conclude that $h(-x) = -h(x) = -x$. Because $h \in H$ was arbitrary, we conclude that $-x \in \mathbb{D}^H$. The proof that for each $x \in \mathbb{D}^H \setminus \{0\}$ the multiplicative inverse x^{-1} is in $\mathbb{D}^H$, too, is similar. ∎

It turns out to be quite challenging to construct a nontrivial example of a group $G(\mathbb{E}/\mathbb{F})$. We will see our first (and only) concrete examples of such groups in Theorem 6.53. Until then we will explore the abstract properties of these groups.

We continue by investigating the groups $G(\mathbb{E}/\mathbb{F})$ when $\mathbb{E}$ is obtained by adjoining a single root of an irreducible polynomial. After all, that is how splitting fields can be obtained.

Theorem 6.31 *Let* $(\mathbb{F}, +, \cdot)$ *be a field, let* $\mathbb{E}$ *be an extension of* $\mathbb{F}$, *let* $p \in \mathbb{F}[x]$ *be an irreducible polynomial of degree* n *and let* $\theta \in \mathbb{E}$ *be a zero of* p. *Then any automorphism* $\sigma \in G(\mathbb{F}(\theta)/\mathbb{F})$ *is completely determined by* $\sigma(\theta)$. *Moreover* $G(\mathbb{F}(\theta)/\mathbb{F})$ *has at most* n *elements.*

Proof. By Theorem 6.20, the set $\{1, \theta, \ldots, \theta^{n-1}\}$ is a basis for $\mathbb{F}(\theta)$ over $\mathbb{F}$. Thus for every $a \in \mathbb{F}(\theta)$ there are $a_0, \ldots, a_{n-1} \in \mathbb{F}$ so that $a = \sum_{j=0}^{n-1} a_j \theta^j$.

Let $\sigma \in G(\mathbb{F}(\theta)/\mathbb{F})$ and let $a \in \mathbb{F}(\theta)$ be arbitrary. Because by definition of $G(\mathbb{F}(\theta)/\mathbb{F})$, σ fixes $\mathbb{F}$, we obtain that

$$\sigma(a) = \sigma\left(\sum_{j=0}^{n-1} a_j \theta^j\right) = \sum_{j=0}^{n-1} \sigma(a_j)\sigma\left(\theta^j\right) = \sum_{j=0}^{n-1} a_j \sigma(\theta)^j.$$

This equation shows that the value of $\sigma(a)$ is uniquely determined by the value of $\sigma(\theta)$. Because $a \in \mathbb{F}(\theta)$ was arbitrary, this means that σ is uniquely determined by $\sigma(\theta)$.

Because the function σ is a field automorphism that fixes the field $\mathbb{F}$ we have that $0 = \sigma(0) = \sigma(p(\theta)) = p(\sigma(\theta))$. That is, the image of θ under σ is another zero of p. Therefore there are at most n possible images for θ under σ, which means there are at most n elements in $G(\mathbb{F}(\theta)/\mathbb{F})$. ∎

The groups $G(\mathbb{E}/\mathbb{F})$ are called Galois groups when $\mathbb{E}$ is a splitting field over $\mathbb{F}$. It turns out that, for irreducible polynomials, Galois groups have a very simple structure (see Theorem 6.34 below). This simple structure will be one of the keys in the investigation of solvability by radicals.

Definition 6.32 *Let* $(\mathbb{F}, +, \cdot)$ *be a field and let* $\mathbb{E}$ *be the splitting field of the polynomial* $p \in \mathbb{F}[x]$. *Then* $G(\mathbb{E}/\mathbb{F})$ *is called the* **Galois group** *of* p.

Notation 6.33 For complicated functions, especially when the preimage or the image or both are functions themselves, the usual function notation can become a bit cumbersome. Therefore, it is common to use the notation $x \mapsto \langle\text{image}\rangle$ instead of the function notation $f(x) := \langle\text{image}\rangle$.

Theorem 6.34 *Let* $(\mathbb{F}, +, \cdot)$ *be a field and let* $\mathbb{E}$ *be the splitting field of the irreducible polynomial* $p \in \mathbb{F}[x]$. *Then the Galois group* $G(\mathbb{E}/\mathbb{F})$ *is isomorphic to a group of permutations of the zeros of* p *in* $\mathbb{E}$. *For a visualization, consider Figure 6.2.*

Proof. Let $\theta_1, \ldots, \theta_k$ be the zeros of p. Let $\sigma \in G(\mathbb{E}/\mathbb{F})$. From $p(\theta_j) = 0$, we obtain that $0 = \sigma(0) = \sigma(p(\theta_j)) = p(\sigma(\theta_j))$, where in the last step we used the fact that σ is a field automorphism that fixes the elements of $\mathbb{F}$. But this means that $\sigma(\theta_j)$ is a root θ_l of p. An argument similar to the proof of Theorem 6.31 (see Exercise 6-36) now shows that σ is uniquely determined by its values on the θ_j. Moreover, $\sigma|_{\{\theta_1,\ldots,\theta_k\}}$ is a permutation of $\{\theta_1, \ldots, \theta_k\}$. The restrictions of the elements of $G(\mathbb{E}/\mathbb{F})$ to $\{\theta_1, \ldots, \theta_k\}$ form a group under composition. The map $\Phi : \sigma \mapsto \sigma|_{\{\theta_1,\ldots,\theta_k\}}$ is injective, because $\Phi(\sigma) = \Phi(\nu)$ implies $\sigma|_{\{\theta_1,\ldots,\theta_k\}} = \nu|_{\{\theta_1,\ldots,\theta_k\}}$, which implies $\sigma = \nu$, because the elements of $G(\mathbb{E}/\mathbb{F})$ are uniquely determined by their values on $\{\theta_1, \ldots, \theta_k\}$. Moreover, for all $\sigma, \nu \in G(\mathbb{E}/\mathbb{F})$ the map Φ satisfies

$$\Phi(\sigma \circ \nu) = \sigma \circ \nu|_{\{\theta_1,\dots,\theta_k\}} = \sigma|_{\{\theta_1,\dots,\theta_k\}} \circ \nu|_{\{\theta_1,\dots,\theta_k\}} = \Phi(\sigma) \circ \Phi(\nu).$$

Thus $G(\mathbb{E}/\mathbb{F})$ is isomorphic to a group of permutations on $\{\theta_1, \dots, \theta_k\}$. ∎

Theorem 6.34 shows that the smallest field extension needed to include all zeros of an irreducible polynomial into a well-behaved algebraic entity leads to a very simple structure, namely, a group of permutations. This is good. In this comparatively simple structure, we will find the promised contradiction to the assumption that a certain equation is solvable by radicals. But just knowing that the Galois group is isomorphic to *some* subgroup of the group of permutations of the zeros is not good enough. Theorem 6.53 will provide the exact Galois groups for certain polynomials. Before we can obtain these exact Galois groups, we must investigate splitting fields in more detail, and we will do so in the next section.

Exercises

6-34. Prove Theorem 6.28.

6-35. Let $\mathbb{E}$ be a field.

    (a) Prove that if $\sigma : \mathbb{E} \to \mathbb{E}$ is a field automorphism, then $\sigma^{-1} : \mathbb{E} \to \mathbb{E}$ is a field automorphism, too.

        Hint. To prove that $\sigma^{-1}(x + y) = \sigma^{-1}(x) + \sigma^{-1}(y)$ use that $x = \sigma(u)$ and $y = \sigma(v)$ for suitable $u, v \in \mathbb{E}$.

    (b) Prove part 1 of Proposition 6.30.

        Hint. Use Theorem 6.28.

6-36. Let $(\mathbb{F}, +, \cdot)$ be a field and let $\mathbb{E}$ be the splitting field of the irreducible polynomial $p \in \mathbb{F}[x]$ of degree n. Let $\theta_1, \dots, \theta_k$ be the zeros of p and let $\sigma \in G(\mathbb{E}/\mathbb{F})$.

    (a) Prove that for every $x \in \mathbb{F}(\theta_1, \dots, \theta_k)$ there are $a_{j_1,\dots,j_k} \in \mathbb{F}$ with $j_1, \dots, j_k \in \{0, \dots, n-1\}$ so that x can be represented as a sum $x = \displaystyle\sum_{j_1,\dots,j_k < n} a_{j_1,\dots,j_k} \theta_1^{j_1} \cdots \theta_k^{j_k}$.

        Hint. $\mathbb{F}(\theta_1, \dots, \theta_k) = \mathbb{F}(\theta_1)(\theta_2) \cdots (\theta_k)$ (see Exercise 6-21).

    (b) Prove that σ is determined by its values on the θ_j.

6-37. Let $\mathbb{E}$ be a field and let $\sigma : \mathbb{E} \to \mathbb{E}$ be a field automorphism.

    (a) Prove that $\sigma(0) = 0$.

        Hint. Uniqueness of the additive identity element.

    (b) Prove that $\sigma(1) = 1$.

6-38. Prove that if G is a finite group, then $S \subseteq G$ is a subgroup of G iff parts 1 and 2 of Theorem 6.28 are satisfied.

6-39. Let $\mathbb{F}$ be a field, let $\mathbb{E}$ be the splitting field of the irreducible polynomial $p \in \mathbb{F}[x]$, let $\mu_1, \dots, \mu_n$ be the zeros of p and for any two μ_i, μ_j let there be an automorphism $\varphi_{ij} \in G(\mathbb{E}/\mathbb{F})$ so that $\varphi_{ij}(\mu_i) = \mu_j$. Prove that $\mathbb{F}$ is the fixed field of $G(\mathbb{E}/\mathbb{F})$.

    Hint. Use that $\mathbb{E} = \mathbb{F}(\mu_1, \dots, \mu_n)$ (see Proposition 6.13) and that there is a basis of $\mathbb{E}$ over $\mathbb{F}$ that consists of products $\mu_1^{k_1} \cdots \mu_n^{k_n}$, where $k_1, \dots, k_n \in \{0, \dots, m - 1 = \deg(p) - 1\}$ (see Theorem 6.20). For elements x for which all coefficients in the expansion are equal to $a \in \mathbb{F} \setminus \{0\}$, use that if x is in the fixed field, then $x - a$ would be in the fixed field and so would $(x - a)^2$, which would imply that some multiple of $\mu_1^m \cdots \mu_n^m$ equals zero.

6.5 Normal Field Extensions

The computation of Galois groups will be simplified by some of the properties of splitting fields presented in this section. First, Theorem 6.35 shows two rather strong properties of splitting fields that are actually equivalent to being a splitting field.

Theorem 6.35 *Let $(\mathbb{F}, +, \cdot)$ be a field of characteristic 0 and let $\mathbb{E}$ be a finite extension of $\mathbb{F}$. Then the following are equivalent.*

1. *$\mathbb{E}$ is the splitting field for a polynomial f of positive degree in $\mathbb{F}[x]$.*

2. *Every irreducible polynomial $p \in \mathbb{F}[x]$ with one zero in $\mathbb{E}$ actually splits in $\mathbb{E}$.*

3. *$\mathbb{F}$ is the fixed field of $G(\mathbb{E}/\mathbb{F})$.*

Proof. For "1⇒2," let $\mathbb{E}$ be the splitting field for a polynomial f of positive degree in $\mathbb{F}[x]$. Let $\mu_1, \ldots, \mu_n \in \mathbb{E} \setminus \mathbb{F}$ be the roots of f that are not in $\mathbb{F}$. Then by Proposition 6.13 we have $\mathbb{E} = \mathbb{F}(\mu_1, \ldots, \mu_n)$.

Let $p \in \mathbb{F}[x]$ be an arbitrary irreducible polynomial with a root $\theta_1 \in \mathbb{E}$. Let θ_2 be an arbitrary root of p that is not equal to θ_1. The root θ_2 lies in some extension of $\mathbb{F}$, but, at this stage, we do not know if it lies in $\mathbb{E}$. By Lemma 6.22 with $\mathbb{F}_1 = \mathbb{F}_2 = \mathbb{F}$ there is an isomorphism $\varphi : \mathbb{F}(\theta_1) \to \mathbb{F}(\theta_2)$ so that $\varphi(\theta_1) = \theta_2$ and φ fixes $\mathbb{F}$.

Moreover, any field in which f splits must contain $\mu_1, \ldots, \mu_n$, so it must contain $\mathbb{E} = \mathbb{F}(\mu_1, \ldots, \mu_n)$. Thus $\mathbb{E} = \mathbb{E}(\theta_1)$ is the splitting field of f over $\mathbb{F}(\theta_1)$ (the splitting field must contain $\mathbb{E}$ and θ_1), and $\mathbb{E}(\theta_2)$ is the splitting field of f over $\mathbb{F}(\theta_2)$ (the splitting field must contain $\mathbb{E}$ and θ_2).

By Lemma 6.23 with $\mathbb{F}_1 = \mathbb{F}(\theta_1)$, $\mathbb{F}_2 = \mathbb{F}(\theta_2)$, $\mathbb{E}_1 = \mathbb{E} = \mathbb{E}(\theta_1)$ and $\mathbb{E}_2 = \mathbb{E}(\theta_2)$ there is an isomorphism $\Phi : \mathbb{E} \to \mathbb{E}(\theta_2)$ that continues φ. In particular, Φ fixes $\mathbb{F}$. But then $[\mathbb{E} : \mathbb{F}] = [\mathbb{E}(\theta_2) : \mathbb{F}]$, which, by Exercise 6-27, because $\mathbb{E} \subseteq \mathbb{E}(\theta_2)$ implies that $\mathbb{E} = \mathbb{E}(\theta_2)$. Because θ_2 was an arbitrary root of p, we conclude that all roots of p are in $\mathbb{E}$, that is, p splits in $\mathbb{E}$.

For "2⇒1," note that if $\{\mu_1, \ldots, \mu_n\}$ is a basis of $\mathbb{E}$ over $\mathbb{F}$, then every μ_k must be a zero of a polynomial in $\mathbb{F}[x]$, because otherwise $\{\mu_k^j : j \in \mathbb{N}_0\}$ would be an infinite $\mathbb{F}$-linearly independent set in $\mathbb{E}$. Thus every μ_k is a zero of an irreducible polynomial in $\mathbb{F}[x]$. By hypothesis, each of these polynomials splits in $\mathbb{E}$. Hence $\mathbb{E}$ is the splitting field of the product $p \in \mathbb{F}[x]$ of these polynomials, because p splits in $\mathbb{E}$ and the splitting field of p over $\mathbb{F}$ must contain $\mathbb{F}$ and $\mu_1, \ldots, \mu_n$.

For "1⇒3," we will prove that 1 *and* 2 imply 3, which is not a problem, because we already know that 1 and 2 are equivalent.

First consider the case that $\mathbb{E} = \mathbb{F}$. In this case $G(\mathbb{F}/\mathbb{F}) = \{\mathrm{id}_\mathbb{F}\}$ and clearly $\mathbb{E} = \mathbb{F}$ is the fixed field of $G(\mathbb{F}/\mathbb{F}) = \{\mathrm{id}_\mathbb{F}\}$.

In case $\mathbb{E} \neq \mathbb{F}$, first recall that, by definition, $\mathbb{F}$ is contained in the fixed field of $G(\mathbb{E}/\mathbb{F})$. We will now show that no further elements of $\mathbb{E}$ are in the fixed field of $G(\mathbb{E}/\mathbb{F})$. Let $\theta \in \mathbb{E} \setminus \mathbb{F}$. Then θ must be a zero of a polynomial in $\mathbb{F}[x]$, because otherwise $\{\theta^j : j \in \mathbb{N}_0\}$ would be an infinite $\mathbb{F}$-linearly independent set in $\mathbb{E}$. But then θ is a zero of an irreducible polynomial q in $\mathbb{F}[x]$. Because $\theta \notin \mathbb{F}$, we have $\deg(q) > 1$. By assumption 2, q splits in $\mathbb{E}$. Let $\theta_1 := \theta$ and let $\theta_2, \ldots, \theta_n$ be the other zeros of

q, which, by Exercise 6-23, must all be distinct. (**So we use characteristic 0 at this stage of the proof.**)

By Lemma 6.22 with $\mathbb{F}_1 = \mathbb{F}_2 = \mathbb{F}$ there is an isomorphism $\varphi : \mathbb{F}(\theta_1) \to \mathbb{F}(\theta_2)$ so that $\varphi(\theta_1) = \theta_2$ and φ fixes every element of $\mathbb{F}$. By Lemma 6.23 with $\mathbb{F}_1 = \mathbb{F}(\theta_1)$, $\mathbb{F}_2 = \mathbb{F}(\theta_2)$, $\mathbb{E}_1 = \mathbb{E}$, $\mathbb{E}_2 = \mathbb{E}$ (recall that, by 1, $\mathbb{E}$ is the splitting field of a polynomial $f = p_1 = p_2$ in $\mathbb{F}[x]$) and φ being the isomorphism, there is an isomorphism $\Phi : \mathbb{E} \to \mathbb{E}$ that continues φ. In particular, $\Phi(\theta) = \varphi(\theta_1) = \theta_2 \neq \theta_1 = \theta$ and Φ fixes every element of $\mathbb{F}$. Hence θ is not in the fixed field of $G(\mathbb{E}/\mathbb{F})$. Because $\theta \in \mathbb{E} \setminus \mathbb{F}$ was arbitrary, the fixed field of $G(\mathbb{E}/\mathbb{F})$ is $\mathbb{F}$.

For "3⇒1," first consider the case that $\mathbb{E} = \mathbb{F}$. In this case $\mathbb{E}$ is the splitting field for $f(x) = x$ over $\mathbb{F}$.

Now consider the case $\mathbb{E} \neq \mathbb{F}$. For each $a \in \mathbb{E}$ let $a_1, \ldots, a_n$ be the elements of $H_a := \{ \nu(a) : \nu \in G(\mathbb{E}/\mathbb{F}) \}$ (by Theorems 6.25 and 6.31 and because $\mathbb{E}$ is a finite extension, this set is finite because $G(\mathbb{E}/\mathbb{F})$ is finite) and let

$$f_a(x) := \prod_{j=1}^{n} (x - a_j) = \sum_{k=0}^{n} b_k x^k \in \mathbb{E}[x].$$

Let $\sigma \in G(\mathbb{E}/\mathbb{F})$. Define $\tilde{\sigma} : \mathbb{E}[x] \to \mathbb{E}[x]$ by

$$\tilde{\sigma}\left(\sum_{k=0}^{m} c_k x^k \right) := \sum_{k=0}^{m} \sigma(c_k) x^k.$$

Then $\tilde{\sigma}$ is an isomorphism from $\mathbb{E}[x]$ to $\mathbb{E}[x]$ (see Exercise 6-40a). Let $a \in \mathbb{E}$. Then for each $\nu \in H_a$ we have that $\sigma(\nu(a)) = \sigma \circ \nu(a) \in H_a$. Therefore, because σ is injective, it permutes the $a_1, \ldots, a_n$. Hence, we can infer

$$\sum_{k=0}^{n} \sigma(b_k) x^k \;=\; \tilde{\sigma}(f_a) = \tilde{\sigma}\left(\prod_{j=1}^{n} (x - a_j) \right) = \prod_{j=1}^{n} \left(x - \sigma(a_j) \right)$$

$$=\; \prod_{j=1}^{n} (x - a_j) = f_a = \sum_{k=0}^{n} b_k x^k,$$

which implies by Corollary 4.59 that $\sigma(b_k) = b_k$ for all k. (**We're "sort of" using that $\mathbb{F}$ has characteristic 0 here, see comment after Corollary 4.59.**) Because $\sigma \in G(\mathbb{E}/\mathbb{F})$ was arbitrary, this means that all b_k are in $\mathbb{F}$, the fixed field of $G(\mathbb{E}/\mathbb{F})$, so $f_a \in \mathbb{F}[x]$. By definition, the polynomial f_a splits in $\mathbb{E}$ and, because $a = \mathrm{id}(a) \in \{a_1, \ldots, a_n\}$, we have $f_a(a) = 0$.

Let $v_1, \ldots, v_m$ be a basis of $\mathbb{E}$ over $\mathbb{F}$. Then $f := \prod_{j=1}^{m} f_{v_k} \in \mathbb{F}[x]$ is a polynomial that splits in $\mathbb{E}$ and which satisfies $f(v_k) = 0$ for all k. Hence the splitting field of f must contain $\mathbb{E}$, which means that $\mathbb{E}$ is the splitting field of f. ∎

Splitting fields are the "normal" field extensions that we are concerned with when investigating solvability by radicals. Hence, to abbreviate terminology, it makes sense to call extensions that satisfy property 2 of Theorem 6.35 "normal field extensions."

Definition 6.36 *Let* $(\mathbb{F}, +, \cdot)$ *be a field and let* $\mathbb{E}$ *be a finite extension of* $\mathbb{F}$. *Then* $\mathbb{E}$ *is called a* **normal extension** *of* $\mathbb{F}$ *iff every irreducible polynomial* $p \in \mathbb{F}[x]$ *that has one zero in* $\mathbb{E}$ *actually splits in* $\mathbb{E}$.

Theorem 6.35 also shows that, with an eye on solvability of radicals, complex numbers are unavoidable.

Example 6.37 $\mathbb{R}$ *is not a normal extension of* $\mathbb{Q}$.

Of course, $\mathbb{R}$ cannot be a normal extension of $\mathbb{Q}$, because $\mathbb{R}$ is not a finite extension of $\mathbb{Q}$ (see Example 6.17). But we can also demonstrate that the property in the definition of normal extensions is violated: The polynomial from Example 6.4 is irreducible in $\mathbb{Q}[x]$ and it has a zero in $\mathbb{R}$, but not all of its zeros are real. So this "all-or-nothing absorption of zeros of irreducible polynomials" is quite a strong property: $\mathbb{R}$ is substantially larger than $\mathbb{Q}$ and $\mathbb{R}$ still does not have this property.   □

We conclude this section with a word on the size of Galois groups.

Theorem 6.38 *Let* $(\mathbb{F}, +, \cdot)$ *be a field of characteristic 0 and let* $\mathbb{E}$ *be a normal extension of* $\mathbb{F}$. *Then* $\left| G(\mathbb{E}/\mathbb{F}) \right| = [\mathbb{E} : \mathbb{F}]$.

Proof. Because $\mathbb{E}$ is a normal extension, it is in particular a finite extension and by Theorem 6.35 **(which requires** $\mathbb{F}$ **to have characteristic 0)** it is the splitting field of some polynomial $f \in \mathbb{F}[x]$. By Proposition 6.13, $\mathbb{E} = \mathbb{F}(\gamma_1, \ldots, \gamma_m)$, where $\gamma_1, \ldots, \gamma_m$ are the zeros of f in $\mathbb{E} \setminus \mathbb{F}$. By Theorem 6.25 there is a $\theta \in \mathbb{E}$ so that $\mathbb{E} = \mathbb{F}(\gamma_1, \ldots, \gamma_m) = \mathbb{F}(\theta)$ and θ is a zero of a polynomial in $\mathbb{F}[x]$. **(We're using that** $\mathbb{F}$ **has characteristic 0 here.)** But then θ also is a zero of an irreducible polynomial of degree n in $\mathbb{F}[x]$, and hence by Theorem 6.20 the set $\left\{ 1, \theta, \ldots, \theta^{n-1} \right\}$ is a basis for $\mathbb{E} = \mathbb{F}(\theta)$ over $\mathbb{F}$. In particular this means that $n = [\mathbb{E} : \mathbb{F}]$.

From Theorem 6.31 we obtain that $\left| G(\mathbb{E}/\mathbb{F}) \right| = \left| G(\mathbb{F}(\theta)/\mathbb{F}) \right| \leq n$. By Exercise 6-23 **(We're using that** $\mathbb{F}$ **has characteristic 0 here.)**, p has n distinct zeros $\theta_1 = \theta, \theta_2, \ldots, \theta_n$. Because $\mathbb{E}$ is a normal extension of $\mathbb{F}$ and $\theta \in \mathbb{E}$, all zeros of p are in $\mathbb{E}$. By Lemmas 6.22 and 6.23, for every $j \in \{1, \ldots, n\}$ there is an automorphism of $\mathbb{E}$ that fixes $\mathbb{F}$ and that maps θ to θ_j. No two of these n automorphisms are equal to each other, so $\left| G(\mathbb{E}/\mathbb{F}) \right| \geq n$. Hence $\left| G(\mathbb{E}/\mathbb{F}) \right| = n = [\mathbb{E} : \mathbb{F}]$.   ■

Remark 6.39 It can be shown that Theorem 6.38 is an equivalence. That is, $\mathbb{E}$ is a normal extension of $\mathbb{F}$ iff $\left| G(\mathbb{E}/\mathbb{F}) \right| = [\mathbb{E} : \mathbb{F}]$. But the proof of the converse of Theorem 6.38 requires a substantial excursion into linear algebra. Because this converse will not be needed for our purposes, we will not take this excursion here.   □

Exercises

6-40. Let $\mathbb{F}$ be a field and let $\sigma : \mathbb{E} \to \mathbb{E}$ be an automorphism. Define the function $\tilde{\sigma} : \mathbb{E}[x] \to \mathbb{E}[x]$ by

$$\tilde{\sigma} \left(\sum_{k=0}^{n} c_k x^k \right) := \sum_{k=0}^{n} \sigma(c_k) x^k$$

(a) Prove that $\tilde{\sigma}$ is a ring isomorphism from $\mathbb{E}[x]$ to $\mathbb{E}[x]$.
   Hint. For $\tilde{\sigma}(pq) = \tilde{\sigma}(p)\tilde{\sigma}(q)$ consider Exercise 4-7(b)i.

(b) Prove that if $p \in \mathbb{F}[x]$ and $\sigma \in G(\mathbb{E}/\mathbb{F})$, then p and $\tilde{\sigma}(p)$ have the same zeros.

(c) Prove that if $p \in \mathbb{E}[x]$ and σ fixes the zeros of p, then p and $\tilde{\sigma}(p)$ have the same zeros.

6-41. Prove that if $\mathbb{F}$ is a field of characteristic 0, $\mathbb{D}$ is a normal extension of $\mathbb{F}$ and $\mathbb{E}$ is a normal extension of $\mathbb{D}$, then $\mathbb{E}$ need *not* be a normal extension of $\mathbb{F}$.

Hint. Use $\mathbb{F} = \mathbb{Q}$, let $\mathbb{D}$ be the splitting field of $p(x) = x^2 - 2$ over $\mathbb{Q}$ and let $\mathbb{E}$ be the splitting field of $q(x) = x^2 - \sqrt{2}$ over $\mathbb{D}$.

6-42. Let $\mathbb{F}$ be a field of characteristic 0 and let $\mathbb{E} \supset \mathbb{F}$ be a normal extension of $\mathbb{F}$. Prove that there is a $w \in \mathbb{E} \setminus \mathbb{F}$ so that $\mathbb{E} = \mathbb{F}(w)$ and so that there is a $k \in \mathbb{N}$ so that $\{1, w, \ldots, w^{k-1}\}$ is a basis of $\mathbb{E}$ over $\mathbb{F}$.

6.6 The Groups S_n

Theorem 6.34 shows that groups, and in particular groups of permutations, are an important tool when investigating solvability by radicals. Therefore we now take a small breather to investigate groups of permutations in more detail. Moreover, Theorem 6.53 will show that, for many polynomials, the Galois group is isomorphic to the whole group of permutations of the roots.

Definition 6.40 *A* **permutation** *is a bijective function* $\sigma : F \to F$ *from a finite set* F *to itself. For* $n \in \mathbb{N}$, *the set* S_n *is the set of all permutations of the set* $\{1, \ldots, n\}$. *Permutations* $\sigma \in S_n$ *are often stated via an assignment table, that is,*

$$\sigma = \begin{pmatrix} 1 & 2 & \cdots & n \\ \sigma(1) & \sigma(2) & \cdots & \sigma(n) \end{pmatrix}.$$

Proposition 6.41 $(S_n, \circ)$ *is a group, called the* **symmetric group** *on* n *elements.*

Proof. See Exercise 4-6b.   ∎

Permutations can be represented in terms of more elementary permutations, called cycles (see Theorem 6.47 below). For permutations, the composition sign is often omitted and we will feel free to do so throughout.

Lemma 6.42 *Let* $\sigma \in S_n$ *be a permutation and let* $k \in \{1, \ldots, n\}$. *Then there is a natural number* m *so that* $\sigma^m(k) = k$.

Proof. Because the set $\{\sigma^j(k) : j \in \mathbb{N}_0\} \subseteq \{1, \ldots, n\}$ is finite, there must be natural numbers $i < j$ so that $\sigma^i(k) = \sigma^j(k)$. After setting $m := j - i$, we obtain $\sigma^m(k) = \sigma^{-i}\sigma^i\sigma^m(k) = \sigma^{-i}\sigma^j(k) = \sigma^{-i}\sigma^i(k) = k$.   ∎

Definition 6.43 *A permutation* $\gamma \in S_n$ *is called a* **cycle** *iff there is a* $k \in \{1, \ldots, n\}$ *so that all elements that are not fixed by* γ *are contained in the set* $\{k, \gamma(k), \gamma^2(k), \ldots\}$. *If* m *is the smallest natural number so that* $\gamma^m(k) = k$, *then the cycle* γ *is also denoted by* $\left(k\gamma(k) \cdots \gamma^{m-1}(k)\right)$ *and called an* m-**cycle**. *The identity function can be considered to be a* 1-*cycle* (k) *for any* $k \in \{1, \ldots, n\}$.

Example 6.44 Working with permutations that are written as cycles takes some getting used to, especially when it comes to compositions. For example, consider the composition (12)(23) of the 2-cycle (12) $\in S_3$ with the 2-cycle (23) $\in S_3$. The best way to compute the composition is to determine where each element of {1, 2, 3} goes. The number 1 is mapped to itself by (23) and then to 2 by (12). The number 2 is mapped to 3 by (23), which is mapped to itself by (12). The number 3 is mapped to 2 by (23), which is mapped to 1 by (12). So, overall, 1 maps to 2, which maps to 3, which maps to 1, or, (12)(23) = (123). □

Although we know that composition of functions is in general not commutative, in some compositions, the order of the factors can be reversed.

Definition 6.45 *Two cycles $\gamma, \delta \in S_n$ are called* **disjoint** *iff for all $k \in \{1, \ldots, n\}$, at most one of $\gamma(k)$ and $\delta(k)$ is not equal to k. Otherwise we say that γ and δ* **intersect**.

Proposition 6.46 *Disjoint cycles commute. That is, if γ and δ are disjoint cycles, then $\gamma\delta = \delta\gamma$.*

Proof. Exercise 6-43a. ∎

Theorem 6.47 *Every permutation in S_n can be represented as a product of disjoint cycles.*

Proof. We prove, by induction on the number s of elements $j \in \{1, \ldots, n\}$ so that $\sigma(j) \neq j$, that every $\sigma \in S_n$ can be represented as a product of disjoint cycles. Clearly for $s = 0$, $\sigma = \mathrm{id}_{\{1,\ldots,n\}} = (1)$.

For the induction step $\{0, \ldots, s-1\} \to s$, let $\sigma \in S_n$ be so that the number of elements $j \in \{1, \ldots, n\}$ so that $\sigma(j) \neq j$ is s. Let $k \in \{1, \ldots, n\}$ be so that $\sigma(k) \neq k$. By Lemma 6.42 there is a smallest $m \in \mathbb{N}$ so that $\sigma^m(k) = k$. Then the set $\{k, \sigma(k), \ldots, \sigma^{m-1}(k)\}$ has exactly m elements (see Exercise 6-44). Let $\gamma := \big(k\sigma(k) \cdots \sigma^{m-1}(k)\big)$ be a cycle with m elements. If $m = s$, then σ is equal to γ. Otherwise, consider $\mu := \gamma^{-1}\sigma$. For all $j \notin \{k, \sigma(k), \ldots, \sigma^{m-1}(k)\}$ we have $\mu(j) = \sigma(j)$ and for all $j \in \{k, \sigma(k), \ldots, \sigma^{m-1}(k)\}$ we have $\mu(j) = j$. Thus μ is a permutation for which the number of elements with $\mu(k) \neq k$ is smaller than for σ. Moreover, because $m < s$, this number is not zero. Hence by induction hypothesis, μ is a product of disjoint cycles $\mu = \gamma_1 \cdots \gamma_c$, and we can assume that none of the cycles is a 1-cycle. Then by induction hypothesis and by construction of μ, none of these cycles $\gamma_1, \ldots, \gamma_c$ contains an element of γ. Thus $\sigma = \gamma\gamma_1 \cdots \gamma_c$ is a product of disjoint cycles. ∎

Among all cycles, the cycles with exactly two elements are special. Every permutation can be represented as a product of cycles with two elements. Note however, that we lose disjointness in such a representation.

Definition 6.48 *A permutation $\tau \in S_n$ is called a* **transposition** *iff it is a 2-cycle.*

Corollary 6.49 *Every permutation in S_n can be represented as a product of (not necessarily disjoint) transpositions.*

Proof. By Theorem 6.47 every permutation in S_n can be represented as a product of cycles. Thus the result is established if we can show that every cycle can be represented as a product of transpositions. Let $\gamma = (\gamma_1\gamma_2 \ldots \gamma_c)$ be a cycle. Then $(\gamma_1\gamma_2 \ldots \gamma_c) = (\gamma_1\gamma_2)(\gamma_2\gamma_3) \cdots (\gamma_{c-2}\gamma_{c-1})(\gamma_{c-1}\gamma_c)$. ∎

The representation of permutations with transpositions shows that certain subgroups of S_n must actually be equal to S_n.

Lemma 6.50 *Let G be a subgroup of S_n that contains the n-cycle $\gamma := (12 \cdots n)$ and the transposition $\tau := (12)$. Then $G = S_n$.*

Proof. We prove by induction on m that for any $k, m \in \{1, \ldots, n\}$ with $k + m \le n$, the transposition $(k(k + m))$ is in G.

Base step $m = 1$. $(12) \in G$ by assumption, and for $k \le n - 1$, the product $\gamma^{k-1}\tau\gamma^{-(k-1)}$ equals $(k(k + 1))$.

Induction step $(m - 1) \to m$. Note that

$$\big(k(k + m)\big) = \big(k(k + (m - 1))\big)\big((k + (m - 1))(k + m)\big)\big(k(k + (m - 1))\big),$$

which is in G by induction hypothesis.

Thus G contains all transpositions and hence $G = S_n$ by Corollary 6.49. ∎

Lemma 6.51 *Let n be prime and let G be a subgroup of S_n that contains an n-cycle and a transposition. Then $G = S_n$.*

Proof. Without loss of generality, we can assume that the n-cycle is $\gamma = (12 \cdots n)$. Moreover, without loss of generality we can assume that the transposition contains 1. Let $k \in \{1, \ldots, n-1\}$ be the smallest number so that $\tau := \big(1(1+k)\big) \in G$. Then for all $m \in \mathbb{N}$ so that $(m+1)k < n$ we have that $\big((1+mk)(1+(m+1)k)\big) = \gamma^{mk}\tau\gamma^{-mk} \in G$. Now, because $2k \le n$, we have $\big(1(1+2k)\big) = \big(1(1+k)\big)\big((1+k)(1+2k)\big)\big(1(1+k)\big) \in G$ and inductively we obtain $\big(1(1 + mk)\big) \in G$ for all $m \in \mathbb{N}$ with $mk < n$. Let j be the largest such number and let z be the smallest positive number that is equivalent to $1 + (j + 1)k$ modulo n. Then $\big((1 + jk)z\big) = \gamma^{jk}\tau\gamma^{-jk} \in G$ and we conclude $(1z) = \big(1(1 + jk)\big)\big((1 + jk)z\big)\big(1(1 + jk)\big) \in G$. But then, by definition of z, we have $z < 1 + k$. Hence, by definition of k, $z = 1$. Then $1 \equiv 1 + (j + 1)k \pmod{n}$, which means that $n \mid (j + 1)k$. But $(j + 1)k < 2n$, so $n = (j + 1)k$. Because n was assumed to be prime, we conclude that $k = 1$.

The result now follows from Lemma 6.50. ∎

Knowing that certain subgroups of S_n must be equal to S_n now allows us to prove that the Galois groups of certain polynomials must be equal to S_n. For prime n, all we need is that the Galois group contains an n-cycle and a transposition.

Theorem 6.52 Cauchy's Theorem. *Let G be a finite group and let p be a prime number that divides $|G|$. Then G has a subgroup H with $|H| = p$.*

Proof. Let S be the set of all p-tuples $(a_1, \ldots, a_p)$ so that $a_1 \cdots a_p = e$ (recall that in general groups we use multiplicative notation and that e is the generic notation for the identity element). Then $|S| = |G|^{p-1}$, because for each of the $|G|^{p-1}$ possible choices for $a_1, \ldots, a_{p-1}$ there is exactly one a_p so that the equation $a_1 \cdots a_p = e$ holds, namely $a_p := a_{p-1}^{-1} \cdots a_1^{-1}$.

For two elements $(a_1, \ldots, a_p), (b_1, \ldots, b_p) \in S$ let $(a_1, \ldots, a_p) \sim (b_1, \ldots, b_p)$ iff $(a_1, \ldots, a_p) = (b_1, \ldots, b_p)$ or there is a number $j \in \{2, \ldots, p\}$ so that we have $(b_1, \ldots, b_p) = (a_j, \ldots, a_p, a_1, \ldots, a_{j-1})$. Then $\sim$ is an equivalence relation. Consider the equivalence classes of $\sim$. If $a_1 = \cdots = a_p$, then the equivalence class of $(a_1, \ldots, a_p)$ has one element. Otherwise the equivalence class of $(a_1, \ldots, a_p)$ has p elements (Exercise 6-45). Let u be the number of 1-element equivalence classes and let v be the number of p-element equivalence classes. Because equivalence classes are disjoint, $u \cdot 1 + v \cdot p = |G|^{p-1}$. But then, because p divides $|G|$, p divides $u = |G|^{p-1} - v \cdot p$. Hence $u > 1$, which means there must be an element $a \neq e$ so that $a^p = e$.

Now $H := \{a^j : j \in \{0, \ldots, p-1\}\}$ is a subgroup of G that has at most p elements and at least two elements. Let k be the smallest natural number so that $a^k = e$. Then k divides p, which implies $k = p$. Therefore $a^i \neq a^j$ for all $i < j$ in $\{0, \ldots, p-1\}$ and hence $|H| = p$. ∎

Theorem 6.53 *Let $p \in \mathbb{Q}[x]$ be an irreducible polynomial of degree n, where n is prime, that has exactly two zeros in $\mathbb{C} \setminus \mathbb{R}$. Then the Galois group of p is isomorphic to S_n.*

Proof. Let $\mathbb{E}$ be the splitting field of p over $\mathbb{Q}$. Let $\theta_1, \ldots, \theta_n$ be the zeros of p. The proof of Theorem 6.34 has shown that $\Phi : \sigma \mapsto \sigma|_{\{\theta_1, \ldots, \theta_n\}}$ is an isomorphism between the Galois group $G(\mathbb{E}/\mathbb{Q})$ and the group of permutations of $\{\theta_1, \ldots, \theta_n\}$. Let θ_1 and θ_2 be the complex zeros of p, which are complex conjugates by the Conjugate Pairs Theorem. The function $\varphi(z) := \overline{z}$ that maps each $z \in \mathbb{E} \subseteq \mathbb{C}$ to its complex conjugate is a field automorphism of $\mathbb{E}$ (see Exercise 6-46) that fixes $\mathbb{E} \cap \mathbb{R}$. Hence $\varphi \in G(\mathbb{E}/\mathbb{Q})$ and φ fixes $\theta_3, \ldots, \theta_n$. Moreover, $\varphi(\theta_1) = \overline{\theta_1} = \theta_2$ and $\varphi(\theta_2) = \overline{\theta_2} = \theta_1$. Hence $\Phi(\varphi) = (\theta_1 \theta_2)$, the transposition that switches θ_1 and θ_2.

Now let θ be any zero of p. By Theorem 6.20 we have that $[\mathbb{Q}(\theta) : \mathbb{Q}] = n$. By Theorem 6.19 we obtain $[\mathbb{E} : \mathbb{Q}] = [\mathbb{E} : \mathbb{Q}(\theta)][\mathbb{Q}(\theta) : \mathbb{Q}] = [\mathbb{E} : \mathbb{Q}(\theta)]n$. Via Theorem 6.38 we conclude that $|G(\mathbb{E}/\mathbb{Q})| = [\mathbb{E} : \mathbb{Q}] = [\mathbb{E} : \mathbb{Q}(\theta)]n$. Because n is prime, Cauchy's Theorem implies that the Galois group $G(\mathbb{E}/\mathbb{Q})$ of the polynomial p contains a subgroup U of size n. By Exercise 6-49, this means that $U = \{\text{id}, \sigma, \sigma^2, \ldots \sigma^{n-1}\}$ for some $\sigma \in G(\mathbb{E}/\mathbb{Q})$. But that means that $\sigma|_{\theta_1, \ldots, \theta_n}$ must be a cyclic permutation of the n zeros of p, because any other permutation μ, being a product of disjoint cycles of shorter length than the prime number n, does not satisfy $\mu^n = \text{id}$ (see Exercise 6-50).

By Lemma 6.51 we conclude that $G(\mathbb{E}/\mathbb{Q})$ is isomorphic to the group of all permutations of the zeros of p. ∎

Example 6.54 *The Galois group of the polynomial in Example 6.4 is isomorphic to S_5.*

Direct consequence of Remark 6.5 and Theorem 6.53. □

Exercises

6-43. Commuting cycles.

    (a) Prove Proposition 6.46.

    (b) Let $\alpha, \beta \in S_n$ be two cycles. Prove that $\alpha\beta = \beta\alpha$ iff α and β are disjoint or there is a $k \in \mathbb{Z}$ so that $\alpha = \beta^k$.

       Hint. First prove that if α and β are not disjoint, then $\alpha(x) \neq x$ iff $\beta(x) \neq x$. Then prove that α commutes with any power of β.

    (c) In S_9, consider $\sigma = (147)(258)(369)$ and $\mu = (123)(456)(789)$. Prove that $\sigma \circ \tau = \tau \circ \sigma$.

       Note. This example shows that two permutations can commute even when no two of the cycles from the representation in Theorem 6.47 commute.

6-44. Let $\sigma \in S_n$ be a permutation, let $k \in \{1, \ldots, n\}$ be so that $\sigma(k) \neq k$ and let $m \in \mathbb{N}$ be the smallest natural number so that $\sigma^m(k) = k$. Prove that the set $\left\{k, \sigma(k), \ldots, \sigma^{m-1}(k)\right\}$ has exactly m elements.

    Hint. The idea for this proof is also used in the proof of Lemma 6.42.

6-45. Let p be a prime number and let $a_1, \ldots, a_p$ be elements of a set S so that at least two of the a_j are not equal to each other. Prove that all of the p-tuples $(a_j, \ldots, a_p, a_1, \ldots, a_{j-1})$ (with $j \in 2, \ldots p$) and $(a_1, \ldots, a_p)$ are distinct.

    Hint. Suppose two are equal and obtain a contradiction to the fact that p is prime.

6-46. Let $\mathbb{E} \subseteq \mathbb{C}$ be the splitting field of the polynomial $p \in \mathbb{Q}[x]$. Prove that if $z = a + ib \in \mathbb{E}$, then $\overline{z} = a - ib \in \mathbb{E}$.

    Hint. Proposition 6.13 and the Conjugate Pairs Theorem.

6-47. Let G be a group and let $H \subseteq G$ be a subgroup. Prove that $a \sim b$ iff $ab^{-1} \in H$ defines an equivalence relation on G.

6-48. **Lagrange's Theorem.** Let G be a finite group and let $H \subseteq G$ be a subgroup of G. Prove that $|H|$ divides $|G|$.

    Hint. Prove that the equivalence classes in Exercise 6-47 all have equally many elements.

6-49. Let G be a finite group so that $n := |G|$ is prime. Prove that $G = \left\{e, a, a^2, \ldots a^{n-1}\right\}$ for some $a \in G$.

    Hint. Use Lagrange's Theorem in a proof by contradiction.

6-50. Let $\mu \in S_n$ be a permutation that can be represented as the product $\mu = \prod\limits_{j=1}^{m} \gamma_j$ of pairwise disjoint cycles γ_j. For each j, let l_j be the number of elements in the cycle γ_j and let $q \in \mathbb{N}$ be the smallest natural number so that $\mu^q = \text{id}$. Prove that $l_j \mid q$ for all $j \in \{1, \ldots, m\}$.

6-51. **Fermat's Little Theorem** revisited.

    (a) Let G be a group and let $a \in G$. Prove that $a^{|G|} = e$.

       Hint. Let $k \in \mathbb{N}$ be the smallest natural number so that $a^k = e$ and use Lagrange's Theorem.

       Note. This result is called **Fermat's Little Theorem**, too.

    (b) Use part 6-51a to prove Theorem 3.108.

    (c) A theorem of Euler. Let $n \in \mathbb{N}$ and let $a \in \mathbb{N}$ have no common factors with n. Prove that $a^{\varphi(n)} = 1$ (mod n), where φ denotes Euler's totient function.

       Hint. Let $\left\{[r_j]_n : j = 1, \ldots, \varphi(n)\right\}$ be the set of equivalence classes of all numbers r between 1 and n with $(r, n) = 1$. Prove that this set is equal to $\left\{[ar_j]_n : j = 1, \ldots, \varphi(n)\right\}$. Then form the products of the classes in each set.

6-52. Give an example of a permutation group S_n and a subgroup G in S_n so that G contains an n-cycle and a transposition, but $G \neq S_n$.

6-53. Explain why the proof of Theorem 6.53 actually establishes that the Galois group of any irreducible polynomial of prime degree n with at least two complex zeros is S_n.

6.7 The Fundamental Theorem of Galois Theory and Normal Subgroups

Proposition 6.30 has shown that if $\mathbb{F}$, $\mathbb{D}$ and $\mathbb{E}$ are nested fields with $\mathbb{F} \subseteq \mathbb{D} \subseteq \mathbb{E}$, then $G(\mathbb{E}/\mathbb{D})$ is a subgroup of $G(\mathbb{E}/\mathbb{F})$. Exercise 6-47 shows that this is good enough to define an equivalence relation on $G(\mathbb{E}/\mathbb{F})$. But, similar to the equivalence classes modulo m, we want to be able to do algebra with these equivalence classes. To do that, we need a slightly stronger property than just being a subgroup. This property, called normality, also arises naturally in the Fundamental Theorem of Galois Theory. To become accustomed with this idea, we first introduce normality for subgroups and then we state the Fundamental Theorem of Galois Theory. After that, the Fundamental Theorem of Galois Theory will give abstract examples of normal subgroups, and Proposition 6.65 will give a concrete example.

We start by introducing a certain type of algebra for sets we would like to utilize.

Definition 6.55 *Let G be a group. For any two sets A, $B \subseteq G$, we define the product $AB := \{ab : a \in A, b \in B\}$. If $A = \{a\}$ we also write aB for $\{a\}B$, and if $B = \{b\}$ we also write Ab for $A\{b\}$.*

So we would like to perform algebra with sets in an element-by-element fashion. This is common practice in mathematics, but it is also our first encounter with this idea. So it may take a little while to get used to this type of algebra. Exercise 6-54 provides some of the properties of this new operation on sets. Next, we focus on the sets with which we would like to do this kind of algebra.

Definition 6.56 *Let G be a group and let $H \subseteq G$ be a subgroup. Then, for $g \in G$, sets of the form $gH = \{gh : h \in H\}$ are called the **left cosets** of H and sets of the form $Hg = \{hg : h \in H\}$ are called the **right cosets** of H.*

By Exercise 6-55 the equivalence classes of the equivalence relation $\sim$ from Exercise 6-47 are the right cosets of the subgroup H. The algebra from Definition 6.55 will be much simpler for cosets if we can move the element of g from the right side of the coset to the left side. But this is not possible for all subgroups. Subgroups for which we can do this are called normal subgroups.

Definition 6.57 *Let G be a group and let $N \subseteq G$ be a subgroup. The subgroup N is called a **normal subgroup** of G, also denoted $N \lhd G$, iff for all $g \in G$ we have that $gN = Ng$.*

Exercise 6-56 shows that left and right cosets need not be equal in general.

Lemma 6.58 *Let G be a group and let $N \subseteq G$ be a subgroup. Then the following are equivalent.*

1. *N is a normal subgroup of G.*

2. *For all $g \in G$ we have that $gNg^{-1} = N$.*

3. *For all $g \in G$ we have that $gNg^{-1} \subseteq N$.*

Proof. Exercise 6-57. ∎

As claimed, normal subgroups allow us to do algebra with their cosets.

Theorem 6.59 *Let G be a group and let $N \lhd G$ be a normal subgroup. Then the operation $gN \circ kN := (g \circ k)N$ turns the set of cosets of N into a group. This group is typically denoted G/N, and it is called the **quotient group** or **factor group**.*

Proof. Exercise 6-58. ∎

Note that the left cosets of a non-normal subgroup H of G need not form a group with the operation $gH \circ kH := (g \circ k)H$, because the operation need not be well-defined (see Exercise 6-59). Further beauty of Theorem 6.59 can be found in the fact that the definition of the group operation matches the operation on sets from Definition 6.55. That is, if we multiply the cosets gN and hN in the indicated "element-by-element" fashion of Definition 6.55, then repeated use of normality of N (as expressed in Exercise 6-60a) proves the following.

$$
\begin{aligned}
(gN)(hN) &= \{gn : n \in N\}\{hm : m \in N\} \\
&= \{gnhm : n, m \in N\} \\
&= \{ghpm : m, p \in N\} \\
&= \{gha : a \in N\} \\
&= ghN.
\end{aligned}
$$

In more abbreviated fashion, using the associativity of the multiplication of sets and the fact that $NN = N$ (see Exercise 6-54c), this computation can also be written as

$$(gN)(hN) = g(Nh)N = ghNN = ghN.$$

In the future, we will freely use the "algebra with sets" as above and the fact that normality allows us to move elements from one side of a coset to the other.

After this introduction, we return our attention to normal field extensions. It is a quick corollary to Theorem 6.35 that normal extensions $\mathbb{E}$ of a field $\mathbb{F}$ are normal extensions of intermediate fields $\mathbb{D}$ with $\mathbb{F} \subseteq \mathbb{D} \subseteq \mathbb{E}$, too.

Corollary 6.60 *(Corollary to Theorem 6.35.) Let $(\mathbb{F}, +, \cdot)$ be a field of characteristic 0, let $\mathbb{E}$ be a normal extension of $\mathbb{F}$ and let $\mathbb{D}$ be an **intermediate field**, that is, a field so that $\mathbb{F} \subseteq \mathbb{D} \subseteq \mathbb{E}$. Then $\mathbb{E}$ is a normal extension of $\mathbb{D}$.*

Proof. By part 1 of Theorem 6.35, $\mathbb{E}$ is the splitting field of a polynomial $f \in \mathbb{F}[x]$. Because $\mathbb{F}[x] \subseteq \mathbb{D}[x]$, $\mathbb{E}$ is the splitting field of a polynomial $f \in \mathbb{D}[x]$ and hence, by Theorem 6.35, $\mathbb{E}$ is a normal extension of $\mathbb{D}$. ∎

The hypothesis that $\mathbb{F}$ has characteristic 0 was needed so that we could apply Theorem 6.35. For the same reason (and because again we will refer to Theorem 6.25), we need $\mathbb{F}$ to have characteristic 0 in our version of the Fundamental Theorem of Galois Theory below. Thus, so far, our need for assuming $\mathbb{F}$ has characteristic 0 can be traced back to this presentation's dependence on Theorem 6.25, Corollary 4.59 and Exercise 6-23. As noted before, Theorem 6.25 and Corollary 4.59 can be obtained without requiring characteristic 0. Moreover, any proofs presented here that utilize Exercise 6-23 can be replaced with proofs that do not rely on this exercise or on characteristic 0. Therefore, even though we did not prove it here, all results so far (except Exercise 6-23) and the Fundamental Theorem of Galois Theory are actually valid for fields $\mathbb{F}$ of any characteristic.

We would now like to know when the intermediate field $\mathbb{D}$ is a normal extension of $\mathbb{F}$. The Fundamental Theorem of Galois Theory tells when this is the case, and it shows the connection between normal extensions and normal subgroups.

Theorem 6.61 Fundamental Theorem of Galois Theory. *Let $(\mathbb{F}, +, \cdot)$ be a field of characteristic 0 and let $\mathbb{E}$ be a normal extension of $\mathbb{F}$. Then the following hold.*

1. *The map $\mathbb{D} \mapsto G(\mathbb{E}/\mathbb{D})$ is a bijective correspondence from the set of intermediate fields $\mathbb{D}$ between $\mathbb{F}$ and $\mathbb{E}$ to the set of subgroups of $G(\mathbb{E}/\mathbb{F})$.*

2. *Let $\mathbb{D}$ be an intermediate field between $\mathbb{F}$ and $\mathbb{E}$. Then $[\mathbb{E} : \mathbb{D}] = \left|G(\mathbb{E}/\mathbb{D})\right|$ and*
$$[\mathbb{D} : \mathbb{F}] = \frac{\left|G(\mathbb{E}/\mathbb{F})\right|}{\left|G(\mathbb{E}/\mathbb{D})\right|}.$$

3. *Let $\mathbb{D}_1$ and $\mathbb{D}_2$ be intermediate fields between $\mathbb{F}$ and $\mathbb{E}$. Then $\mathbb{F} \subseteq \mathbb{D}_1 \subset \mathbb{D}_2 \subseteq \mathbb{E}$ iff $\{id\} \subseteq G(\mathbb{E}/\mathbb{D}_2) \subset G(\mathbb{E}/\mathbb{D}_1) \subseteq G(\mathbb{E}/\mathbb{F})$, where in each case the central inclusion is proper.*

4. *Let $\mathbb{D}$ be an intermediate field between $\mathbb{F}$ and $\mathbb{E}$. Then $\mathbb{D}$ is a normal extension of $\mathbb{F}$ iff $G(\mathbb{E}/\mathbb{D})$ is a normal subgroup of $G(\mathbb{E}/\mathbb{F})$. In this case, $G(\mathbb{D}/\mathbb{F})$ is isomorphic to $G(\mathbb{E}/\mathbb{F})/G(\mathbb{E}/\mathbb{D})$.*

See Figure 6.3 for a visualization.

Proof. For part 1, first recall that, by part 2 of Proposition 6.30, if $\mathbb{D}$ is an intermediate field so that $\mathbb{F} \subseteq \mathbb{D} \subseteq \mathbb{E}$, then $G(\mathbb{E}/\mathbb{D})$ is a subgroup of $G(\mathbb{E}/\mathbb{F})$.

To see that the correspondence $\mathbb{D} \mapsto G(\mathbb{E}/\mathbb{D})$ is one to one, let $\mathbb{D}_1$ and $\mathbb{D}_2$ be intermediate fields, that is, $\mathbb{F} \subseteq \mathbb{D}_1 \subseteq \mathbb{E}$ and $\mathbb{F} \subseteq \mathbb{D}_2 \subseteq \mathbb{E}$, for which the equality $G(\mathbb{E}/\mathbb{D}_1) = G(\mathbb{E}/\mathbb{D}_2)$ holds. By Corollary 6.60, $\mathbb{E}$ is also a normal extension of $\mathbb{D}_1$ and of $\mathbb{D}_2$. By part 3 of Theorem 6.35, this means that $\mathbb{D}_1$ is the fixed field of $G(\mathbb{E}/\mathbb{D}_1)$

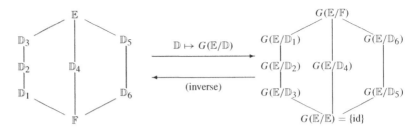

Figure 6.3 The Fundamental Theorem of Galois Theory provides a bijection between the subfields of a normal extension $\mathbb{E}$ that contain $\mathbb{F}$ and the subgroups of the Galois group $G(\mathbb{E}/\mathbb{F})$. Both this bijection and its inverse reverse order and preserve normality.

and that $\mathbb{D}_2$ is the fixed field of $G(\mathbb{E}/\mathbb{D}_2)$. From $G(\mathbb{E}/\mathbb{D}_1) = G(\mathbb{E}/\mathbb{D}_2)$ we infer that $\mathbb{D}_1 = \mathbb{D}_2$. Hence the correspondence is one to one.

To see that every subgroup of $G(\mathbb{E}/\mathbb{F})$ is of the form $G(\mathbb{E}/\mathbb{D})$ for an intermediate field $\mathbb{D}$, let H be a subgroup of $G(\mathbb{E}/\mathbb{F})$. Let $\mathbb{D}^H$ be the fixed field of H (see part 3 of Proposition 6.30). Clearly $H \subseteq G\left(\mathbb{E}/\mathbb{D}^H\right)$ is a subgroup of $G\left(\mathbb{E}/\mathbb{D}^H\right)$. To show that the two are equal, we show that both must have the same number of elements. By Corollary 6.60, $\mathbb{E}$ is also a normal extension of $\mathbb{D}^H$. Therefore by Proposition 6.13 and Theorem 6.25, there is a $\theta \in \mathbb{E}$ so that $\mathbb{E} = \mathbb{D}^H(\theta)$. Now by Theorem 6.38 we have $\left|G\left(\mathbb{E}/\mathbb{D}^H\right)\right| = \left[\mathbb{E} : \mathbb{D}^H\right] = \left[\mathbb{D}^H(\theta) : \mathbb{D}^H\right]$. Let $H = \{\sigma_1, \ldots, \sigma_h\}$. Then $h = |H| \leq \left|G\left(\mathbb{E}/\mathbb{D}^H\right)\right| = \left[\mathbb{E} : \mathbb{D}^H\right]$. For the reverse inequality, consider the polynomial $p(x) = \prod_{j=1}^{h}\left(x - \sigma_j(\theta)\right)$. For all $\sigma \in H$, with $\tilde{\sigma}$ as in Exercise 6-40, we have

$$
\begin{aligned}
\tilde{\sigma}\big(p(x)\big) &= \tilde{\sigma}\left(\prod_{j=1}^{h}\left(x - \sigma_j(\theta)\right)\right) = \prod_{j=1}^{h}\tilde{\sigma}\left(x - \sigma_j(\theta)\right) \\
&= \prod_{j=1}^{h}\left(x - \sigma\big(\sigma_j(\theta)\big)\right) = \prod_{k=1}^{h}\left(x - \sigma_k(\theta)\right) = p(x),
\end{aligned}
$$

because $\{\sigma \circ \sigma_j : j = 1, \ldots, h\} = H$. But this means that the coefficients of p are fixed by H, so $p \in \mathbb{D}^H[x]$. Hence θ is a zero of a polynomial of degree h, and $\left\{1, \theta, \ldots, \theta^h\right\}$ is $\mathbb{D}^H$-linearly dependent. So $\left|G\left(\mathbb{E}/\mathbb{D}^H\right)\right| = \left[\mathbb{E} : \mathbb{D}^H\right] = \left[\mathbb{D}^H(\theta) : \mathbb{D}^H\right] \leq h$, which implies $\left|G\left(\mathbb{E}/\mathbb{D}^H\right)\right| = h = |H|$. We conclude $H = G\left(\mathbb{E}/\mathbb{D}^H\right)$ (recall Theorem 3.54).

For part 2, first note that, by Corollary 6.60, $\mathbb{E}$ is also a normal extension of $\mathbb{D}$. Thus by Theorem 6.38 we have $\left|G(\mathbb{E}/\mathbb{D})\right| = [\mathbb{E} : \mathbb{D}]$, as well as $\left|G(\mathbb{E}/\mathbb{F})\right| = [\mathbb{E} : \mathbb{F}]$. Now by Proposition 6.19 we have

$$
[\mathbb{D} : \mathbb{F}] = \frac{[\mathbb{E} : \mathbb{F}]}{[\mathbb{E} : \mathbb{D}]} = \frac{\left|G(\mathbb{E}/\mathbb{F})\right|}{\left|G(\mathbb{E}/\mathbb{D})\right|}.
$$

For part 3, we first prove the "$\Rightarrow$" direction. So let $\mathbb{F} \subseteq \mathbb{D}_1 \subset \mathbb{D}_2 \subseteq \mathbb{E}$, where the central inclusion is proper. Every automorphism that fixes a field also fixes the subfields of said field. Thus $\{\mathrm{id}\} = G(\mathbb{E}/\mathbb{E}) \subseteq G(\mathbb{E}/\mathbb{D}_2) \subseteq G(\mathbb{E}/\mathbb{D}_1) \subseteq G(\mathbb{E}/\mathbb{F})$. To see that the containment of $G(\mathbb{E}/\mathbb{D}_2)$ in $G(\mathbb{E}/\mathbb{D}_1)$ is proper, note that, by part 1, $\mathbb{D}_1 \neq \mathbb{D}_2$ implies $G(\mathbb{E}/\mathbb{D}_2) \neq G(\mathbb{E}/\mathbb{D}_1)$.

For the "$\Leftarrow$" direction, let $\mathbb{D}_1$ and $\mathbb{D}_2$ be intermediate fields between $\mathbb{F}$ and $\mathbb{E}$ so that $\{\mathrm{id}\} \subseteq G(\mathbb{E}/\mathbb{D}_2) \subset G(\mathbb{E}/\mathbb{D}_1) \subseteq G(\mathbb{E}/\mathbb{F})$, where the central inclusion is proper. By Corollary 6.60, $\mathbb{E}$ is also a normal extension of $\mathbb{D}_1$ and of $\mathbb{D}_2$. Hence, by part 3 of Theorem 6.35, $\mathbb{D}_1$ is the fixed field of $G(\mathbb{E}/\mathbb{D}_1)$ and $\mathbb{D}_2$ is the fixed field of $G(\mathbb{E}/\mathbb{D}_2)$. Because $G(\mathbb{E}/\mathbb{D}_2) \subset G(\mathbb{E}/\mathbb{D}_1)$, every element of $\mathbb{D}_1$ is fixed by every element of $G(\mathbb{E}/\mathbb{D}_2)$. Hence $\mathbb{D}_1 \subseteq \mathbb{D}_2$. By part 1, we must have $\mathbb{D}_1 \neq \mathbb{D}_2$, which concludes the proof of part 3.

For part 4, we first prove the "$\Rightarrow$" direction. So let $\mathbb{D}$ be a normal extension of $\mathbb{F}$. By Lemma 6.58, we must prove that for all $\sigma \in G(\mathbb{E}/\mathbb{D})$ and all $\gamma \in G(\mathbb{E}/\mathbb{F})$ we have that $\gamma \sigma \gamma^{-1} \in G(\mathbb{E}/\mathbb{D})$. By definition of $G(\mathbb{E}/\mathbb{D})$, this is equivalent to showing that for all $\sigma \in G(\mathbb{E}/\mathbb{D})$ and all $\gamma \in G(\mathbb{E}/\mathbb{F})$ the automorphism $\gamma \sigma \gamma^{-1}$ fixes every element of $\mathbb{D}$.

Let $d \in \mathbb{D}$ and let $v \in G(\mathbb{E}/\mathbb{F})$. By Theorem 6.25, $d \in \mathbb{E}$ is a zero of an irreducible polynomial $p(x) = \sum_{j=0}^{n} p_j x^j$ in $\mathbb{F}[x]$. Then

$$0 = v(0) = v\left(\sum_{j=0}^{n} p_j d^j\right) = \sum_{j=0}^{n} v\left(p_j\right) v\left(d^j\right) = \sum_{j=0}^{n} p_j v(d)^j = p(v(d)),$$

that is, $v(d)$ is a zero of p, too. Because $\mathbb{D}$ is a normal extension of $\mathbb{F}$ and $d \in \mathbb{D}$, we conclude that $v(d) \in \mathbb{D}$. Hence for all $d \in \mathbb{D}$ and all $v \in G(\mathbb{E}/\mathbb{F})$, we have that $v(d) \in \mathbb{D}$.

But then for all $\gamma \in G(\mathbb{E}/\mathbb{F})$ and $\sigma \in G(\mathbb{E}/\mathbb{D})$ we obtain the following for every $d \in \mathbb{D}$. Because $\gamma^{-1}(d) \in \mathbb{D}$, we obtain $\sigma \gamma^{-1}(d) = \sigma\left(\gamma^{-1}(d)\right) = \gamma^{-1}(d)$, and then $\gamma \sigma \gamma^{-1}(d) = \gamma\left(\sigma \gamma^{-1}(d)\right) = \gamma\left(\gamma^{-1}(d)\right) = d$. Because $d \in \mathbb{D}$ was arbitrary, $\gamma \sigma \gamma^{-1}$ fixes every element of $\mathbb{D}$. So $\gamma \sigma \gamma^{-1} \in G(\mathbb{E}/\mathbb{D})$ and hence $G(\mathbb{E}/\mathbb{D})$ is normal in $G(\mathbb{E}/\mathbb{F})$.

For the "$\Leftarrow$" direction of part 4, let $G(\mathbb{E}/\mathbb{D})$ be a normal subgroup of $G(\mathbb{E}/\mathbb{F})$. We first establish that $G(\mathbb{D}/\mathbb{F})$ is isomorphic to $G(\mathbb{E}/\mathbb{F})/G(\mathbb{E}/\mathbb{D})$, which also proves the last sentence in part 4. We will make extensive use of the multiplication of sets as in Definition 6.55.

Let $\mu \in G(\mathbb{D}/\mathbb{F})$ be fixed. Then (see Exercise 6-61) μ can be extended to an automorphism $\gamma_\mu \in G(\mathbb{E}/\mathbb{F})$ so that $\mu = \gamma_\mu|_{\mathbb{D}}$. But this extension is not unique. Let $\gamma_\mu, \delta_\mu \in G(\mathbb{E}/\mathbb{F})$ be so that $\mu = \gamma_\mu|_{\mathbb{D}} = \delta_\mu|_{\mathbb{D}}$. Then $\delta_\mu^{-1} \gamma_\mu|_{\mathbb{D}} = \mu^{-1}\mu = \mathrm{id}_{\mathbb{D}}$, that is, $\delta_\mu^{-1} \gamma_\mu \in G(\mathbb{E}/\mathbb{D})$. Hence

$$\delta_\mu G(\mathbb{E}/\mathbb{D}) = \delta_\mu \left(\delta_\mu^{-1} \gamma_\mu\right) G(\mathbb{E}/\mathbb{D}) = \left(\delta_\mu \delta_\mu^{-1}\right) \gamma_\mu G(\mathbb{E}/\mathbb{D}) = \gamma_\mu G(\mathbb{E}/\mathbb{D}).$$

Therefore the function $\Phi : G(\mathbb{D}/\mathbb{F}) \to G(\mathbb{E}/\mathbb{F})/G(\mathbb{E}/\mathbb{D})$ that maps each $\mu \in G(\mathbb{D}/\mathbb{F})$ to the left coset $\gamma_\mu G(\mathbb{E}/\mathbb{D}) \in G(\mathbb{E}/\mathbb{F})/G(\mathbb{E}/\mathbb{D})$, where $\gamma_\mu|_{\mathbb{D}} = \mu$, is well-defined. We claim that this function is the desired isomorphism.

First note that for all $\mu, v \in G(\mathbb{D}/\mathbb{F})$ we have $\gamma_{\mu \circ v} G(\mathbb{E}/\mathbb{D}) = (\gamma_\mu \gamma_v) G(\mathbb{E}/\mathbb{D})$, because $\gamma_{\mu \circ \nu}|_{\mathbb{D}} = \mu \circ \nu = \gamma_\mu|_{\mathbb{D}} \circ \gamma_\nu|_{\mathbb{D}} = (\gamma_\mu \gamma_\nu)|_{\mathbb{D}}$. Therefore

$$
\begin{aligned}
\Phi(\mu \circ \nu) &= \gamma_{\mu \circ \nu} G(\mathbb{E}/\mathbb{D}) = (\gamma_\mu \gamma_\nu) G(\mathbb{E}/\mathbb{D}) \\
&= \gamma_\mu \gamma_\nu G(\mathbb{E}/\mathbb{D}) G(\mathbb{E}/\mathbb{D}) = \gamma_\mu G(\mathbb{E}/\mathbb{D}) \gamma_\nu G(\mathbb{E}/\mathbb{D}) = \Phi(\mu)\Phi(\nu).
\end{aligned}
$$

Hence Φ is a homomorphism. To prove that Φ is an isomorphism, we will construct its inverse.

Let $\gamma \in G(\mathbb{E}/\mathbb{F})$ be fixed and let $\sigma \in G(\mathbb{E}/\mathbb{D})$ and $d \in \mathbb{D}$ be arbitrary. Because $G(\mathbb{E}/\mathbb{D})$ is normal in $G(\mathbb{E}/\mathbb{F})$, there is a $\rho \in G(\mathbb{E}/\mathbb{D})$ so that $\sigma \gamma = \gamma \rho$. Therefore $\sigma\big(\gamma(d)\big) = \gamma \rho(d) = \gamma(d)$, that is, $\gamma(d)$ is fixed by every $\sigma \in G(\mathbb{E}/\mathbb{D})$. By Corollary 6.60 and part 3 of Theorem 6.35, the fixed field of $G(\mathbb{E}/\mathbb{D})$ is $\mathbb{D}$, so $\gamma(d) \in \mathbb{D}$. Because $d \in \mathbb{D}$ was arbitrary, γ maps $\mathbb{D}$ to $\mathbb{D}$. Because the same argument applied to γ^{-1} shows that γ^{-1} maps $\mathbb{D}$ to $\mathbb{D}$, we obtain that for every $\gamma \in G(\mathbb{E}/\mathbb{F})$, the restriction $\gamma|_{\mathbb{D}}$ is an automorphism of $\mathbb{D}$ that fixes every element of $\mathbb{F}$.

Now let $\delta \in \gamma G(\mathbb{E}/\mathbb{D})$. Then there is a $\sigma \in G(\mathbb{E}/\mathbb{D})$ so that $\delta = \gamma \sigma$. Therefore $\delta|_{\mathbb{D}} = \gamma \sigma|_{\mathbb{D}} = \gamma|_{\mathbb{D}}$. Hence the function $\Psi : G(\mathbb{E}/\mathbb{F})/G(\mathbb{E}/\mathbb{D}) \to G(\mathbb{D}/\mathbb{F})$ that maps each $\gamma G(\mathbb{E}/\mathbb{D}) \in G(\mathbb{E}/\mathbb{F})/G(\mathbb{E}/\mathbb{D})$ to $\gamma|_{\mathbb{D}} \in G(\mathbb{D}/\mathbb{F})$ is well-defined. Moreover, for all $\gamma G(\mathbb{E}/\mathbb{D}), \delta G(\mathbb{E}/\mathbb{D}) \in G(\mathbb{E}/\mathbb{F})/G(\mathbb{E}/\mathbb{D})$ we have

$$
\begin{aligned}
\Psi\big(\gamma G(\mathbb{E}/\mathbb{D}) \delta G(\mathbb{E}/\mathbb{D})\big) &= \Psi\big(\gamma \delta G(\mathbb{E}/\mathbb{D})\big) = \gamma \delta|_{\mathbb{D}} \\
&= \gamma|_{\mathbb{D}} \delta|_{\mathbb{D}} = \Psi\big(\gamma G(\mathbb{E}/\mathbb{D})\big) \Psi\big(\delta G(\mathbb{E}/\mathbb{D})\big).
\end{aligned}
$$

Hence Ψ is a homomorphism.

It is now easy to verify that Φ and Ψ are inverses of each other, which proves that Φ is the desired isomorphism.

The above already establishes the last claim of part 4. To conclude the proof of the "$\Leftarrow$" part of part 4, we must prove that $\mathbb{D}$ is a normal extension of $\mathbb{F}$. By part 3 of Theorem 6.35, it suffices to prove that $\mathbb{F}$ is the fixed field of $G(\mathbb{D}/\mathbb{F})$. Let $d \in \mathbb{D}$ be fixed by all automorphisms $\mu \in G(\mathbb{D}/\mathbb{F})$ and let $\gamma \in G(\mathbb{E}/\mathbb{F})$. By the above, $\gamma|_{\mathbb{D}} \in G(\mathbb{D}/\mathbb{F})$. Hence $\gamma(d) = \gamma|_{\mathbb{D}}(d) = d$. Because $\mathbb{E}$ is normal over $\mathbb{F}$ and because $\gamma \in G(\mathbb{E}/\mathbb{F})$ was arbitrary, we conclude that $d \in \mathbb{F}$. Therefore the fixed field of $G(\mathbb{D}/\mathbb{F})$ is $\mathbb{F}$, and $\mathbb{D}$ is normal over $\mathbb{F}$ by part 3 of Theorem 6.35. ∎

The Fundamental Theorem of Galois Theory shows that if we can find a root tower made of normal extensions, then we can generate a nested sequence of normal subgroups $\{\mathrm{id}\} = G_0 \triangleleft G_1 \triangleleft \cdots \triangleleft G_m = G(\mathbb{E}/\mathbb{F})$ in the Galois group. Although the exact proof of unsolvability by radicals will be more technical yet, this is a promising idea: We will show in the following that in S_n such nested sequences of normal subgroups may not be very long. Aside from this progress regarding unsolvability by radicals, the results below provide examples of normal subgroups that are more concrete than those in part 4 of Theorem 6.61.

The representation of permutations via transpositions splits the permutations in S_n into two classes: Those permutations that are a product of an even number of transpositions and those permutations that are the product of an odd number of transpositions. Proposition 6.62 below shows that no permutation can be both.

Proposition 6.62 *Let $\sigma \in S_n$ be a permutation. Then σ cannot be represented both as a product of an even number of transpositions and as a product of an odd number of transpositions.*

Proof. Suppose for a contradiction that $\sigma \in S_n$ can be represented as a product $\tau_1 \cdots \tau_{2k}$ of an even number of transpositions and as a product $\tilde{\tau}_1 \cdots \tilde{\tau}_{2l+1}$ of an odd number of transpositions. Then the identity can be represented as the product $\tilde{\tau}_{2l+1} \cdots \tilde{\tau}_1 \tau_1 \cdots \tau_{2k} = \sigma^{-1}\sigma = $ id of an odd number of transpositions. Let $m \in \mathbb{N}_0$ be the smallest number so that the identity can be represented as a product of $2m + 1$ transpositions. Let the transpositions $\nu_1, \ldots, \nu_{2m+1}$ be so that id $= \nu_1 \cdots \nu_{2m+1}$, so that, among all representations of the identity with $2m + 1$ transpositions, the largest number that occurs in any of the ν_j is as small as possible, and so that the number of transpositions in which this number occurs is as small as possible, too.

Let $b \in \{1, \ldots, n\}$ be the largest number that occurs in any of the transpositions ν_j. Because $\nu_1 \cdots \nu_{2m+1} = $ id, there must be at least two transpositions ν_j that contain b. Because disjoint cycles commute, and because for all $a, c \in \{1, \ldots, n\}$ we have that $(bc)(ac) = (abc) = (ca)(ab)$, it is possible to rewrite the product so that two transpositions that contain b are adjacent, so that the number of factors is still $2m + 1$, and so that the number of occurrences of b does not increase: Go through the product from right to left until you find the second transposition that contains b. Use the equality $(bc)(ac) = (abc) = (ca)(ab)$ repeatedly to push the second rightmost occurrence of b back until it is adjacent to the rightmost occurrence of b.

By choice of m, the two adjacent transpositions that contain b cannot be equal, because then the identity could be represented as a product of $2m - 1$ transpositions. Hence the two adjacent transpositions that contain b are (ab) and (bc) for some $a \neq c$. But now, because $(ab)(bc) = (abc) = (ca)(ab)$, the two adjacent transpositions that contain b can be rewritten in such a way that the total number of bs in the product $\nu_1 \cdots \nu_{2m+1}$ is reduced by 1. This is a contradiction to the choice of $\nu_1, \ldots, \nu_{2m+1}$. ∎

Now the following definition is sensible.

Definition 6.63 *A permutation is called **even** iff it can be represented as a product of an even number of transpositions. A permutation is called **odd** iff it can be represented as a product of an odd number of transpositions.*

Not only do permutations "split" into even and odd permutations, the even permutations form a subgroup of S_n.

Proposition 6.64 *The even permutations form a subgroup of S_n, called the **alternating group** A_n on n elements.*

Proof. Exercise 6-62. ∎

Trivially, for any group G with identity element e, the sets $\{e\}$ and G are normal subgroups. The alternating group is an important example of a nontrivial normal subgroup.

Proposition 6.65 *For $n > 1$, the alternating group A_n is a normal subgroup of the symmetric group S_n.*

Proof. By Proposition 6.62, every permutation in S_n is either even or odd. Moreover, for every $\sigma \in S_n$, the functions $\nu \mapsto \sigma\nu$ and $\nu \mapsto \nu\sigma$ are bijections on S_n. If σ is even, these bijections map even permutations to even permutations and odd permutations to odd permutations. If σ is odd, they map odd permutations to even permutations and even permutations to odd permutations. But then, if σ is even, we have $\sigma A_n = A_n = A_n\sigma$ and, if σ is odd we have that $\sigma A_n = S_n \setminus A_n = A_n\sigma$. Thus A_n is a normal subgroup of S_n. ∎

Moreover, although the alternating group has many subgroups (for every transposition τ, the set $\{\tau, \mathrm{id}\}$ is a group), for $n \geq 5$ the alternating group A_n has no nontrivial *normal* subgroups. This will be crucial for our proof that there are polynomial equations that are not solvable by radicals.

Definition 6.66 *A group is called* **simple** *iff it has no nontrivial normal subgroup.*

Lemma 6.67 *Any even permutation of $n \geq 3$ elements is a product of 3-cycles.*

Proof. Any even permutation is a product of pairs of transpositions. We are done if we can show that any product of two transpositions can be represented as a product of 3-cycles. So consider two transpositions τ and μ. If $\tau = \mu = (ab)$, note that $\tau\mu = (ab)(ab) = \mathrm{id} = (abc)(cba)$. If τ and μ overlap in one element, say $\tau = (ab)$ and $\mu = (bc)$, note that $(ab)(bc) = (abc)$. If τ and μ are disjoint, say $\tau = (ab)$ and $\mu = (cd)$, note that $(ab)(cd) = (acb)(acd)$. Thus any product of two transpositions is a product of 3-cycles. Hence any even permutation is a product of 3-cycles. ∎

Theorem 6.68 *A_n is simple for all $n \geq 5$.*

Proof. Let $N \lhd A_n$ be a normal subgroup that is not equal to $\{\mathrm{id}\}$.

If N contains the 3-cycle (abc) we argue as follows: Let $x, y \notin \{a, b, c\}$ be two distinct elements of $\{1, \ldots, n\}$. Then one of

$$\gamma := \begin{pmatrix} 1 & 2 & 3 & 4 & 5 & \cdots & n \\ a & b & c & x & y & \cdots & \end{pmatrix}$$

(where we simply complete the assignment beyond 5 to somehow get a permutation) or

$$\delta := (xy)\gamma = \begin{pmatrix} 1 & 2 & 3 & 4 & 5 & \cdots & n \\ a & b & c & y & x & \cdots & \end{pmatrix}$$

is even. But then one of the containments $N \ni \gamma^{-1}(abc)\gamma = (123)$ or $N \ni \delta^{-1}(abc)\delta = (123)$ holds. Either way, $(123) \in N$ and a similar argument proves that all 3-cycles are in N (Exercise 6-63), which means that $N = A_n$.

Thus, we are done if we can prove that N contains a 3-cycle. Let $\alpha \in N \setminus \{\mathrm{id}\}$ and consider the representation of α as a product of disjoint cycles.

Case 1: The representation of α as a product of disjoint cycles contains an r-cycle $(ijkl\ldots)$ with $r \geq 4$. Then $\alpha = (ijkl\ldots)\gamma$, where $\gamma = \mathrm{id}$ or γ is a product of cycles that do not share elements with $(ijkl \ldots)$. Let $\beta := (ijk)$. Then N contains the product $\beta\alpha^{-1}\beta^{-1}$ and hence

$$
\begin{aligned}
N \ni{}& \alpha\left(\beta\alpha^{-1}\beta^{-1}\right) \\
={}& \left(\alpha\beta\alpha^{-1}\right)\beta^{-1} \\
={}& \left((ijkl\ldots)\gamma(ijk)\gamma^{-1}(\ldots lkji)\right)(kji) \\
={}& \left((ijkl\ldots)(ijk)(\ldots lkji)\right)(kji) \\
={}& (jkl)(kji) \\
={}& (ilj)
\end{aligned}
$$

So by the above, because N contains a 3-cycle, $N = A_n$.

Case 2: The representation of α as a product of disjoint cycles contains a 3-cycle (ijk) and all cycles in this representation have 2 or 3 elements. If α is equal to the 3-cycle (ijk), then there is nothing to prove. Otherwise there are two distinct elements $l, m \notin \{i, j, k\}$, a cycle $(lm\cdot)$ that is either a transposition or a 3-cycle disjoint from (ijk), and a permutation γ that equals the identity function or consists of cycles that contain none of the elements of (ijk) and $(lm\cdot)$ so that we have $\alpha = (ijk)(lm\cdot)\gamma$. Let $\beta := (ijl)$. Then N contains the product $\beta\alpha^{-1}\beta^{-1}$ and hence

$$
\begin{aligned}
N \ni{}& \alpha\left(\beta\alpha^{-1}\beta^{-1}\right) \\
={}& \left(\alpha\beta\alpha^{-1}\right)\beta^{-1} \\
={}& \left((ijk)(lm\cdot)\gamma(ijl)\gamma^{-1}(\cdot ml)(kji)\right)(lji) \\
={}& \left((ijk)(lm\cdot)(ijl)(\cdot ml)(kji)\right)(lji) \\
={}& (jkm)(lji) \\
={}& (ilkmj)
\end{aligned}
$$

Now by case 1, N contains a 3-cycle and hence $N = A_n$.

Case 3: The representation of α as a product of disjoint cycles contains only transpositions. Then there are distinct elements $i, j, k, l \in \{1, \ldots, n\}$ and a permutation γ that equals the identity function or is a product of transpositions that contain none of the numbers i, j, k, l so that $\alpha = (ij)(kl)\gamma$. Moreover, there is an element $m \in \{1, \ldots, n\} \setminus \{i, j, k, l\}$. Let $\beta := (ikm)$. Then N contains the product $\beta\alpha^{-1}\beta^{-1}$ and hence (using that, this time, $\gamma(m) = \alpha(m)$)

$$
\begin{aligned}
N \ni{}& \alpha\left(\beta\alpha^{-1}\beta^{-1}\right) \\
={}& \left(\alpha\beta\alpha^{-1}\right)\beta^{-1} \\
={}& \left((ij)(kl)\gamma(ikm)\gamma^{-1}(lk)(ji)\right)(mki) \\
={}& \left((ij)(kl)(ik\alpha(m))(lk)(ji)\right)(mki)
\end{aligned}
$$

$$
\begin{aligned}
&= \quad (jl\alpha(m))(mki) \\
&= \quad \begin{cases} (ijlmk) & \text{if } \alpha(m) = m, \\ (jl\alpha(m))(mki) & \text{if } \alpha(m) \neq m. \end{cases}
\end{aligned}
$$

In either case, by case 1 or case 2, N contains a 3-cycle and hence $N = A_n$.

Therefore, in all cases, the normal subgroup $N \lhd A_n$ is equal to A_n and hence A_n is simple. ∎

Exercises

6-54. Let G be a group and consider the multiplication of sets from Definition 6.55.

  (a) Prove that this multiplication of sets is an associative binary operation on the power set $\mathcal{P}(G)$.

  (b) Prove that $\{e\}$ is a neutral element for this multiplication of sets.

  (c) Prove that if $H \subseteq G$ is a subgroup of G, then $HH = H$.

6-55. Let G be a group and let $H \subseteq G$ be a subgroup. Prove that the equivalence classes of the relation $\sim$ of Exercise 6-47 are the right cosets of H.

6-56. Let G be a group and let $H \subseteq G$ be a subgroup. Prove that the left cosets need not be equal to the right cosets when H is not normal.

  Hint. Use $G = S_4$, $H = \left\{ \text{id}, (12) \right\}$ and $g = (123)$.

6-57. Prove Lemma 6.58.

6-58. Prove Theorem 6.59.

6-59. Let G be a group and let $H \subseteq G$ be a subgroup. Prove that the operation $gH \circ kH := (g \circ k)H$ on the left cosets of H need not be well-defined when H is not normal.

  Hint. Use $G = S_4$, $H = \left\{ \text{id}, (12) \right\}$, $g = (123)$, $\tilde{g} = (123)(12)$, $h = (124)$, and $\tilde{h} = (124)(12)$.

6-60. Let G be a group and let $N \subseteq G$ be a subgroup.

  (a) Prove that N is normal iff for all $g \in G$ and all $n \in N$ there is an $m \in N$ so that $gn = mg$.

  (b) Prove that N is normal iff for all $g \in G$ and all $n \in N$ we have $gng^{-1} \in N$.

6-61. Let $\mathbb{F} \subseteq \mathbb{D} \subseteq \mathbb{E}$ be fields so that $\mathbb{E}$ is a normal extension of $\mathbb{F}$. Prove that every automorphism $\gamma' \in G(\mathbb{D}/\mathbb{F})$ can be extended to an automorphism $\gamma \in G(\mathbb{E}/\mathbb{F})$, that is, prove that there is an automorphism $\gamma \in G(\mathbb{E}/\mathbb{F})$ with $\gamma|_{\mathbb{D}} = \gamma'$.

  Hint. Lemma 6.23.

6-62. Prove Proposition 6.64.

  Hint. Theorem 6.28.

6-63. Let $n \geq 5$ and let $N \lhd A_n$ be a normal subgroup. Prove that if $(123) \in N$, then $N = A_n$.

6-64. Let $m \in \mathbb{N} \setminus \{1\}$ and let $m\mathbb{Z} := \{mk : k \in \mathbb{Z}\}$.

  (a) Prove that $m\mathbb{Z}$ is a normal subgroup of $(\mathbb{Z}, +)$.

  (b) Prove that $\mathbb{Z}/m\mathbb{Z}$ is isomorphic to $(\mathbb{Z}_m, +)$.

6-65. Let G be a group and let $H \subseteq G$.

  (a) Define H^{-1}.

  (b) Prove that H is a subgroup iff $H \neq \emptyset$, $HH \subseteq H$ and $H^{-1} \subseteq H$.

  (c) Prove that H is a subgroup iff $H \neq \emptyset$, $HH = H$ and $H^{-1} = H$.

6-66. Prove that the odd permutations cannot form a subgroup of S_n.

6-67. Let G be a group.

    (a) Let $M \lhd G$ and $N \lhd G$ be normal subgroups of G. Prove that MN is a normal subgroup of G, too.

    (b) Use $G = S_4$ and $H = \{ \text{id}, (12) \}$ and $K = \{ \text{id}, (23) \}$ to prove that the product HK of two subgroups need not be a subgroup.

    (c) Prove that if $\mathcal{N}$ is a family of normal subgroups of G, then the set N of all finite products of elements of normal subgroups in $\mathcal{N}$ is a normal subgroup of G, too.

    (d) Let $H \subseteq G$ be a subgroup of G. Prove that there is a largest (with respect to set containment) normal subgroup of G that is contained in H.

    Hint. Part 6-67c.

6-68. Let G be a group.

    (a) Let $\mathcal{N}$ be a family of normal subgroups of G. Prove that $\bigcap \mathcal{N}$ is a normal subgroup of G.

    (b) Let $H \subseteq G$ be a subgroup of G. Prove that there is a smallest (with respect to set containment) normal subgroup of G that contains H.

    Hint. Part 6-68b.

6-69. **Galois connections.** Let $(P, \leq)$ and $(Q, \leq)$ be ordered sets and let $f : P \to Q$ and $g : Q \to P$ be functions so that $x \leq y$ (in P) implies $f(x) \geq f(y)$ and so that $a \leq b$ (in Q) implies $g(a) \geq g(b)$. Then (f, g) is called a **Galois connection** or a **Galois correspondence** iff for all $x \in P$ we have that $gf(x) \geq x$ and for all $a \in Q$ we have that $fg(a) \geq a$.

    (a) Prove that if (f, g) is any Galois connection, then $fgf = f$ and $gfg = g$.

    (b) Prove that both restrictions $f|_{g[Q]}$ and $g|_{f[P]}$ are bijective and inverses of each other.

    (c) Prove that, for each $q \in f[P]$, the set $f^{-1}(q)$ has a largest element and that, for each element $p \in g[Q]$, the set $g^{-1}(p)$ has a largest element.

    (d) Prove that, for each $p \in P$, we have that $gf(p)$ is the smallest element of the set of all $x \in g[Q]$ with $x \geq p$. Then state and prove a similar result for each $q \in Q$.

    (e) Prove that, for each $p \in P$, we have that $f(p)$ is the largest element $q \in Q$ so that $g(q) \geq p$. Then state and prove a similar result for each $q \in Q$.

    (f) Let $\mathbb{F}$ be a field of characteristic 0 and let $\mathbb{E}$ be a normal extension of $\mathbb{F}$. Let P be the set of intermediate fields between $\mathbb{F}$ and $\mathbb{E}$ ordered by set inclusion $\subseteq$. Let Q be the set of subgroups of the Galois group $G(\mathbb{E}/\mathbb{F})$ ordered by set inclusion. Let $f : P \to Q$ be defined by $f(\mathbb{D}) = G(\mathbb{E}/\mathbb{D})$ and let $g : Q \to P$ be equal to f^{-1}.

    Prove that (f, g) is a Galois connection.

    (g) Let $\mathbb{F}$ be a field of characteristic 0 and let $\mathbb{E}$ be a normal extension of $\mathbb{F}$. Let P be the set of intermediate fields between $\mathbb{F}$ and $\mathbb{E}$ ordered by set inclusion $\subseteq$. Let Q be the set of subgroups of the Galois group $G(\mathbb{E}/\mathbb{F})$ ordered by set inclusion. For $f : P \to Q$, let $f(\mathbb{D})$ be the largest normal subgroup of $G(\mathbb{E}/\mathbb{F})$ that is contained in $G(\mathbb{E}/\mathbb{D})$ (see Exercise 6-67d). For $g : Q \to P$, let $g(H)$ be the fixed field of the smallest normal subgroup of $G(\mathbb{E}/\mathbb{F})$ that contains H (see Exercise 6-68b).

    Prove that (f, g) is a Galois connection.

    (h) Let $\mathbb{F}$ be a field of characteristic 0 and let $\mathbb{E}$ be a normal extension of $\mathbb{F}$. Let P be the set of intermediate fields between $\mathbb{F}$ and $\mathbb{E}$ ordered by reversed set inclusion $\supseteq$. Let Q be the set of subgroups of the Galois group $G(\mathbb{E}/\mathbb{F})$ ordered by reversed set inclusion. For $f : P \to Q$, let $f(\mathbb{D})$ be the smallest (with respect to regular inclusion) normal subgroup of $G(\mathbb{E}/\mathbb{F})$ that contains $G(\mathbb{E}/\mathbb{D})$. Let $g : Q \to P$ map every subgroup H of $G(\mathbb{E}/\mathbb{F})$ to its fixed field $\mathbb{D}^H$.

    Prove that (f, g) is a Galois connection.

6.8 Consequences of Solvability by Radicals

With the Fundamental Theorem of Galois Theory we now close in on the unsolvability of certain polynomial equations by radicals: If we can find a root tower in which each new field is a normal extension of the previous one, we can start analyzing nested sequences of normal subgroups of Galois groups. Moreover, we know that Galois groups are isomorphic to groups of permutations (see Theorem 6.34), that some of them are equal to S_n (see Theorem 6.53) and that groups of permutations do not have many normal subgroups (see Theorem 6.68 and Lemma 6.76 below). But even that is not quite enough yet, because splitting fields, can be generated by adjoining a single element (see Theorem 6.25), so that "short" root towers are still a theoretical possibility. Therefore we need extensions so that the factor groups from part 4 of the Fundamental Theorem of Galois Theory are especially nice. Theorem 6.70 below shows that such extensions exist, and the rest of this section investigates the concept of solvability of groups, which is inspired by it.

A key ingredient to our discussion are solutions of the equation $x^n - 1 = 0$. These numbers are also called **roots of unity**. By Theorem 5.64, the field $\mathbb{C}$ contains all n^{th} roots of unity. If our starting field does not contain these roots for a given n, then we could always first adjoin them before we continue. (Indeed, this is done in the proof of Theorem 6.74 below.)

Definition 6.69 An nth root of unity r is called a **primitive** n^{th} **root of unity** iff every n^{th} root of unity is a power of r.

Primitive n^{th} roots of unity exist. For example, the number $e^{i\frac{2\pi}{n}}$ is one such primitive root of unity.

Theorem 6.70 Let $\mathbb{F}$ be a field of characteristic 0 that contains all n^{th} roots of unity, let $f \in \mathbb{F}$ and let w be a root of $x^n - f = 0$. Then $\mathbb{F}(w)$ is the splitting field of the polynomial $x^n - f$. Moreover, the Galois group $G\big(\mathbb{F}(w)/\mathbb{F}\big)$ is isomorphic to a subgroup of $\mathbb{Z}_n$. In particular, the Galois group $G\big(\mathbb{F}(w)/\mathbb{F}\big)$ is commutative.

Proof. Because the result is trivial for $f = 0$, we can assume that $f \neq 0$.

Let r be a primitive n^{th} root of unity. Then the elements $1, r, r^2, \ldots, r^{n-1}$ are the n distinct n^{th} roots of unity. Hence the elements $w, rw, r^2w, \ldots, r^{n-1}w$, which are all in $\mathbb{F}(w)$, are the n distinct roots of the polynomial $x^n - f$. Therefore $\mathbb{F}(w)$ is the splitting field of the polynomial $x^n - f$.

Now let $\sigma \in G\big(\mathbb{F}(w)/\mathbb{F}\big)$. Then $\sigma(w) = wr^k$ for some $k \in \mathbb{Z}$ (see proof of Theorem 6.34), which is unique modulo n. Define the function $\Phi : G\big(\mathbb{F}(w)/\mathbb{F}\big) \to \mathbb{Z}_n$ by $\Phi(\sigma) := [k]_n$, where k is the unique element of $\{0, \ldots, n-1\}$ so that $\sigma(w) = wr^k$.

To see that the function Φ is a homomorphism, let $\sigma, \mu \in G\big(\mathbb{F}(w)/\mathbb{F}\big)$. Then there are $j, k \in \{0, \ldots, n-1\}$ so that $\mu(w) = wr^j$ and $\sigma(w) = wr^k$. Therefore we conclude $\mu \circ \sigma(w) = \mu\big(wr^k\big) = \mu(w)\mu\big(r^k\big) = \mu(w)r^k = wr^jr^k = wr^{j+k}$ and hence $\Phi(\mu \circ \sigma) = [j + k]_n = [j]_n + [k]_n = \Phi(\mu) + \Phi(\sigma)$.

Moreover Φ is injective, because $\Phi(\sigma) = [0]_n$ implies $\sigma(w) = w$, and hence, by Theorem 6.31 that $\sigma = \mathrm{id}$ (use Exercise 6-70 to conclude that Φ is injective).

Hence the Galois group $G\big(\mathbb{F}(w)/\mathbb{F}\big)$ is isomorphic to a subgroup of $\mathbb{Z}_n$, and because $\mathbb{Z}_n$ is commutative, so are its subgroups. ∎

We need $\mathbb{F}$ to have characteristic 0 in Theorem 6.70 because we need n distinct roots of $x^n\text{-}f=0$. In a field of characteristic p, so that p divides n, we would not get n distinct roots. This means that every result that relies on Theorem 6.70 truly needs characteristic 0 (or something rather technical) in the hypothesis. Hence, for the remaining results in this chapter, the hypothesis that $\mathbb{F}$ has characteristic 0 is quite common in abstract algebra, too.

A converse to Theorem 6.70 is proved in Exercise 6-74f. Theorem 6.70 shows that there is a chance for root towers in which the factor groups of the Galois groups $G(\mathbb{F}_j/\mathbb{F}_{j-1})$ of consecutive entries would be commutative. By the Fundamental Theorem of Galois Theory, if we can get the $\mathbb{F}_j$ to be normal extensions of each other, this would mean that the factor groups $G(\mathbb{F}_q/\mathbb{F}_{j-1})/G(\mathbb{F}_q/\mathbb{F}_j)$ are commutative. Translated to groups, this motivates the following definition.

Definition 6.71 *Let G be a finite group. Then G is called* **solvable** *iff there is a nested sequence of subgroups $\{e\} = G_0 \lhd G_1 \lhd \cdots \lhd G_m = G$ so that G_j/G_{j-1} is commutative for $j = 1, \ldots, m$.*

Lemma 6.72 *Let G be a finite group and let $N \lhd G$. If G is solvable, then G/N is solvable.*

Proof. Let $\{e\} = G_0 \lhd G_1 \lhd \cdots \lhd G_m = G$ be a nested sequence of normal subgroups so that G_j/G_{j-1} is commutative for $j = 1, \ldots, m$. The most natural idea is to attempt to work with factor groups "G_j/N," but, because N need not be contained in G_j, we must work with groups G_jN/N instead. Of course this means we must first establish all the requisite properties once more.

Let $j \in \{0, \ldots, m\}$. We first prove that the set G_jN is a subgroup of G. Clearly $G_jN \neq \emptyset$, because $e \in G_jN$. Now let $x, y \in G_jN$. Then there are $g_x, g_y \in G_j$ and $n_x, n_y \in N$ so that $x = g_xn_x$ and $y = g_yn_y$. Because N is normal in G, there in an $n \in N$ so that $n_xg_y = g_yn$. Therefore

$$xy = (g_xn_x)(g_yn_y) = g_x(n_xg_y)n_y = g_x(g_yn)n_y = (g_xg_y)(nn_y) \in G_jN.$$

Finally, for $x^{-1} = n_x^{-1}g_x^{-1}$ the normality of N guarantees that there is an $n' \in N$ so that $x^{-1} = n_x^{-1}g_x^{-1} = g_x^{-1}n' \in G_jN$. Thus G_jN is a subgroup of G.

Because N is normal in G, it is normal in each G_jN, and we can form the quotient groups G_jN/N. Note that if $xN \in G_jN/N$, then there are a $g_j \in G_j$ and an $n \in N$ so that $xN = g_jnN = g_jN$. So each element of G_jN/N is of the form g_jN for some $g_j \in G_j$.

Now let $j \in \{1, \ldots, m\}$ be fixed. We claim that $G_{j-1}N/N$ is normal in G_jN/N. Because $G_{j-1}N/N \subseteq G_jN/N$, it is clear that $G_{j-1}N/N$ is a subgroup of G_jN/N. Now let $g_jN \in G_jN/N$ and let $g_{j-1}N \in G_{j-1}N/N$. Then, by normality of N in G, we have $g_jN = Ng_j$, by normality of G_{j-1} in G_j, there is a $g'_{j-1} \in G_{j-1}$ so that $g_jg_{j-1} = g'_{j-1}g_j$, and, by normality of N in G once more, $Ng'_{j-1} = g'_{j-1}N$. But then

$(g_j N)(g_{j-1} N) = N g_j g_{j-1} N = N g'_{j-1} g_j N = (g'_{j-1} N)(g_j N)$, which proves that $(g_j N)(G_{j-1} N / N) \subseteq (G_{j-1} N / N)(g_j N)$. The reversed inclusion is proved similarly. Thus for all $x \in G_j N / N$ we have that $x(G_{j-1} N / N) = (G_{j-1} N / N)x$. Hence the subgroup $G_{j-1} N / N$ is normal in $G_j N / N$.

To prove that $(G_j N / N)/(G_{j-1} N / N)$ is commutative, first note that the algebra of sets that we use when working with cosets also works when the cosets are sets of sets (as they are in this case). Let $g_j, h_j \in G_j$. Then, because G_j / G_{j-1} is commutative,

$$g_j G_{j-1} h_j = g_j G_{j-1} G_{j-1} h_j = g_j G_{j-1} h_j G_{j-1} = h_j G_{j-1} g_j G_{j-1}$$
$$= h_j G_{j-1} G_{j-1} g_j = h_j G_{j-1} g_j,$$

and hence, for every $g_{j-1} \in G_{j-1}$ there is an element $f_{j-1} \in G_{j-1}$ so that we have $g_j g_{j-1} h_j = h_j f_{j-1} g_j$. Therefore

$$
\begin{aligned}
g_j(G_{j-1} N / N)h_j &= \{g_j g_{j-1} n N h_j : g_{j-1} \in G_{j-1}, n \in N\} \\
&= \{g_j g_{j-1} N h_j : g_{j-1} \in G_{j-1}\} \\
&= \{g_j g_{j-1} h_j N : g_{j-1} \in G_{j-1}\} \\
&= \{h_j f_{j-1} g_j N : f_{j-1} \in G_{j-1}\} \\
&= \{h_j f_{j-1} N g_j : f_{j-1} \in G_{j-1}\} \\
&= h_j(G_{j-1} N / N)g_j.
\end{aligned}
$$

Now recall that the elements of each $G_j N / N$ are of the form $x N$ with $x \in G_j$. Then

$$
\begin{aligned}
g_j N(G_{j-1} N / N)h_j N(G_{j-1} N / N) &= N\big(g_j(G_{j-1} N / N)h_j\big)N(G_{j-1} N / N) \\
&= N\big(h_j(G_{j-1} N / N)g_j\big)N(G_{j-1} N / N) \\
&= h_j N(G_{j-1} N / N)g_j N(G_{j-1} N / N).
\end{aligned}
$$

Because $g_j N(G_{j-1} N / N)$ and $h_j N(G_{j-1} N / N)$ were arbitrary elements of the factor group $(G_j N / N)/(G_{j-1} N / N)$, we conclude that $(G_j N / N)/(G_{j-1} N / N)$ is commutative.

Therefore, $\{e N\} = G_0 N / N \lhd G_1 N / N \lhd \cdots \lhd G_m N / N = G / N$ is a nested sequence of subgroups as in the definition of solvability. Hence G / N is solvable. ∎

Back to field extensions: We will now establish that for every extension with a root tower, there is a normal extension with a root tower.

Theorem 6.73 *Let $(\mathbb{F}, +, \cdot)$ be a field of characteristic 0 and let the nested sequence $\mathbb{F} = \mathbb{F}_1 \subseteq \mathbb{F}_2 \subseteq \cdots \subseteq \mathbb{F}_q$ be a root tower for $\mathbb{F}_q$ over $\mathbb{F}$. Then there is normal extension $\mathbb{E}$ of $\mathbb{F}$ that contains $\mathbb{F}_q$ and that has a root tower over $\mathbb{F}$.*

Proof. The proof is an induction on $q = [\mathbb{F}_q : \mathbb{F}]$, the length of the root tower. Clearly, the base step $q = 1$ is trivial ($\mathbb{F}$ is the splitting field of $x - 1$).

For $q \geq 2$, let $\mathbb{F}'_{q-1}$ be a normal extension of $\mathbb{F}$ that contains $\mathbb{F}_{q-1}$ and that has a root tower over $\mathbb{F}$ (for $q = 2$ let $\mathbb{F}'_1 := \mathbb{F}_1$). By assumption, $\mathbb{F}_q = \mathbb{F}_{q-1}(w_q)$ for some w_q with $w_q^{n_{q-1}} \in \mathbb{F}_{q-1}$. Define the polynomial

$$f(x) := \prod_{\sigma \in G(\mathbb{F}'_{q-1} / \mathbb{F})} \left(x^{n_{q-1}} - \sigma\left(w_q^{n_{q-1}}\right) \right).$$

For all φ in the Galois group $G(\mathbb{F}'_{q-1}/\mathbb{F})$ we have

$$\tilde{\varphi}(f) = \prod_{\sigma \in G(\mathbb{F}'_{q-1}/\mathbb{F})} \left(x^{n_{q-1}} - \varphi\sigma\left(w_q^{n_{q-1}}\right)\right) = \prod_{\gamma \in G(\mathbb{F}'_{q-1}/\mathbb{F})} \left(x^{n_{q-1}} - \gamma\left(w_q^{n_{q-1}}\right)\right) = f,$$

which means that the coefficients of f are in the fixed field $\mathbb{F}$ of $G(\mathbb{F}'_{q-1}/\mathbb{F})$. Hence $f \in \mathbb{F}[x]$.

Let $\mathbb{E}$ be the splitting field of f over $\mathbb{F}'_{q-1}$. Because $\mathbb{F}'_{q-1}$ is a normal extension of $\mathbb{F}$, it is the splitting field of a polynomial $g \in \mathbb{F}[x]$ (in case $q = 2$, use $g(x) = x - 1$). The polynomial $fg \in \mathbb{F}[x]$ splits in $\mathbb{E}$ and any field in which fg splits must contain $\mathbb{F}'_{q-1}$ as well as the elements that must be adjoined to it to obtain $\mathbb{E}$. Hence $\mathbb{E}$ is the splitting field of fg over $\mathbb{F}$, which means that $\mathbb{E}$ is a normal extension of $\mathbb{F}$.

Moreover, by definition of f, $\mathbb{E}$ is obtained from $\mathbb{F}'_{q-1}$ by adjoining n_{q-1}^{st} roots. Hence $\mathbb{E}$ has a root tower over $\mathbb{F}'_{q-1}$. Appending this root tower to the root tower for $\mathbb{F}'_{q-1}$ over $\mathbb{F}$ produces a root tower for $\mathbb{E}$ over $\mathbb{F}$.

Finally, w_q is a root of f. Hence $\mathbb{E}$ must contain $\mathbb{F}_q = \mathbb{F}_{q-1}(w_q)$. ∎

For normal extensions with root towers, we get solvable Galois groups.

Theorem 6.74 *Let $\mathbb{F}$ be a field of characteristic 0 and let $\mathbb{E}$ be a normal extension with a root tower over $\mathbb{F}$. Then the Galois group $G(\mathbb{E}/\mathbb{F})$ is solvable.*

Proof. Let $\mathbb{F} = \mathbb{F}_1 \subseteq \mathbb{F}_2 \subseteq \cdots \subseteq \mathbb{F}_q = \mathbb{E}$ be a root tower for $\mathbb{E}$ over $\mathbb{F}$, let $n := n_1 \cdots n_{q-1}$ and let θ be a primitive n^{th} root of unity. Then for all $j \in \{2, \ldots, q\}$ we have that $\mathbb{F}_j(\theta) = \mathbb{F}_{j-1}(\theta)(w_j)$ and the power $\theta^{\frac{n_1 \cdots n_{q-1}}{n_{j-1}}} \in \mathbb{F}_{j-1}(\theta)$ is a primitive n_{j-1}^{st} root of unity. Thus, by Theorem 6.70, $\mathbb{F}_j(\theta) = \mathbb{F}_{j-1}(\theta)(w_j)$ is a normal extension of $\mathbb{F}_{j-1}(\theta)$ and it has a commutative Galois group $G\left(\mathbb{F}_j(\theta)/\mathbb{F}_{j-1}(\theta)\right)$.

$\mathbb{E}$ is the splitting field of a polynomial $g \in \mathbb{F}[x]$. Then $\mathbb{E}(\theta)$ is the splitting field of the polynomial $g(x)\left(x^n - 1\right)$. Hence $\mathbb{E}(\theta)$ is a normal extension of $\mathbb{F}$ and, by Corollary 6.60, of each $\mathbb{F}_j(\theta)$.

Now consider the root tower $\mathbb{F} \subseteq \mathbb{F}(\theta) = \mathbb{F}_1(\theta) \subseteq \mathbb{F}_2(\theta) \subseteq \cdots \subseteq \mathbb{F}_q(\theta) = \mathbb{E}(\theta)$. For each $j \in \{2, \ldots, q\}$ we have by the Fundamental Theorem of Galois Theory that $G\left(\mathbb{E}(\theta)/\mathbb{F}_j(\theta)\right)$ is a normal subgroup of $G\left(\mathbb{E}(\theta)/\mathbb{F}_{j-1}(\theta)\right)$ and we have that $G\left(\mathbb{E}(\theta)/\mathbb{F}_{j-1}(\theta)\right)/G\left(\mathbb{E}(\theta)/\mathbb{F}_j(\theta)\right)$ is isomorphic to $G\left(\mathbb{F}_j(\theta)/\mathbb{F}_{j-1}(\theta)\right)$, which is commutative. Moreover, $\mathbb{F}(\theta)$ is a normal extension of $\mathbb{F}$ (it's the splitting field of the polynomial $x^n - \theta^n = x^n - 1$) and $G\left(\mathbb{E}(\theta)/\mathbb{F}\right)/G\left(\mathbb{E}(\theta)/\mathbb{F}(\theta)\right)$ is isomorphic to $G\left(\mathbb{F}(\theta)/\mathbb{F}\right)$. An argument similar to the proof of Theorem 6.70 (see Exercise 6-71) shows that this group is commutative. Hence $G\left(\mathbb{E}(\theta)/\mathbb{F}\right)$ is solvable.

By hypothesis, $\mathbb{E}$ is a normal extension of $\mathbb{F}$, and, by the Fundamental Theorem of Galois Theory, we obtain that $G(\mathbb{E}/\mathbb{F})$ is isomorphic to $G\left(\mathbb{E}(\theta)/\mathbb{F}\right)/G\left(\mathbb{E}(\theta)/\mathbb{E}\right)$, which, by Lemma 6.72, is solvable. ∎

A combination of the preceding results leads to the fact that polynomial equations that are solvable by radicals will produce solvable Galois groups for the polynomial.

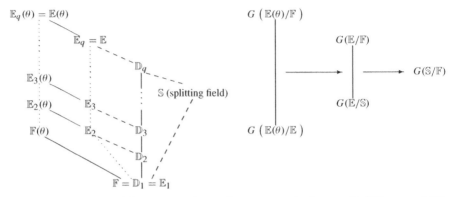

Figure 6.4 Summary of the constructions of root towers in the proof of Theorem 6.75. Solid connections indicate that the higher field is obtained by adjoining an mth root of a number to the lower field. Dotted connections, such as between $\mathbb{E}_{j-1}$ and $\mathbb{E}_j$, indicate that the higher field is obtained from the lower field via a root tower. Dashed connections, such as those for the splitting field $\mathbb{S}$, need not be induced by adjoining roots or via root towers. By Theorem 6.73, the top $\mathbb{D}_q$ of the root tower over $\mathbb{F}$ has an extension $\mathbb{E}$ so that $\mathbb{E}$ contains $\mathbb{D}_q$, and hence the splitting field $\mathbb{S}$, so that $\mathbb{E}$ is a normal extension of $\mathbb{F}$ and so that $\mathbb{E}$ has a root tower, too. The proof of Theorem 6.73 reveals that the root tower for $\mathbb{E}$ is related to that for $\mathbb{D}_q$ as indicated: Some fields $\mathbb{E}_j$ in the root tower for $\mathbb{E}$ are extensions of the corresponding fields $\mathbb{D}_j$ and consecutive fields $\mathbb{E}_j$ are linked by root towers themselves. After adjoining an appropriate primitive root of unity θ to $\mathbb{F}$ (see proof of Theorem 6.74), we see that the Galois group of $G\big(\mathbb{E}(\theta)/\mathbb{F}\big)$ is, essentially by Theorem 6.70 and the Fundamental Theorem of Galois Theory, solvable. All fields, except for $\mathbb{D}_2, \ldots, \mathbb{D}_q$, are normal extensions of $\mathbb{F}$. Repeated use of the isomorphism from part 4 of the Fundamental Theorem of Galois Theory and Lemma 6.72, indicated on the right, shows that $G(\mathbb{E}/\mathbb{F})$ and $G(\mathbb{S}/\mathbb{F})$ are solvable.

Theorem 6.75 *Let* $(\mathbb{F}, +, \cdot)$ *be a field of characteristic* 0 *and let* $p \in \mathbb{F}[x]$ *be so that* $p(x) = 0$ *is solvable by radicals. Then the Galois group of* p *is solvable.*

Proof. Let $\mathbb{S}$ be the splitting field of p over $\mathbb{F}$. Because $p(x) = 0$ is solvable by radicals, there is a root tower over $\mathbb{F}$ whose top field contains all roots of p. So by Theorem 6.73 there is a normal extension $\mathbb{E}$ of $\mathbb{F}$ that contains $\mathbb{S}$ and that has a root tower $\mathbb{F} = \mathbb{F}_1 \subseteq \mathbb{F}_2 \subseteq \cdots \subseteq \mathbb{F}_q = \mathbb{E}$. By Theorem 6.74, $G(\mathbb{E}/\mathbb{F})$ is solvable. By the Fundamental Theorem of Galois Theory, $G(\mathbb{S}/\mathbb{F})$ is isomorphic to $G(\mathbb{E}/\mathbb{F})/G(\mathbb{E}/\mathbb{S})$. By Lemma 6.72, $G(\mathbb{S}/\mathbb{F})$ is solvable. Figure 6.4 provides a summary of the details of the various constructions. ∎

A converse to Theorem 6.75 is proved in Exercise 6-78e, which is the most challenging and, if all supporting exercises are solved, too, the lengthiest exercise in this text.

Exercises

6-70. Let G, H be groups and let the function $\Phi : G \to H$ be a homomorphism. Prove that Φ is injective iff $\Phi(g) = e_H$ implies $g = e_G$.

6-71. Let $\mathbb{F}$ be a field of characteristic 0 and let θ be an nth root of unity. Prove that $G\left(\mathbb{F}(\theta)/\mathbb{F}\right)$ is commutative.

6-72. Prove that every subgroup of $(\mathbb{Z}_m, +)$ is isomorphic to some group $(\mathbb{Z}_n, +)$.

6-73. **Dedekind's Theorem.** Let $\mathbb{F}$ be a field and let $\sigma_j : \mathbb{F} \to \mathbb{F}$, $j = 1, \ldots, n$ be pairwise distinct nonzero field homomorphisms. Prove that if $\lambda_1, \ldots, \lambda_n \in \mathbb{F}$ and $\sum_{j=1}^{n} \lambda_j \sigma_j = 0$, then $\lambda_1 = \cdots = \lambda_n = 0$.

Hint. Induction on n. In the induction step, use an $a \in \mathbb{F}$ so that $\sigma_{n+1}(a) \neq \sigma_1(a)$ and subtract $\sigma_1(a) \sum_{j=1}^{n+1} \lambda_j \sigma_j(\cdot) = 0$ from $\sum_{j=1}^{n+1} \lambda_j \sigma_j(a \cdot) = 0$. Conclude that, by induction hypothesis, we have that $\lambda_{n+1} \left(\sigma_{n+1}(a) - \sigma_1(a) \right) = 0$.

6-74. Let $\mathbb{F}$ be a field of characteristic 0. Let $\mathbb{E}$ be a normal extension of $\mathbb{F}$ so that the Galois group $G(\mathbb{E}/\mathbb{F})$ is isomorphic to a group $(\mathbb{Z}_n, +)$ and so that $\mathbb{F}$ contains all n^{th} roots of unity.

(a) Let $\sigma \in G(\mathbb{E}/\mathbb{F})$ be so that $G(\mathbb{E}/\mathbb{F}) = \left\{ \text{id}, \sigma, \ldots, \sigma^{n-1} \right\}$ and let r be a primitive n^{th} root of unity. Prove that there is a $c \in \mathbb{E}$ so that $b := \sum_{j=0}^{n-1} r^j \sigma^j(c) \neq 0$.

Hint. Use Exercise 6-73.

(b) Prove that $\sigma\left(b^n\right) = b^n$.

Hint. First prove that $\sigma(b) = r^{-1}b$.

(c) Let $a := b^n$. Prove that $a \in \mathbb{F}$ and that b is a root of $x^n - a$.

Hint. Use part 6-74b and the definition of Galois groups.

(d) Let $f \in \mathbb{F}[x]$ be the minimal polynomial of b. Prove that $\deg(f) \geq n$.

Hint. Use that $\tilde{\sigma}(f) = f$ (where $\tilde{\sigma}$ is defined as in Exercise 6-40) and that $\sigma(b) = r^{-1}b$ to generate n distinct roots of f.

(e) Conclude that the minimal polynomial of b is $f(x) = x^n - a$.

(f) Prove that $\mathbb{E} = \mathbb{F}(b)$, which is the splitting field of $x^n - a$.

Hint. Compare $[\mathbb{E} : \mathbb{F}]$ and $\left[\mathbb{F}(b) : \mathbb{F} \right]$.

6-75. Let G and H be groups and let $\varphi : G \to H$ be a group homomorphism.

    (a) Prove that $\big\{ g \in G : \varphi(g) = e \big\}$ is a normal subgroup of G.

    (b) Prove that if φ is surjective, then $G / \big\{ g \in G : \varphi(g) = e \big\}$ is isomorphic to H.

6-76. Let G be a group, let $N \lhd G$ be a normal subgroup of G and let $H \subseteq G$ be a subgroup with $N \subseteq H$.

    (a) Prove that if H is a normal subgroup of G, then G/H is isomorphic to $(G/N)/(H/N)$.
    Hint. Use Exercise 6-75.

    (b) Prove that if G/N is commutative, then H is a normal subgroup of G.

    (c) Prove that if G/N is commutative, then G/H and H/N are commutative groups.

6-77. Prove that for every solvable group G there is a nested sequence $\{e\} = G_0 \lhd G_1 \lhd \cdots \lhd G_n = G$ so that for all $j \in \{1, \ldots, n\}$ the size $\big| G_j / G_{j-1} \big|$ is a prime number p_j and hence the group G_j / G_{j-1} is isomorphic to $\mathbb{Z}_{p_j}$.

    Hint. Take a nested sequence $\{e\} = G_0 \lhd G_1 \lhd \cdots \lhd G_n = G$ so that $\big| G_j / G_{j-1} \big|$ is not prime. Use Cauchy's Theorem to obtain a subgroup S of G_j / G_{j-1} with p elements, where p is prime. Write S as $H G_{j-1}$, where $|H| = p$ and use it to prove that $H G_{j-1}$ is a normal subgroup of G_j / G_{j-1} so that $G_j / H G_{j-1}$ is commutative. Use Exercise 6-76.

6-78. Let $\mathbb{F}$ be a field of characteristic 0, let $\mathbb{E}$ be a normal extension with solvable Galois group $G(\mathbb{E}/\mathbb{F})$, let $\{e\} = G_m \lhd G_{m-1} \lhd \cdots \lhd G_1 \lhd G_0 = G(\mathbb{E}/\mathbb{F})$ be a nested sequence of normal subgroups so that for all $j \in \{1, \ldots, m\}$ the group G_{j-1}/G_j is isomorphic to $(\mathbb{Z}_{p_j}, +)$, where p_j is a prime number (such a nested sequence exists by Exercise 6-77), and let $\mathbb{F}_j := \mathbb{E}^{G_j}$ be the fixed field of G_j.

    (a) Use the Fundamental Theorem of Galois Theory to prove that each $\mathbb{F}_j$ is a normal extension of $\mathbb{F}_{j-1}$ so that $G(\mathbb{F}_j/\mathbb{F}_{j-1})$ is isomorphic to G_{j-1}/G_j.

    (b) Let $n := \big| G(\mathbb{E}/\mathbb{F}) \big|$ and let θ be a primitive n^{th} root of unity. Prove that $\mathbb{F}_j(\theta)$ is a normal extension of $\mathbb{F}_{j-1}$ and of $\mathbb{F}_{j-1}(\theta)$.

    (c) Let $\psi : G \big(\mathbb{F}_j(\theta)/\mathbb{F}_{j-1} \big) \to G(\mathbb{F}_j/\mathbb{F}_{j-1})$ be defined by $\psi(\sigma) := \sigma|_{\mathbb{F}_j}$. Prove that ψ is a group homomorphism and that the restriction $\psi|_{G(\mathbb{F}_j(\theta)/\mathbb{F}_{j-1}(\theta))}$ is injective (use Exercise 6-70). Conclude that $G \big(\mathbb{F}_j(\theta)/\mathbb{F}_{j-1}(\theta) \big)$ is isomorphic to a group $(\mathbb{Z}_{k_j}, +)$ with $k_j \mid p_j \mid n$.

    (d) Prove that $\mathbb{F}_{j-1}(\theta)$ contains a k_j^{th} root of unity and conclude that the $\mathbb{F}_j(\theta)$ are a root tower for $\mathbb{E}(\theta)$ over $\mathbb{F}$.
    Hint. Use Exercise 6-74.

    (e) Prove that if $\mathbb{F}$ is a field of characteristic 0 and the Galois group of $p \in \mathbb{F}[x]$ is solvable, then the equation $p(x) = 0$ is solvable by radicals.
    Hint. Use Exercises 6-77 and 6-78d.

6.9 Abel's Theorem

Now we are ready to prove that there are quintics that are not solvable by radicals. By Example 6.54, there is a a fifth order polynomial whose Galois group is S_5. Theorem 6.75 says that the Galois group of any polynomial p so that the equation $p(x) = 0$ is solvable by radicals must be solvable in the sense of Definition 6.71. But solvability means there is a nested sequence $\{e\} = G_0 \lhd G_1 \lhd \cdots \lhd G_m = G$ of subgroups so that G_j/G_{j-1} is commutative for $j = 1, \ldots, m$, and Theorem 6.68 shows that there may be very few such nested sequences in S_5. Theorem 6.77 below establishes that S_n is not solvable for any $n \geq 5$.

Lemma 6.76 *Let G be a group and let H and N be normal subgroups of G. Then $H \cap N$ is a normal subgroup of N.*

Proof. As an intersection of two subgroups, $H \cap N$ is a subgroup, too. Now let $x \in N$. Because H is normal in G, we have that $xHx^{-1} = H$. That is, for all $h \in H$ we obtain $xhx^{-1} \in H$. Let $y \in H \cap N$. By what we just proved, $xyx^{-1} \in H$. Moreover, because N is a group, we have $xyx^{-1} \in N$. Thus $xyx^{-1} \in H \cap N$ holds for all $y \in H \cap N$. Because $x \in N$ was arbitrary, by Lemma 6.58 we conclude that $H \cap N$ is a normal subgroup of N. ∎

Theorem 6.77 *For every natural number $n \geq 5$, the group S_n is not solvable.*

Proof. Let $n \geq 5$. We first prove that $\{\mathrm{id}\}$, A_n and S_n are the only normal subgroups of S_n. Let $N \lhd S_n$ be a normal subgroup of S_n that is not equal to $\{\mathrm{id}\}$. Then, by Lemma 6.76, $N \cap A_n$ is a normal subgroup of A_n. Because, by Theorem 6.68, A_n is simple, we conclude that $N \cap A_n \in \{\{\mathrm{id}\}, A_n\}$.

Suppose for a contradiction that $N \cap A_n = \{\mathrm{id}\}$. Then all nontrivial permutations in N are odd. Let $\sigma \in N \setminus \{\mathrm{id}\}$. Then $\sigma^2 \in N$ must be even, and thus it must be the identity. Because any odd permutation is a composition of disjoint cycles, the square of each cycle in the representation of σ must be the identity. But the only cycles whose square is the identity are transpositions. Therefore all nontrivial elements of N are products of an odd number of disjoint transpositions.

Without loss of generality, let $\tau \in N \setminus \{\mathrm{id}\}$ be so that (12) is one of the transpositions that make up τ. Then $\tau \neq (12)$: Indeed, otherwise the normality of N implies that $(13) = (32)(12)(23) \in N$, and because N is a subgroup we infer that the even permutation $(123) = (13)(12)$ is in N, which cannot be. Thus τ is a product of an odd number (greater than 1) of disjoint transpositions. Without loss of generality assume that $\tau = (12)(34)\delta$, where δ is a product of an odd number of disjoint transpositions that do not contain any of the numbers $1, 2, 3, 4$. Then, because N is normal, $(13)(24)\delta = (23)(12)(34)\delta(23) \in N$. Because N is a subgroup, $(14)(23) = (12)(34)(24)(13) = (12)(34)\delta\delta^{-1}(24)(13) \in N$. But $(14)(23)$ is an even permutation, which is a contradiction to $N \cap A_n = \{\mathrm{id}\}$.

Thus $N \cap A_n \neq \{\mathrm{id}\}$ and hence $N \cap A_n = A_n$. This means that N is either equal to A_n, or N contains an odd permutation. In case N contains an odd permutation, then, because N contains A_n, N must contain a transposition. Without loss of generality, assume that $(12) \in N$. Then, because N is a normal subgroup, for all $a, b \in \{3, 4, \ldots, n\}$ we have that $(ab) = (1a)(2b)(12)(a1)(b2) \in N$, $(1b) = (2b)(12)(b2) \in N$ and $(a2) = (1a)(12)(a1) \in N$. This means that N contains all transpositions. Hence, if N contains an odd permutation, then $N = S_n$.

We have shown that any normal subgroup N of S_n is one of $\{\mathrm{id}\}$, A_n and S_n. But neither $S_n/\{\mathrm{id}\}$, which is isomorphic to S_n, nor $A_n/\{\mathrm{id}\}$, which is isomorphic to A_n, is commutative. Thus neither of the only two possible nested sequences $\{\mathrm{id}\} \lhd S_n$ and $\{\mathrm{id}\} \lhd A_n \lhd S_n$ of nested normal subgroups from $\{\mathrm{id}\}$ to S_n satisfies the definition of solvability. Thus S_n is not solvable. ∎

Theorem 6.77 was the last piece to solve the puzzle why there is no quintic formula.

Theorem 6.78 Abel's Theorem *or* **Abel-Ruffini Theorem**. *There is a fifth order polynomial p so that the equation $p(x) = 0$ is not solvable by radicals. In particular, this means that there is no "quintic formula."*

Proof. By Example 6.54, the Galois group of the polynomial p from Example 6.4 is isomorphic to S_5. By Theorem 6.77, S_5 is not solvable. If $p(x) = 0$ was solvable by radicals, we would have a contradiction to Theorem 6.75. Hence $p(x) = 0$ cannot be solvable by radicals.   ∎

In conclusion, we may ask what Abel's Theorem actually does for us. In general, results that show that a certain problem cannot be solved will allow mathematicians to focus on problems that *can* be solved. For example, when it was proved that $\sqrt{2}$ was not rational (see Proposition 5.24), a new number system, the real numbers, was needed. Abel's Theorem says that radicals and field operations are not enough to solve polynomial equations. This insight opens the investigation to numerical approximation techniques, as well as other, more complicated operations. For a modern consequence of Abel's Theorem that investigates computability with a computer, consider [22]. Historically, the fact that there are quantities that are not "easily" computable, as well as Galois' approach to the problem had another profound impact on mathematics: The proof showed for the first time that very abstract arguments can lead to useful and powerful results. Ever since, abstraction has been a standard, helpful tool in mathematics. For more on the history of solvability by radicals, consider [25].

Definition 6.79 *In honor of Abel, commutative groups are also called* **abelian groups**. *In honor of Galois, the work that developed from his proof, which was used to write this chapter, is called* **Galois Theory**.

Exercises

6-79. Find a seventh order polynomial p so that the equation $p(x) = 0$ is not solvable by radicals.

6-80. Prove that for every prime degree $n \geq 5$ there is an irreducible polynomial p so that the equation $p(x) = 0$ is not solvable by radicals.

Chapter 7

More Axioms

After Chapter 6 provided an undoubtedly deep "capstone experience," we conclude our presentation of the fundamentals of mathematics with three more axioms for set theory. Of these three, the Axiom of Choice (see Section 7.1) is most widely used throughout mathematics. The Axiom of Replacement (see Section 7.2) and the Continuum Hypothesis (see Section 7.3) are mostly needed in set theory, when extending the ideas of counting and of "standard sizes" for sets. We will therefore focus primarily on the Axiom of Choice. For the two other axioms, we present the essentials, leaving more detailed study to specialized courses and texts in set theory.

7.1 The Axiom of Choice, Zorn's Lemma, and the Well-Ordering Theorem

The Axiom of Choice (especially in its "Zorn's Lemma incarnation," see Theorem 7.6 below) is an indispensable tool for a large variety of existence proofs throughout mathematics. This section provides the Axiom of Choice and some of its consequences that are accessible from this text.

Axiom 7.1 *The* **Axiom of Choice***. Let* $\{A_i\}_{i \in I}$ *be an indexed family of sets. Then there is a function* $f : I \to \bigcup_{i \in I} A_i$ *so that* $f(i) \in A_i$ *for all* $i \in I$*. The function* f *is also called a* **choice function**.

On the surface, the Axiom of Choice looks very natural. Given a family of sets $\{A_i\}_{i \in I}$, it guarantees that (via the choice function) we can *simultaneously* choose exactly one element from each set. The amazing thing about the Axiom of Choice is that, although it is definitely needed in many parts of mathematics, it also has some entirely counterintuitive consequences. The **Banach-Tarski paradox**, which can be

Fundamentals of Mathematics: An Introduction to Proofs, Logic, Sets, and Numbers.
By Bernd S. W. Schröder.

proved using the Axiom of Choice (although we will not do so here), states that there is a way to take a solid ball of radius 1 and partition it into finitely many pieces, which can subsequently be reassembled into a solid ball of radius 2. From a physical point of view, this is completely ridiculous. Therefore, the Banach-Tarski paradox is one of the reasons why there will always be debate about the Axiom of Choice.[1]

Nonetheless, as noted above, the Axiom of Choice is needed to guarantee the existence of various elements in mathematics. Before we continue to the corresponding proofs, let us familiarize ourselves with choice functions. For example, because n-tuples clearly are a finitary notion, choice functions are the only general way to form the product of infinitely many sets.

Definition 7.2 *Let $\{A_i\}_{i \in I}$ be a family of sets. The* **product** $\prod_{i \in I} A_i$ *of the sets A_i is defined as the set of all choice functions $f : I \to \bigcup_{i \in I} A_i$ with $f(i) \in A_i$ for all $i \in I$.*

For finite index sets, Definition 7.2 and the natural generalization of Definition 2.44 to cartesian products with more than two factors produce isomorphic structures. So, for finite index sets, the two constructions can be used interchangeably. For infinite families, the Axiom of Choice implies that the set $\prod_{i \in I} A_i$ is not empty. Conversely, the demand that for every family $\{A_i\}_{i \in I}$ of sets, the product is not empty implies the Axiom of Choice. This is a typical situation for the Axiom of Choice: Many of its consequences are actually equivalent to the Axiom of Choice. That means that, no matter how reasonable these consequences sound and no matter how much we need them, if we want to use the consequences, we must accept the Axiom of Choice (and with it, things like the Banach-Tarski paradox).

Choice functions also allow a sweeping generalization of the notion of distributivity for unions and intersections of sets.

Theorem 7.3 *Intersection and union are* **completely distributive**. *Let $\{J_i\}_{i \in I}$ be a family of index sets and let $\{C_{ij}\}_{i \in I, j \in J_i}$ be a family of sets. Then the following hold.*

1. $\displaystyle \bigcap_{i \in I} \bigcup_{j \in J_i} C_{ij} = \bigcup_{f \in \prod_{i \in I} J_i} \bigcap_{i \in I} C_{if(i)}.$

2. $\displaystyle \bigcup_{i \in I} \bigcap_{j \in J_i} C_{ij} = \bigcap_{f \in \prod_{i \in I} J_i} \bigcup_{i \in I} C_{if(i)}.$

Proof. For part 1, first let $x \in \bigcap_{i \in I} \bigcup_{j \in J_i} C_{ij}$. Then, for every index $i \in I$, there is a j_i with $x \in C_{ij_i}$. That is, for each $i \in I$, the set $A_i = \{j \in J_i : x \in C_{ij}\}$ is not empty. Let $g : I \to \bigcup_{i \in I} A_i$ be a choice function. Then the element x satisfies $x \in \bigcap_{i \in I} C_{ig(i)} \subseteq \bigcup_{f \in \prod_{i \in I} J_i} \bigcap_{i \in I} C_{if(i)}$. Conversely, let $x \in \bigcup_{f \in \prod_{i \in I} J_i} \bigcap_{i \in I} C_{if(i)}$. Then there is a choice function $f \in \prod_{i \in I} J_i$ so that $x \in C_{if(i)}$ for all $i \in I$. But then $x \in \bigcup_{j \in J_i} C_{ij}$ for every $i \in I$, which means $x \in \bigcap_{i \in I} \bigcup_{j \in J_i} C_{ij}$.

---

[1]The author's personal way out of this problem is to note that the Banach-Tarski paradox is not (and cannot be) proved constructively. Therefore our physical intuition must be suspended when we consider it. Once we do that, the Banach-Tarski paradox is not a paradox at all, because the frame of reference in which it would be one is gone.

For part 2, first let $x \in \bigcup_{i \in I} \bigcap_{j \in J_i} C_{ij}$. Then there is an index $i_0 \in I$ so that we have $x \in \bigcap_{j \in J_{i_0}} C_{i_0 j}$. Hence, for every choice function $f \in \prod_{i \in I} J_i$, the containments $x \in \bigcap_{j \in J_{i_0}} C_{i_0 j} \subseteq C_{i_0 f(i_0)} \subseteq \bigcup_{i \in I} C_{if(i)}$ hold. Therefore $x \in \bigcap_{f \in \prod_{i \in I} J_i} \bigcup_{i \in I} C_{if(i)}$. For the reverse inclusion, we prove that if x is not an element of the set on the left side, then x is not in the set on the right side. So let $x \notin \bigcup_{i \in I} \bigcap_{j \in J_i} C_{ij}$. Then for every $i \in I$ there is a $j_i \in J_i$ so that $x \notin C_{ij_i}$. Define the choice function $g \in \prod_{i \in I} J_i$ by $g(i) := j_i$. Then $x \notin \bigcup_{i \in I} C_{ig(i)}$ and hence $x \notin \bigcap_{f \in \prod_{i \in I} J_i} \bigcup_{i \in I} C_{if(i)}$. ∎

Having familiarized ourselves with choice functions, we turn to the possibly most frequently used consequence of the Axiom of Choice: Zorn's Lemma. We first need some terminology for ordered sets.

Definition 7.4 *Let X be an ordered set. A totally ordered subset C of X is also called a **chain**. An element $m \in X$ so that for all $x \in X$ we have that $m \leq x$ implies $m = x$ is called a **maximal element** of X.*

To motivate Zorn's Lemma, let us consider an ordered set in which every chain has an upper bound. We start a construction with a single element. Because that element is a chain, it has an upper bound. If it has a **strict upper bound**, that is, an upper bound that is not an element of the (in this case, singleton) set itself, we add this strict upper bound to get a two element chain. We continue constructing ever larger chains: At every stage of the construction, we have a chain, which must have an upper bound. If there is no strict upper bound for the chain, then the upper bound must be a maximal element. Otherwise, we add a strict upper bound to the chain and continue. Of course this process runs into trouble "after infinitely many steps." But this trouble can be circumvented by forming the union of the chains constructed so far. The process becomes hard to visualize, but it *seems* reasonable. Because the process cannot continue indefinitely, it should ultimately yield a maximal element. As constructive as the above motivation may sound, we need the Axiom of Choice to get past the problem that we encounter after infinitely many steps. Moreover, Exercise 7-2 shows that the existence of maximal elements in ordered sets in which every chain has an upper bound implies the Axiom of Choice. So, although it is easy to say that we somehow "muddle past infinity" above, it is exactly this step which is not possible without invoking the Axiom of Choice.

As we now prove Zorn's Lemma, Lemma 7.5 gives a set theoretical version of what we want to prove, and Theorem 7.6 provides Zorn's Lemma itself.

Lemma 7.5 *Let X be a set, and let $Z \subseteq \mathcal{P}(X)$ be a nonempty set of subsets of X, ordered by set containment $\subseteq$ and with the following properties.*

1. *For every set $C \in Z$, every subset of C is an element of Z.*

2. *For every chain (with respect to set containment) $\mathcal{C} \subseteq Z$, the union $\bigcup \mathcal{C}$ of $\mathcal{C}$ is an element of Z.*

Then Z has a maximal element with respect to set containment.

Proof. Index every element of $\mathcal{P}(X) \setminus \{\emptyset\}$ with itself to obtain the indexed family $\{i\}_{i \in \mathcal{P}(X) \setminus \{\emptyset\}}$. The union of this family is X and, by the Axiom of Choice, there is a choice function $f : \mathcal{P}(X) \setminus \{\emptyset\} \to X$. That is, there is a function $f : \mathcal{P}(X) \setminus \{\emptyset\} \to X$ so that $f(A) \in A$ holds for all $A \in \mathcal{P}(X) \setminus \{\emptyset\}$.

For each $C \in Z$, define the set $E_C := \{x \in X \setminus C : C \cup \{x\} \in Z\}$ and let

$$g(C) := \begin{cases} C \cup \{f(E_C)\}; & \text{if } E_C \neq \emptyset, \\ C; & \text{if } E_C = \emptyset. \end{cases}$$

If $M \in Z$ satisfies $g(M) = M$, then there is no element $x \in X \setminus M$ so that $M \cup \{x\} \in Z$, which means that M is maximal in Z. Hence the proof will be done once we find an $M \in Z$ with $g(M) = M$.

A subset $T \subseteq Z$ will be called a tower iff

1. $\emptyset \in T$, and

2. If $C \in T$, then $g(C) \in T$, and

3. If $\mathcal{C} \subseteq T$ is a chain in T, then $\bigcup \mathcal{C} \in T$.

Note how the idea of a tower reflects our introductory discussion after Definition 7.4. The set Z contains at least one tower, because Z itself is a tower. Moreover, the intersection of any set of towers is a tower, too. The proof is quick. Mentally fill it in. Let T_0 be the intersection of all towers that are contained in Z. Then T_0 is not empty, because $\emptyset \in T_0$.

Call an element $C \in T_0$ comparable iff for all $A \in T_0$ we have $A \subseteq C$ or $C \subseteq A$. We ultimately want to prove that all elements of T_0 are comparable. To do so, we will show that the comparable elements in T_0 form a tower.

First note that, clearly, $\emptyset$ is a comparable set.

Now let $C \in T_0$ be a fixed comparable set. We want to show that $g(C)$ is comparable, too. To do so, consider the set $U := \{A \in T_0 : A \subseteq C \text{ or } g(C) \subseteq A\}$. We will prove that U is a tower, which implies that $U = T_0$, which in turn implies that $g(C)$ is comparable. Clearly, $\emptyset \in U$. Now let $A \in U$. Because C is comparable, we have $A = C$ or $A \subset C$ or $C \subset A$. In case $A = C$, we have $g(A) = g(C) \supseteq g(C)$, which means $g(A) \in U$. In case $A \subset C$, note that, because C is comparable, we have $g(A) \subseteq C$ or $C \subset g(A)$. But the strict containment $C \subset g(A)$ would mean that (by $A \subset C$) C has at least one more element than A and $g(A)$ has at least one more element than C. We would conclude that $g(A)$ has at least two more elements than A, which is impossible. Thus, in case $A \subset C$ we must have $g(A) \subseteq C$, which means $g(A) \in U$. In the last case, $C \subset A$, we note that $A \not\subseteq C$. Thus, by definition of U, $g(C) \subseteq A \subseteq g(A)$ and $g(A) \in U$. Finally, let $\mathcal{A} \subseteq U$ be a chain. If $C \supseteq A$ for all $A \in \mathcal{A}$, then $C \supseteq \bigcup \mathcal{A}$ and $\bigcup \mathcal{A} \in U$. Otherwise, there is an $A \in \mathcal{A}$ so that $C \subset A$. But then $A \not\subseteq C$, which implies $g(C) \subseteq A \subseteq \bigcup \mathcal{A}$ and hence $\bigcup \mathcal{A} \in U$. Thus $U \subseteq T_0$ is a tower. By definition of T_0, $T_0 \subseteq U$ and hence $U = T_0$. Thus for all $A \in T_0$ we have $A \subseteq C \subseteq g(C)$ or $g(C) \subseteq A$. We conclude that if $C \in T_0$ is comparable, then $g(C)$ is comparable, too.

Now let $\mathcal{C} \subseteq T_0$ be a chain of comparable elements and let $A \in T_0$. If there is a $C \in \mathcal{C}$ with $A \subseteq C$, then $A \subseteq C \subseteq \bigcup \mathcal{C}$. Otherwise for all $C \in \mathcal{C}$ we have $C \subseteq A$, which means $\bigcup \mathcal{C} \subseteq A$. Consequently, if $\mathcal{C} \subseteq T_0$ be a chain of comparable elements, then the union $\bigcup \mathcal{C}$ is comparable.

The above shows that the set of comparable elements in T_0 is a tower. Now, because T_0 is the intersection of all towers, we conclude that every element of T_0 is comparable. By definition of comparable elements, we conclude that T_0 is a chain. Therefore, because T_0 is a tower, we have $\bigcup T_0 \in T_0$. By definition of towers, we infer that $g\left(\bigcup T_0\right) \in T_0$, which means that $g\left(\bigcup T_0\right) \subseteq \bigcup T_0$. Moreover, by definition of g, $g\left(\bigcup T_0\right) \supseteq \bigcup T_0$ and we conclude $g\left(\bigcup T_0\right) = \bigcup T_0$.

Hence $M := \bigcup T_0$ is the desired maximal element. ∎

Theorem 7.6 Zorn's Lemma. *Let X be a nonempty ordered set so that every chain in X has an upper bound. Then X has a maximal element.*

Proof. Let Z be the set of all chains in X, ordered by inclusion. If $C \in Z$, then every subset of C is in Z, too. Moreover, the union of every chain in Z is again an element of Z (see Exercise 7-1). By Lemma 7.5, Z has a maximal element M with respect to inclusion. This set M has an upper bound m in X (with respect to the order on X). Hence $M \cup \{m\}$ is a chain in X, that is, $M \cup \{m\} \in Z$. But M is maximal in Z, so $m \in M$. Now let $x \in X$ satisfy $x \geq m$. Then $M \cup \{x\} \in Z$. Using maximality of M in Z once more, we obtain $x \in M$. But then $m \geq x$, and we conclude $m = x$. Therefore, m is maximal in X. ∎

Zorn's Lemma is always invoked in the same way. When a maximal element with respect to a certain order is needed, we prove that every chain in the order has an upper bound. The order often is set containment and the upper bound often is the union.

Of course, the use of Zorn's Lemma need not be very direct. To construct an entity with certain properties, we often must first construct a set of objects that should contain the entity we seek. Then we must construct an order that satisfies the hypothesis of Zorn's Lemma. And finally, we must prove that the maximal element from Zorn's Lemma really is the entity we were looking for.

The following result gives a first indication how Zorn's Lemma can be used. It states that the disjoint union of an infinite set with itself is equivalent (recall Definition 2.59) to the original set. This is a natural extension of the fact that the union of two countable sets is countable (see Theorem 4.65). Note that it is standard practice, if two sets A and B are not guaranteed to be disjoint, but are desired to be disjoint, to consider $A \times \{0\}$ and $B \times \{1\}$ to force disjointness.

Theorem 7.7 *Let A be an infinite set. Then $\left(A \times \{0\}\right) \cup \left(A \times \{1\}\right)$ is equivalent to A.*

Proof. First note that $A \times \{0, 1\} = \left(A \times \{0\}\right) \cup \left(A \times \{1\}\right)$. This will abbreviate notation. Let $\mathcal{F}$ be the set of all bijective functions $f : X \times \{0, 1\} \to X$, where X is a subset of A. The set $\mathcal{F}$ is not empty, because, by Theorem 4.65, it contains all the bijective functions $f : X \times \{0, 1\} \to X$, where $X \subseteq A$ is countable. Now recall that, by Definition 2.49, functions are sets of ordered pairs. This means that $\mathcal{F}$ is ordered by set inclusion. Moreover for any chain $\mathcal{C}$ of functions in $\mathcal{F}$ we can form the union $u := \bigcup \mathcal{C}$, and it will be a bijective function $u : X_u \times \{0, 1\} \to X_u$ for some subset $X_u \subseteq A$ (Exercise 7-3, plus a short argument for domain and range). Clearly, this union is an upper bound for C in the order $\subseteq$ on $\mathcal{F}$. Thus the hypotheses of Zorn's Lemma are satisfied.

Let $h : X \times \{0, 1\} \to X$ be a maximal element of $\mathcal{F}$, as guaranteed by Zorn's Lemma. Suppose for a contradiction that $A \setminus X$ contains a countably infinite set C. Let $b : C \times \{0, 1\} \to C$ be a bijective function. Then the function $t : h \cup b$ is a bijective function between $(X \cup C) \times \{0, 1\}$ and $X \cup C$, that is, $t \in \mathcal{F}$. Moreover clearly $h \subset t$, which contradicts the maximality of h.

Therefore $A \setminus X$ cannot be infinite. If $A \setminus X = \emptyset$, then the function h is the desired bijection between $A \times \{0, 1\}$ and A. Finally consider the case that $A \setminus X \neq \emptyset$. By the above, $A \setminus X$ is finite. Let $C \subseteq X$ be a countably infinite subset of X. Let $R \subseteq C$ be an $|A \setminus X|$-element subset of C. Then $C \setminus R$ is still countably infinite. Let $p : h^{-1}[C] \to C \setminus R$ be a bijective function and let $q : (A \setminus X) \times \{0, 1\} \to A \setminus X \cup R$ be a bijective function. Then $t := \left(h \setminus h|_{h^{-1}[C]} \right) \cup p \cup q$ is the desired bijective function with domain $A \times \{0, 1\}$ and range A. ∎

Note how the last part in the proof of Theorem 7.7 shows how careful we must be with the orders used in proofs that invoke Zorn's Lemma. The function t has a larger domain and a larger range than h, but it does not extend h. Hence h can be maximal with respect to the containment order and it can still not be the function we are looking for.

We would also like to prove that, for infinite sets A, the product $A \times A$ is equivalent to A. But this result requires the Cantor-Schröder-Bernstein Theorem, which will be proved in Section 7.3. Hence we leave further discussions on the sizes of infinite sets to Section 7.3. The proof that every set can be well-ordered is another good example of how Zorn's Lemma is used.

Definition 7.8 *Let S be a set and let $\leq \subseteq S \times S$ be an order relation. Then $\leq$ is called a **well-order** (relation) iff it is a total order and every nonempty subset of S has a smallest element with respect to $\leq$.*

Example 7.9 Theorem 3.51 shows that $\mathbb{N}$ is well-ordered. □

Theorem 7.10 Well-Ordering Theorem. *Every set can be well-ordered. That is, for every set S, there is a well-order relation $\leq \subseteq S \times S$.*

Proof. Let X be the set of all well-order relations $\leq \subseteq D \times D$, where D is a subset of S. This set is not empty, because any one element set can trivially be well-ordered. For any two well-order relations $\leq_1 \subseteq D_1 \times D_1$ and $\leq_2 \subseteq D_2 \times D_2$ in X define $\leq_1 \sqsubseteq \leq_2$ iff $D_1 \subseteq D_2$, every $d_2 \in D_2 \setminus D_1$ is a strict $\leq_2$-upper bound of D_1, and $\leq_2 |_{D_1 \times D_1} = \leq_1$. Then $\sqsubseteq$ is an order relation on X (Exercise 7-4).

Moreover, if $C \subseteq X$ is a chain in X, let $\leq := \bigcup C$. As a union of a chain of order relations, $\leq$ is an order relation: Reflexivity is trivial. Let D be the domain of the relation $\leq$. For antisymmetry, let $x, y \in D$ be so that $x \leq y$ and $y \leq x$. Then there is a $\leq_x \in C$ so that x is in the domain of $\leq_x$ and there is a $\leq_y \in C$ so that y is in the domain of $\leq_y$. Let $\leq'$ be the larger of the two in the order $\sqsubseteq$ on C. Then the domain D' of $\leq' \in C$ contains both x and y. Hence $x \leq' y$ and $y \leq' x$, which implies $x = y$. Transitivity is proved similarly.

To see that $\leq$ is a strict upper bound of $\mathcal{C}$, let $\leq' \in \mathcal{C}$ and let D' be the domain of $\leq'$. Clearly, $D' \subseteq D$. Let $d \in D \setminus D'$ and let $d' \in D'$. Then there is a $\leq'' \in \mathcal{C}$ with domain D'' so that $\leq' \sqsubseteq \leq''$ and $d \in D'' \setminus D'$. But then $d \geq'' d'$, which means $d > d'$. Hence d is a strict $\leq$-upper bound of D'. Finally, because $\leq = \bigcup \mathcal{C}$, $\leq |_{D' \times D'} = \leq'$. **Note that this paragraph does _not_ establish that $\leq$ is an upper bound of $\mathcal{C}$, because we still do not know if $\leq \in X$. But this paragraph does establish that the domain D' of every $\leq' \in \mathcal{C}$ is an "initial segment" of D under $\leq$ and that the restriction of $\leq$ to D' is $\leq'$.**

To see that $\leq \in X$, let $A \subseteq D$ be a nonempty subset of D. Then there is a $\leq' \in \mathcal{C}$ with domain D' so that $A \cap D' \neq \emptyset$. Because $\leq'$ is a well-order, $A \cap D'$ has a $\leq'$-smallest element a. Because $\leq |_{D' \times D'} = \leq'$, a is the $\leq$-smallest element of $A \cap D'$. Because all elements of $D \setminus D'$ are $\leq$-strict upper bounds of D', a is the $\leq$-smallest element of A. Therefore (see Exercise 7-5) $\leq$ is a well-order, and the preceding paragraph shows that $\leq$ is a $\sqsubseteq$-upper bound of $\mathcal{C}$.

By Zorn's Lemma, X has a $\sqsubseteq$-maximal element $\leq$. Then $\leq$ is a well-order with domain D. Suppose for a contradiction that $D \neq S$ and let $s \in D \setminus S$. Define $\leq'$ to be an order relation on $D \cup \{s\}$ so that $\leq' |_{D \times D} = \leq$ and so that s is a strict $\leq'$-upper bound of D. Then $\leq' \in X$ is a strict $\sqsubseteq$-upper bound of $\leq$, contradicting the maximality of $\leq$. Hence $\leq$ must be a well-order for S.   ∎

Exercise 7-6 shows that the Well-Ordering Theorem, too, is equivalent to the Axiom of Choice.

Exercises

7-1. Let X be an ordered set. Let $\mathcal{C} \subseteq \mathcal{P}(X)$ be a set of chains in X so that $\mathcal{C}$ itself is a chain with respect to set inclusion. Prove that $\bigcup \mathcal{C}$ is a chain in X.

7-2. Use Zorn's Lemma to prove the Axiom of Choice.

   Note. This proves that Zorn's Lemma and the Axiom of Choice are equivalent. Hence either one can be used as an axiom for set theory.

7-3. Let A, B be sets and let $\mathcal{F}$ be a set of functions $f : X \to B$, where $X \subseteq A$. Then $\mathcal{F}$ is ordered by set inclusion. Let $\mathcal{C} \subseteq \mathcal{F}$ be a chain.

   (a) Prove that $f := \bigcup \mathcal{C}$ is a function $f : Y \to B$ for some $Y \subseteq B$.

   (b) Prove that if every function in $\mathcal{C}$ is injective, then $f := \bigcup \mathcal{C}$ is injective, too.

   (c) Prove that if $A = B$ and every function $g \in \mathcal{C}$ satisfies $\mathrm{dom}(g) = \mathrm{rng}(g)$, then $f := \bigcup \mathcal{C}$ satisfies $\mathrm{dom}(f) = \mathrm{rng}(f)$, too.

7-4. Prove that the relation $\sqsubseteq$ from the proof of Theorem 7.10 is an order relation on X.

7-5. Let S be a set and let $\leq \subseteq S \times S$ be an order relation. Prove that if every nonempty subset of S has a smallest element with respect to $\leq$, then $\leq$ is a total order.

7-6. Use the Well-Ordering Theorem to prove the Axiom of Choice.

   Note. This proves that the Well-Ordering Theorem and the Axiom of Choice are equivalent. Hence either one can be used as an axiom for set theory.

7-7. Prove that every vector space has a base.

   Hint. Apply Zorn's Lemma to the set of linearly independent subsets.

7-8. Let $\mathbb{F}$ be a field and let $\mathbb{E}$ be an extension of $\mathbb{F}$. Prove that $\mathbb{E}$ has a basis over $\mathbb{F}$.

   Hint. Zorn's Lemma.

7-9. A **filter** $\mathcal{F}$ on a set S is a nonempty set of nonempty subsets of S so that if $F, G \in \mathcal{F}$, then $F \cap G \in \mathcal{F}$ and if $F \in \mathcal{F}$ and $F \subseteq H$, then $H \in \mathcal{F}$. An **ultrafilter** $\mathcal{U}$ on S is a filter $\mathcal{U}$ so that for all filters $\mathcal{F}$ on S we have that $\mathcal{U} \subseteq \mathcal{F}$ implies $\mathcal{U} = \mathcal{F}$.

　(a) Let A be a set and let $S \subseteq A$ be a nonempty subset. Prove that $\mathcal{F}_S := \{X \subseteq A : S \subseteq X\}$ is a filter. Then prove that if $S = \{a\}$ is a singleton set, then $\mathcal{F}_{\{a\}}$ is an ultrafilter.

　(b) Prove that $\{A \subseteq \mathbb{R} : \mathbb{R} \setminus A$ is finite $\}$ is a filter, but not an ultrafilter.

　(c) Prove that every filter $\mathcal{F}$ on a set S is contained in an ultrafilter $\mathcal{U}$ on S.

　　Note. This statement is also known as the **Ultrafilter Axiom**, the **Ultrafilter Theorem** or the **Ultrafilter Lemma**. Unlike Zorn's Lemma and the Well-Ordering Theorem, it is not equivalent to the Axiom of Choice.

　(d) Prove that a filter is an ultrafilter iff for all $A \subseteq S$ we have $A \in \mathcal{F}$ or $S \setminus A \in \mathcal{F}$.

7.2　Ordinal Numbers and the Axiom of Replacement

The Peano Axioms for the natural numbers (see Theorem 2.64) facilitate nothing more than the fundamental idea of counting. Further ideas were introduced in Chapter 3 to abbreviate repeated counting (addition), to abbreviate repeated addition of the same number (multiplication) and to compare sizes (the comparability relation). Thereafter, in Chapters 4 and 5, the integers, the rational numbers, the real numbers, and the complex numbers extended our respective "current" number system to facilitate more of the familiar algebraic operations. At the end of this process we had gone from the Peano Axioms and counting to number systems that are sophisticated enough to investigate algebra (and we analyzed the solution of polynomial equations in detail in Chapter 6) and calculus/analysis. This is good, because it could rightly be claimed that most modern mathematics rests upon algebra and analysis.

But there is another way to extend the natural numbers, and it is connected to the idea of counting rather than the algebraic operations. The natural numbers, including zero to make the set theory simpler, facilitate the counting of finite entities. But we get stuck at infinity. Or do we? Definition 3.53 introduced infinity as a concept and as the symbol ∞. Theorem 2.60 shows that infinity "comes in different sizes," and Sections 4.6 and 5.6 give concrete examples of infinite sets that "have different sizes." The idea for ordinal and cardinal numbers is to assign these infinite sizes a "numerical value." The only way to do that is to "count past infinity," and this section will show how counting past infinity can be achieved.[2] The Well-Ordering Theorem is a pretty good indication that that it should be possible to "count past infinity." In a well-ordered set W, every element x that is not the largest element of W has an immediate successor $x' := \min\{w \in W : w > x\}$, and, of course, this immediate successor would be the next "number" in our counting. Moreover, every infinite well-ordered set contains a copy of the natural numbers (see Exercise 7-10). We could leave it at that, but it would be nice to have a certain standard set of numbers, such as $\mathbb{N}$, for counting, rather than

---

[2]The author's daughters, aged around 7 and 11 at the time, once were competing who could find the largest number. The conversation went from "Ten," "Twenty," etc. to larger numbers until the 11-year old came up with "Infinity," which she had heard about in school. She was sure she had won. The 7-year old countered "Infinity plus one," and the 11-year old looked to the author to settle the competition, saying "Daddy, she can't do that." Of course the author did not give details, but the reply started "Well, technically …"

an amorphous mass of isomorphic well-ordered sets. In the following, we present these "standard numbers," thus answering question 11 on page x.

To see how counting can be extended past $\mathbb{N}$ in a standard fashion that includes $\mathbb{N}$ as constructed in Theorem 2.64, we need to analyze the natural numbers and zero once more. With digits representing numbers, we see that

$$
\begin{aligned}
0 &= \emptyset \\
1 &= \{\emptyset\} = \{0\} \\
2 &= \big\{\emptyset, \{\emptyset\}\big\} = \{0, 1\} \\
3 &= \big\{\emptyset, \{\emptyset\}, \{\emptyset, \{\emptyset\}\}\big\} = \{0, 1, 2\} \\
&\ \vdots
\end{aligned}
$$

The important thing to realize here is that every natural number contains all the natural numbers before it as elements *and* as strict subsets.

Proposition 7.11 *Consider* $\mathbb{N}_0$, *where* $\mathbb{N}$ *is constructed as in Theorem 2.64, and* $0 := \emptyset$ *(see footnote to Definition 3.53). Then every* $n \in \mathbb{N}_0$ *is so that for all* $m \in n$ *we have* $m = \{k \in n : k \subset m\}$.

Proof. Induction on n. The base step $n = 0$ is trivial, because $0 = \emptyset$ has no elements.

Induction step $n \to n' = n \cup \{n\}$: Let $m \in n'$.

If $m \in n$, then by induction hypothesis we have $m = \{k \in n : k \subset m\} \subseteq n$. Thus $n \not\subset m$ and so, because $n' \setminus n = \{n\}$, we infer that $m = \{k \in n' : k \subset m\}$.

If $m \notin n$, then $m = n = \{k \in n : k \subseteq n\}$, because it was proved in the proof of Theorem 2.64 that every element of $n \in \mathbb{N}$ is a subset of n, a claim that trivially holds for 0, too. Moreover, see Exercise 7-11, the containment in the above claim can be sharpened to strict containment, so $m = n = \{k \in n : k \subset n\}$. Finally, because $n' \setminus n = \{n\}$ and because n is not a strict subset of itself, $m = n = \{k \in n' : k \subset n\}$. ∎

Coming back to set theory, the key property for the natural numbers as a counting system is given in Proposition 7.11. Therefore, the natural way to extend counting systems is to define them as sets of sets that have this property.

Definition 7.12 *An* **ordinal number** *is a set* α *of sets that is well-ordered by set inclusion so that for each* $\beta \in \alpha$ *we have that* $\beta = \{\gamma \in \alpha : \gamma \subset \beta\}$.

Clearly, by Proposition 7.11 every natural number is an ordinal number. But, with $\omega := \mathbb{N}_0$, it is easy to see that ω is an ordinal number, too. This means that ordinal numbers allow us to continue the counting process beyond finite numbers. Moreover, by Exercise 7-12, any two ordinal numbers will always satisfy a containment one way or the other. Therefore, we actually end up with a "linear" counting process.

In the setting of ordinal numbers, it is customary to call the natural numbers ω. With ω being the last letter in the Greek alphabet, the notation indicates the end of finite counting. Another name for ω is the **first infinite ordinal number**. This name indicates that, now that "infinity"/ω is tangible, we can try to count beyond it. Proposition 7.13 shows how.

Proposition 7.13 *Let* α *be an ordinal number and define the* **successor** *of* α *to be* $\alpha' := \alpha \cup \{\alpha\}$. *Then* α' *is an ordinal number, too.*

Proof. Exercise 7-13.   ∎

So with ω being the first infinite ordinal, or (countable) "infinity," ω' counts one beyond ω. Hence ω' is also denoted $\omega + 1$. Numbers $\omega + 2$, etc. are defined similarly. This process runs into problems once we have constructed all $\omega + n$. No axiom of set theory guarantees that the collection of all the $\omega + n$ is a set once more. For example, we cannot use the Axiom of Unions, because we have no guarantee that the totality of the singleton sets $\{\omega + n\}$ forms a set. Here is where the Axiom of Replacement comes in.

Axiom 7.14 Axiom of Replacement. *Let A be a set. If $S(a, b)$ is a sentence such that for each $a \in A$ there is an element b that satisfies $S(a, b)$, then there exists a set S that contains these elements b. In particular, $\{b : S(a, b), a \in A\}$ is a set. The name of the axiom comes from the fact that each element of A is replaced with the element(s) b for which $S(a, b)$ holds.*

For our attempt to "count twice infinity," let $A := \omega = \mathbb{N}_0$, let $C(x) := x \cup \{x\}$, let C^n denote the n-fold application of the function C and let $S(n, b) := [b = C^n(\omega)]$. Then, via the Axiom of Replacement, we have that $\{C^n(\omega) : n \in \mathbb{N}_0\}$ is a set. It contains $\omega + 0 = \omega = C^0(\omega)$, $\omega + 1 = \omega' = C^1(\omega)$, ..., $\omega + n = C^n(\omega)$, etc. Moreover, the set $2\omega := \omega \cup \{C^n(\omega) : n \in \mathbb{N}_0\}$ is an ordinal number (Exercise 7-14).

In this fashion, ever larger ordinal numbers (3ω, 4ω, ..., ω^2, etc.) can be constructed. But there is no set that consists of all ordinal numbers: Suppose for a contradiction that there was such a set. If this set had a maximum m, then the successor $m' = m \cup \{m\}$ would be an ordinal number that is not in the set. If the set did not have a maximum (with respect to set inclusion), then the union of all its elements would be an ordinal number, too, and would not be in the set. This is called the **Burali-Forti paradox**.

We should also note that the ordinal numbers are "totally ordered" by set inclusion. Of course, because the ordinal numbers do not form a set, inclusion is formally not an order relation here. But reflexivity, antisymmetry and transitivity are trivial, and any two ordinal numbers α and β will always satisfy $\alpha \subseteq \beta$ or $\beta \subseteq \alpha$ (see Exercise 7-12). So we have everything we need for a total order relation, except that we cannot consider all ordinal numbers within the safe confines of a set. This is usually not a problem: When we work with orderings, we only compare two elements at a time.

Exercises

7-10. Let W be an infinite well-ordered set. Prove that there is a bijective function φ from $\mathbb{N}$ to a subset $M \subseteq W$ so that $m \leq n$ implies $\varphi(m) \leq \varphi(n)$ and so that every $w \in W \setminus M$ is a strict upper bound of the set M.

7-11. Consider $\mathbb{N}_0$, where $\mathbb{N}$ is constructed as in Theorem 2.64, and $0 := \emptyset$ (see footnote to Definition 3.53). Prove that if $n \in \mathbb{N}_0$, then every element of n' is a *strict* subset of n'.

7-12. Let α and β be two ordinal numbers. Prove that $\alpha \subseteq \beta$ or $\beta \subseteq \alpha$.
   Hint. Assume $\alpha \nsubseteq \beta$ and consider the smallest element of α that is not contained in β.

7-13. Prove Proposition 7.13.

Hint. First prove that every element of an ordinal number is a strict subset of the ordinal number, too.

7-14. For any x, let $C(x) := x \cup \{x\}$, and let C^n denote the n-fold application of this operation. Prove that the set $2\omega := \omega \cup \{C^n(\omega) : n \in \mathbb{N}_0\}$ is an ordinal number.

7-15. Let A be a set.

(a) Prove that the superstructure over A is a set, too.

(b) Explain why the Axiom of Replacement is needed to prove part 7-15a.

(c) Explain why the superstructure itself (as a set) is not part of the model for set theory given in the discussion after Definition 2.62.

7-16. Let W be a well-ordered set. Prove that there is an ordinal number α and a bijective function $\varphi : W \to \alpha$ so that $x \leq y$ iff $\varphi(x) \subseteq \varphi(y)$.

Hints. Use Zorn's Lemma, extending functions that are defined on down-sets of W.

7-17. The **Axiom of Substitution** states the following. Let A be a set. If $S(a, b)$ is a sentence such that for each $a \in A$ the set $\{b : S(a, b)\}$ can be formed, then there exists a function F with domain A such that $F(a) = \{b : S(a, b)\}$ for all $a \in A$.

Prove that the Axiom of Replacement and the Axiom of Substitution are equivalent.

7-18. When working with ordinal numbers, it can be handy to know that no set is an element of itself. (We proved a similar result just for natural numbers in the proof of Theorem 2.64.) It turns out that this simple and intuitive statement actually is close to another axiom of set theory. The **Axiom of Regularity** states the following. Every nonempty set A contains an element B which is disjoint from A.

Use the Axiom of Regularity to prove that no set is an element of itself.

Hint. Suppose for a contradiction that $C \in C$ and apply the Axiom of Regularity to $A := \{C\}$.

7.3 Cardinal Numbers and the Continuum Hypothesis

Just as the natural numbers provide the standard sizes for finite sets, we want to use ordinal numbers to define "standard sizes" for infinite sets. These "standard sizes" are called cardinal numbers. Because it is possible to do arithmetic with finite numbers, we also want to introduce an arithmetic with infinite cardinal numbers. To start, let us explore the arithmetic for the sizes of finite sets.

Theorem 7.15 *Let A, B and C be finite sets. Then the following hold.*

1. If $A \cap B = \emptyset$, then $|A \cup B| = |A| + |B|$.

2. $|A \cup B| = |A| + |B| - |A \cap B|$.

3. $|A \cup B \cup C| = |A| + |B| + |C| - |B \cap C| - |A \cap B| - |A \cap C| + |A \cap B \cap C|$.

4. $|A \times B| = |A| \cdot |B|$.

5. With A^B denoting the set of all functions from B to A, we have $\left|A^B\right| = |A|^{|B|}$.

Proof. For part 1, a bijective function between $A \cup B$ and $|A| + |B|$ is constructed in Exercise 7-19a.

Parts 2-4 require the representation of the sets in question as unions of disjoint sets. For example, for part 2 the argument is the following.

$$
\begin{aligned}
|A \cup B| + |A \cap B| &= \left|A \cup (B \setminus A)\right| + |A \cap B| \\
&= |A| + |B \setminus A| + |A \cap B| \\
&= |A| + \left|B \setminus (A \cap B)\right| + |A \cap B| \\
&= |A| + |B|
\end{aligned}
$$

The argument for part 3 is no different in principle, but it is a little longer.

$$
\begin{aligned}
|A \cup B \cup C| \\
= \ & \left|A \cup (B \cup C)\right| \\
= \ & |A| + |B \cup C| - \left|A \cap (B \cup C)\right| \\
= \ & |A| + |B| + |C| - |B \cap C| - \left|(A \cap B) \cup (A \cap C)\right| \\
= \ & |A| + |B| + |C| - |B \cap C| - \left[|A \cap B| + |A \cap C| - \left|(A \cap B) \cap (A \cap C)\right|\right] \\
= \ & |A| + |B| + |C| - |B \cap C| - |A \cap B| - |A \cap C| + |A \cap B \cap C|
\end{aligned}
$$

The proof of part 4 is left to Exercise 7-19b.

Finally, the proof of part 5 is an induction on $|B|$ (see Exercise 7-19c). ■

Parts 2 and 3 of Theorem 7.15 are special cases of the Principle of Inclusion and Exclusion, which is proved in Exercise 7-20. This result is often used in counting arguments.

Example 7.16 *At a certain school, 57 students lettered in academics, track, or band. Of these students, 53 lettered in academics, 19 lettered in track, 27 lettered in band, 13 lettered in academics and band, 18 lettered in track and band, and 5 lettered in academics, track and band. How many students lettered in academics and track? How many students lettered in academics and track, but not in band?*

Let A be the set of students who lettered in academics, let T be the set of students who lettered in track, and let B be the set of students who lettered in band. Then by the above, $|A| = 53$, $|T| = 19$, $|B| = 27$, $|A \cap B| = 13$, $|T \cap B| = 18$, $|A \cap T \cap B| = 5$, and $|A \cup T \cup B| = 57$. Therefore,

$$
\begin{aligned}
|A \cap T| &= |A| + |T| + |B| - |A \cap B| - |T \cap B| + |A \cap B \cap C| - |A \cup B \cup C| \\
&= 53 + 19 + 27 - 13 - 18 + 5 - 57 = 16.
\end{aligned}
$$

So 16 students lettered in academics and track (and possibly in band) and 11 students lettered in academics and track, but not in band. □

After this "warm-up," we are ready to tackle sizes of infinite sets and the arithmetic that comes with these sizes. It is easy to see that $\omega + 1$ and ω are equivalent sets (recall Theorem 4.65). Therefore, not every ordinal number will be a standard size for sets. Instead we use the smallest ordinal number of every possible size. By the Well-Ordering Theorem and by Exercise 7-16, such an ordinal number does exist.

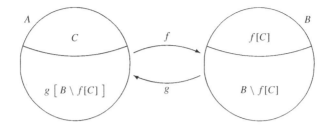

Figure 7.1 The idea for the proof of the Cantor-Schröder-Bernstein Theorem is to partition A into two sets C and $A \setminus C$ so that $A \setminus C = g[B \setminus f[C]]$. Bringing A to the right side in this equation produces a fixed point equation for C as well as the function used in the proof.

Definition 7.17 A **cardinal number** *is an ordinal number* α *so that for all ordinal numbers* β *that are equivalent to* α *(in the sense of Definition 2.59) we have* $\alpha \subseteq \beta$.

Definition 7.18 *For every infinite set* S, *we define the* **cardinality** $|S|$ *of* S *to be the unique cardinal number* α *that is equivalent to* S.

To analyze the arithmetic of cardinal numbers, we must understand how sizes of sets can be ordered. Because arbitrary sets do not necessarily share any (much less all) of their elements, comparison via containment is not an option. But it would make sense to say that a set A is "smaller than or equal to" another set B iff there is an injective (but not necessarily surjective) function from A to B. This relation is easily seen to be reflexive and transitive on any set of sets. Antisymmetry has a surprisingly sophisticated proof (see Theorem 7.19 below). Note that, just as for ordinal numbers, we cannot quite call this way to compare sets an order relation, because there is no set of all sets.

Theorem 7.19 Cantor-Schröder-Bernstein Theorem. *Let* A *and* B *be sets so that there is an injective function* $f : A \to B$ *and an injective function* $g : B \to A$. *Then there is a bijective function* $h : A \to B$.

Proof. Define the function $F : \mathcal{P}(A) \to \mathcal{P}(A)$ on the power set of A by setting $F(X) := A \setminus g[B \setminus f[X]]$ for all $X \subseteq A$. Then $X \subseteq Y$ implies $f[X] \subseteq f[Y]$, which implies $B \setminus f[X] \supseteq B \setminus f[Y]$, which implies $g[B \setminus f[X]] \supseteq g[B \setminus f[Y]]$, which implies $F(X) = A \setminus g[B \setminus f[X]] \subseteq A \setminus g[B \setminus f[Y]] = F(Y)$.

Now let $C := \bigcup \{H \in \mathcal{P}(A) : H \subseteq F(H)\}$. The union is well-defined, because $\emptyset \subseteq F(\emptyset)$. Let $c \in C$. Then there is an $H \in \mathcal{P}(A)$ so that $c \in H \subseteq F(H) \subseteq F(C)$. Hence $C \subseteq F(C)$. But then $F(C) \subseteq F(F(C))$, and by definition of C, this means $F(C) \subseteq C$. Thus $C = F(C)$.

Now $C = F(C) = A \setminus g[B \setminus f[C]]$, which implies $g[B \setminus f[C]] = A \setminus C$, which implies $B \setminus f[C] = g^{-1}[A \setminus C]$. Hence $g^{-1}\big|_{A \setminus C}$ is a bijective function from $A \setminus C$ onto $B \setminus f[C]$. Define $h : A \to B$ by $h|_C := f|_C$ and $h|_{A \setminus C} := g^{-1}\big|_{A \setminus C}$. Then

$h|_C : C \to f[C]$ and $h|_{A \setminus C} : A \setminus C \to B \setminus f[C]$ are bijective, which means that h is surjective. To prove that h is injective, let $x, y \in A$ be so that $x \neq y$. If $x, y \in C$, then $h(x) = f(x) \neq f(y) = h(y)$ and if $x, y \notin C$, then $h(x) = g^{-1}(x) \neq g^{-1}(y) = h(y)$. If neither of the above is the case, we can assume without loss of generality that $x \in C$ and $y \notin C$. In this case, $h(x) \in f[C]$ and $h(y) \in B \setminus f[C]$, which means $h(x) \neq h(y)$.

Thus the function $h : A \to B$ is bijective. For an illustration of the construction, see Figure 7.1.   ∎

Using the Cantor-Schröder-Bernstein Theorem and the fact that for any two sets A and B there must be an injective function from A to B or from B to A (Exercise 7-21), we can prove that every product of an infinite set with itself is equivalent to the set itself.

Theorem 7.20 *Let A be an infinite set. Then $A \times A$ is equivalent to A.*

Proof (sketch). The proof is an application of Zorn's Lemma to the set $\mathcal{F}$ of all bijective functions $f : X \times X \to X$, where $X \subseteq A$. To prove that a maximal element $g : Y \times Y \to Y$ of $\mathcal{F}$ must be a bijective function from $A \times A$ to A, assume that Y is not equivalent to A. By Exercise 7-21 and Theorem 7.19, there must be an injective function from Y to $A \setminus Y$. Then let $Z \subseteq A \setminus Y$ be equivalent to Y and expand g to a function from $(Y \cup Z) \times (Y \cup Z)$ to $Y \cup Z$, using Theorem 7.7 to establish that $Y \times Z \cup Z \times Z \cup Z \times Y$ is equivalent to Z. The reader will supply all necessary details in Exercise 7-22.   ∎

Similar to the arithmetic for sizes of finite sets (see Theorem 7.15), the arithmetic of cardinal numbers mirrors the algebra of sets, too. Note that for the sum of cardinal numbers, we use products with "dummy factors" to assure that the sets involved will be disjoint.

Definition 7.21 Cardinal arithmetic. *Let α and β be cardinal numbers and let A and B be sets with $|A| = \alpha$ and $|B| = \beta$. We define*

1. $\alpha + \beta := |A \times \{0\} \cup B \times \{1\}|$

2. $\alpha\beta := |A \times B|$

3. $\alpha^\beta := |A^B|$, *where* $A^B := \{f : f \text{ is a function from } B \text{ to } A\}$.

The algebra for infinite cardinal numbers is surprisingly simple. By Exercise 7-12, for any two cardinal numbers, we have $\alpha \subseteq \beta$ or $\beta \subseteq \alpha$. So we define $\max\{\alpha, \beta\} := \beta$ iff $\alpha \subseteq \beta$ and $\max\{\alpha, \beta\} := \alpha$ iff $\beta \subseteq \alpha$.

Theorem 7.22 *Let α, β be cardinal numbers so that one of α and β is infinite. Then $\alpha + \beta = \max\{\alpha, \beta\}$ and $\alpha\beta = \max\{\alpha, \beta\}$.*

Proof. Both results are consequences of Theorems 7.7 and 7.20 and the Cantor-Schröder-Bernstein Theorem (see Exercise 7-23).   ∎

Theorem 7.23 *Let α, β and γ be cardinal numbers. Then*

1. $\alpha^{\beta}\alpha^{\gamma} = \alpha^{\beta+\gamma}$

2. $\alpha^{\gamma}\beta^{\gamma} = (\alpha\beta)^{\gamma}$

3. $\left(\alpha^{\beta}\right)^{\gamma} = \alpha^{\beta\gamma}$

Proof. Let A, B, C be pairwise disjoint sets with $|A| = \alpha$, $|B| = \beta$ and $|C| = \gamma$. Then $A^{B\cup C}$ is equivalent to the set $A^{B} \times A^{C}$ via the map that maps every $f \in A^{B\cup C}$ to $(f|_{B}, f|_{C})$. Hence $\alpha^{\beta+\gamma} = \left|A^{B\cup C}\right| = \left|A^{B} \times A^{C}\right| = \alpha^{\beta}\alpha^{\gamma}$.

The proofs of the remaining parts are similar. The reader can provide them in Exercise 7-24. ∎

The exponentiation of cardinal numbers brings us to a final (for this text), mysterious feature of set theory. The **first infinite cardinal** is called $\aleph_0$. (The symbol $\aleph$ is the Hebrew letter **aleph**.)[3] This cardinal is, of course, the familiar ordinal number ω. Because any set of cardinal numbers is well-ordered (after all, it is a set of ordinal numbers and these sets are well-ordered), there is a **second infinite cardinal**, which is called $\aleph_1$. We have seen only one set-theoretical way to obtain sets with larger cardinality: Theorem 2.60. In the notation of cardinal arithmetic, $|\mathcal{P}(X)| = 2^{|X|}$, because the subsets of X are in bijective correspondence with the functions from X to $2 = \{0, 1\}$ (for each $A \in \mathcal{P}(X)$, set $\chi_A(x) := 1$ if $x \in A$ and $\chi_A(x) := 0$ if $x \notin A$). So the question beckons: Is $2^{\aleph_0} = \aleph_1$? The answer is quite surprising. Set theory can be formulated in a consistent way if the equality holds, and it can be formulated in a consistent way, if it does not. The more common assumption seems to be that the equality holds. Demanding this equality is called the Continuum Hypothesis, which is the last axiom we present.

Axiom 7.24 *The* **Continuum Hypothesis.** $2^{\aleph_0} = \aleph_1$.

Because set theory can be formulated in a valid way with and without the Continuum Hypothesis, we also say that the Continuum Hypothesis is **independent** of the other axioms of set theory. The Axiom of Choice as well as the Axiom of Replacement are independent of the other axioms, too. Detailed investigation of this notion of independence is best left to a specialized text or course in set theory. The interested reader can find the proofs in [6].

Exercises

7-19. Prove the remaining parts of Theorem 7.15.

    (a) Prove part 1 of Theorem 7.15.

    (b) Prove part 4 of Theorem 7.15.

    (c) Prove part 5 of Theorem 7.15.

---

[3]This is the third name that we have for this set: $\aleph_0 = \omega = \mathbb{N}_0$. Depending on the context, one or another of these names will be used.

7-20. The **Principle of Inclusion and Exclusion**. Let $A_1, \ldots, A_n$ be finite sets. Prove that

$$\left| \bigcup_{j=1}^{n} A_j \right| = \sum_{k=1}^{n} (-1)^{k-1} \sum_{j_1 < j_2 < \cdots < j_k} \left| A_{j_1} \cap A_{j_2} \cap \cdots \cap A_{j_k} \right|.$$

7-21. Let A and B be sets. Prove that there must be an injective function from A to B or from B to A.

   Hint. Consider the set of all injective functions from A to B. The maximal element from Zorn's Lemma will be bijective, but it may not be totally defined.

7-22. Give a fully detailed proof of Theorem 7.20.

7-23. Let α, β be cardinal numbers so that one of α and β is infinite.

   (a) Prove that $\alpha + \beta = \max\{\alpha, \beta\}$.

   (b) Prove that $\alpha\beta = \max\{\alpha, \beta\}$.

7-24. Finish the proof of Theorem 7.23.

   (a) Prove part 2 of Theorem 7.23.

   (b) Prove part 3 of Theorem 7.23.

7-25. Prove that if α, β are finite cardinals greater than 1 and γ is an infinite cardinal, then $\alpha^\gamma = \beta^\gamma$.

   Hint. Use that $\alpha^\gamma = \left| \prod_{c \in C} A \right|$ and Theorem 7.22.

7-26. Revisiting the Cantor-Schröder-Bernstein Theorem. Let A and B be sets so that there is a surjective function $f : A \to B$ and a surjective function $g : B \to A$. Prove that there is a bijective function $h : A \to B$. You may use the Axiom of Choice.

7-27. Explain why it is not possible to assign 30 people to three groups of 15 so that any two groups have at most 4 members in common.

7-28. Explain why it is not possible to assign 30 people to three groups of 15 so that any two groups have at least 11 members in common.

7-29. In everyday usage, numbers are used as cardinals (to indicate sizes), as ordinals (for precedence/order), and for identification (such as social security numbers, etc.). For each of the following uses of numbers, indicate if the number was used as a cardinal, an ordinal, or for identification.

   (a) Our 7 points beat your 3 points.

   (b) My bank account is empty.

   (c) Please send us your checking account number.[4]

   (d) Our best high jumper just cleared 6 feet.

   (e) Out team's top 2 mile runner finished in 12 minutes, your team's top 2 mile runner finished in 13 minutes.

   (f) Does my daughter's 3 foot height meet the requirements for this ride?

   (g) I need to lose 8 pounds.

7-30. Prove that $|\mathbb{R}| = \left| \mathcal{P}(\mathbb{N}) \right|$.

   Hint. Construct an injective function from $(0, 1)$ to the set of all sequences of zeros and ones (similar idea as in Theorem 5.50) and show that the set of elements not in the range of this function has at most the cardinality of the range itself. Then show that $(0, 1)$ is equivalent to $\mathbb{R}$ and use Exercise 5-49.

---

[4]Never respond to e-mails that include such requests.

Appendix A

Historical Overview and Commentary

Mathematics is presented in linear fashion to safeguard against circular reasoning. But mathematics rarely develops linearly. It is therefore quite interesting to find out what concepts were known when and how they influenced other developments. This appendix gives a brief overview of how and when various concepts arose, as well as some idea about the people who discovered them. The references back to the text will hopefully serve to reinforce the results themselves, too.

I should say right here that I am not a historian. So what follows is pieced together from various references, such as [7], [14], [19], and [25], as well as from my memory of historical remarks heard in lectures. It is not meant to be a comprehensive historical account that goes back to the primary references. Mixed in are anecdotes that I sometimes tell in class to allow the brain to relax amid the often demanding mathematics, as well as (hopefully appropriate) attempts at social commentary to illustrate that mathematics as well as history are ultimately made by people with all their ingenuity, but also with all their faults. The perspective is European, because, ultimately, Europe is the place in which most of the significant breakthroughs happened, but also because I am originally European. I hope you'll enjoy my first attempt at extended prose.

Use this appendix as light reading when the mathematics gets a little deep. Think about any of my comments that you may consider enlightening or provocative. (If they are, the intent is to provoke thought, not to make people take my point of view.) But use an actual history text if you are interested in further details and more background. For a nice text that views history through the focus of prime numbers, consider [8].

Fundamentals of Mathematics: An Introduction to Proofs, Logic, Sets, and Numbers.
By Bernd S. W. Schröder.
Copyright © 2010 John Wiley & Sons, Inc.

A.1 Ancient Times: Greece and Rome

Much of what we consider mathematics and mathematical reasoning goes back to the (rightly) revered philosophers of ancient Greece, whose names are especially entwined with geometry, triangles, and straightedge and compass constructions. Historians can point towards evidence that various pieces of mathematics were known earlier. For example, quadratic equations were solved in ancient Babylon and the quadratic formula (see Theorem 5.63) was known to the Babylonians. There also seems to be evidence that Pythagoras may not have proved what we know as Pythagoras' Theorem, or that, at the very least, he was not the first to do so.

But the salient and original feature of the Greek's pursuit of mathematics is that it established the foundations of logical reasoning (see Chapter 1) and that it rigorously followed these rules of reasoning. The Greeks may not have worked with formal logic and truth tables. But they adhered to the rules that govern logical discourse and they established the idea that results should be derived from axioms that are accepted as true. The first significant organized summary for this approach may be Euclid's *Elements*. In it, a set of postulates for geometry is collected and then further results (such as Example 1.52) are derived from them. These results are fundamental and they still are the subject of high school geometry. But what is even more fundamental is that logical deduction connects the results in an ironclad fashion. If we accept the axioms, then all their consequences must be true. This is exactly the strength of mathematical reasoning.

At the same time, the Greeks, being human after all, were hesitant to rethink parts of their philosophy, even when it became clear that these parts were unsustainable. Therefore, clashes between what people wanted to be and what actually *is* were unavoidable. For example, the Greeks adhered to a notion that was summarized by Pythagoras as *"All is number."* By numbers, the Greeks specifically meant *natural* numbers. Consequently, everything was expected to be expressible in terms of (ratios of) natural numbers. Now, "everything" includes lengths of sides of triangles, such as the familiar right triangle for which the sides that make the right angle both have length 1. Eventually it was proved that $\sqrt{2}$, the length of the third side of this triangle, is not rational (see Example 1.57 and Proposition 5.24). The philosopher Hippasus, who either proved the result or disclosed the secret to some uninitiated people, was drowned for his deed. Unfortunately, the history of mankind serves as evidence that such behavior is possible in human beings in any time period. We can name examples from Galileo's significant problems with the church to humorous stories about administrators being worried because half the people in their charge performed below the group's median.

From a mathematical point of view, the square root of 2 may well be the first paradox, or even crisis, in the history of mathematics: Reasonable looking assumptions were used in logical fashion, and the result was a contradiction. The problem was deeper than just completing a simple proof by contradiction. Similar to how Russell's Paradox is presented in Section 1.8, the assumptions with which the Greeks started the proof that $\sqrt{2}$ is irrational, including the dogma that all numbers are rational, were *all considered to be true*, and there was no deliberate attempt to debunk any of them. The problems with irrational numbers would only be completely resolved by the end of the nineteenth century. The philosopher Eudoxos (408?-355?B.C.) created a theory of proportion that included irrational numbers. It remained the definitive work on this subject until the end of the nineteenth century. Moreover, once one gets past the different way of writing mathematics, it was quite close to what we now consider real numbers. Considering how sophisticated the ideas surrounding irrational numbers are, it may be surprising that the Greeks did not work with negative numbers. In fact, avoidance of negative numbers persisted until well into the sixteenth century.

With other problems, the Greeks simply were stuck. For example, they knew how to bisect an angle with compass and straightedge, but they could not find a compass and straightedge construction to trisect an angle. Of course, the reason is that no such construction exists (see Exercises 6-5, 6-16, 6-31, 6-32 and 6-33), but the main tools for the proof, such as cartesian coordinates, were not available. Plus, it is a quantum leap to start thinking about proving that a construction is *not* possible. Proving that a construction is possible is comparatively "easy": Once the construction is spelled out, we show that it works, and that's it. But how do we prove that a construction is impossible? We can't list all the possible sequences of operations with straightedge and compass and show that none of them work! Chapter 6 certainly shows that non-existence proofs require a significant leap into abstraction that was not available at the time.

Similarly, the Greeks were close to the concept of the limit. Their proof for the formula for the area of a circle is essentially a limit argument. But the Greeks never were comfortable with infinity. Proofs that would have been easier using infinity (many of them by the aforementioned Eudoxos) were instead completed by using exhaustion arguments that culminated in two proofs by contradiction, but which avoided infinity. Infinity was even avoided in the language. The Greeks knew the set of prime numbers is not finite (see Exercise 4-59) and they used the proof that is given today. But instead of saying the set is infinite, they left it at saying that there are not finitely many prime numbers.

In some cases, the Greeks' avoidance of infinity even led to wrong results. The "paradox" of Achilles and the turtle states that if the warrior Achilles ran a race against a turtle and gave the turtle a head start, then he would never catch up to the turtle. The "argument" is that by the time Achilles reaches the turtle's starting point, the turtle has moved ahead a little. By the time he reaches that point, the turtle has moved ahead again, and so on, indefinitely. So he can never catch up. Of course the Greeks had observed runners catching up to each other, so they knew something was wrong. Hence they considered it a "paradox." From a modern point-of-view, "Achilles and the turtle" is a false argument, not a paradox. The times it takes Achilles to catch up get shorter and shorter and their (infinite) sum is finite. Thus Achilles catches up in finite time and then passes the turtle. But this was exactly the leap the Greeks did not make. If something had infinitely many parts, their avoidance of infinity sometimes kept them from the correct conclusion.

From the author's layman's perspective, there are many key ideas and even attitudes in history, which may seem simple from a modern point of view, but without which other truly monumental developments would have been impossible. The above are some examples and more will be pointed out in this essay. Such situations seem to show the distinctly human element that influences our discovery of mathematical and natural laws. Nonetheless, as we examine such key changes in attitude, we should be careful to not dismiss the development as simple, just because we have learned it in school. Anyone who has ever been stuck doing a proof that turned out to be "simple" after all will know better than that.

The Greeks pursued natural science, too. Archimedes possibly was the most famous Greek philosopher to tackle applications, such as, for example, the principle of buoyancy. But the Greeks were mostly philosophers, not engineers. The Roman empire ultimately supplanted the Greek culture and the transition was often violent. Archimedes was killed when the Romans took Syracuse in a surprise attack in 212B.C. The Greeks were celebrating a feast in honor of the goddess Artemis, and wine as well as negligence made the job comparatively easy on the Romans. As the story goes, Archimedes was working on a proof when a Roman soldier came to take him to the commander Marcellus. When Archimedes, 75 years old and deep in thought, asked the soldier to "not disturb my circles," the soldier slew Archimedes with his sword.

The ascendancy of the Roman empire was not based on military might alone. The Romans had a keen ability to apply scientific principles, whether they understood them or not, and they were indeed capable engineers. Roman weapons technology was superior in the ancient world and some of the arches that supported their aquaeducts stand to this day. So it cannot be said that the Romans were intellectually inferior to the Greeks. They simply had no interest in theory. Things like Caesarian ciphers (see introduction to Section 3.9) were invented, they worked, and that was it. This approach works in a non-technological, non-scientific world. But it ultimately runs into limitations because deeper ideas that could lead to better inventions will forever be hidden.

We can only speculate how successful a *perfect*[1] merger of the Greeks' affinity for theory and the Romans' ingenuity for applications would have made a combined Greco-Roman empire. The Greeks were close to the concept of the limit, and the Romans were pragmatic enough to accept something like infinity. So a combination of approaches could have led to calculus. Archimedes did some very sound physics, and combining these ideas with the Romans' knack for applications might have triggered an industrial revolution of sorts. So, once more we have a situation in which attitudes may have prevented progress. Then again, there would have still been problems with the cumbersome Roman number system. But maybe the Greeks could have eventually found a way around that. (And, clearly, this line of thought could be spun on and on.)

This essay is focused on science and mathematics, with other aspects, sometimes ironically and deliberately, understated. Therefore I will forgo any obvious extended comments regarding how my enthusiasm for these significant early achievements is tempered by my dismay about the dark sides of either culture. Both relied heavily on slave labor and punishments as well as ways to obtain confessions were simply barbaric. Unfortunately, these activities were not limited to the ancient world. The human affinity for cruelty is well documented throughout all of history.

---

[1] Historically, the Romans did use Greek inventions and employed Greek scholars. But the use was strictly utilitarian, which is much less than perfect.

A.2 The Dark Ages and First New Developments

Perhaps surprisingly, the Roman empire was not defeated by better science and technology, but by complacency (on the part of the Romans) and brute force (on the part of the teutonic barbarians who are the author's ancestors). So whereas the succession of Greece by Rome replaced pure science and philosophy with pure applications, after the fall of the Roman empire there was, well, nothing of significance in science and mathematics for a long time. During these dark ages, scientific activity, if it can be called that, was focused on preserving ancient knowledge, rediscovering it, but also on active ways to suppress what early Christianity considered "heathen" activity. On these grounds the last remaining Greek schools of philosophy were closed in 529A.D., effectively ending nearly a thousand years of steady, if slow, scientific and mathematical discovery.

Unsurprisingly, not much of mathematical (or scientific) significance could be reported for the next thousand years. Europeans clung to what was left of the knowledge of the Romans and the Greeks, including the rather cumbersome Roman numerals, the absence of a number zero and a strong suspicion of negative quantities. Of course, mathematics did not vanish. Numbers were needed in the trading of goods, and accounting was a regular activity (for those who owned anything worth accounting for). The people who did this accounting were the mathematicians of that age. But mathematics did not progress during the dark ages. Leonardo of Pisa, Italy, better known as Fibonacci (1175?–1250?), may be the one bright light during this time. He was the first European to extensively use arabic numerals in a treatise on arithmetic and algebra. It still took a long time for arabic numerals to become more widely used. Part of this slow development can be ascribed to the, to put it mildly, substantial animosity and suspicion between Christian Europe and Muslim Arabia. But part of the reason why Roman numerals hung on may also simply be human nature. After all, there is no dire tension between continental Europe on one side and Britain and the United States on the other, and yet the British unit system persists even though the metric system is more convenient. So, this may be an instance where the main problem with human nature was not denial or suspicion, but simply too much inertia.

Mathematics became exciting again in 1530 and the location was Italy once more. That year, there was a mathematical contest between Niccolò Fontana Tartaglia (1500-1557)[2] and Antonio Fiore. The contest consisted of solving problems posed by the other contestant. The deciding factor was that Tartaglia had figured out how to solve arbitrary cubic equations (see Theorems 5.65 and 5.66), whereas Fiore could only solve cubic equations of the form of Theorem 5.66, based on a method he learned from his teacher Scipione del Ferro (1465-1526).

Gerolamo Cardano (1501-1576) ultimately convinced Tartaglia to to tell him the method, but Tartaglia asked Cardano to not reveal it and to give him time to publish the result himself. Several years passed and Tartaglia did not publish his result. Meanwhile Cardano found out about del Ferro's work. Ultimately Cardano published work that could be attributed to del Ferro, but Tartaglia was enraged nonetheless. To exact

---

[2]"Tartaglia" means "stammerer." Early in life, Niccolò Fontana sustained a stab through the jaw and palate that made it impossible for him to speak normally.

vengeance, Tartaglia challenged Cardano to a mathematical contest, but Cardano declined. The challenge was eventually accepted by Cardano's student Lodovico Ferrari (1522-1565), who had discovered a way to solve quartic equations (see Theorems 5.67 and 5.68). Because quartic equations were part of the challenge, Tartaglia lost.

So, on one hand, we have a warning tale about not leaving good results unpublished too long. On the other hand, it is advisable to not try to publish every triviality. Every productive mathematician walks this line and we all may have erred on one side or the other. We are also led to the desire to give proper credit for results. There are several good mathematical results to which several mathematicians can lay rightful claim. This leads to theorems being named after multiple people, such as the Cantor-Schröder-Bernstein Theorem, which will be examined later in this essay. But the method to solve cubic equations is a curious example indeed. It is due to del Ferro and Tartaglia, but because it first occurred in publications by Cardano, it is commonly called "Cardano's formula." At least the Ferrari method for quartic equations is rightly named after Ferrari. This is not the only time history has somewhat overlooked Tartaglia. Tartaglia had a way to compute binomial coefficients that we call Pascal's triangle (see Theorem 3.76) after Blaise Pascal (1623-1662), even though it's quite obvious that if Tartaglia knew it, Pascal could not have been the original inventor. Pascal's triangle is also called Tartaglia's triangle, especially in Italy. Yet neither man was the first to know the formula, as the triangle was depicted 500 years earlier in India and in Iran. The author does not know if Tartaglia learned the formula or rediscovered it. Either way, at least in this case Tartaglia is not the original party to whom credit is due. In fact, for Pascal's triangle (similar to Pythagoras' Theorem) it does not seem clear at all to whom credit is due.

After about the year 1500, the pace of development of mathematics quickened in Europe. Negative numbers, which were suspect to Tartaglia, Cardano, Ferrari et. al., became more and more accepted and notation slowly improved, too. By 1500, formulas were written in in-line format: No exponents, subscripts, etc., and powers were expressed by writing the variable the appropriate number of times until René Descartes (1596-1650) introduced exponents. Among the ingenious discoveries that we do not touch upon in this text are the introduction of cartesian coordinates by René Descartes, which then led to the possibly most monumental mathematical discovery ever, calculus, which was discovered independently by Sir Isaac Newton (1643-1727) and Gottfried Wilhelm Leibniz (1646-1716). This text does not touch upon calculus/analysis, because it can only be done justice in a separate, and rather large, text.[3] But calculus required the solution of many complicated equations and so eventually complex numbers (see Section 5.7) found more and more acceptance. Johann Carl Friedrich Gauss' (1777-1855) endorsement of (at least) the imaginary unit i possibly was the final word in a long discussion. So it took over a thousand years to accept negative quantities, but then it took mere hundreds of years to accept computations with imaginary numbers. That may still seem slow, but it indicates a fundamental shift in mathematicians' mindsets. Abstract notions were still abstract, but they gained acceptance. Plus, the idea that notation can help clarify ideas started to take hold. Descartes' introduction of superscripts to denote powers was another step that is small from a modern point of

---

[3] Interested readers can find the author's take on the subject in [30].

view, but significant nonetheless.

Along the way, mathematicians became used to computing with irrational entities, such as $\sqrt{2}$, without truly understanding them. So, a new abstract pragmatism took hold. In Tartaglia's times, the significant qualms with negative numbers led to equations being rewritten so that quantities on both sides were positive. Moreover, negative solutions were substituted into equations, negative signs were pulled out, the equation was rewritten to have only positive coefficients, and then the absolute value of the solution was presented as a solution of the rewritten equation. This substantial amount of work just to avoid notions that seemed uncomfortable was replaced with the attitude that if something, such as irrational or complex numbers, works, then it may as well be used. I'm pretty sure part of this development must have been driven by the amazing success of calculus, but I cannot substantiate details.

A.3 There is No Quintic Formula: Abel and Galois

The increasing acceptance of what from a modern point of view may only seem slightly abstract concepts, but which were nonetheless significant steps, probably helped in making the leap to attempt something that looks impossible at first: To prove that there is no fifth order analogue of the formulas for equations of order 1 (if you want to call that a formula), 2, 3, and 4 (see Chapter 6). Attempts to derive a quintic formula had failed time and again. In fact, Niels Henrik Abel (1802-1829), a Norwegian, as well as Evariste Galois (1811-1832), a Frenchman, both started their pursuits with an incorrect "quintic formula" and only pursued their respective proofs once they found a mistake in their respective formulas. The first to really attempt the previously unthinkable proof that a quintic formula does not exist was Paolo Ruffini (1765-1833), an Italian, who offered several proofs, all of which were flawed. Ruffini's best proof is very close to the proof that Abel conceived, not knowing of Ruffini's work. But Ruffini's proof had a significant technical gap, which Abel was mindful to close in his proof. Because of Ruffini's role in this story, Abel's Theorem is also called the Abel-Ruffini Theorem.

Calculus certainly will always be connected to the names of Newton and Leibniz, who introduced calculus as an overarching framework and who produced significant methods in this theory. But mathematicians before Newton and Leibniz had tackled the tangent problem and had made contributions that must have helped along with the birth of calculus. The proofs by Abel and Galois came, comparatively, out of nowhere: There was little to no preparatory work available to them and the idea itself feels quite unthinkable. This comment should not in any way be taken as an attempt to diminish the genius of Newton and Leibniz, but rather as an attempt to emphasize Abel's and Galois' singular achievement.

Unfortunately, the times were not right to appropriately receive Abel's and Galois' ideas. Both men led brief, tragic lives and their achievements were only properly honored after their deaths. Part of the reason may be that calculus was still a fresh invention and much work was to be done developing it. Given such an exciting option, it may be understandable that contemporary mathematicians did not pay enough heed to Abel's and Galois' work. Part of the reason may also have been that lesser individuals did submit incorrect attempts at proving that there is no quintic formula to famous mathematicians. (Hence the earlier note to not try to publish too quickly. A reputation is easily spoiled and hard to regain.) For example, Gauss supposedly tossed Abel's paper aside, unread, exclaiming, "Here is another of these monstrosities." But nonetheless, Abel and Galois would have deserved better.

Abel spent his life in poverty. His youth, which apparently was happy nonetheless, took place in a small village in Norway. Abel had seven siblings and did mathematics amid the the noise and distraction that can be caused by a large family in small quarters. When Abel was 18, his father died and Abel had to provide for his mother and his six siblings. At that time, Abel had studied the works of Newton, Euler and Lagrange and he subsided on small grants from the Norwegian government and taking on private students. In 1825, at age 23, Abel received a grant to travel and study in France and Germany for a year. Armed with copies of his work, among them the proof that there is no quintic formula, he traveled to see the giants of his time. His reception was lukewarm at best, as the anecdote about Gauss in the preceding paragraph shows. The only mathematician who recognized Abel's contribution was August Leopold Crelle (1780-1855), the founder of the *Journal für die reine und angewandte Mathematik* which, for the period of Crelle's life, is also known simply as *Crelle's Journal*. Because Crelle recognized Abel's genius, Abel frequently published in Crelle's Journal. Abel's reception in France was similarly lukewarm, with Cauchy losing Abel's manuscript and not being able to locate it until requested to do so (by the Norwegian government!) a year after Abel's death. Cauchy may have been a bit unenthusiastic about proofs that there is no quintic formula, because he may have believed that Ruffini's proof was correct, but that did not alleviate Abel's hardship. By 1827, Abel had contracted tuberculosis, but he refused to believe it. He returned first to Germany and then to Norway, where he died of tuberculosis in 1829. In an irony of fate, two days after Abel's death, Crelle wrote to Abel to inform him that he had finally managed to find him a position at the University of Berlin. This position would have allowed Abel to live at least a modest life and to finally marry his fiancee of many years.

Whereas Abel apparently was a humble genius, Galois was described as a difficult student. Part of this can be explained by poor conditions that Galois found in schools and by rather strange requirements in the schools of his day. Galois was not afraid to speak out about such things (and no one should be), but it did not endear him to teachers. Part of the explanation is Galois' undoubted genius. He had developed his theory of equations, which mathematicians would still fill in and expand to new applications a hundred years after his death, by the age 17! (That's an exclamation sign, not 17 factorial, which would still not be enough time for the author to create anything that would even approach Galois' work.) He did complicated work in his head, and often only communicated results. This approach may not even work well with patient and highly educated teachers, and it certainly bears its own perils with lesser minds. Interestingly, Cauchy played an unfortunate role in Galois' life, too. Galois gave a manuscript to Cauchy to be submitted on Galois' behalf to the French Academy. But Cauchy forgot to submit the manuscript and lost the abstract. Failing the entrance exams to the Ecole Polytechnique several times further frustrated Galois. His failure here was not for lack of intelligence. But Galois did not provide many steps in his solutions and he flatly refused to study certain subjects. So even as a much lesser mathematician, the author feels compelled to say that although Galois was brilliant, he was unwise at times. Eventually, Galois became entangled in revolutionary activities that landed him in jail. Soon after his release, political enemies rigged a duel with him and Galois foresaw that he would not survive it. The night before the duel he wrote up what became the centerpiece of his legacy, a letter to a friend, which contained at least an outline of his ideas. Every so often there are marginal notes stating "I have not time," indicating either his urgency, or, if parts were written earlier, his impatience with details he considered trivial. The duel ended with Galois being shot and left to die. Eventually he was brought to a hospital where his brother spent Galois' last hours with him. In these hours, Galois asked this of his brother. "Don't cry. It takes all my courage to die at twenty."

As a non-mathematical aside, duels were part of the times back then, but they were not always the honorable affairs depicted in cloak-and-dagger movies. In fact, both in Germany and in France, many duels devolved into legalized assassinations. Pretenses of injured honor are easy to invent, and showing cowardice by avoiding a duel was considered worse than death by many. First of all, the concept of honor, even when in some instances deeply misguided, was deeply rooted in all individuals. So there were strong personal inhibitions against fleeing. But fleeing and starting anew somewhere else was also virtually impossible in the nineteenth century. If you were poor, you did not have the funds, if you had the funds, they were usually impossible to liquidate on short notice, papers took a long time to come by, and being an outsider in any part of the world was not easy. The practice of duels became so bad that in France and Germany duels were eventually outlawed. In Germany, a ritualized form of combat – the *Mensur* – took the place of actual duels: The Mensur was fought with swords, held above your head, protective head gear was worn to protect all but certain areas of the face, and the fight would stop as soon as one party, who would be declared the winner, drew first blood, typically with a gash to the face. The loser was treated and no lives were lost. But the treatment of the wound often was for the express purpose of leaving a really tough looking scar. Such scars were worn proudly, despite the fact that they indicated that you lost. Being of distinctly non-noble background, the author never understood this practice, until a friend indicated the obvious fact that at least people would not be killed in this type of fight.

But sometimes smart (and hardened) people found creative ways to even beat this system. In this paragraph we may actually descend into what is merely a tall tale, but I read a synopsis of the story in a newspaper in the 1980s. After World War One, the story goes, a veteran — he was not a mathematician, but that's not the point — was challenged by a university student, who had not seen combat, to one of these ritualized duels. But the traditional rules for duels were still in place. The challenged party had the right to choose the weapons. Everyone chose swords, but this veteran had seen too much to deal with rituals. He chose mortars, as he had serviced one during the war. Needless to say, the duel never happened.

Coming back to mathematics, it is an eerie coincidence that a third, brilliant individual whose name is associated with the unsolvability of the quintic by radicals also died very young. Ferdinand Gotthold Max Eisenstein (1823-1852) proved several amazing results. Among other things, he creatively used what we now call Eisenstein's Irreducibility Criterion (Theorem 6.3, the result was actually proved by Schönemann).

Although Eisenstein's name suggests physical strength,[4] he was actually of poor health. Like Galois, he fell in with revolutionary circles and was imprisoned, which affected his health. Like Abel, he died of tuberculosis, and funding that might have provided for care to prevent his death was only obtained too late to make a difference.

A.4   Understanding Irrational Numbers: Set Theory

Mathematics continued to develop after Abel and Galois, and certainly much of the development was geared towards making calculus rigorous. Said feat was eventually achieved with the introduction of limits by Augustin-Louis Cauchy (1789-1857), the Frenchman who intersected the lives of Abel and Galois in such unfortunate ways, but who was brilliant and who is nonetheless and rightly well-respected, and Karl Weierstrass (1815-1897), a German. But to work productively with limits, certain sequences, called Cauchy sequences, must have limits. These limits must be numbers. As noted earlier, people had become accustomed to certain irrational numbers, such as roots of prime numbers, but they worked with them in a rather mechanical fashion. From a modern point of view, most numbers were still undiscovered (see Corollary 5.54 and Exercise 6-9). So to truly get a notion of limits, it was necessary to define what real numbers are. Richard Dedekind (1831-1916) was unsatisfied with certain arguments involving limits and in 1871 he looked to "geometrically evident facts" to make the argument rigorous. Instead of talking about the real numbers as "not having gaps" (which is to this day a good, albeit for mathematics not formal enough, description) he noted that if numbers on "the line" are partitioned into two sets so that all numbers in one set lie to the left of all numbers of the other set, then there is exactly one real number in the middle. He apologized for the triviality of his observation, but he proceeded to produce the construction of the real numbers from the rational numbers that is given here in Definition 5.30. The only difference is that he insisted that the numbers were not equal to what we now call Dedekind cuts, but that the numbers were what created the Dedekind cuts.

So Dedekind's approach used a naïve notion of sets (in this context, naïve is used to distinguish it from formal set theory, not as a value judgment) and it is assumed that there are rational numbers. This is not an unreasonable assumption. Later, in 1899, the Italian mathematician Giuseppe Peano (1858-1932) gave an axiomatic description of the natural numbers (see Theorem 2.64) as well as a rigorous construction of the integers and the rational numbers (see Chapters 3 and 4 and Sections 5.1-5.3, although he probably did not talk about groups, rings, fields and ordered sets, as these notions were only about to emerge in full force). In this fashion, mathematics has come back to its Grecian roots. "All is number" once more.

But wait, if we look at the approach presented here, then "every number is a set," which consequently means that "all is set," not number. Set theory, though undoubtedly fundamental, entered the mathematical conscience later than the ideas of numbers, calculus or the unsolvability of the quintic by radicals. The development of set theory is primarily connected with the name of the German mathematician Georg Ferdinand

---

[4]Literally translated, it means "Ironstone," and the author would not want to be on the wrong side of anyone with such a name.

Ludwig Phillip Cantor (1845-1918). It seems most ideas of set theory can be traced back to Cantor, including one-to-one functions (Definition 2.52), the fact that there are more real numbers than there are rational numbers (Corollary 5.52) and the associated diagonal argument that is used to this day (see Figure 5.4), well-ordered sets (see Definition 7.8), the fact that no set can have the same size as its power set (see Theorem 2.60), and the Continuum Hypothesis (see Axiom 7.24), which he tried to, but could not, prove. Interestingly, Cantor only attempted to prove what we call the Cantor-Schröder-Bernstein Theorem (see Theorem 7.19), but did not succeed. So did the German mathematician Ernst Schröder (1841-1902, and the author is not related to him) who stated it earlier, with a proof that was flawed, too. It was Cantor's student Felix Bernstein (1878-1956) who finally supplied a correct proof in his dissertation. Cantor's result that no set is equivalent to its power set, which consequently gives rise to an infinitude of sizes for infinite sets, resulted in bitter disputes with several great mathematicians of his day. Cantor certainly and rightly stood by his guns, although he temporarily needed to be treated for depression at the *Nervenheilanstalt Halle*. A famous colleague supposedly commented upon hearing of Cantor's plight that "This is what happens when you think about that abstract stuff." Later in life, accolades were rightly showered upon Cantor for his contributions to mathematics, which may make for a slightly happier ending to this essay. Cantor at least earned some fruits for his labors. But because of World War One, Cantor, too, died in poverty.

At this stage my knowledge of the history becomes desperately thin. But any time I have the choice to study more of the history of mathematics or more mathematics itself, mathematics wins. And it feels great. So with apologies for the obviously abrupt ending, let me tell what I know.

Set theory itself was axiomatized by Ernst Friedrich Ferdinand Zermelo (1871-1953), a German, and the axiom system was improved by Abraham Halevi (Adolf) Fraenkel (1891-1955), an Israeli, and by Thoralf Albert Skolem (1887-1963), a Norwegian. In honor of Zermelo and Fraenkel, the customary axioms for set theory are called the Zermelo-Frankel axioms. Skolem has multiple constructions and results in logic named after him. The authors is unsure how Russell's Paradox (see Section 1.8) fits into this story, but obviously it must have played a role.

Obviously there are more stories to tell about the people who have created the mathematics we learn, use and expand today. Among them are tragic stories about what happened to great mathematicians during World War Two, such as the ones woven into the content of [16], or the final act of defiance or desperation of Felix Hausdorff (1868-1942). Hausdorff was one of the founders of topology and he wrote a classic text on set theory. When he saw that the Nazis were downstairs to take them away, he and his wife committed suicide. But there are also more hopeful stories of mathematicians helping to turn the tide of history (see Section 3.9).

The author hopes this essay has shown that the discovery of mathematics, despite the rigor and purity of the discipline, is shaped by people who should be admired for their genius and their courage, even if it is only duplicated to a lesser degree within ourselves.

Conclusion and Outlook

This text was meant to lay a foundation for advanced mathematics classes, that is, for proof classes. Chapters 1-5 did that by analyzing logic, sets and the familiar number systems in a rigorous light. In doing so, we became accustomed to proofs as well as to standard ideas and constructions. These first five chapters should have prepared you as well as possible for advanced classes, and they certainly provide sufficient material for a one term course.

Chapter 6 was a tour de force through Galois Theory. It certainly showed how a natural, but hard, question leads to sophisticated, but, in retrospect, natural, ideas and theorems. It also gives an idea of the challenges that lie ahead. Do not be discouraged if Chapter 6 felt difficult. It *is* difficult, and you will get used to this level of difficulty. Chapter 7 completed our exposure to set theoretical concepts that working mathematicians are most likely to encounter.

So what is next? After an introduction such as this one, there are three subjects that are usually tackled from an advanced point-of-view .

Algebra involves the further detailed study of groups, rings and fields. The results in Chapter 6 are often presented in less compressed form in a first (two term) algebra course. Some algebra courses also put emphasis on **number theory** and they investigate primality and various concepts closely related to numbers. Applications include complexity of algorithms, encryption, and the connection of group theory to theoretical physics.

Linear algebra studies vector spaces and the transformations between them. Some courses on linear algebra are also called **matrix theory**, because matrices feature prominently in them. Applications include fast solutions of systems of equations and matrix factorizations that are needed in the numerical and theoretical analysis of differential equations.

Analysis is the one topic that we have barely touched in this text. With all due respect to algebra and linear algebra, analysis probably is the broadest of these three subjects. Briefly speaking, it is calculus with proofs. Its further abstract development leads to branches of mathematics, such as topology, that are almost entirely hidden from elementary classes, as well as to branches, such as functional analysis, that are both highly theoretical and highly applied. Scientific applications include all of theoretical physics, which is a substantial field indeed, as well as scientific computation.

You are ready for courses in all of the above subjects, and it is likely that your curriculum requires them. I hope your experience with this text will ease the transition into the abstractions to come. They are beautiful in and of themselves, and they are immensely useful to help us model and understand nature. Enjoy!

Bibliography

[1] Z. Adamowicz, P. Zbierski (1997), *Logic of Mathematics: A Modern Course of Classical Logic*, J. Wiley and Sons, Inc., Hoboken, NJ

[2] A. Albert (1943), An Inductive Proof of Descartes' Rule of Signs, *Amer. Math. Monthly* 50, 178–180

[3] R. Bjork (1994), Memory and Metamemory Considerations in the Training of Human Beings, in J. Metcalfe and A. Shimamura (eds.), Metacognition: Knowing about knowing, MIT Press, Cambridge, MA, 185–205

[4] J.Bransford, R. Sherwood, N. Vye, J. Rieser (1986), Teaching Thinking and Problem Solving, American Psychologist, October issue

[5] N. Carter and K. Monks, Lurch open srouce mathematics validation software, available at http://lurch.sourceforge.net/

[6] P. J. Cohen (2008), *Set Theory and the Continuum Hypothesis*, Dover Publications, Inc., Mineola, NY

[7] C. Dodge (1969), *Sets, Logic and Numbers*, Prindle, Weber & Smith, Incorporated, Boston, London, Sydney

[8] M. du Sautoy (2003), *The Music of the Primes*, HarperCollins Publishers, New York

[9] J. Ewing (2007), Paul Halmos: In His Own Words, *Notices of the American Mathematical Society* 54, 1136–1144

[10] Free Software Foundation (2008), GNU Multiple Precision Arithmetic Library, available at http://gmplib.org/

[11] M. R. Garey and D. S. Johnson (1979), *Computers and intractability: A guide to the theory of NP-completeness*, Freeman, San Francisco

Fundamentals of Mathematics: An Introduction to Proofs, Logic, Sets, and Numbers.
By Bernd S. W. Schröder.
Copyright © 2010 John Wiley & Sons, Inc.

[12] P. R. Halmos (1974), *Naive set theory*, Undergraduate Texts in Mathematics, Springer Verlag, New York

[13] M. Hazewinkel (ed.), Encyclopaedia of Mathematics, available at `http://eom.springer.de/`

[14] H. Heuser (1983), *Lehrbuch der Analysis, Teil 2 (2. Auflage)*, B. G. Teubner, Stuttgart

[15] A. Hurd and P. Loeb (1985), *An Introduction to Nonstandard Real Analysis*, Academic Press, Orlando, FL

[16] P. Lax (2002), *Functional Analysis*, J. Wiley and Sons, Inc., Hoboken, NJ

[17] R. Leckie (1987), *Delivered from Evil – The Saga of World War II*, Harper & Row, New York

[18] B. G. Lipták (2005), *Instrument Engineers' Handbook: Process control and optimization*, CRC Press, Boca Raton, FL

[19] J. Maxfield and M. Maxfield (1971), *Abstract Algebra and Solution by Radicals*, W. B. Saunders Company, Philadelphia, London, Toronto

[20] K. Meyberg (1980), *Algebra, Teil 1 (2. Auflage)*, Hanser Verlag, München, Wien

[21] K. Meyberg (1976), *Algebra, Teil 2*, Hanser Verlag, München, Wien

[22] R. Miller (2008), Computable Fields and Galois Theory, *Notices of the AMS* 55(7), 798–807, available at `http://www.ams.org/notices/200807/tx080700798p.pdf`

[23] M. A. Nielsen (2002), Rules for a Complex Quantum World: Qubits Explained, Scientific American, October issue

[24] `http://nobelprize.org/nobel_prizes/physics/laureates/1921/`

[25] P. Pesic (2003), *Abel's Proof*, MIT Press, Cambridge, MA, London, England

[26] P. Phipps, J. Maxwell, C. Rose (2007), 2007 Annual Survey of the Mathematical Sciences, *Notices of the American Mathematical Society* 54, 253–263, available at `http://www.ams.org/notices/200802/tx080200253p.pdf`

[27] G. Pólya (1945), *How to Solve It: A New Aspect of Mathematical Method*, Princeton University Press, Princeton, NJ

[28] R. Rivest, A. Shamir, L. Adleman (1978), A Method for Obtaining Digital Signatures and Public-Key Cryptosystems, *Communications of the ACM* 21, 120-126

[29] F. Ruskey, C. D. Savage, and S. Wagon (2006), The Search for Simple Symmetric Venn Diagrams, *Notices of the American Mathematical Society* 53, 1304–1311, available at
http://www.ams.org/notices/200611/fea-wagon.pdf

[30] B. Schröder (2007), *Mathematical Analysis – A Concise Introduction*, J. Wiley and Sons, Inc., Hoboken, NJ

[31] B. Schröder (2010), *Mathematical Analysis, Vol. 2 – Applied Abstract Analysis*, in preparation, inquire with author

[32] B. Schröder (2009), Videos on the Fundamentals of Mathematics, available at
http://www2.latech.edu/~schroder/fund_videos.htm

[33] S. Y. Yan (2009), *Primality Testing and Integer Factorization in Public-Key Cryptography*, Springer Verlag, Heidelberg

[34] M. Tommila (2008), Calculator Applet (arbitrary precision calculator), available at http://www.apfloat.org/apfloat_java/applet/calculator.html

[35] D. Velleman (2008), Proof Designer, available at
http://www.cs.amherst.edu/~djv/pd/pd.html

[36] Wikipedia, Archon: The Light and the Dark, available at
http://en.wikipedia.org/wiki/Archon_(computer_game)

Index

Printed and bound by CPI Group (UK) Ltd, Croydon, CR0 4YY

27/10/2024

14580252-0002